Physical Mathematics

Edited by
Lucy Flynn

Willford Press
www.willfordpress.com

Published by Willford Press,
118-35 Queens Blvd., Suite 400,
Forest Hills, NY 11375, USA

ISBN: 978-1-68285-591-1

Cataloging-in-Publication Data

Physical mathematics / edited by Lucy Flynn.
p. cm.
Includes bibliographical references and index.
ISBN 978-1-68285-591-1
1. Mathematical physics. 2. Physics. 3. Mathematics. I. Flynn, Lucy.

QC20 .P49 2019
530.15--dc21

For information on all Willford Press publications
visit our website at www.willfordpress.com

Willford Press

Contents

Preface

Every book is a source of knowledge and this one is no exception. The idea that led to the conceptualization of this book was the fact that the world is advancing rapidly; which makes it crucial to document the progress in every field. I am aware that a lot of data is already available, yet, there is a lot more to learn. Hence, I accepted the responsibility of editing this book and contributing my knowledge to the community.

Physical mathematics is a branch of theoretical mathematics that has been developed for finding solutions of complex physical problems of quantum gravity, string theory, super symmetry, etc. The subject builds on the advanced theoretical dimensions in physics and complex mathematical abstraction in algebra, group theory and topology. This book discusses the fundamentals as well as modern approaches of both physics and mathematics that are relevant to the growth of this discipline. It also unfolds the innovative dimensions of analysis in physical mathematics that are being explored. A number of latest researches have been included to keep the readers up-to-date with this emerging field of study.

While editing this book, I had multiple visions for it. Then I finally narrowed down to make every chapter a sole standing text explaining a particular topic, so that they can be used independently. However, the umbrella subject sinews them into a common theme. This makes the book a unique platform of knowledge.

I would like to give the major credit of this book to the experts from every corner of the world, who took the time to share their expertise with us. Also, I owe the completion of this book to the never-ending support of my family, who supported me throughout the project.

Editor

How SI Units Hide the Equal Strength of Gravitation and Charge Fields

Michael Lawrence*

Maldwyn Centre for Theoretical Physics, Cranfield Park, Burstall, Suffolk, UK

Abstract

This paper shows that there are deeper symmetries within physics than are currently recognised. The use of SI units in their existing form hides that gravity is not the weakest force. The paper shows through symmetry arguments that Planck's constant h and the Gravitational constant G are both dimensionless ratios when dimensional analysis is used at property levels deeper than mass, length and time. The resultant adjustments shown to be needed for SI units produce much simpler sets of units which also solve the issue of why magnetic field H and magnetic inductance B have not previously had the same units. The result shows that gravitational and charge fields have the same strengths when considered in fractional adjusted-Planck values. By showing that h and G are dimensionless, they can be understood to be unit-dependent ratios which can be eliminated from all equations by merging them within new adjusted SI units. The implications are that mass and charge sizes, and distance, are not the properties which separate quantum and classical gravitational systems. The equivalence of gravitational and inertial mass is also shown. The new type of dimensional analysis shows how to uncover any law of nature or universal constant and that the current set of properties of nature is missing two from the set, whose dimensions and units can be inferred.

Keywords: Symmetry; Gravitational constant; Planck constant; Planck units; SI units; Dimensionality; Properties; Parameters; Ratios; Field strength

Introduction

The paper by Mohr et al. [1] explains the current state, where SI units are being bought more into the quantum measurement realm. The paper by Duff et al. [2] includes a broad and varied introduction to the problems of fundamental units and also covers their relationship with SI units. The issue is not new [3], but has missed the deeper implications on the relative strength of gravitational and charge fields.

To paraphrase Okun [4] "The use of fundamental units h and c in SI has introduced greater accuracy in some of the units, but some electromagnetic units are based on pre-relativistic classical electrodynamics and so their measurement is not as accurate as other units. The use of permeability and permittivity spoils the perfection of the special relativistic spirit and, whilst this is useful for engineers, it results in the four physical properties D, H and E, B having four different dimensions". It is only by starting with the most basic, symmetrical and simple physical maximal sized set of Planck type units - and maintaining the integrity of the relationships within that set by not stretching property space unequally - that it is possible to see that the electromagnetic and mechanical properties are misaligned versus each other and that the current value of permeability (and thus permittivity) results in a further misalignment. A new form of dimensional analysis underpins this and allows both mechanical and electromagnetic properties to be treated on an identical basis. It addresses Okun's concerns in that the pair magnetic inductance B, and magnetic field H are shown to have the same units, separated only by the new dimensionless ratio $\sqrt{|G|}$ which replaces permeability. For electric field $\xi\_$ and electric displacement field D the relationship factor is the permittivity $\varepsilon\_$, equal to c^{-2} in DAPU units, as explained below, meaning that D is an energy. There appears to be no current work in this field, trying to investigate physics by simplifying the measurement system and units used. The author has privately experienced the complete reverse view, that no new physics can be uncovered from simplifying units. This paper is a riposte. On the foundations of the changes to SI units, it is possible to show that gravitational and charge fields have the same strengths when considered in fractional adjusted-Planck values.

Methodology

This paper shows, using very simple manipulation of formulae based around Planck units, that both h and G can be eliminated from all formulae as being dimensionless ratios – numbers set by the choice of units – and that this implies that gravitational and charge fields have the same strength. This paper is not directed at simply changing units, as in the case of the misalignments existing in current SI units, but shows how clearing up and simplifying those units enables hidden deeper relationships between properties to be uncovered.

Significance and Objectives

The significance of the paper lies mainly in the reinterpretation of h and G which undermines current notions of where the quantum world ends and the classical world starts. Without the misguided emphasis on large/small distances and masses and the differential strength of charge and gravity fields, it is not clear what properties define where the quantum world becomes classical and vice versa. The paper sets out from the premise based on symmetry that our physical properties are built on deeper dimensions, and that identifying those dimensions will give us better tools to explain what we observe. The first issue that needs investigation is why SI units do not work consistently together across mechanical and electromagnetic properties. The solution is to adjust the SI value of charge and split Planck's constant h and the Gravitational constant G between mass and distance parameters, rather than just mass alone, plus a small redefinition of the value of permeability. But to do this requires understanding of the dimensionality of all the properties – meaning how they are related to each other and to a

***Corresponding author:** Michael Lawrence, Maldwyn Centre for Theoretical Physics, Cranfield Park, Burstall, Suffolk, UK
E-mail: lawrence@maldwynphysics.org

single base property which can be used to define triple-adjusted Planck units (TAPU) and their new SI values. The splitting of the adjustments between mass, charge and distance properties is novel because usually G is considered only to be mass-specific and h has never before been subject to elimination in this manner. Also the Planck charge is usually taken to be the electronic charge size, rather than the larger size implied through symmetry by the Planck mass, but here it is the latter that is used as the ultimate maximum size, with the two different sizes leading to two different sets of adjusted-Planck units. The result shows that only some fundamental constants are really constant, and that those which are, are actually ratios whose size is set by the choice of units. The objective is to show that the strength of gravitational and charge fields are the same for fractional adjusted-Planck values. This implies on the small scale that within the nucleus the actions of gravity probably are as important as charge and that possibly the strong force is gravity in disguise, working with charge. And on the large scale that the universe may not be as large as currently calculated.

Equations Used

It might be reasonably asked why the simple rearrangements of Planck mass $M_p = \sqrt{\hbar c / G}$ into triple-adjusted Planck mass $M_T = M_p\sqrt{G/\hbar} = \sqrt{c}$ and Planck length $L_p = \sqrt{\hbar G / c^3}$ into triple-adjusted Planck length $L_T = L_p / \sqrt{\hbar G} = \sqrt{c^{-3}}$ as the base cases used here should need so much explanation. This will become clear below. Since only the dimensionality, explained later, of each property in these Planck-size based equations is what matters initially, the use of c rather than v for velocity is not an issue. Each property in the Planck formulae takes its appropriate and accepted initial Planck value, apart from charge Q which here is the larger size, rather than the actual charge size that is experimentally observed which is $\sqrt{\alpha / 2\pi}$ smaller. Amongst the issues are the units of h and G, which are not immediately obviously dimensionless ratios, the deeper dimensionalities of properties which allow the new maximum value of mass and minimum value of length to be described in terms of powers of c and the parallel treatments of two sizes of Planck charge based on either an observational basis or on a symmetry argument.

The symmetry argument is that the foundation Planck size for charge is not the electron charge observed, but a TAPU size Q_T related to the TAPU massonly by c. The underlying symmetry has been hidden by two misalignments within SI units. No indications of the accuracy of any property values are given in the paper because the main final values are all powers of $\sqrt{c}$ which is defined in SI units as being exact. The only factor remaining within the paper with any experimental error is the value of the dimensionless fine structure constant α. That it is dimensionless suggests that it really is a constant.

Foundations

All the equations in the paper use only Planck values, unless specifically mentioned otherwise. The Planck, or adjusted-Planck, values are call 'maximal' in that they represent either the largest (eg velocity, c) or smallest (eg distance, $L_{p)}$ that is possible for that property. The Planck unit sets are eventally based in TAPU form on the maximal values using either Q_T as explained below for the 'larger' set and q_{eT} for the 'smaller' set. The most basic two formulae for defining a Planck unit sized system are the gravitational and charge force equations $F = GM^2/L^2 = Q^2c^2/L^2$ and the quantum angular momentum equation $h = McL$. The normal usage of the latter is to define a Planck mass M_p and Planck Length L_p such that $\hbar = M_p c L_p$ and $M_p = \sqrt{\hbar c / G}$. The transformation is to replace the Planck set by the TAPU set, such that

$$M_T = M_p\sqrt{G/\hbar} = \sqrt{c} \text{ and}$$

$$L_T = L_p / \sqrt{\hbar G} = \sqrt{c^{-3}}$$

The force equation provides the simple relationship that the Planck mass M_p and theoretical larger Planck charge Q_p are related such that $M_p\sqrt{G} = Q_p c$. Since the latter equation does not include L_p it is not immediately apparent that there is a need to adjust Q_p into Q_T by using $\sqrt{h}$ and $\sqrt{2\pi}$ so that $M_T = Q_T c$ if the latter factors are distributed in the same way as $\sqrt{G}$. As mentioned, this stretches property space equally along the mass and length properties, rather than just the mass property as is usually done when trying to eliminate G [5]. It is also possible to define a useful intermediate adjustment that retains h in order to provide simplified SI values that can be compared with observable measurements, such that the DAPU set consists of:

$$M_* = M_p\sqrt{2\pi G},$$

$$Q_* = Q_p\sqrt{2\pi}$$

$$L_* = L_p\sqrt{2\pi / G}$$

with $h = M_* c L_*$, where Q_* is the DAPU charge. This is the maximum charge based on symmetry with the maximum mass and is not the electron charge, which is considered later.The result is the foundation of a DAPU property set and units based on

$$h = M_* c L_* \tag{1}$$

And

$$F_* L_*^2 = M_*^2 = Q_*^2 c^2 = hc \tag{2}$$

which excludes G. The dimensionality of G will be shown to be zero later. This is the most basic set of Planck properties that can be devised using two universal constants h and c . However, as shown before, only c is required in the maximal TAPU set.

The relationship between M_* and Q_* is simply $M_* = Q_* c$ with the deeper relationships $M_* = \sqrt{hc}$ and $Q_* = \sqrt{h/c}$.

Equivalence

Considering inertial and gravitational mass, the starting point is the simple DAPU relationship

$$F_* L_*^2 = M_*^2 = Q_*^2 c^2 = M_* c^2 L_* = hc \tag{1}$$

Here now there is no need to differentiate between the M of the gravitational side of the equation and the M of the inertial side because the treatment of both M's is identical and the result independent of G. The subsuming of G within the mass and distance units eliminates the difference between gravitational and inertial masses, since there is no longer any purely gravitational mass. This is not equivalent to making G=1 because the effect of subsuming G into M_* and L_* is to stretch current property space into the more symmetric DAPU properties space, which does not occur when simply setting G=1. The result of eliminating L_* is also that the field strength of any fractional charge q_f/Q_* is equal to the same strength of gravitational field for an equal fractional mass m_f/M_*, the actual factor between the two being c. To maintain the topology and symmetry of the base property space requires that the two properties M and L are stretched proportionately together. Provided Q is treated in the same way as M, it will stay symmetric. Any non-symmetric stretching results in an asymmetric set of properties and will require the use of factors such as $\sqrt{\alpha / 2\pi}$ in the relationships between the stretched properties.

SI Units and Dapu

The above two relationships hold in the system in DAPU units, but unfortunately in SI units the first misalignment becomes apparent. To align the charge and mass side of the Planck equation in SI units requires that the base unit size Planck charge is altered by the factor $\sqrt{1\times10^{-7}}$ relative to the mass side since $GM_P^2/L_P^2 = Q_P^2c^2(1\times10^{-7})/L_P^2$ in SI units.

To identify this difference, each equation in future may, where it might otherwise confuse, be identified either as being in DAPU or SI units, so that

$$Q_* = M_*/c(DAPU) = M_*\sqrt{1\times10^{7}}/c(SI) \tag{4}$$

It is useful for display purposes, as will be used liberally later, to define a factor

$$d = \sqrt{\alpha/2\pi} \tag{5}$$

which represents the ratio $d = q_{e*}/Q_*$, where q_{e*} is the DAPU size of the electronic charge.

The second SI misalignment appears when comparing electromagnetic and mechanical SI units that have material content requiring permeability or permittivity. The use of permeability u_* as $4\pi\times10^{-7}$ causes the factor $4\pi\times10^{-7}/\sqrt{|G|} = 1/6.501$ to appear in some properties when compared with what their DAPU based value should be. This arises from some properties whose SI units may mix electromagnetic and mechanical properties within their definition, such as the Farad. So the second SI re-alignment is to define u_* to be equal to the ratio $\sqrt{|G|}$ rather than the usual $4\pi\times10^{-7}$, which relegates $|G|$ from gravitational to permeability use, so that it represents a measure of the strength of interactions within materials, not between masses. It will be shown below that u_* and $\sqrt{|G|}$ both have the same units, in that they are both dimensionless. The value of permittivity also needs to be adjusted to maintain the value of its product with permeability. The result is that the proposed new adjusted-SI units (NSI) which should be used are either the same as the normal SI units or are different to normal SI units by a power of either the ratio $\sqrt{|G|}$, the $\sqrt{1\times10^{-7}}$ factor, the 6.501 factor or a combination of these. Wherever there is a factor q_{e*} used, the same power of $\sqrt{1\times10^{-7}}$ is used. Where there is no q_{e*} or u_* factor, the NSI and SI values are the same. In this paper, where the current SI unit is adjusted by a power of the $\sqrt{1\times10^{-7}}$ factor, the property unit has a cedilla above it Û, or as a subscript in the tables thus $U_\wedge$. So the SI unit Watts, W becomes $W^\wedge$ in NSI where $W^\wedge = \sqrt{1\times10^{-7}}W$ Note that NSI units include h, but will be changed to Brand New SI units on the elimination of h later when considering values in TAPU units. Because most of the property examples used in this paper do not have any specific material dependence, as would be the case for the magnetic field H, there is no use of permeability u_* or permittivity ε_* within most of the property examples given, except to show that Magnetic Field H and magnetic inductance B have the same underlying units. For the examples used here, there are no complications of additional 6.501 usage or identification of double adjusted SI units, other than in the permittivity ε_* and capacitance c_*, where the SI unit the Farad F is adjusted by that factor to be $_F^{\#}$ in NSI with $F^{\#} = F/6.501$. So the adjustment of SI units to make them self-consistent across both mechanical and electromagnetic properties, and to ensure that they have the same overall shape in property space as the underlying DAPU units, allows the direct comparison of all properties in either DAPU or NSI units, with the only difference being the actual number value in each set of units. For the Q_* set of properties, in DAPU the maximal values are always one multiplied by the combination of h and c representing that property, except where $\sqrt{|G|}$ is needed. For the q_{e*} set of properties, the maximal values are always powers of d multiplied by the h, c combination, again except for $\sqrt{|G|}$. For both these sets, the NSI values are shown in Supplimetary Tables 1 and 2, with translation factors between units in Supplimetary Table 3. The SI values should be multiplied by the factors in the appropriate column to produce the DAPU values of that property.

Dimensionality of *h* and *G*

The new dimensionality analysis goes deeper than considering properties in terms of mass, length and time by uncovering a dimension in which adjusted-Planck sizes of mass, length and time are themselves only powers of a single underlying property. The subsuming of G within the DAPU mass M_*, and the DAPU length L_* would seem to ignore the units of G, effectively treating G as being without units. This is not the case since G has units of $m^3Kg^{-1}s^{-2}$ but it is necessary to show that, based on Planck sizes, these units cancel completely to leave only a ratio. A consideration of the standard laws of nature and the fundamental constants through a form of dimensional analysis shows that if each property at its Planck size is assigned an appropriate dimensionality, every fundamental constant, other than c, will have a total dimensionality of zero, or to state the reverse – every property that has dimensionality of zero is a fundamental constant. The dimensional analysis consists of solving for a basis vector in vector Planck property space which produces zeroes of dimension for four important constants of nature, h, G, Permeability (u) and Boltzmann's constant K_B. Using h and G in the analysis may appear circular, but the analysis supports their use. It also shows that Boltzmann's constant, like h and G, is simply a ratio that can be discarded in the correct units and that there may exist other properties, as yet unrecognized, that correspond to missing dimensionalities. The dimensionalities of the main SI, NSI, DAPU or TAPU properties in terms of a hypothetical dimension Y that emerge from the consideration are:

Mass M Y Velocity $c = Y^{+2}$

Length $L_* = Y^{-3}$ Energy $E_* = Y^{+5}$

Charge $Q_* = Y^{-1}$ Time $T_* = Y^{-5}$

$h = Y^0$ $G = Y^0$

The units of G are $m^3kg^{-1}s^{-2} = Y^{-9}Y^{-1}Y^{+10} = Y^0$ dimensionality and h has units of $m^2kgs^{-1} = Y^{-6}Y^{+1}Y^{+5} = Y^0$ dimensionality. So the units of both h and G are actually irrelevant because they represent fundamental constants with zero dimensionality. Similarly Boltzmann's constant has units of $JK^{-1} = Y^5Y^{-5} = Y^0$ dimensionality as well. Thus adjusting the Planck mass to the DAPU mass, and Planck length to DAPU length, involves only multiplying or dividing by the ratio $\sqrt{|G|}$ and $\sqrt{2\pi}$ as dimensionless numbers, and does not affect the dimensionality of the units of mass, charge or length, other than changing the sizes of the base Planck mass, charge and distance units. This stretches the current property space into the more symmetric DAPU property space, which is different to treating G to be equal to one, which does not affect the current property space topology at all. The same analysis can be done for permeability to give units of $u_* = NA^{-2} = m^{-1}kgs^{-2}(\sqrt{kgms^{-1}})^{-2} = Y^0$ dimensionality which shows that the replacement of u_* by $\sqrt{|G|}$ does not affect the units used because they are both dimensionless.

Producing Laws of Nature

This hypothetical dimensionality tool can be used to produce any law of nature by creating equations where the dimensionalities are equal on both sides.One example from the tables would be $F=Ma$, where force is Y^{+8} and is equal to the product of mass Y^{+1} and acceleration Y^{+7}, so that both sides have Y^{+8} dimensionality. Another example would be the product of volume and viscosity which produces Y^0 on one side and could represent a new constant of nature on the other. To produce a constant of nature, aside from c the minimum that is required is that it has Y^0 dimensionality. In this instance, there is no need for a new constant since the product of volume and viscosity is equal to h, through $V_{*}\eta_{*}=h$ in DAPU. However, producing laws of nature through dimensional analysis of Planck unit sizes does not provide the exact relationship between the fractional Planck property values, because these depend on the specific context in which the properties are being considered. An example would be the kinetic energy of a particle in motion $E_{ke}=(\gamma_v-1)mc^2 \cong 0.5mv^2$ compared with the rest mass energy of the same particle $E_{rm}=mc^2$. Dimensionally, at Planck unit sizes, these two formulae exhibit the same relationships between mass, energy and velocity but as fractional Planck values they describe different specific aspects of that relationship.

Values of the Q_* Set of Properties

Table 1 provides a list of the main Q_* property set and their NSI values at their maximal Planck sizes. The set is produced by starting with the base property space M, h, L, c and Q and extending through the use of standard formulae to find each additional property value in this 'larger' set. The column headed 'NSI units' means that where the current electromagnetic SI units appear they have been adjusted by a power of the factor $\sqrt{1\times10^7}$ mentioned earlier and their use is denoted by a cedilla above the unit or $F^{\#}$ describes the SI unit F adjusted by the 6.501 factor. Note that the factor d does not appear in Supplimetary Table 1 because these values are all based on the DAPU charge Q_*

Von Klitzing and Josephson Constants

The discovery that the von Klitzing constant $R_k=h/q_e^2$ [5] the Josephson constant $K_J=2q_e/h$ [6] could be measured directly has improved the precision of measurement of h and some SI electromagnetic units [7]. It is unfortunate that the misalignment of SI units between mechanical and electromagnetic properties has not been addressed before. These two experimentally measured 'smaller' Planck unit constituents can only easily be shown to be members of that set if the current misalignment of SI units is corrected initially into New SI units (NSI) and then finally into Brand New SI units (BNSI). This is shown in both formulaic and numerical comparisons.What emerges from the q_e set are values in the new fundamental units for R_k and K_j. These two constants are members of the set of q_{e*} units, as should be expected, although K_j appears inversely and twice the anticipated size. From these two observable constants (which are not universal constants because their dimensionalities are not equal to zero) all the other q_{e*} set of adjusted-Planck property values can be constructed as power combinations. The dimensional analysis used to subsume G and h is employed to show that R_k can be considered as equivalent to a velocity, and that many of the electromagnetic properties can similarly be considered equivalent to mechanical properties. This invites a reinterpretation of not just R_k and K_j, but of all electromagnetic properties. The measured value of R_k is shown to equate to a speed greater than light speed. Although it is not clear whether this increased maximum velocity applies to either physical objects, the media through which the physical objects travel or patterns created by subluminal physical objects, this can be experimentally tested. The experimentally observed value of R_k probably implies that a minimum electron velocity is required in order to pass through resistive materials.

R_k and K_j - Members of the q_{e*} Property Set whose Values can be Measured Directly

The maximal value for Resistance R_{e*} is equal to the von Klitzing constant R_k,

$$R_{e*}=R_k(DAPU) \tag{6}$$

and the value of the Magnetic Flux φ_{e*} is equal to twice the inverse of the Josephson constant K_j,

$$\varphi_{e*}=(2/K_J)(DAPU) \tag{7}$$

Supplimetary Table 3 shows that the NSI values of R_k and K_j are identical to R_{e*} and $2/\varphi_{e*}$ when translated into DAPU units by multiplying by the factor 1×10^{-7} for R_k $(2.58128076\times10^4\Omega(SI))$ and $\sqrt{1\times10^{-7}}$ for K_J $(4.835870\times10^{14}HV^{-1}(SI))$.

Values of the q_{e*} Set of Properties

In DAPU the value of each property in Supplimetary Table 1 is one multiplied by the constants factor containing h and c, except where $\sqrt{|G|}$ is needed. To arrive at the maximal real values that can be found experimentally, the list needs to be adjusted to use q_{e*} instead of Q_* since we do not observe Q_* charges usually. As before, the base property space is extended using standard formulae to produce the maximal values in this new 'smaller' set. The maximal values in NSI units of some properties under this limitation are listed in Supplimetary Table 2. Note that the power of the factor d is inversely proportional to the dimensionality of every property.

Properties, Physical Constants and Laws of Nature

All the properties in Supplimetary Tables 1 and 2 have been produced using standard relationships and formulae. It is interesting to observe that some properties on the mechanical side have identical size and dimension partners on the electromagnetic side, for example mass M_* and magnetic flux φ_*. To ensure that the above values can be understood properly, the following series of relationships at the Q_* level can be culled from standard laws and the results confirmed to be correct using their NSI values in Supplimetary Table 1 as:

$$F_*=(M_*/L_*)^2=(\varphi_*/L_*)^2=M_*a_*=\varphi_*B_*=i_*^2 \tag{8}$$

It is also possible to use the same relationships at the q_{e*} level, using the property values from Supplimetary Table 2 thus:

$$F_{e*}=(M_{e*}/L_{e*})^2=(\varphi_{e*}/L_{e*})^2=\varphi_{e*}B_{e*}=i_{e*}^2 \tag{9}$$

Since the values of some electromagnetic properties are identical to the values of some mechanical properties, it suggests that mechanical formulae could be used with electromagnetic properties substituted instead, and vice versa.

One example would be the simple $L_{e*}=v_{e*}T_{e*}=\angle_{e*}$ which suggests that in some way electromagnetic inductance is equivalent to a mechanical distance. Were this only done in SI units, the mix of mechanical and electromagnetic properties would not show that the properties were interchangeable because of the misalignment of those two types of property in the SI units system. The Tables show that most electromagnetic properties can be reinterpreted in terms of mechanical properties. It requires a complete reinterpretation of what is understood by the terms magnetic inductance (acceleration), magnetic flux (mass), inductance (distance), current density (mass density) and other electromagnetic properties.

Equivalence of Electromagnetic and Mechanical Properties in Experiments

The new law of nature mentioned earlier, producing Planck's constant *h* as the product of DAPU volume V_* and viscosity η_*, together with the equivalence in DAPU units of viscosity η_* and electric field ξ_*, provide two interesting possibilities, one already experimentally hinted at. Firstly, that any fundamental physical framework based on a single fundamental particle of one volume size, which combines with others in a composite structure and moves against a background viscosity, would have similar viscosity acting on the motion of every such component particle. This would be equivalent to the action of air resistance on a skydiver, providing a terminal velocity for all such particles. The same type of action on such fundamental particles could be the underlying reason for the terminal velocity that we describe as light speed, the irreversible arrow of time as energy is always lost in motion to overcome viscosity and could also provide an additional redshift factor to the passage of photons, almost completely directly related to their distance travelled, reducing the real size and expansion rate of the universe. Correspondingly, where such viscosity is not present, there will be no maximum velocity and non-locality could result. It may be that the presence of viscosity produces a relativistic environment and an absence of viscosity produces a quantum environment. Secondly, and having potential experimental justification, is that viscosity η_* and electric field ξ_* could be the same property in different disguises. A recent paper [8] mentioned that the 'stickiness' of spiders' silk could be turned on and off through the application of an electric field. If such stickiness and viscosity are related, then this would show directly how viscosity is related to electric field and vice versa. This effect would not the same as the creation of magnetorheological fluids [9] with dual fluids, but would be describing a deeper level of equivalence.

Triple-adjusted Planck Units

Having reintroduced *h* earlier in order to show clearly the link between the q_{e*} set of property maximal values and R_k and K_J, it is now useful to eliminate it again to produce the most simple definitions possible of mass and charge, that is the TAPU definitions

$$M_T = M_* / \sqrt{h} = \sqrt{c}$$

$$Q_T = Q_* / \sqrt{h} = 1/\sqrt{c}$$

$$L_T = L_* / \sqrt{h} = \sqrt{c^{-3}}$$

and to show their simple relationships to all other properties through a new ratio $\vartheta = \sqrt{c/d} = \sqrt{2\pi c/\alpha}$.

The base formulae are now:

$$1 = M_T c L_T \tag{10}$$

and

$$F_T L_T^2 = M_T^2 = Q_T^2 c^2 = c \tag{11}$$

It is now considered here what it means to have those properties, also described as parameters, as ratios of ϑ. The starting point is to consider how each of the parameters could be most simply described in terms of the product the normal length, velocity and time parameters (L *v* T) and respectively ϑ^1 (mass *m*) and ϑ^{-1} (charge *q*) parameters. This is done to understand better what the electromagnetic properties represent when considered as mechanical properties. This analysis is the reversal of the way that the description of the properties was parameterised into powers of *c* and *d*, and now ϑ The new TAPU sets are based around the X_T set $M_T = \sqrt{c}$ and $Q_T = 1/\sqrt{c}$ and the X_{eT} set $m_{eT} = \sqrt{c/d} = \vartheta^1$ and $q_{eT} = \sqrt{d/c} = \vartheta^{-1}$.

It is also worth noting how the current equation relating energy and time, instead of position and momentum in the original Heisenberg relationship (11), in DAPU was $E_* T_* = h$ and now becomes $E_T T_T = 1$ in TAPU.

Comparisons and Unit Foundations

Supplimetary Tables 1 and 2 should be compared with Supplimetary Table 4 for understanding. The q_{eT} set is the observable set of TAPU parameters which can be compared with the maximal Q_T TAPU set. Although the Q_T set is described as maximal because it is based on all adjusted Planck unit sizes, it does contain smaller values when ϑ takes positive powers.

Note that the L*v*T groups used may not correspond to the normally accepted set due to the inclusion of *m* or *q* in every parameter formula. It is clear from a comparison of Table 4 columns 1-3 and 4-6 that the same grouping of L*v*T parameters with mass *m* and with the product *qc* can be described identically. The two sets have the same powers of ϑ which should make the properties the same. So, for example, Shear Viscosity (η) and Electric Field (ξ) appear to be the same properties, and Acceleration (a)seems equivalent to Magnetic Inductance (*B*) The accepted definitions of the electromagnetic properties are therefore shown to be incorrect. They should all be adjusted by the extra *c* factor. One difficulty in considering the alignments across all possible powers of ϑ is that there are gaps where no known properties exist for that power of ϑ, at powers ϑ^{15} and . These gaps are properties that we have not yet realised actually exist. Doubtless they will be uncovered experimentally in due course, although it is not clear what set of parameters or units would best describe them since there are many different ways to produce their dimensionalities [10]. The simplest set has been used in Supplimetary Table 4. The best possible descriptions for these two properties would be: for the ϑ^{15} property 'Kinetic Intensity' since it can be formed from the product of velocity and intensity and for the ϑ^{-8} property 'Inverse Force'.

Brand New SI Units

In translating between DAPU units used above in Supplimetary Tables 1 and 2 and TAPU units used in Supplimetary Table 5, it is helpful to show the adjustments to each of the properties in the parameter sets. The results are displayed in Supplimetary Table 5 which combines the two parameter sets and shows both the BNSI values of the TAPU parameters and their values in terms of ratios of *c*, or of *c* and $d = \sqrt{\alpha/2\pi}$ or $\vartheta = \sqrt{c/d}$. The changes can be split into six groupings, where X_T / X_* is the relationship between the TAPU units in BNSI and the DAPU units in NSI when eliminating *h* content with the description of the units in Supplimetary Table 5 given as BNSI units (*h*-adjusted).

The parameters Mass (*m*), Magnetic Flux ϕ, Charge-mass (*qc*), Momentum (*mv*), Energy (*E*), Temperature (*K*), Charge (*q*), Distance (*L*), Inductance (*Z*), Capacitance (*C*) and Time (*T*) hange in the form $X_T = X_* / \sqrt{h}$.

The parameters Angular Frequency (*w*), Frequency(*f*), Acceleration (*a*), Magnetic Inductance (*B*), Magnetic Field (*H*), Electric Field (ξ) and Viscosity(η) change in the form $X_T = X_* \sqrt{h}$. The parameters Velocity (*v*), Resistance (*R*), Current (*i*) Action (*m/L*) Potential Difference (V) Force (*F*) Power(*P*), Conductance (ζ) and Permittivity (ε) remain in the form $X_T = X_*$.

The parameters Moment (mL), and Area (A) change in the form $X_T = X_* / h$.

The parameters Mass Density (ρ), Current Density (J), Pressure (p) and Energy Density (ψ) change in the form $X_T = X_* h$. The parameter Volume (V) changes in the form $X_T = X_* / h^{3/2}$.

Discussion

Why is the action of charge so strong compared with gravity? The answer is that the strength of action of both is identical. It is the relative sizes in which each occur that starts the confusion and then the gravitational constant that hides the situation further. The latter is caused by the inconsistencies in SI units and lack of understanding of the underlying deeper dimensions in nature.

On the subject of the TAPU interpretation of properties what, for example, does it mean that the maximal value of the TAPU of observable adjusted-Planck unit energy is ϑ^5 whilst that of mass is ϑ^1?

This tells us that regardless of the relative size of the electronic charge in the q_{eT} set to its maximum value in the Q_T set, the relationship between the maximal values of the two adjusted-Planck unit properties energy and mass in terms of one being the fifth power of the other will always be the same, only the actual measurable value in whatever units are used will differ, dependent on the value of α. The laws of nature would be constructed in the same way regardless of the relative sizes of G, h,c and α

It is also possible to infer that the underlying reason for the value of the fine structure constant must be motional, since it is part of the ratio $\vartheta = \sqrt{2\pi c / \alpha}$. Because the relationship is inverse, it does not necessarily mean that a is a translational velocity, instead it could be linked to rotational or spinning motion.The total dimensionality of any object is based on the observation that there must be at least 16 + 9 + 1 = 26 dimensions existing to accommodate all the properties that we currently observe, even if we do not have names for either the mechanical or electromagnetic properties at some values of powers of ϑ, where they have not yet been recognized to exist. Note that, other than for m and q parameters, the formulae used to provide the appropriate powers of for each parameter in Supplimetary Table 4 do not use the target parameter in the formula, so velocity v does not have v in its formula, for example .It is now clear that the use of h, G and the omission of the $\sqrt{1\times10^7}$ and 6.501 factors in SI units serve to hide the underlying symmetry within the current set of Planck units. Only in their final TAPU form in BNSI units is it clear that the set of TAPU units have adjusted-Planck unit property values TAPU=Y^x with $14 \geq x \geq -9$ where for the larger set $Y = \sqrt{c}$, with the smaller set having $Y = \sqrt{2\pi c / \alpha}$. Whilst the elimination of h and G provides advantages in terms of simplification of units and improved understanding of how properties are related, it undermines the idea that the quantum realm belongs to small distances and small masses, and that the classical relativistic world belongs to large distances and large masses. Since the paper shows that there is no difference in field strengths for identical fractional Planck values of mass and gravity, it asks the question why quantum effects are seen in the world of the small and not in the world of the large. The answer appears to be that nature prefers to balance out the larger effects first. So the naturally occurring fractional Planck size of charge is significantly larger than the normal fractional Planck size of mass of any of the basic building blocks of matter. The preference is to reduce the effect of charge first, even though this may increase the amount of mass. The primary example is the neutralising of the charges on a proton and electron to form a neutron. The existence of positive and negative units of charge enables the balancing.

So as the mass size of grouping particles increases towards equality with the field strength of a unit of charge on these masses, the existence of unitised positive and negative charges allows the net charge effect to become the easier one to balance. The attractive-only gravitational field then becomes the stronger overall as mass increases, but has no ability to balance because there is no negative gravitational effect. So below a certain size of mass, unitised and balanceable systems will exist, where gravity plays a subsidiary role – even though its field strength is the same as that of charge its actual strength is much smaller. Above a certain size of mass, gravity will dominate because its actual strength then exceeds that of individual charges. This does not mean that charge fields do not play a role in gravitational systems, nor that gravity does not act in charge balanced systems, only that the relative effect will be small at either end of the scale. There ought to exist at the size where the two forces balance in actual strength, some systems where the gravity and charge actions both need to be considered equally in their dynamics. The final output in Supplimetary Table 5 is to display all the Q_T property set as powers of only $\sqrt{}$ and all the q_{eT} property set as powers of only $\sqrt{2\pi c / \alpha}$. This highlights how the adjusted-Planck sized properties are linked and dependent and shows that the laws of nature would be constructed in the same way regardless of the relative sizes of G, h, c and α .The dimensional analysis enables new laws to be constructed and new constants of nature to be uncovered, although it is not clear that there are any of the latter needed since c is all that is required to generate all the Q_T fundamental property set. However, since c is not strictly a fundamental constant, have dimensionality Y^2, the local value of the maximal adjusted-Planck properties will depend on the local value of c.

Conclusion

This paper presents new ways of understanding the relationships between properties whilst undermining the current interpretation of where the quantum and classical worlds diverge because the strength of gravitational and charge fields are equivalent. The novel insights and predictions include:

- If our current units are simplified and corrected for two misalignments, the underlying symmetry of the maximal values of all properties can be seen.
- The reinterpretation of h and G implies that size and distance are not the properties which separate quantum and classical gravitational systems.
- The reinterpretation of the gravitational constant G as a dimensionless ratio and its relegation from gravitational to permeability use as a ratio enables it to represent a measure of the strength of interactions within materials not between masses.
- The reinterpretation of G eliminates the need to test the equivalence of gravitational and inertial masses.
- The strength of equal fractional adjusted-Planck sized charge and gravitational fields has been shown to be equal.
- The fundamental constants h and G have zero values for dimensionality and can be eliminated from all equations by appropriate adjustment of SI units because they are only dimensionless ratios.
- The adjustment of SI units results in the same units for

magnetic inductance B and magnetic field H , separated only by the dimensionless ratio $\sqrt{|G|}$ which replaces permeability. For electric field ξ_{-} and electric displacement field D the relationship is the permittivity ε_{-} equal to c^{-2} in TAPU units, meaning D is an energy property.

- To correctly understand the relationships between properties the fundamental constant G needs to be split equally between both mass and distance properties and h equally between both mass and charge properties on the one hand and distance on the other.
- There is a self-contained and consistent new Planck unit set of maximal Q_T based properties from which all observed values can be produced and easily combined in equations.
- There is a self-contained and consistent new Planck unit set of electron charge-size q_{eT} based properties can be produced, some of which are directly observable in experiments.
- All properties can be displayed in terms of only c for the Q_T property set and in terms of only c and α for the q_{eT} set (other than permeability, permittivity, H and other material properties which have $|G|$ content), which was previously considered impossible.
- There exists a new hypothetical dimensionality analysis that can be used to describe adjusted-Planck unit property dimensions and to uncover any law of nature or any universal constants.
- All that is required to produce a law of nature is to create an equation where the adjusted-Planck unit dimensionalities are equal on both sides.
- To produce a constant of nature, aside from c, the minimum that is required is that it has Y^0 dimensionality.
- That most of the Q_T and q_{eT} property sets can be described solely in terms of ratios of the R_k and K_j (and d for the Q_T set) and so will benefit from the precision of measurement of these two properties.
- That the experimentally observed value of R_k probably implies that a minimum electron velocity is required sin order to pass through resistive materials.
- That most electromagnetic properties can be reinterpreted in terms of mechanical properties. It requires a complete reinterpretation of what is understood by the terms magnetic inductance (acceleration), magnetic flux (mass), inductance (distance), current density (mass density) and other electromagnetic properties. One possible experimental verification exists in equating viscosity and electric field.
- That the reinterpretation of R_k and $K_j/2$ with their current excellent precision of measurement, should enable increased accuracy in the estimation of the values of other adjusted-Planck unit properties and fundamental constants identified as novel composite functions of R_k and $K_j/2$.
- A universal method of discovering laws of nature that applies regardless of any stretching of property space. A unit with $q_{eT}/Q_T \neq \sqrt{\alpha/2\pi}$ would still have the same relationships between adjusted-Planck unit properties although the numerical values of the results would be different.
- Physics can be better understood when stripped to its bare essentials using a better tool set consisting of a repaired system of SI units, which are currently misaligned across the electromagnetic and mechanical properties. By adjusting SI units to be self-consistent and consistent with TAPU units, greater clarity will ensue.
- The adjustments necessary to align and make SI units self-consistent and also consistent with the simplicity of TAPU units have been proposed, producing a system of Brand New SI units.
- The new dimensional analysis shows that the current set of properties is missing two from the set, whose dimensions and probable units can be inferred and are suggested be called 'Kinetic Intensity' and 'Inverse Force'.

References

1. Mohr P (2010) Recent progress in fundamental constants and the International System of Units. Third workshop on Precision Physics and Fundamental Physical Constants.
2. Duff MJ, Okun LB, Veneziano G (2002) Trialogue on the number of fundamental constants. JHEP.
3. Fiorentini G, Okun L, Vysotsky M (2001) Is G a conversion factor or a fundamental unit.
4. Okun LB (2003) Fundamental units: physics and metrology. "Astrophysics, Clocks and Fundamental Constants.
5. Klitzing KV, Dorda G, Pepper M (1980) New method for high-accuracy determination of the fine-structure constant based on quantized Hall resistance. Physical Review Letters 45: 494-497.
6. Josephson BD (1974) The discovery of tunnelling supercurrents.Rev Mod Phys 46: 251-254.
7. Mohr PJ,Taylor BN, David B (2008) CODATA Recommended Values of the Fundamental Physical Constants: 2006. Rev Mod Phys 80: 633-730.
8. Tao H (2012) Silk-Based Conformal, Adhesive, Edible Food Sensors. Advanced Materials 24: 1067-1072.
9. Andrade EN, Dodd C (1946) The Effect of an Electric Field on the Viscosity of Liquid Proceedings of the Royal Society of London. Series A, Mathematical and Physical Sciences 187: 296-337.
10. Heisenberg W (1930) The Physical Principles of Quantum Theory. Chicago: University of Chicago Press.

2

The Last Challenge of Modern Physics

Michaud A*

SRP Inc Educational Research Service Quebec, Canada

Abstract

Synthesis of the current state of research on the conversion processes involving electromagnetic energy and mass, and description of an expanded space geometry that may help resolve many of the remaining issues.

Keywords: Hamiltonian; Trispatial geometry; Trispatial LC equations; Quantum of action; Quantum of induction; Adiabatic energy induction; 3-Spaces geometry

Objective Physical Reality

Over the course of the 20th century, a whole bunch of particles has been identified, or "defined", that have mostly been categorized as part of the Standard Model of particle physics. The Standard Model is the superset of all particles that are deemed to exist and serve as material for the construction of the material universe, which is the foundation of objective physical reality.

They can be regrouped into numerous subsets: virtual particles, unstable complex particles, unstable elementary particles, stable complex particles, stable elementary particles, and finally, neutrinos. We will examine each of these subsets at the general level.

But before proceeding, let us examine the tools that we have at our disposal to identify and verify the physical existence of these elementary particles.

Destructive *vs.* Non-Destructive Scattering

Verifying the existence of elementary particles can be achieved only by colliding them with each other. Their trajectories, deflected during such encounters, can be recorded by various means to be subsequently studied and interpreted. In fact, the recorded trace of the deflected trajectories of colliding particles is the only proof out of any doubt of the physical existence of these particles.

Elementary particle do not interact during such collisions like hard solid objects as could be expected from our macroscopic perspective, but like elastic "objects" due to their common electromagnetic nature. They can interact either electrically according to the well known inverse square law of distance between such particles, or magnetically according to the less familiar inverse cube magnetic interaction law between the same particles [1]. From our macroscopic perspective, they behave as if they electrically "attracted" or "repelled" each other according to the inverse square law.

The closer they come to each other, the more strongly they seem to electrically repel each other if their electric charges are of the same sign, and the more strongly they seem to electrically attract each other if they have opposite electric signs. Similarly, the closer they come to each other, the more strongly they will magnetically repel each other if they interact in parallel spin alignment, and the more strongly they will magnetically attract each other if they interact in anti-parallel spin alignment.

Exploratory high energy collisions of fundamental particles can be carried out in two different ways, which are the non-destructive collision mode or the destructive collision mode, and the absence of a clear description in textbooks of the difference between both methods has been the source of widespread confusion.

Non-Destructive Scattering

Non-destructive scattering was used for a short period of time during the second half of the 1960s to explore the only two stable composite particles in existence, the proton and the neutron, which also are the only components of all existing atomic nuclei. Since it had previously been confirmed that they occupy a measurable volume in space, this hinted at the possibility that they could have an internal structure involving smaller particles, and consequently, that they might not be elementary. The proton was discovered in 1919 by Ernest Rutherford and the neutron in 1932 by James Chadwick.

Non-destructive collisions with neutrons and protons (nuclei of hydrogen and deuterium atoms captive in water molecules, for example) involves colliding them with electrons or positrons sufficiently accelerated with magnets thus, increasing their momentum sustaining kinetic energy, to enter the nucleon structures, but with insufficient energy to knock the components making up their internal structure out of their structure.

Contrary to protons and neutrons, electrons and positrons seem not to occupy any measurable volume in space and always behave as if they were point-like in the mathematical sense each time they are involved in collisions. They are considered "elementary" because the most energetic non-destructive head-on collisions between 2 electrons are, for example, the closer they come to each other's "point-like center" before rebounding, without any unbreachable limit at some distance from this center having been reached. They were then the ideal projectiles to attempt resolving the enigma of the internal structure of protons and neutrons. The electron was discovered in 1856 by Joseph Thompson and the positron in 1932 by Carl Anderson.

Electrons or positrons that met no obstacle inside nucleons crossed completely through the volume occupied by nucleons, but had their trajectories deflected to various degrees depending on how close they came to the inner components of nucleons, which were thus directly detected for the first time. Some of these incident electrons or positrons

***Corresponding author:** Michaud A, SRP Inc Educational Research Service Quebec, Canada, E-mail: srp2@srpinc.org

were very strongly deflected, some even directly backscattered when their trajectories happened to be in direct line with one of these inner components.

Careful analysis revealed that these inner components are electrically charged like electrons and positrons, because their deflected trajectories all obeyed the same deflection law that governs collisions between two electrons or two positrons, that is, the inverse square Coulomb law [2].

The closer the incident particles came to these inner components during these flybys, the more strongly their trajectories were deflected. The incident negative electrons were attracted to the positively charged inner components and repelled by the negatively charged inner components, while positive positrons were attracted to the negatively charge inner components and repelled by the positively charged inner components. On final analysis, the scattering patterns revealed by the deflected trajectories led to the confirmed discovery that only two different scatterable elementary particles charged in opposition existed inside protons and neutrons.

The positive component was named up quark, possessing 2/3 of the charge of the positron, and the negative component was named down quark, possessing 1/3 of the charge of the electron [2]. This is how the discovery was made that the inner scatterable structure of the proton is made of 2 up quarks and 1 down quark (uud), while that of the neutron is made of 1 up quark and 2 down quarks (udd).

It was discovered furthermore, that the up quark was only marginally more massive than the electron and that the down quark was only marginally more massive than the up quark ([3], p. 11-6). Let us also note that the addition of the fractional charges of these inner elementary particles directly explains the measurable electric charges of the proton and the neutron: +2/3 +2/3 -1/3=+1 for the proton and 2/3 -1/3 -1/3=0 for the neutron.

Destructive Scattering

Finding no other scatterable components inside nucleons, destructive scattering started being used towards the end of the 1960s, and has been used to higher and higher energy levels ever since. This method may involve the liberation of the carrying energy of two same sign particles, such as two electrons, that occurs when two such particles collide head-on against each other, causing this carrying energy to escape as highly energetic bremmsstrahlung photons as their motion is brutally stopped momentarily, or the physical destruction of two opposite sign elementary particles, such as an electron and a positron, when these particles ultimately meet, which causes the energy of which their rest masses are made to convert to free energy (electromagnetic photons), thus causing them to cease existing under their initial form, on top of their carrying energy also being released as electromagnetic photons.

When the destructive level is reached during such collisions, huge amounts of free electromagnetic energy are liberated as the incident bullet (an electron, for example) and the up or down quark that it directly collides against are converted to energy. The total amount of liberated energy is made up of all of the kinetic energy sustaining the momentum of the incoming electron, plus all of the adiabatic energy of the quark being scattered against [4], plus the energy that made up the rest masses of the quark and of the electron involved if they also convert.

Whenever such a huge amount of free electromagnetic energy is liberated, it immediately recongeals back into all sorts of transient metastable hyper-excited massive particles that are generically named "partons". The larger the amount of energy liberated during one such collision, the more massive the transient particles momentarily created will be, generally way more massive than the colliding particles.

During such destructive collisions, there may also be cases when the up or down quark involved will be ejected without having been destroyed, that is, without being converted to energy. Let us note that this barely diminishes the total amount of energy liberated, since the energy making up the rest masses of both types of quarks and of the incident electron or positron is very small with respect to that sustaining the momentum of the incoming particle and the stabilized adiabatic energy induced in the quark concerned at the moment of impact [3].

Note that the momentum sustaining energy of the incoming particle and the stabilized adiabatic energy induced in each up and down quark that are captive of the inner structure of nucleons, is kinetic energy in excess of their invariant rest mass energy, and unless specific identification is required, both types will generally be referred to in this text as "carrying energy" or "carrier-photon", for reasons that will become clear further on.

It is a fact that up and down quarks have never been observed moving freely after having been ejected while *still displaying the same characteristics that they have when inside nucleons.* This doesn't mean however that they have not been observed *displaying different characteristics after having been ejected*, which is an as of yet unexplored possibility that may well have prevented experimentalists from recognizing them as the same particle.

For example, if up and down quarks were in reality positrons and electrons whose masses and charge characteristics had been warped into these altered states by the stresses imposed by these most energetic least action equilibrium states that electrons and positrons could reach if they were the actual material that Nature used to build nucleons [4], then as one of them is scattered out of a nucleus without being destroyed, it would of course recover its normal electron or positron characteristics as soon as it escapes these warping stresses, which might directly explain why free moving up and down quarks have never been observed.

All partons produced during destructive collisions almost instantly decay into a cascade of transient states, the last stage of which always is one or other or a combination of the stable particles set, that is, electron, positron, proton, neutron and residual photons. All of these decay sequences have been thoroughly analyzed and are available in reference [3].

The more energetic the incident electron or positron will be, the more energy will be liberated as it destructively collides with one of the up or down quarks inside a nucleon, allowing more and more massive metastable partons to fleetingly appear before almost instantly decaying as previously described.

Even the recently much hyped about Higgs boson belongs to this parton category, actually the most massive parton yet detected as an up or down quark in an incident proton directly and destructively scattered against one of the up or down quarks of another proton at the LHC facility.

Four of the first partons that were long-lived enough to be detected in the 1970s were given the names charm quark, strange quark, bottom quark and top quark, because they seemed to satisfy the most popular theory of the time, even though, like all other partons, they all almost immediately decay into one or other of the stable particles subset.

Unfortunately, all of these short-lived partons are useless as far as describing normal matter in the universe is concerned, because they can exist only outside of protons and neutrons since they are created by means of this type of destructive scattering. Under no circumstance could they ever be identified within proton or neutron structures via non-destructive scattering.

This verified fact did not prevent the physics community from classifying these short-lived metastable massive states as part of the Standard Model, in an apparent endless search for more and more of these transient massive energy states, even though they obviously cannot be part of the stable material structures in the universe.

The same restriction applies to the variety of "virtual" particles that were "defined" such as gluons and "virtual photons" for example, which are mathematical concepts conjured up as mathematical artifacts to make the mathematical description of particles' interactions easier in currently popular theories.

A clear distinction must also be made between real electromagnetic photons, that are scatterable against electrons, and whose trajectories can be deflected by gravity [5], and Quantum Electrodynamics "virtual photons" which are mathematical metaphors conceived of by Richard Feynman [6] to make easier the calculation of interactions between elementary particles.

Moreover, the QED virtual photon metaphor unfortunately bundles together two fundamentally very different aspects of the relation between particles, that is, the Coulomb force proper and the "motion sustaining" or "momentum sustaining" kinetic energy induced by this force, which, when combined with the presence of the word "photon", induces a high level of confusion with "real electromagnetic photons" that are made of only free moving kinetic energy, as analyzed in reference [7], and as will be put in perspective further on.

In Nature, unstable partons also occur as fleeting massive states such as the various configurations of π and K mesons as well as hyperons, the latter being unstable complex particles still more massive than protons and neutrons, and as a few elementary unstable particles such as muon and tau, that have life expectancies never exceeding a few fractions of a second.

They are created as fleeting byproducts of cosmic radiation colliding with the nuclei of atoms in celestial bodies or their atmosphere, or as by-products of high energy particles' interaction in stars' coronas [8,9] and inside the permanently exploding star masses [10].

Let us note here that what has come to be generically named "cosmic radiation" is mostly made up of protons that are many orders of magnitude more energetic than can be achieved even in the LHC accelerator, which means that they can theoretically create fleeting partons more massive yet than the recently detected Higgs boson when they collide with other particles.

Just as in high energy accelerators, the final end product of the practically instantaneous decay of these naturally occurring partons is always a stable particle belonging to the stable massive particles subset already mentioned, besides photons and neutrinos.

The positron, known to be the anti-particle of the electron, is totally identical to the electron except for the sign of its charge [11], but does not become part of stable atoms contrary to the electron because it readily reverts to various electromagnetic photon states as soon as it individually interacts with an electron, also converting the electron to electromagnetic photon energy in this process which is named positronium decay.

Positrons being the anti-particles of electrons, they are viewed in the physics community as "antimatter" with respect to electrons, which are thus viewed as "normal" matter. There is incidentally a century old assumption that the universe is made almost entirely of "normal matter" (a concept that also includes protons and neutrons), and endless speculation is still ongoing as to why so few "antimatter" is to be found, which is deemed to directly contradict the principle of symmetry.

This issue may be completely resolved simply by considering that when the three charged scatterable elementary components of protons and neutrons are taken into account instead of the protons and neutrons themselves, which are not elementary, there exists by structure in the universe the exact same amount of normal matter and anti-matter, that is, the same number of negatively charged elementary particles and positively charged elementary particles [9,12,13].

Let us now examine the various particles subsets.

Virtual Particles

We can include in this subset "virtual photons", which are a mathematical metaphor that Feynman proposed in 1949 [6] to introduce the notion of quantization of the interaction between charged particles, that allowed using the simpler static Lagrangian calculation method instead of the more elaborate Hamiltonian method to account for interactions between elementary charged particles. These virtual photons regroup in a single concept the Coulomb force and the intensity of the amount of related energy, a method easier to mathematically manipulate than the Hamiltonian which on its part more readily accounts for the infinitesimally progressive nature of energy variation.

Let us also include here gluons, which also are pseudo-quantized mathematical metaphors, but this time, of the presumably also progressive, but still not fully explored interaction at play between the charged inner components of nucleons in the frame of Quantum Chromodynamics; a yet to be explored interaction, one of whose laws can only be the Coulomb interaction, since up and down quarks are electrically charged.

What allows us to clearly distinguish these metaphorical virtual particles from real particles is the fact that it is impossible to prove their physical existence by means of the only available method, which is by colliding them directly with particles of the stable set.

In other words, all virtual particles turn out to be, without exception, simple mathematical concepts.

Unstable Complex Particles

Here we find various configurations of π and K mesons as well as hyperons, and the Higgs boson which are unstable complex particles still more massive than protons and neutrons, with life expectancy never exceeding a few fractions of a second.

What is remarkable about all unstable complex particles, which all are partons produced only in high energy accelerators, or as fleeting by-products of cosmic radiation, is that without exception as already mentioned, the final end product of their decay is systematically one or other, or a combination of the only known stable particles subset, that is, electrons, positrons, protons, neutrons and photons.

Consequently, these unstable complex particles could all be considered as only temporary hyper-energetic metastable states of the fundamental stable particles subset.

Unstable Elementary Particles

Here we find the various quarks, except the up and down quarks of course, and also all elementary partons, which, as already discussed quickly decay to end up as one or other particle of the stable particles subset.

In this category, we also find the muon particle, which is a second generation parton since it is typically generated from meson decay, which are first generation partons, and the tau particle, which is a first generation parton produced by destructive head-on electron-positron collisions, first observed at the SLAC facility in the 1970s. Both particles always leave behind a single electron as a solitary massive by product of their decay, besides neutrinos and occasionally a few gamma photons.

In a certain way, both particles muon and tau can be considered temporary unstable hyper-massive states of electrons that quickly decay to ultimate electron stable rest mass state by neutrino pairs emissions. The mechanics of electronic, muonic and tauic neutrino pair's emission in the trispatial space geometry that soon will be described is analyzed in reference [14].

Of course, anti-muon and anti-tau leave behind a solitary positron instead of an electron.

Stable Complex Particles

In this category, we find only the proton, which is totally stable, and the neutron, that becomes totally stable when associated to protons in atomic nuclei *(Although there are limit cases of neutron instability in some unstable nuclei).*

Neutrons, although totally stable when part of nuclei, with the previously mentioned restriction, become unstable when isolated, with a half-life of about 16.88 min. When they decay, they leave behind two totally stable particles, a proton and an electron plus a pair of neutrinos [14].

Stable Elementary Particles

In this very special stable elementary particle subset, we find a single elementary boson, the electromagnetic photon, and four fermions, which are the electron, the positron (which is the antiparticle of the electron), and finally, the up quark and the down quark.

These particles are considered "elementary", because absolutely all non-destructive collision experiments that were ever carried out with them, even the most energetic up to, but short of destructive, reveal that they behave in all circumstances as point-like particles.

This point-like behavior is characterized by the experimental fact that no identifiable unbreachable limit is reached even during the most energetic non-destructive head-on collisions between 2 electrons, for example, however close they come to each other's "point-like center" before rebounding. This gave us the formal proof that they are not made of smaller interacting particles as is the case with protons and neutrons.

They are considered stable, because unless they are physically converted back to electromagnetic energy, they have an unlimited life span. A stable particle is considered destroyed if affected by a collision in such a way that it ceases to exist under the form that it previously had, either by combining with another particle, as is the case for electromagnetic photons when they are "absorbed" by electrons, communicating part of all of its momentum sustaining energy to the electron, or, in the case of the four stable elementary fermions, by converting to electromagnetic photon state during specific collision patterns previously described. In very special circumstances, electrons and positrons are known to release some of their energy as neutrinos [14].

Something peculiar can be observed about these stable elementary particles. It is the fact that, except for the electromagnetic photon, they all have spin 1/2, and that they all possess a signed electrical charge.

The case of the electromagnetic photon is very special, in the sense that despite the fact that it behaves point-like at all times like the four stable fermions, its spin is equal to 1, which is an unmistakable telltale identifying particles made of two elements, and that it is electrically neutral and deemed to be massless.

Louis de Broglie elaborated a most promising hypothesis to help explain these special characteristics of the photon. Having analyzed them in light of the verified aspects of the various pertaining theories, he eventually concluded that the only way for an electromagnetic photon to satisfy at the same time Bose-Einstein's statistic and Planck's law, and to perfectly explain the photoelectric effect while obeying Maxwell's equations and conforming to the symmetry property of complementary corpuscles in Dirac's Hole Theory, would be for it to be made not of one corpuscle, but of two corpuscles, or half-photons, that would be complementary, like the electron is complementary to the positron in Dirac's Hole Theory [15].

This conclusion mandates the association of charges (possibly unsigned) to each half-photon, and consequently to the photon itself, which would account for the apparent unsigned electric aspect of its electromagnetic nature. This hypothesis resulted in a clear description of the internal dynamic structure of the de Broglie photon in the trispatial space geometry, as described in reference [7].

What is remarkable about all stable elementary particles is that without exception, their objective physical existence can be experimentally verified by colliding them with other particle of the same subset.

Presently, it could even be considered that at the fundamental level, physical objective reality can only be made of the whole collection of these discrete stable electromagnetic particles in constant electromagnetic interaction, whose existence can be physically proven through mutual collisions, and of the whole collection of their continuous mutual electromagnetic interactions.

Neutrinos

Neutrinos are a still unresolved issue in particle physics. We know since the early 1920s that part of the energy of a decaying neutron seems to completely vanish when it decays into a proton and an electron, by the observed fact that the sum of the energies making up the masses of the resulting electron and proton, plus the energy sustaining the momentum of the escaping electron, is almost always (but not always) less than the total energy of the neutron rest mass before decay.

The amount of lost energy appears directly dependent on the

velocity of the escaping electron. It seems that in some limit cases, the electron escapes with a velocity sufficient for no loss to be measurable while at the other extreme the loss is maximized when the electron escapes with very low velocity.

Fermi proposed the hypothesis that this unaccounted for energy must be carried away by some sort of new particle that we were then unable to physically detect, and that he proposed to name "neutrino". Even if the variability of the energy loss at the level of each individual neutron decay occurrence was observed, the limit cases for which no energy was lost did not induce a re-questioning of the concept of an evacuating particle being involved, even if there was no energy lost in these cases.

Particles muon and tau also seem to lose their excess mass in a similar manner, leaving behind an isolated electron as the only massive detectable end product of their decay, besides occasional gamma photons, the process always involving the apparent "disappearance" of an amount of energy.

Even after close to a century of research and experimentation, we still have not been able to physically detect neutrinos by colliding them with particles of the stable particles subset in a directly verifiable manner, although the definition of "direct verification" was eventually extended to include observed indirect phenomena that only the existence of neutrinos can explain. A possible coherent explanation to the neutrino conundrum is explored in reference [14].

The Stable Matter of the Universe

Let us now examine more closely the very restricted set of stable elementary particles of which all atoms making up all bodies in the universe are made.

It has been clearly established that scatterable up and down quarks are associated in groups of 3 to form the nucleons (protons and neutrons) of which all existing atomic nuclei are made. The various elements of the periodic table and all of their isotopes are made of all possible combinations of these nucleons. On their part, electrons settle in the various electronic resonance states that define their possible orbitals about atomic nuclei and thus define each atom's measurable volume.

When a photon is absorbed by an electron in an atom, this excess energy forces the electron to leave its rest resonance state to move to an orbital further away from the nucleus that minimally matches the increased energy that it just absorbed, or to even completely escape from the atom if the added energy is sufficient to allow complete escape.

Electromagnetic photons are generated when such over-energized electrons in atoms lose such excess energy under the form of an electromagnetic photon as they fall back towards the nucleus until they ultimately reach the least action resonance orbital closest to the nucleus that they can possibly reach, that is, the rest orbital, or "least action" orbital for this electron in this atom. Electromagnetic photons can also be produced when nucleons in nuclei lose excess energy in a similar manner, and when nucleons are captured by nuclei.

The Nature of Stable Elementary Particles

Given that all unstable particles turn out to be only extremely short lived hyper-energetic states of stable particles, from this point on, we will restrict our discussion to only the stable particles subset, assuming of course, that all laws applying to stable particles, also apply to unstable particles.

As Maxwell was in the process of integrating into a coherent whole the discoveries of Gauss, Ampere and Faraday on the various aspects of electricity and magnetism, he eventually understood and mathematically explained that light had to be an electromagnetic phenomenon that could move in space only at a very specific and invariant velocity, as he concluded that the light that reaches us from the stars had to be caused by the interaction of an electric aspect of the energy interacting orthogonally with a magnetic aspect of the same energy, and that the energy that we perceived as light was moving in space perpendicularly to a plane determined by the orthogonal relation between these two electric and magnetic aspects.

He perceived light as a wave whose "surface", or wavefront, propagated in spherical expansion at the speed of light away from its point of origin in a medium that he conceived of and name the "ether". But from analyzing Wien's experimental results on the black body however, Planck demonstrated mathematically that this "wave" could not be a continuous phenomenon at the fundamental level contrary to Maxwell's conclusion, but appeared to rather be a discontinuous phenomenon.

Einstein confirmed Planck's hypothesis in 1905, with his photoelectric proof. Further confirmations were subsequently provided by Compton and Raman. These separate light quanta were later named "photons".

Doubt was no longer allowed. At the submicroscopic level, free energy of all electromagnetic frequencies could be experimentally verified as consisting of innumerable discrete individual electromagnetic photons, each moving at the speed of light, each being produced only by de-excitation of electrons reaching an orbital closer to the nuclei of atoms somewhere in the universe or by de-excitation of up or down quarks inside nucleons inside atomic nuclei, or by nucleons being captured by nuclei.

A little later, de Broglie hypothesized that electrons also had to be electromagnetic in nature and consequently also had to have a frequency, which was then experimentally confirmed by Davisson and Germer.

Proof that Photons and Electrons are made of the Same Substance

A further step was then taken when Frédéric Joliot and Irène Curie demonstrated experimentally in 1933 that any photon of energy 1.022 MeV or more can de-couple into an electron-positron pair when it is caused to graze the nucleus of an atom [16], which left no doubt whatsoever as to the close relationship between the energy of massless electromagnetic photons and the energy that makes up the rest mass of electrons and positrons.

Moreover, electron-positron pair creation during close flyby of two photons, at least one of which exceeding the 1.022 MeV minimum energy threshold without any atomic nuclei being close by, was experimentally confirmed by Kirk McDonald et al. with experiment #e144, at the Stanford Linear Accelerator in 1997 [17]

On the other hand, we already knew that there existed a direct link between the energy that an electron accumulates due to the Coulomb force as it accelerates between the electrodes of a Coolidge tube, for example, and that this energy is of the same nature as that of which photons are made, because after an electron has left the cathode and has accelerated through the vacuum of the tube, an electromagnetic photon is evacuated in the X-ray frequency range at the very moment when the electron brutally slows down, as it is captured in electromagnetic least action equilibrium on an orbital of an atom of the anode.

We know though experimental verification that the energy quantum of this electromagnetic photon is exactly equal to the amount of kinetic energy that sustained the momentum of the electron at the very moment of its capture, just prior to the release of this photon. We also know that the photon is released at the very moment of capture, because the origin of the emission is clearly established as being the point of capture of the electron.

Consequently, we have since the 1930s the formal experimental proof that it is possible to convert to electromagnetic photon state the kinetic energy sustaining the momentum of a moving electron, an energy that it accumulates through Coulomb acceleration, and to convert electromagnetic photons of energy 1.022 MeV or more to pairs of massive electron/positron.

To complete this cycle, it has been experimentally proven that when an electron and a positron are made to interact in a sufficiently small volume of space, they always end up mutually capturing in a metastable system named positronium, that will quickly decay until the particles collide and then completely revert to electromagnetic photon state energy. The same conversion to electromagnetic photon state is also observed when any pair of opposite electric signs particles are made to collide.

In short, we have the experimental proof that the "substance" that sustains the momentum of moving electromagnetic particles, that of which massless photons are made and that of which the rest mass of electrons and positrons are made, can only be the same "substance", that is, pure kinetic energy, despite the also established facts that photons appear massless and that electrons and positrons appear massive and display a variety of other apparently conflicting characteristics, such as opposite electric sin.

Coming briefly back to the issue of neutrinos, theoretical considerations stemming from de Broglie's conclusions regarding the internal structure of photons, and by extension, to that of electrons and positrons, lead to think that the energy associated to neutrinos, when muons or tau de-energize or when neutrons decay, could be energy that would de-quantify into space as simple free kinetic energy through a process reverse of that observed when momentum sustaining kinetic energy is induced in electrons by the Coulomb force as in the Coolidge tube example previously mentioned. This possibility is explored in a separate reference [14].

The Electromagnetic Mechanics of Elementary Particles

In the subset of stable scatterable elementary charged and massive particles, only two massive particles other than the electron and the positron have been identified. They are the up and down quarks.

Since they are electrically charged and massive just like electrons and positron, the possibility that they may also be made of the same kinetic energy "substance" is far from implausible. It is indeed a practical certainty, since their energy has been liberated as electromagnetic energy for decades by destructive scattering in numerous high energy accelerators.

But to this date, the process that would allow understanding how they integrate into the sequence of conversion processes that comprises conversion of momentum sustaining kinetic energy into electromagnetic photons, followed by conversion of electromagnetic photons into pairs of massive electrons and positrons, and re-conversion of electron-positron pairs back to electromagnetic photon state, that we have just put in perspective, has not been identified and described since their discovery in 1968 at the SLAC facility.

Since understanding this last conversion process would have at long last have given us mastery of the complete sequence of fundamental electromagnetic mechanics of the stable elementary particles subset, one can of course wonder why this possibility seems to not have been explored.

So, the following question comes to mind:

"Why has no attempt been made to identify and describe this last remaining missing process since the confirmation of the physical existence of up and down quarks?"

This issue constitutes indeed the last challenge of modern physics, since its resolution would finally put at our disposal the complete sequence of energy transformation processes that seem to be possible at the submicroscopic level. So before attempting to resolve this issue, the research philosophy that prevailed during the past century needs to be put in perspective.

The Wave Function and the Real State of Physical Systems

At the 1927 Survey Congress, Quantum Mechanics was adopted as the most fundamental theory for dealing with elementary particles and atoms. Twenty five years later, in 1952, Einstein had this to say about quantum theory:

"I have no doubt that quantum theory (more precisely "Quantum Mechanics") is the most perfect theory compatible with experience, inasmuch as its description is made to rest on the concept of the material point and potential energy as being elementary concepts. But what I find unsatisfactory in the theory resides elsewhere, in the interpretation that is given to the "ψ-function". In any case, this is at the origin of my conception of a thesis which is categorically rejected by most current theoreticians:

There is something like "the real state" of a physical system, that exists objectively, independently of any observation or measure, and that can in principle be described by physics description means.

Now, there is no doubt that the ψ-function is a manner of description of a "real state". The question is then to determine if this description of a real state is complete or incomplete."

Albert Einstein [18].

Sixty five years after Einstein passed away in 1955, this issue still raises heated debates that are quickly put to rest, given the obvious success of Quantum Mechanics in providing utterly precise information about the probability amplitude of momentum, position, and many other physical properties of particles.

What Einstein lamented in fact, was that the wave-function could not give a clear description of moving elementary particles. He felt that this should eventually be possible by some means yet to be discovered, and that no stone should remain unturned in the search for a clearer picture of elementary particles. The very properties of the wave function, however, are such that there seems to be no way to clarify further the description of moving particles by means of amending Quantum Mechanics.

As emphasized by Einstein, though, this "real state" of elementary particles that exists objectively, independently of any observation or measure, and that the ψ-function describes only vaguely when they are in motion, is also known to be related to "real energy" that possesses known electromagnetic properties, which are not completely integrated into QM.

For example, although the wave function is the ideal tool to explore the various least action resonance orbital states of electrons in atoms, it does not allow separating the invariant rest mass energy of electrons from the energy which is adiabatically induced in them as a function of the inverse square of the distance from other charged particles, when electrons are translationally immobilized in resonance states, and that sustains their momentum when free moving.

Maxwell's Electromagnetic Wave Theory

Maxwell's theory, on its part, deals with the electromagnetism aspect of this "real energy", but has also not yet satisfactorily jumped the gap from treating electromagnetic energy as a featureless energy density per unit volume or a featureless energy flow per unit surface, to treating it by adding the energy of localized moving electromagnetic photons enclosed in a unit volume or flowing through a unit surface, that would take localization into account and would represent just as well all observed electromagnetic phenomena at the macroscopic level, while also accounting for energy quanta localization at the submicroscopic level, and thus possibly eventually link up with Quantum Mechanics.

This is due to the fact that electromagnetic energy as theorized by Maxwell is described as a continuous wave phenomenon propagating in an underlying "ether", a concept that can hardly be done without from the continuous wave perspective, but which is not directly reconcilable with the concept of localized electromagnetic quanta moving separately, that would self-propel without any need for a supporting medium such as the "ether".

So, from the electromagnetism perspective, there also seemed to be no clear path leading to a clearer description of localized photons from Maxwell's more general electromagnetic theory. And similarly to the case of the success of Quantum Mechanics, even with Maxwell's rather nondescript continuous wave concept approach in the background, his equations nevertheless allow the most precise calculation of all aspects of electromagnetic energy that are useful at our macroscopic level.

A glimmer of hope remained, however, regarding Maxwell's equations proper, when considered separately from his wave theory. Louis de Broglie, who discovered the link between discrete quantum states and resonance states that inspired Schrödinger his wave equation, who then introduced the wave function and gave birth to Wave Mechanics, afterward enriched by Heisenberg and Feynman, upgrading it to full fledged Quantum Mechanics, also concluded in the early 1930's, that a permanently localized photon following a least action trajectory can satisfy at the same time Bose-Einstein's statistic and Planck's Law, perfectly explain the photoelectric effect while obeying Maxwell's equations, while remaining totally conform to the properties of Dirac's Hole Theory of complementary corpuscles symmetry, if it involves two particles, or half-photons of spin 1/2 ([15], p. 277).

The solution he subsequently elaborated in the 1930s, and 1940s by means of the wave function, although interesting, was not convincingly conclusive despite his best efforts, presumably because the ψ-function really cannot be reconciled with any description of a permanently localized moving photon.

Expanding the Space Geometry

Confronted with difficulties inherent in defining this concept of a localized double-particle photon by means of the psi-function, he eventually concluded in 1936 that it was impossible to exactly represent elementary particles in the reference frame of a continuous three dimensional space:

"*... the non-individuality of particles, the exclusion principle and exchange energy are three intimately related enigmas; all three are tied to the impossibility of exactly representing elementary physical entities within the frame of continuous three dimensional space (or more generally of continuous four dimensional space-time). Some day maybe, by escaping from this frame, will we better grasp the meaning, still quite cryptic today, of these major guiding principles of the new physics.*" ([15], p. 273).

Further analysis of the currently augmented pool of data and accumulated knowledge now allows establishing a clearly Maxwell equation's compliant electromagnetic description of the inner structure of localized electromagnetic photons, in line with de Broglie's hypothesis, and also of localized massive electromagnetic elementary particles, in the frame of the electromagnetic mechanics of elementary particles that can be defined in such an expanded space geometry.

This new space geometry was summarily proposed at Congress-2000 held at St Petersburg State University in July of 2000 [19,24], and the seminal considerations that gave rise to this expanded space geometry are exposed in reference [7].

This is of course not the first attempt at resolving the remaining issues of particle physics by considering higher dimensionality levels of spacetime, the most notable being the eleven dimensions M-theory, that apparently opens too many possibilities to easily identify a completely coherent foundation for particle physics.

Various approaches have been explored in these attempts, most of them involving compactification, which consists in defining extra dimensions that would not be significant from our macroscopic 3+1 spacetime perspective (3 spatial dimensions + time), but that become mathematically usable the deeper we explore towards the submicroscopic level. Various flavors explored 9+1, 25+1, 10+1 space time geometries and others. The reverse direction was also explored, involving that our spacetime be a 3+1 sub-spacetime belonging to a higher dimensional super-spacetime, which gave rise to brane theories. All attempts however, fundamentally involve multidimensional "single" spacetimes with various numbers of spatial dimensions plus one time dimension.

There is, however, one aspect of 4D Minkowski 3+1 spacetime that elicits universal agreement and is mathematically easy to deal with. It is the fact that all 4 dimensions of 3+1 spacetime are orthogonal to each other.

A close look at the stable set of elementary electromagnetic particles reveals also that orthogonality is also a fundamental characteristic of electromagnetic energy, and that there is universal agreement also that the momentum in space of an electromagnetic quantum is orthogonal to the electric aspect of the same quantum which is itself orthogonal to its magnetic aspects, which is the recognized triple orthogonality fundamental in electromagnetism.

Given the fact that increasing the number of dimensions in a single space exponentially increases the complexity, this obvious parallel between the orthogonal structures of both Minkowski's 3+1 spacetime and the electromagnetic structure common to all electromagnetic quanta gave rise to the idea that linking the electromagnetic orthogonality of energy to the orthogonal structure of the "space" concept, might reduce the mathematical complexity of the resulting model.

Then came into being the idea of separating the various orthogonal aspects of energy quanta among 3 orthogonal spaces that would co-exist and act as communicating vessels by means of a "junction area" or "junction-point", which junction-point would be the point-like behaving scatterable "object" that we identify as moving in normal space.

Thus, as described in reference [7], the momentum sustaining energy of an electromagnetic particle would be located in its own separate 3D space (X-space or normal space), and the energy of the same particle that oscillates between orthogonal electric and magnetic states would now oscillate between two other separate spaces, which would be a second 3D space (Y-space or electrostatic space) where energy displays electric characteristics and a third orthogonal 3D space (Z-space or magnetostatic space) where energy displays magnetic characteristics.

The orthogonal inner dimensions of each of these spaces can then be identified as X-x, X-y, X-z, Y-x, Y-y, Y-z and finally Z-x, Z-y and Z-z, all uniquely identified, the orthogonality of all three spaces being structurally established by means of assuming that the minor x-axes of all three spaces would be parallel to the conventional direction of motion of energy in normal space in plane wave treatment. A new superset of major **IJK** unit vectors would then identify globally each space while local minor unit vectors **ijk** would conserve their traditional function in each space.

This perspective immediately sheds new light on the issue of the sign of electric charges, given that they would henceforth "live" within Y-space. The electric charge of elementary particles can now be represented as a vector with a negative, positive or null sign in Y-space. The charge of the electron would then amount to momentum in the negative direction along the Y-x axis, that of the positron to momentum in the positive direction along the Y-x axis, and the null sign of de Broglie's half-photons' charges would become explainable by these charges oscillating in opposite directions on the Y-y/Y-z plane perpendicularly to the Y-x axis, as put in perspective in reference [7].

Such a trispatial structure also raises the question of the function of time in this new geometry. Would we be dealing with three 3-dimensional spaces plus time 3 × (3+1), or with a single 3-spaces complex plus time (3 × 3)+1?

Consistency mandates here that time would run at the same "speed", so to speak, for the various dynamic aspects of a given electromagnetic quantum. So, it also mandates that the passage of time would also be common to all possible electromagnetic quanta, each "living" in such space complexes, so (3 × 3)+1 appeared to be the best option.

But since a parameter common to all elements in a set cannot by definition be itself an element of that set, by very nature this parameter belongs to the reference frame of that set (it is an element of the superset), whatever other elements, if any, could belong to that superset. This hints at the possibility that time, which apparently progresses at a presumably constant "velocity", would be more fundamental than space. This issue is analyzed in reference [19].

Another clue that comes in support of this possibility is the fact that electromagnetic energy is induced strictly as a function of the "distance" between charged particles (the inverse square law), and NOT as a function of the time elapsed, because even when not supporting momentum, the adiabatic carrying energy induced in charged particles as a function of the inverse square of the distance between them remains adiabatically induced in them even when they are captive in electromagnetic equilibrium states that prevent translation, irrespective of the passage of "time" [4].

Defining a Distance Based Quantum of Action

At first glance, this idea seems to paradoxically go counter the fact that Planck's quantum of action h=6.626068759E-34 j·s (joules/second), that underlies quantum physics, is time based. However, there exists a distance based corresponding quantum of action not currently used in quantum physics.

This constant emerges from the fact that not only the frequency, but also the wavelength of a free moving electromagnetic quantum (a photon) depends solely on the amount of energy of this quantum. When relating this energy to its wavelength, the simple fact that a photon possessing twice the energy of another, requires a distance twice shorter in space to complete its cycle, is sufficient in and of itself to demonstrate that the photon's energy locally behaves as a totally incompressible material.

Given that the speed of light is constant in vacuum, it can thus be forcefully asserted that the quantity of energy constituting a photon's energy quantum is inversely proportional to the distance it must travel in vacuum for one cycle of its wavelength to be completed, which can be represented by $E=1/\lambda$.

This means that the product $E\cdot\lambda$ is a constant. Analyzing the various constants based definitions of energy reveals that by isolating these two variables in a new definition of energy defined in reference ([20], equation (11)), such a constant can be defined from the familiar set of known electromagnetic constants and the absolute wavelength of an energy quantum (λ), instead of Planck's quantum of action and its frequency:

$$E = hf = \frac{e^2}{2\varepsilon_0 \alpha \lambda} \tag{1}$$

Isolating product $E\lambda$ on the left side of this equation, leaving only the set of constants on the right side, then allowed defining this distance based quantum of action from the same set of known electromagnetic constants in reference ([21], equation (17)), where it was named the *electromagnetic intensity constant*:

$$H = E\lambda = \frac{e^2}{2\varepsilon_0 \alpha} = 1.98644544\mathrm{E}-25\,\mathrm{j \cdot m\,(joules \cdot meter)} \tag{2}$$

Dividing this constant by the speed of light (c), we then have the surprise of obtaining Planck's time based quantum of action from the same set of electromagnetic constants, which reveals that H=hc directly relates Planck's constant to electromagnetism:

$$h = \frac{H}{c} = 6.62606876\mathrm{E}-34\,\mathrm{j \cdot s(joules \cdot second)} \tag{3}$$

Incidentally, we indeed observe that combining equations (2) and (3) allows defining Planck's time based quantum of action from the same set of electromagnetic constants:

$$h = \frac{e^2}{2\varepsilon_0 \alpha c} = 6.626068757\mathrm{E}-34\,\mathrm{J \cdot s} \tag{4}$$

Close analysis shows that Planck's quantum of action is time based only due to the fact that it is equal to energy corresponding to 1 orbit that an electron would run about a hydrogen atom nucleus if it was free to so translate at the mean distance from the nucleus at which the psi-function averages out for the rest orbital of a hydrogen atom.

It was Louis de Broglie who discovered this relation as he observed that Planck's constant was exactly equal to the product of the electron Bohr orbit momentum by the length of the Bohr orbit, an orbit whose radius exactly matches the mean distance that which the probabilistic density of the psi function reaches maximum for the hydrogen atom ground state. Since the hydrogen rest orbital resonance state is key to establishing all other electronic orbital resonance states, this explains why Planck's quantum of action based QM provides so precise information about electronic orbitals:

$$h = m_o v \lambda_B = 6.62606876E\text{-}34 \text{ j.s} \tag{5}$$

Strangely, this precise Bohr atom electron momentum based definition of Planck's constant discovered by de Broglie is nowhere to be found in the formal literature, nor is any definition correlated to electromagnetic constants, neither at NIST nor in the CRC Handbook of Chemistry & Physics [3].

Even the obvious definition of h from equation (4) obtained from the known electromagnetic constants set derived from equation (1) is also nowhere to be found, which implies that h seems to still considered a measured constant, not a derived constant.

Since the Bohr orbit is λ_B=3.32491846E-10 meter long, the total amount of translational energy induced at the Bohr orbit can be obtained by multiplying Planck's quantum of action by the number of times this distance needs to be traveled in 1 second at the Bohr orbit classical velocity (v=2187691.253 m/s) for the total amount of the Bohr ground state energy to be accounted for (h multiplied by v/λ_B) which is why Planck's constant is related to time:

$$E_B = \frac{vh}{\lambda_B} = 4.359743808E - 18\,j \quad (27.21138346 \text{ eV}) \tag{6}$$

The reason why Planck's constant can be so precisely defined from the non-relativistic velocity calculated for the Bohr radius is precisely because the Bohr radius is obtained from the Coulomb equation, which allows calculating the correct amount of adiabatic energy induced at the real mean hydrogen atom rest orbital, thus the correct amount of electromagnetic energy corresponding to one orbital cycle.

The outcome is that dividing an amount of electromagnetic energy by Planck's constant provides the exact electromagnetic frequency of this amount of energy

$$f = \frac{E_B}{h} = 6.579683921E15 \text{ Hz} \tag{7}$$

and dividing the speed of light (c) by this electromagnetic frequency provides the electromagnetic wavelength of this amount of energy:

$$\lambda = \frac{c}{f} = 4.55633525E - 8\text{m} \tag{8}$$

which is the established procedure for calculating electromagnetic wavelength and frequency of energy quanta.

But from equation (2), dividing *the electromagnetic intensity constant* by the amount of energy induced at the Bohr orbit also provides the same absolute wavelength:

$$\lambda = \frac{H}{E_B} = 4.556335252E - 8\text{m} \tag{9}$$

Consequently, the Bohr ground state energy can be obtained from the distance based quantum of action and the absolute wavelength of the carrying energy induced at the Bohr orbit:

$$E_B = \frac{H}{\lambda} = 4.359743808E - 18\,j \tag{10}$$

Which disconnects fundamental energy calculation from any need to use the Bohr atom ground state orbit parameters, and rather relates it to electromagnetic parameters, and shows that energy calculation can be disconnected from the flow of time.

Separating the Carrying Energy of a Particle from the Energy of Its Rest Mass

One interesting outcome of the new definition of energy revealed by equation (1) is that it allows in ref. [20] to define local electric and magnetic fields to represent the energy of localized individual photons with the wavelength of an electromagnetic photon as the only variable, all other parameters being the well known set of electromagnetic constants:

$$\mathbf{E} = \frac{\pi e}{\varepsilon_0 \alpha^3 \lambda^2} \qquad \mathbf{B} = \frac{\mu_0 \pi e c}{\alpha^3 \lambda^2} \tag{11}$$

Interestingly, the same equations can directly represent the electric and magnetic fields of the rest mass energy of an electron by using the electron Compton wavelength:

$$\mathbf{E} = \frac{\pi e}{\varepsilon_0 \alpha^3 \lambda^2} \qquad \mathbf{B} = \frac{\mu_0 \pi e c}{\alpha^3 \lambda^2} \tag{12}$$

Having established in references [7,20] that the carrying energy of a particle such as the electron has the same electromagnetic structure as that of a free moving photon, this provided the opportunity to unify equations (11) and (12) to build relativistic field equations for the moving electron having the wavelength of the carrying energy and that of rest mass energy of the particle as the only variables. Simple addition and simplification of the magnetic fields parameters of both carrying energy and rest mass energy of the electron does provide the correct unified equation [20]:

$$\mathbf{B} = \frac{\pi \mu_0 e c}{\alpha^3} \frac{\left(\lambda^2 + \lambda_C^{\ 2}\right)}{\lambda^2 \lambda_C^{\ 2}} \tag{13}$$

But combining their electric fields turns out to be much more complex, because as mentioned previously, in the trispatial space structure, the charge of electrons is related to momentum in the negative direction along Y-x axis while the electric aspect of its carrying energy can only be momentum presumably due to oscillation on the Y-y/Y-z plane of the electromagnetic half of the carrying energy quantum.

Pending the eventual development of a specific integration procedure that would mathematically resolve this relation in Y-space, the issue can be indirectly resolved by redefining the relativistic velocity parameter v in equation **E**=v**B** stemming from the Lorentz force equation, to involve only the wavelengths of the energies of the carrying energy and of the rest mass energy of the particle [20,22]:

$$v = c\frac{\sqrt{4\lambda\lambda_C + \lambda_C^{\ 2}}}{2\lambda + \lambda_C} \tag{14}$$

So, by multiplying equation (14) defining the value of v, by relativistic equation (13) defining the value of **B**, the following relativistic electric fields equation complementary to magnetic fields equation (13) can be obtained for the moving electron [21]:

$$\mathbf{E} = \frac{\pi e}{\varepsilon_0 \alpha^3} \frac{(\lambda^2 + \lambda_C^2)\sqrt{\lambda_C(4\lambda + \lambda_C)}}{\lambda^2 \lambda_C^2 \quad (2\lambda + \lambda_C)} \tag{15}$$

From equations (13) and (15), any relativistic electron velocity can now be calculated from the wavelength of its carrying energy and the rest mass energy wavelength of the electron as the only variables, with the usual equation v=**E**/**B**.

The Trispatial LC Equation for Permanently Localized Photons in the 3-Spaces Geometry

The next equation is a trispatial LC equation developed in reference [7] showing the momentum sustaining half of a free moving photon's energy located in X-space as it propels its other half, which is "translationally inertly" oscillating between Y-space and Z-space. Since the only energy that can sustain longitudinal momentum in space is located in X-space, this second half of a photon's energy is translationally inert within Y-spaces and Z-spaces along their x axes:

$$E\vec{I}\,\vec{i} = \left(\frac{hc}{2\lambda}\right)_X \vec{I}\,\vec{i} + \left[\begin{array}{l} 2\left(\dfrac{e^2}{4C}\right)_Y (\vec{J}\,\vec{j}, \vec{J}\,\overleftarrow{j})\cos^2(\omega t) \\ +\left(\dfrac{L\,i^2}{2}\right)_Z \overleftrightarrow{K}\,\sin^2(\omega t) \end{array}\right] \tag{16}$$

where

$$C = 2\varepsilon_0 \alpha\lambda \quad L = \frac{\mu_0 \alpha\lambda}{8\pi^2} \quad i = \frac{2\pi ec}{\alpha\lambda} \quad \omega = \frac{2\pi c}{\alpha\lambda} \tag{17}$$

This even split of a photon's energy between an amount of momentum sustaining energy, propelling an equal amount of electromagnetic energy transversely oscillating within two perpendicularly oriented mutually orthogonal 3D spaces, is what explains in this space geometry why the speed of light can only be constant in vacuum [7].

For simplicity, this oscillating structure allows observing that the two half-photons of Louis de Broglie's hypothesis (two electric charges) are shown as oscillating along the Y-y axis. Given that in the case of a photon, no motion is possible along the perpendicular Y-x axis in this space geometry, this provides a possible explanation to the observed null value of the electric charges presumed to exist in electromagnetic photons in de Broglie's hypothesis, since that in this space geometry the minus sign of the electron charge is related to momentum sustaining energy being oriented in the negative direction along the Y-x axis, while the positive sign of the positron charge is related to the momentum sustaining energy being oriented in the positive direction along this axis. This will be made more obvious with equations (20) and (21) that define the trispatial LC equations of electrons and positrons.

Replacing the inductance and capacitance representations by their equivalent electric and magnetic fields representations shown as equations (11) allows observing them oscillating from one state to the other within the Y-space/Z-space complex in the trispatial space geometry in relation with the momentum sustaining energy of the particle in normal X-space:

$$E\vec{I}\,\vec{i} = \left(\frac{hc}{2\lambda}\right)_X \vec{I}\,\vec{i} + \left[\begin{array}{l} 2\left(\dfrac{\varepsilon_0 \mathbf{E}^2}{4}\right)_Y (\vec{J}\,\vec{j}, \vec{J}\,\overleftarrow{j})\cos^2(\omega t) \\ +\left(\dfrac{\mathbf{B}^2}{2\mu_0}\right)_Z \overleftrightarrow{K}\,\sin^2(\omega t) \end{array}\right] V \tag{18}$$

Where V is the related *theoretical stationary isotropic volume* that the incompressible oscillating kinetic energy quantum would occupy if it was immobilized as a sphere of isotropic density, as defined in ref. [20]:

$$V = \frac{\alpha^5 \lambda^3}{2\pi^2} \tag{19}$$

The Trispatial LC Equations Describing the Rest Masses of the Electron and the Positron

The trispatial LC equations derived in reference [11], describe the inner circulation of the energy constituting the invariant rest masses of the electron and positron after the decoupling of the 1.022 MeV mother photon. The trispatial LC equation for the electron is thus:

$$E\vec{0} = m_e c^2 \vec{0} = \left[\frac{hc}{2\lambda_C}\right]_Y \vec{J}\,\overleftarrow{i} + \left(\begin{array}{l} 2\left[\dfrac{(e')^2}{4C_C}\right]_X (\vec{I}\,\vec{j}, \vec{I}\,\overleftarrow{j})\cos^2(\omega t) \\ +\left[\dfrac{L_C i_C^2}{2}\right]_Z \overleftrightarrow{K}\,\sin^2(\omega t) \end{array}\right) \tag{20}$$

And for the invariant rest mass of a positron:

$$E\vec{0} = m_e c^2 \vec{0} = \left[\frac{hc}{2\lambda_C}\right]_Y \vec{J}\,\vec{i} + \left(\begin{array}{l} 2\left[\dfrac{(e')^2}{4C_C}\right]_X (\vec{I}\,\vec{j}, \vec{I}\,\overleftarrow{j})\cos^2(\omega t) \\ +\left[\dfrac{L_C i_C^2}{2}\right]_Z \overleftrightarrow{K}\,\sin^2(\omega t) \end{array}\right) \tag{21}$$

where λ_c is the electron Compton wavelength.

These representations allow observing that the electric momentum sustaining half of the particle's energy located in Y-space is oriented in the negative direction along the Y-z axis for the electron and in the positive direction for the positron. We can also observe that no energy remains to induce any momentum along the X-x axis of normal space since the energy now oscillating between Z-space and X-space can now only oscillate on the X-y/X-z plane due the constraints of the decoupling process [11], which is oriented perpendicularly to axis X-x, which is the only direction that allows momentum to be expressed as a velocity in plane wave treatment in the trispatial geometry. This oscillation is represented here as being aligned along the X-y axis.

The Trispatial LC Equations Describing a Moving Electron in the Trispatial Geometry

Equations (13) and (15) previously established the inner structure of the relativistic electric and magnetic fields of a moving electron, whose velocity can then be calculated with equation v=**E**/**B**.

In Table 1 equations (16) and (20) are used to provide a trispatial LC representation of the same electron moving at relativistic velocity, by using the fields representations of equations (11) for the carrying energy and the fields representation of equation (12) for the rest mass energy of the electron. It can be observed that only possible momentum sustaining energy has to be located in normal X-space along X-x axis.

Pending the potential development some more advanced integration means to unify further these equations, this table seems to be the next best unifying presentation of the electron in motion in the trispatial space complex.

The Last Challenge

Now that moving electromagnetic photons and massive electrons and positrons have been described in the trispatial space geometry, time has come to address the issue of the last two remaining members of the stable set, the up and down quarks, which are the only charged and massive scatterable elementary components of all atomic nuclei, and that up to now have not been linked to the series of kinetic energy

	Direct kinetic energy in X normal-space	Energy located in Y and Z spaces making up the inert mass of the particle
Rest mass energy (m_0c^2)		$\left\{\left(\frac{\varepsilon_0 E_e^2}{2}\right)_Y \vec{J}\overleftarrow{i}+\left(\frac{B_e^2}{2\mu_0}\right)\overleftrightarrow{K}\right\}V_{me}$
Carrying energy K	$\left(\frac{hc}{2\lambda}\right)_X \vec{I}\vec{i}$	$\left\{2\left(\frac{\varepsilon_0 E_k^2}{4}\right)_Y (\vec{J}\overleftarrow{i}\vec{J}\overleftarrow{i})+\left(\frac{B_K^2}{2\mu_0}\right)\overleftrightarrow{K}\right\}V_K$
Total relativistic mass energy (mc^2)		$\left[\left\{\left(\frac{\varepsilon_0 E_e^2}{2}\right)_Y \vec{J}\overleftarrow{i}+\left(\frac{B_e^2}{2\mu_0}\right)_z \overleftrightarrow{K}\right\}V_{m_o}+V_K\left(\frac{B_{eK}^2}{2\mu_0}\right)\overleftrightarrow{K}\right]$

Table 1: Combined fields equations of the moving electrons and its carrier-photon.

Table of the energies contained in the effective masses of quarks Up and Down, estimated on the assumptions that unit charge would be a measure of decoupling distance of electron/positron pairs in electrostatic space			
Particle	$r'=a_0\,\alpha$	$E=K/r^2$	$\lambda=hc/E$
Electron	r'_e=3.861592641E-13 m	0.5109989027 MeV	2.426310215E-12 m
Quark up	r'_{eu}=2.574395094E-13 m	1.149747531 MeV	1.078360096E-12 m
Quark down	r'_{ed}=1.287197547E-13 m	4.598990173 MeV	2.69590021E-12 m

Table 2: Calculated effective rest mass energies of up and down quarks.

transformation processes that unites the other members of the set.

Since up and down quarks "live" in the nuclei of atoms, quantum of action constants h and H that are quite appropriate to calculate the momentum sustaining energy of elementary particles, are not appropriate to deal with energy induction, since this energy is induced as a function of the inverse square of the distance separating any two electrically charged particles, which implies use a radial, or axial, distance with respect to the wavelength, even with their proper definitions from the known electromagnetic constants set (ref: equations (2) and (4)).

As observed with equation (6), calculation of the Bohr ground state energy makes no direct reference to the distance between the electron and the nucleus, and calculates the correct amount of energy strictly from orbital considerations that are fundamentally perpendicular to the direction of energy induction.

What is required is a constant acting axially, that is, perpendicularly to the plane on which the translational motion of an electron can be expressed, which is representable by the Hamiltonian.

Such an appropriate energy induction constant can be defined from the Coulomb equation, since this equation effectively calculates the energy induced at the Bohr orbit as a function of the inverse square of the actual distance separating the Bohr orbit from the central proton. We can thus write that at distance r_B the energy induced will be:

$$E_B = F_B r_B = \frac{e^2}{4\pi\varepsilon_0 r_B} = 4.359743805\text{E}-18\ \text{Joules} \tag{22}$$

which matches the energy calculated with equation (6) from orbital considerations and with equation (10) from electromagnetic considerations.

This precise quantity of kinetic energy is permanently and adiabatically induced at the mean hydrogen rest orbital [4] and does not depend on the time elapsed as previously highlighted. The only possible way for this amount of energy to vary is for the distance between the electron and the proton to vary.

The required *electrostatic energy induction constant*, that we will name K and that could be seen as the *"quantum of induction"*, can be established in two different manners, the fist method stems from the analysis of the manner in which a photon of energy 1.022 MeV or more can decouple into a pair of electron-positron in the 3-spaces geometry as established in reference [11], and the second method consists in simply multiplying equation (22) by r_B squared:

$$K = E_B \cdot r_B^{\ 2} = \frac{e^2 \cdot r_B}{4\pi\varepsilon_0} = 1.220852596\text{E}-38\,\text{j}\cdot\text{m}^2 \tag{23}$$

With this constant, it is possible to enter the hydrogen nucleus "vertically" or "axially", so to speak, by varying the distance r between two charged particles in equation $E=K/r^2$, and so establish the exact amounts of adiabatic energy induced in each of the inner components of the proton and the neutron (Table 2), thus allowing us to finally establish coherent trispatial LC equations for the up and down quarks and their carrier-photons, as analyzed in reference [23].

Indeed, dealing with axial energy induction in atomic structures seems to be the only way that such a space geometry can be explored, which induces acute awareness of the adiabatic levels of energy permanently induced in all massive particles making up massive objects and which is not representable by the Hamiltonian when translational motion, thus momentum, is prevented from being expressed by the translationally immobilizing electromagnetic equilibrium states that they generally are captive in. The issue of axial adiabatic energy induction in atoms is analyzed in reference [4].

This analysis highlighted the surprising fact that although the physics community has been aware since Coulomb that energy is induced as a function of the inverse square of the distance between charged particles, and since the beginning of the 20th century that charged particles organize axially in atomic structure, Classical Mechanics, Relativistic Mechanics, Quantum Electrodynamics, Electromagnetic Theory and Quantum mechanics all apparently still deal with energy "horizontally" so to speak, as witnessed by the fact that the Hamiltonian, basic to quantum physics, and stemming from a reformulation of Classical Mechanics, can fundamentally represent energy only if it involves the momentum of a "moving" particle, which causes it to be unable, for example, to represent the adiabatic amount of 27.2 eV energy induced at the hydrogen ground state if the electron is translationally immobilized by the local electromagnetic equilibrium state.

Clear awareness of adiabatically stabilized energy in atomic structures also sheds new light on gravitation and on how the collected data on spacecrafts' hyperbolic trajectories, systematic so-called anomalous spacecrafts' flybys accelerations, and systematic so-called anomalous rotation slowdown of all spacecrafts, can be interpreted [9,10,25-28].

The Fractional Charges of Up and Down Quarks

In Y-space, the sign intensity of the positive and negative electric charges of electrons and positrons is tied to the distance from the trispatial junction-point in Y-space (r' in Table 2) at which their electric energy expresses their momentum in opposite directions parallel to the Y-x axis. The diminished charges of the up and down quark are thus related in the trispatial geometry to the precise shorter distances that the stress of their equilibrium states forces them to express their momentum in opposite directions parallel to this axis within the structure of nucleons [23] See parameter r' shown in Table 2.

In the trispatial geometry, momentum that cannot be expressed as a "velocity", is expressed as a "measurable pressure" in the direction of application of the Coulomb force in X-space in the case of the least action electromagnetic equilibrium states [4], and is expressed as a "measurable intensity" of the electric charge of a particle in Y-space [7,23]. Their related increased rest masses are similarly related to these shorter distances as a function of the previously mentioned axial inverse square law of distance from the trispatial junctions [23] shown in Table 3.

The Trispatial LC Equations of the Up and Down Quarks

The outcome of this axial exploration of the inner structure of nucleons came in support of the possibility that up and down quarks could simply be positrons and electrons whose masses and charge characteristics would be warped into these altered states by the stresses imposed by these most energetic least action equilibrium states that electrons and positrons can reach in nature [4,23].

The trispatial LC equation for the up quarks is:

$$m_U=\frac{E_U}{c^2}=\frac{1}{c^2}\left\{\begin{array}{l} S_U\left[\dfrac{hc}{2\lambda_U}\right]_Y \\ +(2-S_U)\left[\begin{array}{l}2\left(\dfrac{(e')^2}{4C_U}\right)_X\cos^2(\omega t)\\ +\left(\dfrac{L_U i_U{}^2}{2}\right)_Z\sin^2(\omega t)\end{array}\right]\end{array}\right\} \tag{24}$$

where λ_u is the wavelength of the energy making up the invariant rest mass of the up quark, and S_u is the up quark *magnetic drift stress constant* [23], with dimensionless value 2/3.

And the trispatial LC equation for the down quark is:

$$m_D=\frac{E_D}{c^2}=\frac{1}{c^2}\left\{\begin{array}{l} S_D\left[\dfrac{hc}{2\lambda_D}\right]_Y \\ +(2-S_D)\left[\begin{array}{l}2\left(\dfrac{(e')^2}{4C_D}\right)_X\cos^2(\omega t)\\ +\left(\dfrac{L_D i_D{}^2}{2}\right)_Z\sin^2(\omega t)\end{array}\right]\end{array}\right\} \tag{25}$$

Where λ_d is the wavelength of the energy making up the invariant rest mass of the down quark, and S_d is the down quark *magnetic drift stress constant* [23], with dimensionless value 1/3.

In both cases, the trispatial LC equations describing the carrying energy of up and down quarks are identical to equation (16) for the permanently localized photon.

Since the three quarks of a proton (uud) as well as those of a neutron (udd) simultaneously translate about two different axes in the tri-spatial geometry [23], that is the coplanar axis Y-z and the normal space X-x axis, there would be need to build 6 tables such as Table 1 to represent each possible configuration of the three quarks whose motion about coplanar Y-z axis would be sustained by their carrier-photons as perceived from X-space, each possessing an energy of approximately 310 MeV [23], Table III), and thee more such Tables to represent each carrier-photon being considered the propelled particle as perceived from Y-space, being propelled by the quarks then acting as their carrier-photons, sustaining their motion about the X-x axis.

Obviously, this set of trispatial LC equations is only an entry step into this space geometry, considering that they seem to already have reached their representational limit with these table representations.

Conclusion

These equations summarize the description of all stable and massive point-like behaving electromagnetic particles that have been experimentally detected at the sub-microscopic level. The trispatial LC descriptions of electron, muon and tau particles before they release momentary excess mass in the form of neutrinos are derived in reference [14].

Similar trispatial LC equations can of course be defined for all point-like behaving scatterable electromagnetic sub-components of all unstable partons that were detected, but their description exceeds the scope of the present paper, and are not required to describe normal matter, since they exist only fleetingly and amount to practically nothing in the universe since their short life span prevents any accumulation of these particles.

These conclusions from the analysis of the manner in which electromagnetic energy is likely to behave in this expanded space geometry are tentative entry level in many respects and may require some re-focusing to better formulations, and may even be overly speculative pending experimental confirmation, which means that deep formal analysis remains to be carried out.

But complete mathematization of axially induced adiabatic energy in atomic structures, and whose existence becomes so obvious in the 3-spaces model, could bring to fruition some important applied physics benefits hinted at in reference [4], that will remain out of reach until such mathematization has been accomplished.

So after having thoroughly explored the momentum based "translational plane" of particle physics, so to speak, mainly by means of the Hamiltonian, the last challenge of modern physics may really be to finally go 3 dimensional to finally integrate the so promising orthogonally oriented adiabatic energy induction process.

	Up quark	Down quark
Rotation diameter	r=r'sin 60º=3.344237326E-13 m	
Rotation radii	2r/3=2.229491551E-13 m	r/3=1.114745775E-13 m
Orbit lengths D=2πr	1.400830855E-12 m	7.004154277E-13 m
Quark masses in kg m=E•1.6E-19/c²	2.049610923E-30 kg	8.198443779E-30 kg

Table 3: Relation between up and down quarks masses and their translational and rotational radii about the Y-z axis and the X-x axis in the trispatial geometry.

References

1. Kotler S, Akerman N, Navon N, Glickman Y, Ozeri R (2014) Measurement of the magnetic interaction between two bound electrons of two separate ions. Nature 510: 376-380.
2. Breidenbach M, Friedman JI, Kendall HW, Bloom ED, Coward DH, et al. (1969) Observed Behavior of Highly Inelastic Electron-Proton Scattering. Phys Rev Lett 23: 935-939.
3. Lide DR (2003) CRC Handbook of Chemistry and Physics. 84th Edition 2003-2004 CRC Press, New York.
4. Michaud A (2016) On Adiabatic Processes at the Elementary Particle Level. J Phys Math 7: 177.
5. Ciufolini I, Wheeler JA (1995) Gravitation and Inertia. Princeton University Press.
6. Feynman R (1949) Space-Time Approach to Quantum Electrodynamics. Phys Rev 76: 769.
7. Michaud A (2016) On De Broglie's Double-particle Photon Hypothesis. J Phys Math 7: 153.
8. Lowrie W (2007) Fundamentals of Geophysics. Second Edition Cambridge University Press.
9. Michaud A (2013) The Corona Effect. International Journal of Engineering Research and Development Volume 7: 01-09.
10. Michaud A (2013) Inside Planets and Stars Masses. International Journal of Engineering Research and Development 8: 10-33.
11. Michaud A (2013) The Mechanics of Electron-Positron Pair Creation in the 3-Spaces Model. International Journal of Engineering Research and Development 6: 36-49.
12. Michaud A (2004) Expanded Maxwellian Geometry of Space. 4th Edition, SRP Books.
13. Michaud A (2016) Electromagnetic Mechanics of Elementary Particles. Scolar's Press.
14. Michaud A (2013) The Mechanics of Neutrinos Creation in the 3-Spaces Model. International Journal of Engineering Research and Development 7: 01-08.
15. De Broglie L (1937) New physics and quanta, Flammarion, 2nd 1993 new Preface.
16. Curie I, Joliot F (1933) Comptes Rendus.
17. McDonald K, Burke DL, Field RC, Horton-Smith G, Spencer JE, et al. (1997) Positron Production in Multiphoton Light-by-Light Scattering. Phys Rev Lett 79: 1626.
18. Einstein A, Schrödinger E, Pauli W, Rosenfeld L, Born M, Joliot-Curie IF, et al. (1953) Louis de Broglie, physicist and thinker. Editions Albin Michel, Paris.
19. Michaud A (2016) The Birth of the Universe and the Time Dimension in the 3-Spaces Model. American Journal of Modern Physics. Special Issue: Insufficiency of Big Bang Cosmology 5: 44-52.
20. Michaud A (2007) Field Equations for Localized Individual Photons and Relativistic Field Equations for Localized Moving Massive Particles. International IFNA-ANS Journal 13: 123-140.
21. Michaud A (2013) The Expanded Maxwellian Space Geometry and the Photon Fundamental LC Equation. International Journal of Engineering Research and Development 6: 31-45.
22. Michaud A (2013) From Classical to Relativistic Mechanics via Maxwell. International Journal of Engineering Research and Development 6: 01-10.
23. Michaud A (2013) The Mechanics of Neutron and Proton Creation in the 3-Spaces Model. International Journal of Engineering Research and Development 7: 29-53.
24. Michaud A (2000) On an Expanded Maxwellian Geometry of Space. Proceedings of Congress-2000, Russia.
25. Anderson JD (2005) Study of the anomalous acceleration of Pioneer 10 and 11. Cornell University Library.
26. Anderson JD, Laing PA, Lau, Liu EL, Nieto MM, et al. (1998) Indications from Pioneer 10/11, Galileo, and Ulysses Data, of an Apparent Anomaleous, Weak, Long-Range Acceleration.Cornell University Library.
27. Nieto, Goldman, Anderson, Lau and Perez-Mercader (1994) Theoretical Motivation for Gravitation Experiments on Ultra low Energy Antiprotons and Antihydrogen. Cornell University Library.
28. Anderson JD, Campbell JK, Nieto MM (2006) The energy transfer process in planetary flybys, astrology. Cornell Library university

Existence and Smoothness of the Navier-Stokes Equation in Two and Three-Dimensional Euclidean Space

Tim Tarver*

Department of Mathematics, Bethune-Cookman University, USA

Abstract

A solution to this problem has been unknown for years and the fact that it hasn't been solved yet leaves a lot of unanswered questions regarding Engineering and Pure Mathematics. Turbulence is a specific topic in fluid mechanics which is a vital part of the course when it comes to real life situations. In two and three dimensional systems of equations and some initial conditions, if the smooth solutions exist, they have bounded kinetic energy. In three space dimensions and time, given an initial velocity vector, there exists a velocity field and scalar pressure field which are both smooth and globally defined that solve the Navier-Stokes equations. There are difficulties in two-dimensions and three dimensions in a possible solution and which have been unsolved for a long time and our goal is to propose a solution in three-dimensions. Lets see if we can relate a couple of courses of pure mathematics to come up with an implication.

Keywords: Navier-Stokes equation; Three-dimensional Euclidean space

Introduction

The set of real numbers R^n can also be identified as the n-dimensional Euclidean Space if we wish to emphasize its Euclidean nature. It is mentioned that the Euler and Navier-Stokes equations describe the motion of a fluid in the Euclidean Space R^n, where n could equal 2 or 3 and that these equations are to be solved for an unknown velocity vector $\vec{u}(x,t)=(\vec{u}_i(x,t))_{1\le i\le n}\in\mathbb{R}^n$ and a pressure $p(x,t)$defined for position $x\in R^n$ and time $t \geq 0$. It is also mentioned that we restrict attention here to incompressible fluids filling all of R^n. The Navier-Stokes equations are given by

$$\frac{\partial \vec{u}_i}{\partial t}+\sum_{j=1}^{n}\vec{u}_j\frac{\partial \vec{u}_i}{\partial x_j}=v\Delta\vec{u}_i-\frac{\partial p}{\partial x_i}+\vec{f}_i(x,t),(x\in\mathbb{R}^n,t\ge 0), \tag{1}$$

and the divergence of the velocity field $\vec{u}$ yields

$$div\,\vec{u}=\sum_{i=1}^{n}\frac{\partial \vec{u}_i}{\partial x_i}=0,(x\in\mathbb{R}^n,t\ge 0) \tag{2}$$

with initial condition yielding,

$$\vec{u}(x,0)=\vec{u}^o(x)(x\in\mathbb{R}^n). \tag{3}$$

Our given $\vec{u}^o(x)$ is said to be a C^∞ divergence-free *vector field* on R^n and $\vec{f}_i(x,t)$ is the components of our given constant applied force. For example, *gravity* is a continuous force. A constant v is a positive coefficient for *viscosity*, and

$$\Delta_x=\sum_{i=1}^{n}\frac{\partial^2}{\partial x_i^2}$$

is the Laplacian in any given space. The *Euler equations* are the previous three with v set equal to zero. Equation 1 is Newton's Second Law of Motion $f=m\vec{a}$ for a fluid element subjected to the external force $\vec{f}=(\vec{f}_i(x,t))_{1\le i\le n}$ and to the forces being created from pressure and friction.

We can rearrange equation 1 to look like Newton's Second Law of a fluid element such that,

$$\frac{\partial \vec{u}_i}{\partial t}+\sum_{j=1}^{n}\vec{u}_j\frac{\partial \vec{u}_i}{\partial x_j}=v\Delta\vec{u}_i-\frac{\partial p}{\partial x_i}+\vec{f}_i(x,t)$$

$$\frac{\partial \vec{u}_i}{\partial t}+\sum_{j=1}^{n}\vec{u}_j\frac{\partial \vec{u}_i}{\partial x_j}-v\Delta\vec{u}_i+\frac{\partial p}{\partial x_i}=\vec{f}_i(x,t)$$

$$\vec{f}_i(x,t)=\frac{\partial \vec{u}_i}{\partial t}+\sum_{j=1}^{n}\vec{u}_j\frac{\partial \vec{u}_i}{\partial x_j}-v\Delta\vec{u}_i+\frac{\partial p}{\partial x_i}.$$

The unit vector $\vec{u}_j$ can be brought onto the other side of the summation. Since $\Delta=\nabla^*\nabla=\nabla^2$ our equation now yields,

$$\vec{f}_i(x,t)=\frac{\partial \vec{u}_i}{\partial t}+\vec{u}_j\sum_{j=1}^{n}\frac{\partial \vec{u}_i}{\partial x_j}-v\nabla^2\vec{u}_i+\frac{\partial p}{\partial x_i}.$$

Setting the divergence of the velocity field equal zero will specify that it is the *incompressible continuity equation*. Since we have initial conditions on the velocity field $\vec{u}$, we could possibly yield initial conditions on the force and scalar fields.

Body

The Navier-Stokes equation for an ideal fluid with zero viscosity states that the acceleration is proportional to the *derivative* of *internal pressure*. As a result, the solutions of the Navier-Stokes equation for a given physical problem must be found with the help of *calculus*. One possible way to solving the N.S. equation is to use the *conservation of mass* with boundary conditions in a system of linear or non-linear equations to produce a solution. In wave mechanics, or wave theory, "waves in one-dimension is said to be called *plane waves*. In two-dimensions, the waves are said to be called *cylindrical waves*. In three-dimensions, the waves are said to be called *spherical waves*" [1]. There is an initial velocity vector $\vec{u}^o(x)$ and a divergence-free vector field $\vec{u}_i$ on R^n. The force field $\vec{f}_i(x,t)$ is the component of a given external applied force i.e. *gravity*. The scalar v is the viscosity and

$$\Delta=\sum_{i=1}^{n}\frac{\partial^2}{\partial x_i^2}$$

is the Laplacian with respect to x in the space variables. The Euler equations are numbers 3, 4, and 5 with the viscosity v set equal to zero.

***Corresponding author:** Tim Tarver, Professor, Department of Mathematics, Bethune-Cookman University, USA, E-mail: ttarver31@gmail.com

The goal here is to prove letter A in the paper that states solutions of the Navier-Stokes equation exist on R^3.

Lets take $v>0$ and $n=3$. Let $\vec{u}^{o}(x)$ be *any* smooth, divergence-free vector field satisfying equation 4 stated in the proposal of C. Fefferman. We will also take $\vec{f}_i(x,t)$ to be equal to zero. Then there exists smooth functions $p(x,t)$ and $\vec{u}_i(x,t)$ on R^3 x$[0,\infty)$ that satisfy equations 1, 2, 3, 6, and 7. Going back to our equation of motion of a fluid element,

$$\vec{f}_i(x,t)=\frac{\partial\vec{u}_i}{\partial t}+\vec{u}_j\sum_{j=1}^{n}\frac{\partial\vec{u}_i}{\partial x_j}-v\nabla^2\vec{u}_i+\frac{\partial p}{\partial x_i}$$

we let $\vec{f}_i(x,t)=0$ such that,

$$\vec{f}_i(x,t)=\frac{\partial\vec{u}_i}{\partial t}+\vec{u}_j\sum_{j=1}^{3}\frac{\partial\vec{u}_i}{\partial x_j}-v\nabla^2\vec{u}_i+\frac{\partial p}{\partial x_i}=0.$$

Since we let $n=3$ we have our unit vector interval $1\le j\le 3$and our $1\le i\le 3$ for the force, velocity vectors, and number of positions in space variables. Thus we have a set of force and velocity vectors. Newton's Second Law states that *The acceleration* $\vec{a}$ *of a body is parallel and directly proportional to a net force* $\vec{f}$ *and inversely proportional to a mass m such that* $\vec{f}=m\vec{a}$. In this case, the acceleration could be defined as the partial derivative of the velocity with respect to time from i to *infinity*. Since the acceleration is the derivative of the velocity and in this case we use partial derivatives with respect to time to describe the nature of the fluid such that,

$$\vec{a}=\frac{\partial\vec{u}_i}{\partial t}$$

where $1\le i\le 3$. The net force is inversely proportional to a corresponding mass and velocity field with respect to time $t\ge 0$ which yields,

$$\{\vec{f}_i(x_i,t)\}=\frac{\partial\{\vec{u}_i\}}{\partial t}+\{\vec{u}_j\}\sum_{j=1}^{3}\frac{\partial\{\vec{u}_i\}}{\partial\{x_j\}}-v\nabla^2\{\vec{u}_i\}+\frac{\partial p}{\partial\{x_i\}}=0.$$

Hence, the sequence yields,

$$\vec{f}_1(x_1,t)=\frac{\partial\vec{u}_1}{\partial t}+\vec{u}_1\sum_{j=1}^{3}\frac{\partial\vec{u}_1}{\partial x_1}-v\nabla^2\vec{u}_1+\frac{\partial p}{\partial x_1}=0$$

$$\vec{f}_2(x_2,t)=\frac{\partial\vec{u}_2}{\partial t}+\vec{u}_2\sum_{j=1}^{3}\frac{\partial\vec{u}_2}{\partial x_2}-v\nabla^2\vec{u}_2+\frac{\partial p}{\partial x_2}=0$$

$$\vec{f}_3(x_3,t)=\frac{\partial\vec{u}_3}{\partial t}+\vec{u}_3\sum_{j=1}^{3}\frac{\partial\vec{u}_3}{\partial x_3}-v\nabla^2\vec{u}_3+\frac{\partial p}{\partial x_3}=0.$$

Different types of partial differential equations often need to be matched with different types of boundary conditions in order for their solutions to *exist* and be *unique*.

Suppose that the force field $\vec{f}$ is not equal to zero. Let's set $\vec{f}$ to be arbitrary. Then, we go back to the initial equation derived to be,

$$f_i(x_i,t)=\frac{\partial\vec{u}_i}{\partial t}+\sum_{=1}\vec{u}_j\frac{\partial\vec{u}_i}{\partial x_j}-v\Delta\vec{u}_i+\frac{\partial}{\partial x_i} \qquad (4)$$

where n equals three. Recall back to Newton's Second Law of Motion, $F=ma$. Now, let's translate this equation into partial derivative terms. We have the acceleration as $\frac{\partial\vec{u}_i}{\partial t}$ with out mass m to equal the force field as follows,

$$\vec{f}_i(x_i,t)=m\frac{\partial\vec{u}_i}{\partial t}.$$

Using the equation of force above and solving for the acceleration, we get

$$\frac{\partial\vec{u}_i}{\partial t}+\sum_{j=1}^{3}\vec{u}_j\frac{\partial\vec{u}_i}{\partial x_j}-v\Delta\vec{u}_i+\frac{\partial p}{\partial x_i}=m[v\Delta\vec{u}_i-\sum_{j=1}^{3}\vec{u}_j\frac{\partial\vec{u}_i}{\partial x_i}-\frac{\partial p}{\partial x_i}+\vec{f}_i(x_i,t)].$$

After plugging in the given functions, we distribute mass inside to obtain

$$\frac{\partial\vec{u}_i}{\partial t}+\sum_{j=1}^{3}\vec{u}_j\frac{\partial\vec{u}_i}{\partial x_j}-v\Delta\vec{u}_i+\frac{\partial p}{\partial x_i}=mv\Delta\vec{u}_i-m\vec{u}_j\sum_{j=1}^{3}\frac{\partial\vec{u}_i}{\partial x_j}-m\frac{\partial p}{\partial x_i}+m\vec{f}_i(x_i,t). \qquad (5)$$

We substitute the force field back in using equation 4, we get

$$\frac{\partial\vec{u}_i}{\partial t}+\sum_{j=1}^{3}\vec{u}_j\frac{\partial\vec{u}_i}{\partial x_j}-v\Delta\vec{u}_i+\frac{\partial p}{\partial x_i}=mv\Delta\vec{u}_i-m\vec{u}_j\sum_{j=1}^{3}\frac{\partial\vec{u}_i}{\partial x_j}-m\frac{\partial p}{\partial x_i}+m[\frac{\partial\vec{u}_i}{\partial t}+\sum_{j=1}^{3}\vec{u}_j\frac{\partial\vec{u}_i}{\partial x_j}-v\Delta\vec{u}_i+\frac{\partial p}{\partial x_i}].$$

After another distribution of the mass into force, it becomes

$$\frac{\partial\vec{u}_i}{\partial t}+\sum_{j=1}^{3}\vec{u}_j\frac{\partial\vec{u}_i}{\partial x_j}-v\Delta\vec{u}_i+\frac{\partial p}{\partial x_i}=mv\Delta\vec{u}_i-m\vec{u}_j\sum_{j=1}^{3}\frac{\partial\vec{u}_i}{\partial x_j}-m\frac{\partial p}{\partial x_i}+m\frac{\partial\vec{u}_i}{\partial t}+m\sum_{j=1}^{3}\vec{u}_j\frac{\partial\vec{u}_i}{\partial x_j}-mv\Delta\vec{u}_i+m\frac{\partial p}{\partial x_i}.$$

Next, algebraic manipulation is necessary to cancel out all like terms to represent,

$$\frac{\partial\vec{u}_i}{\partial t}+\vec{u}_j\sum_{j=1}^{3}\frac{\partial\vec{u}_i}{\partial x_j}-v\Delta\vec{u}_i+\frac{\partial p}{\partial x_i}=m\frac{\partial\vec{u}_i}{\partial t}$$

$$\vec{u}_j\sum_{j=1}^{3}\frac{\partial\vec{u}_i}{\partial x_j}-v\Delta\vec{u}_i+\frac{\partial p}{\partial x_i}=m\frac{\partial\vec{u}_i}{\partial t}-\frac{\partial\vec{u}_i}{\partial t}$$

$$\vec{u}_j\sum_{j=1}^{3}\frac{\partial\vec{u}_i}{\partial x_j}-v\Delta\vec{u}_i+\frac{\partial p}{\partial x_i}=\frac{\partial\vec{u}_i}{\partial t}(m-1). \qquad (6)$$

I will put this into a scenario for the application of waves. Recall throwing a rock in a pond and recognize the ripple effect. I see those as mini waves ripping across the pond. Look at the rock as force into the water with created waves as a result. There could be a pressure developed as a result of a hand toss.

Let's take a look at the the equation would look like if we set $\vec{f}_i(x_i,t)$ equal to zero as given. Refer back to equation five. If we set the force field equal to zero, we would get

$$\frac{\partial\vec{u}_i}{\partial t}+\vec{u}_i\sum_{\partial\vec{u}_i}^{\partial x_j}-v\Delta\vec{u}_i+\frac{\partial p}{\partial x_i}=mv\Delta\vec{u}_i-m\vec{u}_j\sum_{j=1}^{3}\frac{\partial\vec{u}_i}{\partial x_j}-m\frac{\partial p}{\partial x_i}$$

$$f_i(x_i,t)=mv\Delta\vec{u}_i-m\vec{u}_j\sum_{=1}\frac{\partial}{\partial x_j}-m\frac{\partial}{\partial x_i}$$

$$0=m[v\Delta\vec{u}_i-\vec{u}_j\sum_{j=1}^{3}\frac{\partial\vec{u}_i}{\partial x_j}-\frac{\partial p}{\partial x_i}].$$

The reason why I decided to go through these derivations is because of the application mentioned above. Let us continue with the scenario of waves. Now, recall the equation of a tangent line from algebra. Let a two-dimensional wave be shaped like a bell curve similar to the normal distribution curve. Since we have a curve there exists a tangent line on all sides of the curve. Imagine the tangents keeping the shape of the wave as the force from the rock penetrates the water.

Now, we can define equations of tangent planes over a three-dimensional wave on a graph. You can use a mathematical program to visualize what I mean and verify. Recall from Calculus, the equation of tangent plane in two and three space variables such as

$$\vec{f}_i(x_i,y_i)=\frac{\partial\vec{f}_i}{\partial x_i}(x-x_0)+\frac{\partial\vec{f}_i}{\partial y_i}(y-y_0)$$

at some initial point (x_0,y_0). Then, the three-dimensional tangent plane yields

$$\vec{f}_i(x_i,y_i,z_i)=\frac{\partial\vec{f}_i}{\partial x_i}(x-x_0)+\frac{\partial\vec{f}_i}{\partial y_i}(y-y_0)+\frac{\partial\vec{f}_i}{\partial z_i}(z-z_0). \qquad (7)$$

at some initial point (x_0,y_0,z_0). What if the partial derivative of any function $\vec{f}_i$ is applied to see if there are infinitely many derivatives implying an infinite force vector field? The force field may not be infinite when we think of the wave application used earlier. Figuratively, we have a derived equation of a tangent plane to look something like

$$\frac{\partial \vec{f}_i}{\partial x_i} = \frac{\partial^2 \vec{f}_i}{\partial x_i^2}(x - x_0) + \frac{\partial^2 \vec{f}_i}{\partial y_i^2}(y - y_0) + \frac{\partial^2 \vec{f}_i}{\partial z_i^2}(z - z_0).$$

What would happen if we referred back to equation four and implemented derivatives with respect to space variables and time. Mathematically, it may look like,

$$\vec{f}_i(x_i, t) = \frac{\partial \vec{u}_i}{\partial t} + \sum_{j=1}^{3} \vec{u}_j \frac{\partial \vec{u}_i}{\partial x_j} - v\Delta \vec{u}_i + \frac{\partial p}{\partial x_i}$$

$$\frac{\partial \vec{f}_i}{\partial x_i} = \vec{u}_j \sum_{j=1}^{3} \frac{\partial^2 \vec{u}_j}{\partial x_j^2} - v\frac{\partial^3 \vec{u}_i}{\partial x_i^3} + \frac{\partial^2 p}{\partial^2 x_i^2}$$

$$= \vec{u}_j \Delta \vec{u}_j - v\frac{\partial^3 \vec{u}_i}{\partial x_i^3} + \frac{\partial^2 p}{\partial^2 x_i^2}. \tag{8}$$

Deriving this equation once more and so forth may imply infinite derivatives towards the divergence free vector field.

Existence

"Consider a quantity of fluid contained in a plane region *S bounded* by a simple closed curve ∂S. The *Law of Conservation of Mass* states that the rate at which fluid pours across ∂S into S must *balance* the rate at which the total amount of fluid in S *increases with time*" [2]. "An existence theorem for our initial-boundary value problem for the wave equation can be proved under a more restrictive hypothesis on the *smoothness* of the initial data" [3]. "Consider the PDE,

$$\frac{\partial \vec{u}_i}{\partial t} + \vec{u}_j \frac{\partial \vec{u}_i}{\partial x_i} = K\vec{u}_i$$

where we have our given domain to be $-\infty<x_i<\infty$, $t>0$, $i=1,2\ldots m$ and C_∞, K are constants. Solving the characteristic equations and the compatibility conditions

$$\frac{dx}{dt} = \vec{u}_j, \frac{d\vec{u}}{dt} = K\vec{u}.$$

The initial manifold t=0is non-characteristic and we will select the initial state of the system described by the function $\vec{u}(x_\alpha, t)$ by the *Cauchy* data

$$\vec{u}(x_i, 0) = \vec{u}(x_i), \infty < x_i < \infty"$$

Consider the equation

$$\sum_{i=1}^{3} \frac{\partial^2 \vec{u}_i}{\partial x_i^2} - \frac{1}{v^2}\frac{\partial^2 \vec{u}_i}{\partial t^2} = 0.$$

Theorem 1 *If* $\vec{u}_1(x) \in C^2$ *and* $\vec{u}_1(x) \in C^2$ *in* $-\infty<x<\infty$, $i=1,2,3$ *then the function*

$$\vec{u}(x,t) = tM(t)\vec{u}_1 + \frac{\partial}{\partial t}(tM(t)\vec{u}_0)$$

where

$$M(t)\vec{u}_i = \frac{1}{4\pi}\int_{\|v\|=1} \vec{u}_i(x + vt)d\omega, i = 0,1$$

belongs to C^2 in $-\infty<(x,t)<\infty$, and is a solution of the *Cauchy* problem

$$\Delta\vec{u} - \frac{1}{v^2}\frac{\partial^2 \vec{u}}{\partial t^2} = 0$$

$$\vec{u}(x,0) = \vec{u}_0(x)$$

$$\frac{\partial \vec{u}}{\partial t}(x,0) = \vec{u}_1(x).$$

$M(t)\vec{u}_i$ denotes the mean value of $\vec{u}_i$ over the sphere with center at x and radius vt in three-dimensional space.

Proof. "Assuming that given $\vec{u}(x,t)$ holds, we first verify whether the initial conditions are satisfied:

$$\vec{u}(x,0) = M(0)\vec{u}_0$$

$$= \frac{1}{4\pi}\int_{\|v\|=1} \vec{u}_0(x)d\omega$$

$$= \vec{u}_0(x)$$

where $M_t(t)\vec{u}_i = \frac{\partial}{\partial t}M(t)\vec{u}_i$ so that

$$M_t(0)\vec{u}_0 = \frac{1}{4\pi}v\sum_{i=1}^{3}\int_{\|v\|=1} \vec{u}_{0\alpha i}(x)v_i d\omega$$

$$= \frac{1}{4\pi}v\sum_{i=1}^{3}\vec{u}_{0\alpha i}(x)\int_{\|v\|=1} v_i d\omega$$

$$= 0.(fromTheGaussIntegralTheorem)$$

It follows that both conditions satisfy the wave equation as well as the partial derivative with respect to t of $tM(t)\vec{u}_0$. Therefore, the given $\vec{u}(x,t)$ is a solution of the three-dimensional wave equation satisfying the given conditions" [4].

"In the *Goursat* problem the data specified on two interesting non-characteristic curves strictly contained in an angle between two characteristics passing through the point of intersection of the curves. Without loss of generality, we can take the point of their intersection to be the origin. Now let,

$$f(x) = F(x) + G(x)$$

$$g(\alpha) = F(\alpha(1+\beta)) + G(\alpha(1-\beta))$$

$$x = 0 and \beta x = vt, 0 < \beta < 1.$$

It is not easy to determine the functions F and G from these relations. From the given data above

$$G\{(1+\beta)x\} - G\{(1-\beta)x\} = f\{(1+\beta)x\} - g(x).$$

Let

$$x = \frac{x}{1+\beta}, 0 < \delta - \frac{1-\beta}{1+\beta} < 1, p(x) = g(\frac{x}{1+\beta}) - f(x),$$

then

$$p(x) = -G(x) + G(\delta x).$$

It follows that

$$p(\delta x) = -G(\delta x) + G(\delta^2 x)$$

and so on. Therefore

$$\sum_{i=0}^{n} p(\delta^i x) = -G(x) + G(\delta^{n+1}x).$$

Since G is continuous and $0<\delta<1$, letting n tend to infinity we get

$$G(x) = G(0) - \sum_{i=0}^{\infty} p(\delta^i x)$$

provided $\sum_{i=0}^{\infty} p(\delta^i x)$ exists. Using the initial data $f(x)= F(x)+G(x)$ we can now find $F(x)$. Hence, the solution is given by

$$\vec{u}(x,t) = f(x+t) + \sum_{i=0}^{\infty} p\{\delta^i(x+vt)\} - \sum_{i=0}^{\infty} p\{\delta^i(x-vt)\}.$$

The functions f and g must be such that $\sum_{i=0}^{\infty} p(\delta^i x)$ *converges* in order for the solution to be valid. It is also unique. The region of determinacy when $f(x)$ and $g(\alpha)$ are specified for the bounds$0 \le x \le a$ and $0 \le \alpha \le b$, is the region bounded by the characteristics through $(a,0)$, $(b, \frac{b\beta}{c})$, and the given segments on $t=0$ and $\beta x=vt$. In this case, the problem is *well-posed*. The solution $\vec{u}$ is given in terms of f and g which when equal to zero, imply that $\vec{u}$ is also zero. If *f* and *g* are arbitrarily small, then $\vec{u}$ is also the same way. Hence, the solutions are unique and *stable* therefore, the problem is well-posed" [4].

"There are equations of a viscous incompressible fluid that are called *stationary* that yield,

$$\vec{u}_1\frac{\partial \vec{u}_1}{\partial x}+\vec{u}_2\frac{\partial \vec{u}_1}{\partial y}=-\frac{1}{\rho}\frac{\partial p}{\partial x}+v\Delta\vec{u}_1,$$

$$\vec{u}_1\frac{\partial \vec{u}_2}{\partial x}+\vec{u}_2\frac{\partial \vec{u}_2}{\partial y}=-\frac{1}{\rho}\frac{\partial p}{\partial y}+v\Delta\vec{u}_2,$$

$$\frac{\partial \vec{u}_1}{\partial x}+\frac{\partial \vec{u}_2}{\partial y}=0$$

can be reduced to an equation in question by defining a *stream function* w such that $\vec{u}_1=\frac{\partial w}{\partial y}$ and $\vec{u}_2=-\frac{\partial w}{\partial x}$ followed by the elimination of the pressure *p* from the first two equations" [5,6]. "Going back to the system of stationary *hydrodynamic* equations we add *F*(*y*) to the first equation such that,

$$\vec{u}_1\frac{\partial \vec{u}_1}{\partial x}+\vec{u}_2\frac{\partial \vec{u}_1}{\partial y}=-\frac{1}{\rho}\frac{\partial p}{\partial x}+v\Delta\vec{u}_1+F(y).$$

The above equation along with the other two equations could describe the *plane flow* of a viscous incompressible fluid under the *action of a transverse force*. Then,

$$f(y)=\frac{\partial F}{\partial y}.$$

Letting *F*(*y*)=*asin*(*λy*) corresponds to *A.N. Kolmogorov's* model, which is *used for describing sub-critical and transitional* (laminar-to-turbulent) flow" [7]. Again, we yield the equations of a viscous incompressible fluid. Only this time we make the equations "*non-stationary* such that,

$$\frac{\partial \vec{u}_1}{\partial t}+\vec{u}_1\frac{\partial \vec{u}_1}{\partial x}+\vec{u}_2\frac{\partial \vec{u}_1}{\partial y}=-\frac{1}{\rho}\frac{\partial p}{\partial x}+v\Delta\vec{u}_1,$$

$$\frac{\partial \vec{u}_2}{\partial t}+\vec{u}_1\frac{\partial \vec{u}_1}{\partial x}+\vec{u}_2\frac{\partial \vec{u}_1}{\partial y}=-\frac{1}{\rho}\frac{\partial p}{\partial x}+v\Delta\vec{u}_1,$$

$$\frac{\partial \vec{u}_1}{\partial x}+\frac{\partial \vec{u}_2}{\partial y}=2a,$$

Describing the motion of a viscous incompressible fluid by two parallel disks moving towards each other is reduced to the given equation. In this case, *a* is the relative velocity of the disks which can also be denoted as $v_r e$ *l*while $\vec{u}_1$ and $\vec{u}_2$ are the horizontal velocity components and $\vec{u}_3=-2az$ is the vertical velocity component. Then a new stream function is defined such that $\vec{u}_1=ax+\frac{\partial w}{\partial y}$ and $\vec{u}_2=ay-\frac{\partial w}{\partial x}$ followed by the elimination of the pressure *p* leading to the equation we are looking for" [5]. "We shall derive a formula expressing the *Law of Conservation of Momentum* in the *x*-direction (one-dimension),(*x*,*y*)-directions (two-dimensions), and the (*x*,*y*,*z*)directions (three-dimensions) "when not only viscosity but also external fields of force, such as gravity are neglected" [2]. The horizontal component of force exerted on a small section *S* of our fluid by the *scalar pressure p* along it boundary ∂S is given by the integral,

$$-\int_{\partial S}pdy=-\iint_S\frac{\partial p}{\partial x}dxdy.$$

Since the velocity field is $dx/dt=\vec{u}$ in one-dimension, the horizontal acceleration d^2x/dt^2 of this particle is equal to

$$\frac{d\vec{u}}{dt}=\frac{\partial \vec{u}}{\partial t}+\vec{u}\frac{\partial \vec{u}}{\partial x}+\vec{v}\frac{\partial \vec{u}}{\partial y}$$

in two-dimensions respectively such that

$$\frac{\partial \vec{u}_i}{\partial t}+\sum_{j=1}^{n}\vec{u}_j\frac{\partial \vec{u}_i}{\partial x_j}=v\Delta\vec{u}_i-\frac{\partial p}{\partial x_i}+\vec{f}_i(x_i,t)$$

$$\frac{d\vec{u}}{dt}=v\Delta\vec{u}_i-\frac{\partial p}{\partial x_i}+\vec{f}_i(x_i,t)$$

and since we let $f_i(x,t)=0$ in the beginning we have now,

$$\frac{d\vec{u}}{dt}=v\Delta\vec{u}_i-\frac{\partial p}{\partial x_i}.$$

In one-dimensional flow, *Euler's equations of motion* reduce to,

$$\frac{\partial \vec{u}}{\partial t}+\vec{u}\frac{\partial \vec{u}}{\partial x}=-\frac{\partial p}{\partial x}.$$

Then in two and three-dimensions, the Euler's equation will yield,

$$\frac{\partial \vec{u}}{\partial t}+\hat{u}\frac{\partial \vec{u}}{\partial x}+\hat{v}\frac{\partial \vec{u}}{\partial y}=-\frac{\partial p}{\partial x\partial y}$$

and

$$\frac{\partial \vec{u}}{\partial t}+\hat{u}\frac{\partial \vec{u}}{\partial x}+\hat{v}\frac{\partial \vec{u}}{\partial y}+\hat{w}\frac{\partial \vec{u}}{\partial z}=-\frac{\partial p}{\partial x\partial y\partial z}$$

Where *u*,*v* and *w* are the unit vectors in the specified directions.

Theorem 2 *"The solution of*

$$\frac{\partial^2\vec{u}}{\partial t^2}-v^2\frac{\partial^2\vec{u}}{\partial x^2}=\vec{f}(x,t)$$

subject to the *boundary conditions*

$$\vec{u}(a,t)=\vec{u}(b,t)=0$$

$$\vec{u}(x,0)=g_1(x)$$

$$\frac{\partial \vec{u}}{\partial t}(x,0)=g_2(x)$$

is

$$\vec{u}(x,t)=\int_0^t\int_a^b G(x,t,\eta,\tau)\vec{f}(\eta,\tau)d\eta d\tau+$$

$$\int_a^b[g_2(\eta)G(x,t,\eta,0)-g_1(\eta)\frac{\partial G}{\partial \tau}(x,t,\eta,0)]dx,$$

Where(*a*,*b*) is(-∞,∞) and *G* is *Green's function* for the wave equation" [8].

The proof is located in the referenced book above. As we know already "mathematically, the partial differential equation to be solved is non-linear and is of *fourth order*, with two, three, or even four independent variables. With these numerical techniques they tend to require very large computer time, tend to lack accuracy due to the non-linearity, and tend to be unstable" [9]. It is said that "when a body moves through a viscous fluid, the Navier-Stokes equations are satisfactorily approximated by the *boundary layer equations* in a narrow region adjacent to the body" [9].

Smoothness

The solution to the three-dimensional wave equation given by $\vec{u}(x,t)$ is of class C^2 for t ≥ 0 when $\vec{u}_0\in C^3(\mathbb{R}^3)$ and $\vec{u}_1\in C^2(\mathbb{R}^3)$. Therefore the solution can be less smooth than the data. There is a possible loss of one derivative. This loss could be due to what happens for *m*>1, where $\vec{u}(x_1,x_2,...,x_m,t)$ in -space variables. For *m*=1the solution is *smooth* for all *t* as the initial data at *t*=*0*. The solution of $\vec{u}(x,t)$ of the three-dimensional wave equation given in Theorem 1 depends on the values of $\vec{u}_0$, $\vec{u}_1$", and the first derivatives of $\vec{u}_0$ on the surface of the sphere of center *x* and radius *vt*.If $\vec{u}_0$ and $\vec{u}_1$ have support in a *closed bounded region* Ω of R^3, for example, if they are oth zero outside of Ω, then at t>0 $\vec{u}(x,t) \neq 0$ at those points x which lie on a sphere of radius *vt* and centered at a point $y\in\Omega$ and $x\in S_{y,yt}$ for some $y\in\Omega$. $S_{y,yt}$ is the sphere with respect to point *y* and radius *vt*. We begin to learn about the development of shock waves from the initial-value problem for $\vec{u}(x,t)$

$$\frac{\partial \vec{u}}{\partial t} + p(\vec{u})\frac{\partial \vec{u}}{\partial x} = 0, -\infty < x < \infty, t > 0,$$

$$\vec{u}(x,0) = \vec{u}_0(x), -\infty < x < \infty,$$

where $p(\vec{u})$ and $\vec{u}(x)$ are $C^{\prime}(\mathbb{R})$ functions of their are arguments, that is, they are *smooth functions*. There are characteristic equations that correspond with the equation above which yield,

$$\frac{dt}{1} = \frac{dx}{p(\vec{u})} = \frac{d\vec{u}}{0}.$$

These equations imply that,

$$\frac{d\vec{u}}{dt} = 0 and \frac{dx}{dt} = p(\vec{u}).$$

The solution of $dx/dt = p(\vec{u})$ represents characteristics of the first above equation along the condition that,

$$\frac{d\vec{u}}{dt} = \frac{\partial \vec{u}}{\partial x}\frac{dx}{dt} + \frac{\partial \vec{u}}{\partial t} = p(u)\frac{\partial \vec{u}}{\partial x} + \frac{\partial \vec{u}}{\partial t} = 0.$$

"The condition means that $\vec{u}$ is *constant* on the characteristics which propagate with *speed* $p(\vec{u})$. The dependence of p and $\vec{u}$ produces a *gradual non-linear distortion* of the wave profile as it propagates. It follows that $p(\vec{u})$ is also *constant* on the characteristics, and therefore must be straight lines in the *(x,t)*-plane with a constant slope of $1/p(\vec{u})$." [5] "If there are two points(ξ,0) and(η,0) with$\xi<\eta$ then the characteristics starting at (ξ,0) and (η,0) will intersect at a *pressure point p(x,t)* for *t*>0. "At the *point of intersection p(x,t)*, the solution of $\vec{u}(x,t)$ has two different values $\vec{u}(\xi)$ and $\vec{u}(\eta)$. This means that $\vec{u}$ is double valued, and hence, the solution is *not unique* at the point of intersection *of the characteristics*. Thus, the solution *must* be *discontinuous* at the point of intersection. The result is that if no two characteristic lines intersect in the *half plane* t>0, *there exists a solution of the initial-value problem* as a *differentiable function* for all *t*>0. This can happen only if the reciprocal of the slope $p(\vec{u})$ is an *increasing function* of the intercept. In other words, the family of characteristics spreads only for *t*>0 and generates a solution of the problem that is at least *as smooth as* $\vec{u}(x)$. Such as solution is called an *expansive* or a *refractive wave*" [5]. "Let the periodic boundary conditions yield,

$$\vec{u}(0,t) = \vec{u}(1,t) = 0 \, for \, t > 0$$

and

$$\frac{\partial \vec{u}}{\partial x}(0,t) = \frac{\partial \vec{u}}{\partial x}(1,t) = 0 \, for \, t > 0$$

and the initial condition be

$$\vec{u}(x,0) = \vec{u}(x) \forall x \in \mathbb{R}^n.$$

"We assume that this initial-boundary problem for an equation possesses a *smooth function* which is uniquely determined by the initial data $\vec{u}$. Two invariants of the problem which are constants of the motion are given [10].

$$I(t) = \int_{\mathbb{R}^n} |\vec{u}(x,t)|^2 \, dx.$$

We consider *any* function $\vec{u}(x,t)$, not necessarily a solution of a certain equation, which is defined for *t*>0 and sufficiently smooth. We will define a *strict solution* for the initial-boundary value problem to be a function $\vec{u}(x,t)$. This velocity is continuous together with its first and second-order derivatives and satisfies a specific PDE for -∞<*x*<∞, *t*>0. The initial and boundary conditions are satisfied in the sense of *equality*. We assume that $\vec{u}(x,t)$ satisfies the given boundary conditions and evolves in time, so that function *I(t)* is a constant in time. A solution $\vec{u}(x,t)$ is said to be *Lagrangian stable* if there exists a constant *C* independent of *t*, but at the same time could be dependent on initial data such that,

$$|\vec{u}(x,t)|_\infty^2 = \max_{x\in\mathbb{R}^n} |\vec{u}(x,t)|^2 \leq C, forall t \geq 0.$$

"A *strict solution* is one of showing the continuity of $\vec{u}(x,t)$ which implies that,

$$\lim_{t\to 0} \vec{u}(x,t) = \vec{u}(x,0) = \vec{u}(x)$$

uniformly for -∞<*x*<∞ and this becomes

$$\lim_{t\to 0} \int_{-\infty}^{\infty} \vec{u}(x,t)dx = \int_{-\infty}^{\infty} \vec{u}(x,0)dx = \int_{-\infty}^{\infty} \vec{u}(x)dx$$

where the initial conditions yield

$$\vec{u}(x,0) = \vec{u}(x)$$

$$\frac{\partial \vec{u}}{\partial t}(x,0) = 0.$$

We have our given velocity field $\vec{u}$ and let our scalar pressure field be *p(x,t)* be equal to a function $g_i(x,t)$. "Let $\vec{u}(x,t)$ be *generalized solution* of the initial conditions if there is a *sequence* of *strict solutions* $\{\vec{u}_i(x,t)\}$ with $\vec{u}_i(x)$,

$p_i(x,t) = g_i(x,t)$ such that

$$\lim_{i\to\infty} \vec{f}_i(x,t) = \vec{f}(x,t), \lim_{i\to\infty} g_i(x,t) = p(x,t)$$

uniformly for -∞<*x*<∞, and

$$\lim_{i\to\infty} \vec{u}_i(x,t) = \vec{u}(x,t),$$

uniformly for -∞<*x*<∞ and $t \geq 0$" [8]. Take the solution of the *plucked string* problem to be generalized as well. The *i-th* partial sum yields,

$$\vec{u}_i(x,t) = \sum_{i=1}^{3} b_i cos(i\pi vt) sin(i\pi x)$$

where $\{\vec{u}_i(x,t)\} = \{\vec{u}_1(x,t), \vec{u}_2(x,t), \vec{u}_3(x,t)\}$ for $1 \leq i \leq 3$ is clearly a strict solution with initial data,

$$\vec{f}_i(x) = \sum_{i=1}^{3} b_i sin(i\pi x), g_i(x) = p_i(x),$$

where $\{\vec{f}_i(x,t)\} = \{\vec{f}_1(x,t), \vec{f}_2(x,t), \vec{f}_3(x,t)\}$. The limit of $\vec{u}$ by the definition of $\vec{u}(x,t)$ and the limit of $\vec{f}$ hold by the *pointwise convergence theorem* for *Fourier Sine Series* and the limit of $g_i(x)$ trivially" [3]. The process of limiting is one way that can explain the continuity of each function. As we can see, $\vec{f}$,*p*, and $\vec{u}$ *converges uniformly* with respect to the domain -∞<*x*<∞ and $t \geq 0$. In this perception, there are partial derivatives of the functions $\vec{u}$ and p that are continuous with initial functions of $\vec{u}$ and *p* that are also continuous. The set of these functions are denoted as the infinite *differentiability class* on the set R^n with time $t \geq 0$. Higher order differentiability classes should correspond to the existence of higher order derivatives. Functions that have derivatives of all orders could be named as *smooth* functions. We must now show that *p* and $\vec{u}$ is *infinitely differentiable* on the set R^n x[0,∞)where as in this case we are letting *n* equal 3 such that R^3 x[0,∞).The notation of C^∞ means that the scalar field p_i and velocity field $\vec{u}_i$ are in a specific type of differentiability class of *smooth functions* if and only if they have derivatives of all orders. We must show that these functions are in this class with respect to the given space R^n x $[0,\infty)$ where we let n equal to

3. When we look at equation 3 we see that we have $\frac{\partial p_i}{\partial x_i}$ where$\{x_i\}$ is the sequence of positions in the *x*-direction such that

$$\frac{\partial p_1}{\partial x_1}$$

$$\frac{\partial p_2}{\partial x_2}$$

$$\frac{\partial p_3}{\partial x_3}$$

.

.

.

$$\frac{\partial p_n}{\partial x_n}$$

If this is the case, there could be a corresponding sequence of *pressure intersection points* $p_i\{(x_i, t)\}$ in a given body from *i* to ∞ such that

$$p_1(x_1,t)$$

$$p_2(x_2,t)$$

$$p_3(x_3,t)$$

.

.

.

$$p_n(x_n,t).$$

Analysis of the Navier-Stokes equation

$$\frac{\partial \vec{u}_i}{\partial t}+\sum_{j=1}^{n}\vec{u}_j\frac{\partial \vec{u}_i}{\partial x_j}=v\Delta\vec{u}_i-\frac{\partial p}{\partial x_i}+\vec{f}_i(x_i,t)$$

The motion of a *non-turbulent Newtonian fluid* could be governed by the equation above where the first term is the time-derivative of any fluid's velocity or the *acceleration* in Newton's Second Law of motion for a fluid element. This element is subjected to an external force that is stated above and also to forces coming from the pressure field and friction in this equation which is known as the *material derivative* such that,

$$\frac{\partial \vec{u}_i}{\partial t}+(\vec{u}*\nabla\vec{u})$$

then

$$\frac{\partial \vec{u}_i}{\partial t}+\vec{u}_j\nabla\vec{u}_i=v\Delta\vec{u}_i-\frac{\partial p}{\partial x_i}+\vec{f}_i(x_i,t)$$

where ($\vec{u}*\nabla\vec{u}$) is a term of convection in fluid mechanics. The equation could also be used to model turbulent flows where the fluid parameter could be interpreted as time averages. The second term could be the velocity in the change of coordinates by the Law of Coordinate Transformation also known as a *contravariant*. The fourth term could be the gradient vector of the scalar pressure field *p* in all space dimensions where *x* is in the real numbers from *i* to *infinity*.

The fifth term represents a sequence of external force vectors $\vec{f}_i$ from *i* to *infinity* that correspond to a mass *m* and a velocity field $\vec{u}_i$ as stated above from Newton's Second Law. As shown earlier, we seen a perception on the scalar field p_i and velocity field u_i being elements of the differentiability class C^∞ with respect to all space dimensions (*x,y,z*) and the time interval $t \geq 0$ on R^3 $x[0,\infty)$. This confirms equation six which states the scalar and velocity field p *and* $\vec{u}_i$ are elements of the infinite differentiability class C^∞ (R^3 $x[0,\infty)$). Equation seven is of bounded energy or *global regularity* that could be expressed as the magnitude or *modulus* of the velocity field squared with respect to all space dimensions and time $t \geq 0$. In non-relativistic wave mechanics, there could exist a wave function $\vec{u}(x,t)$ of a particle that satisfies a certain wave equation where,

$$\frac{\partial \vec{u}_i}{\partial t}=\frac{-h^2}{2m}(\frac{\partial}{\partial x}+\frac{\partial}{\partial y}+\frac{\partial}{\partial z})\vec{u}_i$$

so that

$$\frac{\partial \vec{u}_i}{\partial t}=\frac{-h^2}{2m}(\frac{\partial}{\partial x}+\frac{\partial}{\partial y}+\frac{\partial}{\partial z})\vec{u}_i$$

$$\frac{\partial \vec{u}_i}{\partial t}=\frac{-h^2}{2m}\nabla\vec{u}_i$$

since $\frac{-h}{2m}$ is constant let it be equal to *v* such that

$$\frac{\partial \vec{u}_i}{\partial t}=v\nabla\vec{u}_i.$$

The velocity vector field $\vec{u}$ can be looked at as the *average velocity* that was differentiated to obtain the *average acceleration*. Now we take the a second derivative of $\vec{u}_i$ such that

$$\frac{\partial^2 \vec{u}_i}{\partial t^2}=v\nabla^2\vec{u}_i$$

which is also known as "*The Simple Wave* equation" [6]. As we stated earlier $\nabla^2=\Delta$ so we can use this to change notation to

$$\frac{\partial^2 \vec{u}_i}{\partial t^2}=v\Delta\vec{u}_i.$$

"This Simple Wave equation can be solved in three dimensions with the initial conditions

$$\vec{u}(x,y,z,0)=\phi(x,y,z) for t\geq 0$$

and

$$\frac{\partial \vec{u}}{\partial t}(x,y,z,0)=\psi(x,y,z)$$

Where $(x,y,z)\in R^3$ and how this method to a solution satisfies *Huygen's Principle*. This method can also be used to solve this wave equation in two-dimensions. To solve this problem in three-dimensions we start with an easier one first. "Let $\vec{u}=0$ so that,

$$\vec{u}(x,y,z,0)=0$$

$$\frac{\partial \vec{u}}{\partial t}(x,y,z,0)=\psi(x,y,z)$$

where Δ is the Laplacian Operator stated earlier in the paper. This problem can be solved by a *Fourier Transform* and has a solution

$$\vec{u}(x,y,x,t)=t\overline{\psi}$$

where $\overline{\vec{u}_1}$ is the *average* of the initial disturbance $\vec{u}_1$ over the sphere of radius *vt* centered at (*x,y,z*). The symbol $\overline{\psi}$ yields the *Fourier Transform*" [7]. "The verbal interpretation of this solution is that *initial* disturbance $\vec{u}_i$ radiates outward spherically (viscosity *v*) at each point, so that after so many seconds, the point (*x,y,z*) will be influenced by those initial disturbances on a sphere (of radius *vt*) around that point. Now let $\vec{u}=\phi$ and $\vec{u}_t=0$ " [7]. "A famous theorem developed by *Stokes* says all we have to do to solve this problem is change the initial conditions to $\vec{u} = o$, $\vec{u}_t = \phi$ and then differentiate this solution with respect to time. So we solve our given simple wave problem to get $\vec{u} = t\phi$ and then differentiate with respect to time. This gives us the solution to the simple wave problem which is,

$$\vec{u} = \frac{\partial}{\partial t}[t\bar{\phi}].$$

For the one-dimensional wave equation the solution of the switch(shift) is

$$\vec{u}(x,t) = \frac{1}{2v}\int_{x-vt}^{x+vt}\phi(s)ds.$$

Differentiating this equation will yield,

$$\vec{u}_t(x,t) = \frac{1}{2}[\phi(x+vt) + \phi(x-vt)]$$

which is the solution to our given wave equation. Knowing this we have the solution to the three-dimensional simple wave equation where $\vec{u} = \phi$ and $\vec{u}_t = \psi$ initially which is just

$$\vec{u}(x,y,x,t) = t\bar{\psi} + \frac{\partial}{\partial t}[t\bar{\phi}]$$

where $\bar{\phi}$ and $\bar{\psi}$ are averages of the functions φ and ψ. This generalization is known as the *Poisson's formula* for the *free-wave equations* in three dimensions" [1].

If a general solution can be formed, then a specific one can be made with a large amount of computer time. With the given stationary and non-stationary equations, we derive them all with our stream function for both stationary and non-stationary equations. We will start at the derivation of

$$\frac{d\vec{u}_1}{dt} = v\Delta\vec{u}_1 - \frac{\partial p}{\partial x_1}.$$

Since we have our defined stream function for the stationary equations, we get

$$\frac{d}{dt}(\frac{\partial w}{\partial y}) = v[\frac{\partial^2}{\partial x_1^2}(\frac{\partial w}{\partial y})] - \frac{\partial p}{\partial x}.$$

If we integrate both side with respect to *x*, we will see that

$$w(x)dydt = vw(x_1^2)dy - p(x).$$

Now integrating both sides with respect to *y* and *t*, we will get

$$w(x,y,t) = w(x_1^2,y,t) - p(x,y,t).$$

Going back to our given non-stationary equations from the beginning and we solve for the acceleration with the force field equaling zero such that,

$$\frac{\partial\vec{u}_i}{\partial t} = -\vec{u}_j\sum_{j=1}^{n}\frac{\partial\vec{u}_i}{\partial x_j} + v\Delta\vec{u}_i - \frac{\partial p}{\partial x_i}.$$

Integrating both sides with respect to *x* and *t*, we arrive at

$$\vec{u}_1(x,t) = -\hat{u}_1\vec{u}_1(x_1,t) + v\vec{u}_1(x_1,t) - p(x_1,t).$$

We could describe this as a viscous velocity with a unit vector $\hat{u}_1$, where *j*=1. Going back to our equation where we solved for our acceleration. Now, we will substitute yet again our stream function *w* where we have our stationary functions with the relative velocity *a* $\vec{u}_1 = \frac{\partial w}{\partial y}$ such that,

$$\frac{\partial}{\partial t}\frac{\partial w}{\partial y} = -\vec{u}_1\frac{\partial}{\partial x_1}\frac{\partial w}{\partial y} + v\frac{\partial^2}{\partial x^2}\frac{\partial w}{\partial y} - \frac{\partial p}{\partial x_i},$$

then

$$\frac{\partial^2 w}{\partial y\partial t} = -\hat{u}_1\frac{\partial^2 w}{\partial x_1\partial y} + v\frac{\partial^3 w}{\partial x^2\partial y} - \frac{\partial p}{\partial x_1}.$$

Now integrating both sides with respect to two space dimensions and time, we get

$$w(x,y,t) = \hat{u}_1 w(x,y,t) + vw(x,y,t)dx - p(x,y,t).$$

Since the pressure p is being eliminated and the stream function exists, the velocity $\vec{u}$ exists [11,12].

Conclusion

Turbulence is a specifc topic in fluid mechanics which is a vital part of the course when it comes to reallife situations. In two and three dimensional systems of equations and some initial conditions, if the smooth solutions exist, they have bounded kinetic energy. In three space dimensions and time, given an initial velocity vector, there exists a velocity field and scalar pressure field which are both smooth and globally defined that solve the Navier-Stokes equations.

References

1. Farlow P, Stanley J (1982) Partial Differential Equations for Scientists and Engineers.Dover Publications, New York.
2. Garabedian PR (1964) Partial Differential Equations. John Wiley and Sons, New York.
3. Paul BN, McGregor K, James L (1966) Elementary Partial Differential Equations. Holden-Day,USA.
4. Renuka R (1984) Partial Differential Equations. Halsted Press, Newyork.
5. Polyanin K, Andrei D Zaitsev R, Valentin F (2004) Non-Linear Partial Differential Equations. Chapman and Hall, London.
6. Loitsyanskiy LG (1996) Mechanics of Liquids and Gases. Begell House, New York.
7. Belotserkovskii OM, Oparin AM (2000) Numerical Experiment in Turbulence. Nauka, Moscow.
8. Mayer H, William MB (1992) Boundary Value Problems and Partial Differential Equations. PWS-Kent Boston.
9. Mitchell AR (1969) Computational Methods in Partial Differential Equations. John Wiley and Sons New York.
10. Debnath H, Lokenath P (1997) Nonlinear Partial Differential Equations for Scientists and Engineers. Birkhauser Boston in Cambridge,USA.
11. Smith RA (1961) Wave Mechanics of Crystalline Solids. Chapman and Hall LTD, London.
12. Henry B (1932) Partial Differential Equations of Mathematical Physics. The Syndics of the Cambridge University Press in New York.

Polignac's Conjecture with New Prime Number Theorem

YinYue Sha*

Dongling Engineering Center, Ningbo Institute of Technology, Zhejiang University, China

Abstract

There are infinitely many pairs of consecutive primes which differ by even number En.Let Po(N, En) be the number of Polignac Prime Pairs (which difference by the even integer En) less than an integer (N+En), Pei be taken over the odd prime divisors of the even integer En less than √(N+En), Pni be taken over the odd primes less than √(N+En) except Pei, Pi be taken over the odd primes less than √(N+En), then exists the formulas as follows:

Po(N, En) ≥ INT {N × (1-1/2) × ∏ (1-1/Pei) × ∏ (1-2/Pni)} - 1

≥ INT {Ctwin × Ke(N) × 2N/(Ln (N+En))^2} - 1

Po(N, 2) ≥ INT {0.660 × 1.000 × 2N/(Ln (N+2))^2} - 1

∏ (Pi(Pi-2)/(Pi-1)^2) ≥ Ctwin=0.6601618158...

Ke(N)=∏((1-1/Pei)/(1-2/Pei))=∏((Pei-1)/(Pei-2)) ≥ 1

where -1 is except the natural integer 1.

Keywords: Twin prime, Polignac prime, Bilateral sieve method

Introduction

In number theory, Polignac's conjecture was made by Alphonse de Polignac in 1849 and states: For any positive even number En, there are infinitely many prime gaps of size En. In other words: There are infinitely many cases of two consecutive prime numbers with difference En [1].

The conjecture has not yet been proven or disproven for a given value of En. In 2013 an important breakthrough was made by Zhang Yitang who proved that there are infinitely many prime gaps of size En for some value of En<70,000,000 [2].

For En=6, it says there are infinitely many primes (p, p + 6). For En=4, it says there are infinitely many cousin primes (p, p + 4). For En=2, it is the twin prime conjecture that there are infinitely many twin primes (p, p + 2) as shown in Figure 1. For En=0, it is the new prime theorem.

The Polignac Prime of Even Integer

For an any even integer En there exists a prime P for which the Polignac number Q=En+P is also prime. The Polignac Prime pairs shall be denoted by the representation En=Q-P=(En+P) - P, where P and Q are primes and prime P{P ≤ Q} is a Polignac prime of even integer En. Looking at the Polignac partition a different way, we can look at the number of distinct representations (or Polignac primes)that exist for En.

For example, as noted at the beginning of this discussion:

2=05 - 03=(2+03) - 03; 2=07 - 05=(2+05) - 05;

2=13 - 11=(2+11) - 11; 2=19 - 17=(2+17) - 17;

2=31 - 29=(2+29) - 29; 2=43 - 41=(2+41) - 41;

2=61 - 59=(2+59) - 59; 2=73 - 71=(2+71) - 71;

where 3, 5, 11, 17, 29, 41, 59 and 71 are Polignac primes of even integer 2.

4=07 - 03=(4+03) - 03; 4=11 - 07=(4+07) - 07;

4=17 - 13=(4+13) - 13; 4=23 - 19=(4+19) - 19;

4=41 - 37=(4+37) - 37; 4=47 - 43=(4+43) - 43;

4=71 - 67=(4+67) - 67; 4=83 - 79=(4+79) - 79;

where 3, 7, 13, 19, 37, 43, 67 and 79 are Polignac primes of even integer 4.

6=11 - 05=(6+05) - 05; 6=13 - 07=(6+07) - 07;

6=17 - 11=(6+11) - 11; 6=19 - 13=(6+13) - 13;

6=23 - 17=(6+17) - 17; 6=29 - 23=(6+23) - 23;

6=37 - 31=(6+31) - 31; 6=43 - 37=(6+37) - 37;

6=47 - 41=(6+41) - 41; 6=53 - 47=(6+47) - 47;

6=59 - 53=(6+53) - 53; 6=67 - 61=(6+61) - 61;

6=73 - 67=(6+67) - 67; 6=79 - 73=(6+73) - 73;

6=89 - 83=(6+83) - 83; 6=103 - 97=(6+97) - 97;

where 5, 7, 11, 13, 17, 23, 31, 37, 41, 47, 53, 61, 67, 73, 83 and 97 are Polignac primes of even integer 6.

It shows that generally the number of distinct representations (or Polignac primes) increases with increasing N.

The Sieve Method about the Polignac Primes

Let En is an any even integer, Ci is a positive integer more not large than N, then exists the formula as follows:

***Corresponding author:** YinYue Sha, Dongling Engineering Center, Ningbo Institute of Technology, Zhejiang University, China
E-mail: shayinyue@qq.com

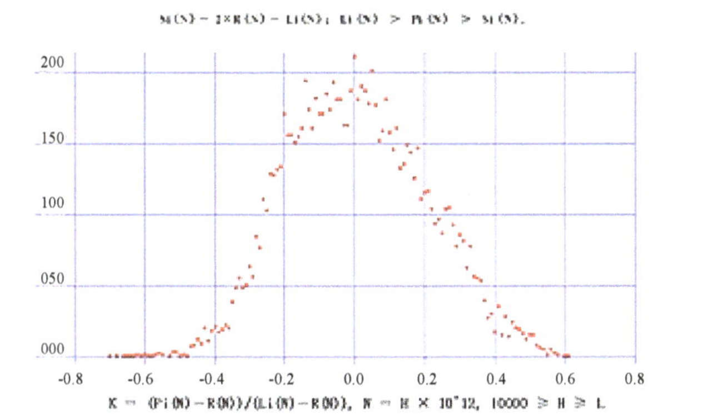

Figure 1: Riemann prime counting function with new prime number theorem.

En=(En+Ci) - Ci (1)

where Ci and En+Ci are two positive integers more not large than N+En.

If Ci and En+Ci any one can be divided by the prime anyone more not large than √(N+En), then sieves out the positive integer Ci; If both Po and En+Po can not be divided by all primes more not large than √(N+En), then both the Po and En+Po are primes at the same time, where the prime Po is a Polignac prime of even integer En.

The Total of Representations of Even Integer

Let En is an any even integer, then exists the formula as follows:

En=(En+Ci)-Ci (2)

where Ci is the natural integer less than N.

In terms of the above formula we can obtain the array as follows:

(En+1, 1), (En+2, 2), (En+3, 3), (En+4, 4), (En+5, 5),..., (En+N, N).

From the above arrangement we can obtain the formula about the total of Polignac numbers of even integer En as follows:

Ci(N, En)=N=Total of integers Ci more not large than N (3)

The Bilateral Sieve Method of Even Prime 2

It is known that the number 2 is an even prime, and above arrangement from (En+1, 1) to (En+N, N) can be arranged to the form as follows:

(En+1, 1), (En+3, 3), (En+5, 5),..., (En+N-X:X< 2, N-X:X< 2).

(En+2, 2), (En+4, 4), (En+6, 6),..., (En+N-X:X< 2, N-X:X< 2),

From the above arrangement we can known that: Because the even integer En can be divided by the even prime 2, therefore, both Ci and En+Ci can be or can not be divided by the even prime 2 at the same time.

The number of integers Ci that Ci and En+Ci anyone can be divided by the even prime 2 is INT (N × (1/2)).

The number of integers Ci that both Ci and En+Ci can not be divided by the even prime 2 is N-INT (N × (1/2))=INT{N-N × (1/2)}=INT{N × (1-1/2)}.

The density of integers Ci that both Ci and En+Ci can not be divided by the even prime 2 (or the ratio of the number of integers Ci that both Ci and En+Ci can not be divided by the even prime 2 to the total of integers Ci more not large than N) as follows:

Si(N, En, 2)=INT (N × (1/2)), Ci(N, En, 2)=N-Si(N, En, 2) (4)

Di(N, En, 2)=Ci(N, En, 2)/(N)=INT{N × (1-1/2)}/N (5)

The Bilateral Sieve Method of Odd Prime 3

It is known that the number 3 is an odd prime, and above arrangement from (En+1, 1) to (En+N, N) can be arranged to the form as follows:

(En+1, 1), (En+4, 4), (En+7, 7),..., (En+N-X:X< 3, N-X:X< 3),

(En+2, 2), (En+5, 5), (En+8, 8),..., (En+N-X:X< 3, N-X:X< 3),

(En+3, 3), (En+6, 6), (En+9, 9),..., (En+N-X:X< 3, N-X:X< 3).

From the above arrangenment we can known that:

If the even integer En can be divided by odd prime 3, then both the Ci and En+Ci can be or can not be divided by odd prime 3 at the same time.

The number of integers Ci that the Ci and En+Ci anyone can be divided by odd prime 3 is INT (N × (1/3)).

The number of integers Ci that both Ci and En+Ci can not be divided by odd prime 3 is N-INT (N × (1/3))=INT {N-N × (1/3)}=INT{N × (1-1/3)}.

The density of integers Ci that both Ci and En+Ci can not be divided by odd prime 3 (or the ratio of the number of integers Ci that both Ci and En+Ci can not be divided by the odd prime 3 to the total of integers Ci more not large than N) as follows:

Sei(N, En, 3) =INT (N × (1/3)), Cei(N, En, 3)=N-Sei(N, En, 3) (6)

Dei(N, En, 3)=Cei(N, En, 3)/(N)=INT{N × (1-1/3)}/N (7)

If the even integer En can not be divided by the odd prime 3, then both Ci and En+Ci can not be divided by the odd prime 3 at the same time, that is the Ci and En+Ci only one can be divided or both the Ci and En+Ci can not be divided by the odd prime 3.

The number of integers Ci that the Ci and En+Ci anyone can be divided by the odd prime 3 is INT(N × (2/3)).

The number of integers Ci that both the Ci and En+Ci can not be divided by the odd prime 3 is N-INT (N × (2/3))=INT {N-N × (2/3)}=INT{N × (1-2/3)}.

The density of integers Ci that both Ci and En+Ci can not be divided by odd prime 3 (or the ratio of the number of integers Ci that both Ci and En+Ci can not be divided by the odd prime 3 to the total of integers Ci more not large than N) as follows:

Sni(N, En, 3)=INT(N × (2/3)), Cni(N, En, 3)=N-Sni(N, En, 3) (8)

Dni(N, En, 3)=Cni(N, En, 3)/(N)=INT{N × (1-2/3)}/N (9)

The Bilateral Sieve Method of Odd Prime 5

It is known that the number 5 is an odd prime, and above arrangement from (En+1, 1) to (En+N, N) can be arranged to the form as follows:

(En+1, 1), (En+06, 06), (En+11, 11),..., (En+N-X:X< 5, N-X:X< 5),

(En+2, 2), (En+07, 07), (En+12, 12),..., (En+N-X:X< 5, N-X:X< 5),

(En+3, 3), (En+08, 08), (En+13, 13),..., (En+N-X:X< 5, N-X:X< 5),

(En+4, 4), (En+09, 09), (En+14, 14),..., (En+N-X:X< 5, N-X:X< 5),

(En+5, 5), (En+10, 10), (En+15, 15),..., (En+N-X:X< 5, N-X:X< 5).

From the above arrangenment we can known that:

If the even integer En can be divided by odd prime 5, then both the Ci and En+Ci can be or can not be divided by odd prime 5 at the same time.

The number of integers Ci that the Ci and En+Ci anyone can be divided by odd prime 5 is INT (N × (1/5)).

The number of integers Ci that both Ci and En+Ci can not be divided by odd prime 5 is N-INT (N × (1/5))=INT {N-N × (1/5)}=INT{N × (1-1/5)}.

The density of integers Ci that both Ci and En+Ci can not be divided by odd prime 5 (the ratio of the number of integers Ci that both Ci and En+Ci can not be divided by odd prime 5 to the total of integers Ci more not large than N) as follows:

Sei(N, En, 5)=INT (N × (1/5)), Cei(N, En, 5)=N-Sei(N, En, 5) (10)

Dei(N, En, 5)=Cei(N, En, 5)/(N)=INT{N × (1-1/5)}/N (11)

If the even integer En can not be divided by the odd prime 5, then both Ci and En+Ci can not be divided by the odd prime 5 at the same time, that is the Ci and En+Ci only one can be divided or both the Ci and En+Ci can not be divided by the odd prime 5.

The number of integers Ci that the Ci and En+Ci anyone can be divided by the odd prime 5 is INT (N × (2/5)).

The number of integers Ci that both the Ci and En+Ci can not be divided by the odd prime 5 is N-INT (N × (2/5))=INT {N-N × (2/5)}=INT{N × (1-2/5)}.

The density of integers Ci that both Ci and En+Ci can not be divided by odd prime 5 (or the ratio of the number of integers Ci that both Ci and En+Ci can not be divided by the odd prime 5 to the total of integers Ci more not large than N) as follows:

Sni(N, En, 5)=INT (N × (2/5)), Cni(N, En, 5)=N-Sni(N, En, 5) (12)

Dni(N, En, 5)=Cni(N, En, 5)/(N)=INT{N × (1-2/5)}/N (13)

The Sieve Function of Bilateral Sieve Method

Let En is an even integer, then exists the formula as follows:

En=(En + Ci) - Ci (14)

where Ci is the natural integer less than N.

In terms of the above formula we can obtain the array as follows:

(En+1, 1), (En+2, 2), (En+3, 3), (En+4, 4), (En+5, 5),..., (En+N, N).

Let Pi be an odd prime less than √(N+En), then the above arrangement can be arranged to the form as follows:

(En+1, 1), (En+Pi+1, Pi+1),..., (En+N-X:X< Pi, N-X:X< Pi),

(En+2, 2), (En+Pi+2, Pi+2),..., (En+N-X:X< Pi, N-X:X< Pi),

(En+3, 3), (En+Pi+3, Pi+3),..., (En+N-X:X< Pi, N-X:X< Pi),

(En+Pi, Pi), (En+2Pi, 2Pi),..., (En+N-X:X< Pi, N-X:X< Pi).

If the even integer En can be divided by the odd prime Pei, then both the Ci and En+Ci can be or can not be divided by the odd prime Pei at the same time.

The number of integers Ci that the Ci and En+Ci anyone can be divided by the odd prime Pei is INT (N × (1/Pei)).

The number of integers Ci that both the Ci and En+Ci can not be divided by the odd prime Pei is N-INT (N × (1/Pei))=INT {N-N × (1/Pei)}=INT{N × (1-1/Pei)}

The density of integers Ci that both the Ci and En+Ci can not be divided by the odd prime Pei (or the ratio of the number of integers Ci that both the Ci and En+Ci can not be divided by the odd prime Pei to the total of integers Ci more not large than N) as follows:

Sei(N, En, Pei)=INT (N × (1/Pei)), Cei(N, En, Pei)=N-Sei(N, En, Pei) (15)

Dei(N, En, Pei)=Cei(N, En, Pei)/(N)=INT{N × (1-1/Pei)}/N (16)

If the even integer En can not be divided by the odd prime Pni, then both the Ci and En+Ci can not be divided by the odd prime Pni at the same time, that is the Ci and En+Ci only one can be divided or both the Ci and En+Ci can not be divided by the odd prime Pni.

The number of integers Ci that the Ci and En+Ci anyone can be divided by the odd prime Pni is INT(N × (2/Pni)).

The number of integers Ci that both the Ci and En+Ci can not be divided by the odd prime Pni is N-INT (N × (2/Pni))=INT {N-N × (2/Pni)}=INT{N × (1-2/Pni)}.

The density of integers Ci that both the Ci and En+Ci can not be divided by the odd prime Pni (or the ratio of the number of integers Ci that both the Ci and En+Ci can not be divided by the odd prime Pni to the total of integers Ci more not large than N) as follows:

Sni(N, En, Pni)=INT(N × (2/Pni)), Cni(N, En, Pni)=N-Sni(N, En, Pni) (17)

Dni(N, En, Pni)=Cni(N, En, Pni)/(N)=INT{N × (1-2/5)}/N (18)

Let Po(N, En) be the number of Polignac Prime Pairs (which difference by the even integer En) less than an integer (N+En), Pei be taken over the odd prime divisors of the even integer En less than √(N+En), Pni be taken over the odd primes less than √(N+En) except Pei, Pi be taken over the odd primes less than √(N+En), then exists the formulas as follows:

Po(N,En) ≥ INT{N × Di(N,En,2) × ∏Dei(N,En,Pei) × ∏Dni(N,En,Pni)}-1

=INT {N × (1-1/2) × ∏ (1-1/Pei) × ∏ (1-2/Pni)} - 1 (19)

where -1 is except the natural integer 1.

The Polignac Prime Theorem

From above we can obtain that:

Let Po(N, En) be the number of Polignac Prime Pairs (which difference by the even integer En) less than an integer (N+En), Pei be taken over the odd prime divisors of the even integer En less than √(N+En), Pni be taken over the odd primes less than √(N+En) except Pei, Pi be taken over the odd primes less than √(N+En), then exists the formulas as follows:

Po(N,En) ≥ INT{N × Di(N,En,2) × ∏Dei(N,En,Pei) × ∏Dni(N,En,Pni)}-1

=INT {N × (1-1/2) × ∏ (1-1/Pei) × ∏ (1-2/Pni)} - 1 (20)

Apply the Prime Number Theorem as follows:

Let Pi(N) be the number of primes less than or equal to N, Pi ($3 \le Pi \le Pm$) be taken over the odd primes less than $\sqrt{N}$, then exists the formulas as follows:

$$Pi(N \mid N \ge 10^4) = INT\{N \times (1-1/2) \times \prod(1-1/Pi) + m + 1\} - 1 \quad (21)$$

$$\ge INT\{N \times (1-1/2) \times \prod(1-1/Pi)\} - 1 \ge INT\{N/Ln(N)\} - 1 \quad (22)$$

$$\prod(Pi(Pi-2)/(Pi-1)^2) \ge Ctwin = 0.6601618158\ldots \quad (23)$$

$$Ke(N) = \prod((1-1/Pei)/(1-2/Pei)) = \prod((Pei-1)/(Pei-2)) \ge 1 \quad (24)$$

From the above and the formula (20) we can obtain the formula as follows:

$$Po(N \mid N \ge 10^4, En) \ge INT\{N \times (1-1/2) \times \prod(1-1/Pei) \times \prod(1-2/Pni)\} - 1 \quad (25)$$

$$\ge INT\{Ctwin \times Ke(N) \times 2N/(Ln(N+En))^2\} - 1 \quad (26)$$

$$\ge Ctwin \times Ke(N) \times 2N/(Ln(N+En))^2 - 2 \quad (27)$$

When the numbe $N \to \infty$, we can obtain the formula as follows:

$$Po(N \mid N \to \infty, En) \ge Ctwin \times Ke(N) \times 2N/(Ln(N+En))^2 - 2 \quad (28)$$

$$\ge 0.660 \times 1.000 \times 2N/(Ln(N+En))^2 - 2 \to \infty \quad (29)$$

The above formula expresses that there are infinitely many pairs of Polignac primes which differ by every even number En.

When the En=2, then there are infinitely many twin primes.

Every Even Integer Greater than Four Can be Expressed as a Sum of Two Odd Primes

Every even integer greater than four can be expressed as a sum of two odd primes, and exists the formula as follows:

$$Gp(N) \ge INT\{Kpc \times Ctwin \times N/(Ln\ N)^2\} - 1 \ge INT\{0.66016 \times N/(Ln\ N)^2\} - 1 \ge 185 >> 1$$

where the Gp(N)be the number of primes P with N-P primes, or, equivalently, the Gp(N) be the number of ways of writing N as a sum of two primes, the N be the even integer greater than 30000.

The proof method of Goldbach's conjecture

The Goldbach's Conjecture is one of the oldest unsolved problems in Number Theory. In its modern form, it states that every even integer greater than two can be expressed as a sum of two primes.

Let N be an even integer greater than 2, and let N=(N-Gp)+Gp, with N-Gp and Gp prime numbers, the Gp{$Gp \le N/2$} be a Goldbach Prime of even integer N. Let Gp(N) be the number of Goldbach Primes of even integer N. The number of ways of writing N as a sum of two prime numbers, when the order of the two primes is important, is thus GP(N)=2Gp(N) when N/2 is not a prime and is GP(N)=2Gp(N)-1 when N/2 is a prime. The Goldbach's Conjecture states that Gp(N) > 0, or, equivalently, that GP(N) > 0, for every even integer N greater than two.

We known that the Goldbach's Conjecture is true for every even integer N no greater than 30000, therefore, we only need to prove that the Goldbach's Conjecture is true for every even integer N greater than 30000, that is: $Gp(N \mid N > 30000) \ge 1$.

TWO: The Sieve Method about the Goldbach Primes

Let N be an even integer greater than 30000, then the even integer N can be expressed to the form as follows:

$$N = (N - Gn) + Gn,\ Gn \le N/2 \quad (1)$$

where Gn be the positive integer no greater than N/2.

Sieve method

Let N-Gn and Gn are two positive integers, if N-Gn and Gn any one can be divisible by the prime P, then sieves the positive integer Gn; if both the N-Gp and Gp can not be divisible by the all primes no greater than $\sqrt{N}$, then both the N-Gp and Gp are primes at the same time, the prime Gp be called the Goldbach Prime of even integer N.

Theorem 1: Let Pc be an odd prime factor of even integer N and no greater than $\sqrt{N}$, then the ratio of the number of integers Gp that both the N-Gp and Gp can not be divisible by the prime Pc to the total of integers Gn no greater than N/2 is follows:

$$R(N,Pc) = INT\{N/2 - N/2/Pc\}/(N/2) = \{INT(N/2) - INT(N/2/Pc)\}/(N/2)$$

Proof: Because Pc is an odd prime factor of even integer N, therefore, both the N-Gn and Gn can or can not be divisible by prime Pc at the same time, then the number of integers Gn that the N-Gn and Gn any one can be divisible by the prime Pc is INT{(N/2)/Pc}, the number of integers Gn that both the N-Gn and Gn can not be divisible by the prime Pc is {INT(N/2) - INT(N/2/Pc)}=INT{N/2-N/2/Pc}, the ratio of the number of integers Gn that both the N-Gn and Gn can not be divisible by the prime Pc to the total of integers Gn no greater than N/2 is follows:

$$R(N,Pc) = \{INT(N/2) - INT(N/2/Pc)\}/(N/2) = INT\{N/2 - N/2/Pc\}/(N/2) \quad (2)$$

Theorem 2: Let Pn be an odd prime no factor of even integer N and no greater than $\sqrt{N}$, then the ratio of the number of integers Gn that both the N-Gn and Gn can not be divisible by the prime Pn to the total of integers Gn no greater than N/2 is follows:

$$R(N,Pn) = INT\{N/2 - N/Pn\}/(N/2) = \{INT(N/2) - INT(N/Pn)\}/(N/2)$$

Proof: Because the Pn is an odd prime no factor of even integer N, therefore, both the N-Gn and Gn can not be divisible by the prime Pn at the same time, that is the N-Gn and Gn only one can be divisible or both the N-Gn and Gn can not be divisible by the prime Pn, then the number of integers Gn that the N-Gn and Gn any one can be divisible by the prime Pn is INT{N/Pn}, the number of integers Gn that both the N-Gn and Gn can not be divisible by the prime Pn is {INT(N/2) - INT(N/Pn)}=INT{N/2 - N/Pn}, the ratio of the number of integers Gn that both the N-Gn and Gn can not be divisible by the prime Pn to the total of integers Gn no greater than N/2 is follows:

$$R(N,Pn) = \{INT(N/2) - INT(N/Pn)\}/(N/2) = INT\{N/2 - N/Pn\}/(N/2) \quad (3)$$

Theorem 3: The integer 2 is an even prime factor of even integer N, the ratio of the number of integers Gn that both the N-Gn and Gn can not be divisible by the even prime 2 to the total of integers Gn no greater than N/2 is follows:

$$R(N,2) = INT\{N/2 - N/2/2\}/(N/2) = \{INT(N/2) - INT(N/2/2)\}/(N/2)$$

Proof: Because the 2 is an even prime factor of even integer N, therefore, both the N-Gn and Gn can be divisible or can not be divisible by the even prime 2 at the same time, then the number of integers Gn that the N-Gn and Gn any one can be divisible by the even prime 2 is INT{N/2/2}, the number of integers Gn that both the N-Gn and Gn can not be divisible by the even prime 2 is {INT(N/2) -

INT(N/2/2)}=INT{N/2 - N/2/2}, the ratio of the number of integers Gn that both the N-Gn and Gn can not be divisible by the even prime 2 to the total of integers Gn no greater than N/2 is follows:

$$R(N,2)=\{INT(N/2) - INT(N/2/2)\}/(N/2)=INT\{N/2 - N/2/2\}/(N/2) \quad (4)$$

Three: The Number of Goldbach Primes of Even Integer

Let Gp(N) be the number of Goldbach primes of even integer N, let Gp(N,Pn) be the number of Goldbach primes no greater than √N, then exists the formulas as follows:

$$Gp(N)=INT\{(N/2) \times R(N,2) \times \Pi R(N,Pci) \times \Pi R(N,Pni)\} + Gp(N,Pni) - 1(\text{if } N\text{-}1 \text{ prime})$$

$$=INT\{(N/2) \times (1-1/2) \times \Pi(1-1/Pci) \times \Pi(1-2/Pni)\}+Gp(N,Pni)-1(\text{if } N\text{-}1 \text{ prime}) \quad (5)$$

Where Pci and Pni are odd primes no greater than √N.

Let Pi(N) be the number of primes less than an integer N, then, be the formula as follows:

$$Pi(N) \equiv INT\{N \times (1-1/P1) \times (1-1/P2) \times \ldots \times (1-1/Pm)+m-1\} \equiv P(N)+Pi(\sqrt{N}) - 1$$

$$Pi(N) \approx Psha(N) \equiv Li(N)- 1/2 \times Li(N^{0.5})$$

$$P(N \geq N \geq 10^9) \geq 2/(1+\sqrt{(1-4/Ln(N))}) \times N/Ln(N) \geq N/(Ln(N)-1)$$

$$P(N \geq N \geq 10^4) \equiv INT\{N \times (1-1/2) \times \Pi(1-1/Pi)\} \geq N/Ln(N) \quad (6)$$

The Proof of Goldbach's Conjecture

Theorem 4: Every even integer greater than 30000 can be expressed as a sum of two odd primes.

Proof: According to the formula (5),

We can obtain the formula as follows:

$$Gp(N)+ 1 \geq INT\{(N/2) \times (1-1/2) \times \Pi(1-1/Pci) \times \Pi(1-2/Pni)\}$$

$$=INT\{(N/2) \times (1-1/2) \times \Pi((Pci-1)/(Pci-2)) \times \Pi(1-2/Pci) \times \Pi(1-2/Pni)\}$$

$$=INT\{(N/2) \times (1-1/2) \times \Pi((Pci-1)/(Pci-2)) \times \Pi(1-2/Pi)\}$$

$$=INT\{(N/2) \times (1-1/2) \times Kpc \times \Pi(1-2/Pi)/\Pi(1-1/Pi)^2 \times \Pi(1-1/Pi)^2\}$$

$$=INT\{(N/2) \times (1-1/2) \times Kpc \times \Pi(1-1/(Pi-1)^2) \times \Pi(1-1/Pi)^2\}$$

$$\geq INT\{(N/2) \times (1-1/2) \times Kpc \times Ctwin \times \Pi(1-1/Pi)^2\} \quad (7)$$

Apply the formula (6), we can obtain the formula as follows:

$$Gp(N \mid N \geq 30000) \geq INT\{(N/2) \times (1-1/2) \times Kpc \times Ctwin \times \Pi(1-1/Pi)^2\} - 1$$

$$\geq INT\{Kpc \times Ctwin \times N/Ln(N)^2\} - 1 \geq INT\{0.66016 \times N/Ln(N)^2\} - 1$$

$$\geq INT\{0.66016 \times (30000)/Ln(30000)^2\}-1=INT\{186.355\ldots\}-1=185 >>1 \quad (8)$$

From above formula (8) we can obtain that:

Every even integer greater than 30000 can be expressed as a sum of two odd primes.

Conclusion

For every even integer En there are infinitely many pairs of Polignac primes which difference by En.

When the En=0, we can obtain New Prime Number Theorem: Let Pi(N)be the number of primes less than or equal to N, for any real number N, the New Prime Number Theorem can be expressed by the formula as follows: Pi(N)=R(N)+ K × (Li(N)- R(N)), 1>K>-1. The Goldbach's Conjecture is a Complete Correct Theorem.

References

1. Tattersall JJ (2005) Elementary number theory in nine chapters. Cambridge University Press, UK.
2. Erica K (2013) Unheralded Mathematician Bridges the Prime Gap. Simons Science News, China.

Convergence and the Grand Unified Theory

Cusack P*

Brealey Drive, Peterborough, 77432, Infrobright, Canada

Abstract

This paper examines the relationship between the four fundamentals forces and shows how, using fluid mechanics, electromagnetism, quantum mechanics and gravity, these forces converge on one solution just as Mathematics does. An equation for the universal processes is provided.

Keywords: Quantumn mechanics; Gravity; Convergence; Fluid mechanics; Schrodiner equation

Introduction

This paper shows how as Mathematics, namely Algebra, Linear Algebra, Geometry and Calculus, converge to one solution, so too does Fluid Mechanics, Quantum Mechanics, Electromagnetism, and Gravity converge to one solution, namely the Superforce. We begin with mathematics convergence and end with Cusack's Universal Equation. Cosmology and Quantum Mechanics are united.

Permeability

We began over 4 years ago with this notion of Permeability. It still stands. It is calculated from the above equation of the circle:

$2x^2=\tan 60°$

$x=\sqrt{2} * \sqrt{3}=[(E^2+t^\wedge 2)^{1/2}] *(t)=2.4495$

F/2.34495=2.667/2.4495=1.08 rads

1.08 rads=62.3o

sin 62.3o=0.886=Permeability

Prime numbers

In the solution to the Reinmann hypothesis, the critical line for Prime Numbers is:

$Y=e^z+Pi$

The first Prime Number=1

$Y=e^1+Pi$

$Y=5.85=1/\sqrt{3}=\tan 60°=\sin 60°/\cos 60°$

From Geometry and the 30°-60°-90° triangle, we can see that R=1.

$x^2+y^2=R^2$

$2x^2=1^2$

$x=1/\sqrt{2}=\sin 45°=\cos 45°$

Universal Young's Modulus=$\pi-e^1=0.4233$=cuz

$cuz=e^1+\pi$

$-e^1=e^z$

$-1=z$

z is the first prime number. Mass is formed at z=-1

$e^{-1}+\pi$

=3.50

$cuz^2=1/81$

Mass

Eigenvectors

Light=space. Space implies G

The rest of the universe follows at once - the crystal.

Time is a vector - an eigen vector. There is no degree of freedom with time. It flows forward and constant. It is Energy converted from PE to KE that changes with time [1-3].

11 Integrals of $\pi=1/11! * \pi^{11}=1356\sim s$

c=d/t=s/t

s=ct

$\pi^{10}+c=10! *\pi^{10}$

$1/t+t=11!=3.99!\sim 0.4=1/c^2$

s=ten ∫ of π=ct=1/c^2* c=c

s=c

s=s'

If we Integrate π 10 times (N=10), we get space.

s=∫0-10 implies π

c=s=2.9947

2.947/2.9979=305

Light is an Eigen value; not an eigenvector. In fact, light must be what an eigen value really is. Light reaches out in a spherical way. It isn't a vector. Time is the eigen vector [4].

c=2.9979=d/t=s/t

***Corresponding author:** Cusack P, Researcher, Software Engineer, Infobright, Canada, E-mail: jcxxxx@163.com

$s/t=(4/3)/\ t=2/9979$ (really, light is a derivative or it is 3 as it approaches max value).

$t=0.4$

$(s/t)\ \delta t=(|E||t|\sin\theta)'\delta t$

$s\ \delta t * 1/t\ \delta t=s\ \delta t - \text{Ln}\ t\ v - \text{Ln}\ t=\sin 1 - \text{Ln}\ 1=\sin 1$

$\sin 1=\delta E/\delta t * (\delta t/\delta t) * \cos t$

DOT PRODUCT=CROSS PRODUCT WHEN $t=1$, $\delta E/\delta t=1$

$\delta E/\delta t=\sin 1/\cos 1$

$E=\int \tan t=\text{Ln}\cos t$

$\sin\sin t=\cos t$

$E=\text{Ln}\ (\cos t)$

$1=\text{Ln}\ (\cos t)$

$e^t=\cos t$

derivative

$e^t=\sin t$

$\sin t=\cos t$

$E=1$

$\sin^{-1}(0.4)=23.57$ degrees

Ln (23.57 degrees)=π=Operator

The universe is an in-homogeneous Lorentz Group. There a ten real parameters.

I think where Einstein was going wrong is that he thought space and time were independent, if I understand relativity correctly. They are not. Space is the cross product of Energy and time vectors and thus completely dependent upon one another. Einstein couldn't find the Superforce because he did not know the universe is being compressed by sin theta. And of course, there is no such thing as a vacuum, at least inside the universe. Quantum Mechanics doesn't need to be recast because Einstein was incorrect. QM fits within my theory perfectly well.

Above is THE solution to Mathematics. It encompasses the 4 branches of Mathematics. It is the simplest way to explain and show the universe [5-6].

Convergence

The 4 branches of math: Algebra, Calculus, Geometry and Matrices are all about convergence. Where do they converge? Calculus converges on $y=y'=e^x$. Algebra converges on the Golden Mean $t^2-t-1=E$ $E=z=-1$ when $t=0$ • Geometry converges on the Area and circumference of a circle $R=2$, π, and on the Cartesian, Spherical, and Cylindrical Coordinates [7,8]. And Matrices converge on the eigenvector=$\sqrt{3}$ and eigenvalue=3 Convergence occurs at the first prime number given by $Y=e^z+\pi$, $z=-1$.

Why The Gravitational Constant Is 6.67?

$\delta^2E/\delta t^2=G$

$\int=\delta^2E/\delta t^2=\int G=0.4233$

$\int G=cuz$

$6.67=0.4233x$

$x=15.75$

$100-15.75=84.24=\sin\theta=\cos\theta$

$\int G=cuz$

$6.67X=0.4233$

X0.0635

$1/X=15.75=1-\sin\theta$

Integrate

$\text{Ln}\ X=1-\cos\theta$

$\text{Ln}\ X=1-0.8415=0.1585$

$X=\theta$

Gravity waves

So with the discovery of "Gravity Waves", predicted by Einstein, I wondered if and how this discovery fits in with my theory. It does.

The collision of the black holes would be akin to a giant sink of the universal fluid. This would cause a draw down of the surrounding fluid. See illustration below.

However, I don't think that is what is happening. What is happening is that the scientists are encountering the edge of the universe. Recall that the universe is egg shaped with axes 1 x 8 x 22 or 3 x 24 x 66 LY. Consider the short axis.

3 LY * 365.25 days/ yr * 24 Hours / Day * 3600 sec/ Hr * 30,000 Km / sec=2.8188 x 10^12 km

If we consider the draw down to be exponential,

$e^2=2.8188$

$\text{Ln}\ e^2=\text{Ln}\ 2.8188=0.4004=t$

$t=1/2\pi$ rads$=0.4$

Clairnaut:

$E''-E=0$

$t^2-t-1=0$ $2t-1=0$ $2=G=E''$

$1/[t*G]=1/(0.4004)(2)=1.249\sim1.25\sim1.3$ BLY

This is the distance that the Black hole collision was thought to have occurred.

So this discovery of Gravity Waves could fit in to my theory.

Eigen Function and Electromagnetism

$t=-0.618+1618=2.2236$

$t^2-t=0.1236$

$t(t-1)=1/81$ $t=1/8=0.01236$

$t=1/81=M$

$F=\sin t=\sin(1/81)=1/\sqrt{2}=\sin 45°=\cos 45°$

As F cycles, a charge like phenomenon dissipates on the Energy parabola (Characteristic Equation or Eigen Equation). This is why we sense quantum mechanics.

$G=\text{Pi}/\ \text{Ln}\ 1.618$

G *Ln 1.618=π

e^{G}* 1.618=e^{π}

1/0.618=$e^{\pi}/e^{G}=e^{\pi-G}$

x=1/(x-1)

x=t

$e^{\pi-G}$=1/t-1

E=1/t

$e^{\pi-G-t}-e^{\pi-G}$=1

Taking Ln of both sides,

π-G-t-π+G=0

-t=0

t=0

1/t E=Infinity

t=0, E=Infinity

Qunatum Packages are released at every cycle of π. Human Perception is freq=1/π=31.8 Hz

When the continuous charge is built up drop by drop over this time, a quantum is released. This is when the Energy of sin+cos is maximum, or its slope is 0. $\partial E/\partial t=0$, t=1

The universe is really continuous is the Spiritual World. Humans however have evolve to sense only the E max. Why?

Q=sin θ t=1. E=1

Integral dQ/dt=Integral Pi

Q=π Q

π=Q/Q=1

=quanta

Quanta=n*1

F=sin θ

$\partial F/\partial t=\cos\theta$

$\theta=\pi/4=1/\rho$

t=1/ρ=E E=ρ=0.127

1602/1273=79.5~80

1/81=0.01234567

The Mass crests over at the quanta. That is what humans perceive. The world outside the human mind is continuous.

From Electromagnetism,

$Jr^2=\cos\gamma$

$\omega=\iint S\cos\gamma/r^2\,\partial a=\iint S\,J\,\partial a$

$J^2 2=\cos\gamma * 1/r^2$

$J^2 * r^2/2=\cos\gamma$

The Orb is 44 cm diameter. So

$[(4/3)^2 * (44/2)^2]/2=\cos\gamma$

$\gamma=\pi/2$

23.704/2=cos γ

118.52=cos γ

γ=83.2°=1.452 rads

An/A=cos θ

<E,t>=|An| I

I=current=1.334

s=|An| (4/3)

An=1

-sin θ=-a=An=1/ cos θ

sin θ(cos θ)=-1

sin θ=-(1/cos θ)

-cos $\theta=-1/2x^2$

-cos=1/ $2x^2$

-2cos ^{3}(t)=1

$\cos^3$ (t)=1/2

cos t=0.7937

t=37.467 degrees=6539 rads=π/ Ln 1.618=G

t=G

Now s=|E||t|cos θ

<E,t>=s=cos θ M=<E,t>

An/A=<E,t>=M

An / A=cos theta An=A cos γ n

An=A<E,t>

1/An=cos θ

1/An=4/3

A=3/4 Spherical

z=rcos θ

z=cos θ=r An

r=An

Everything in existence is relative to the number 1, even scalars.

Hilbert Space

The Unique vector that is the convergence of orthogonal subspaces is given by:

f (x)=Σ f(x) k=1 to ∞

x=f=x=^(∞+1)/ (∞+1) $xx^{\infty/\infty}$=0.99999~1

The universe converges upon 1, where the fraction meets the multiple.

[INTRODUCTION TO SPECTRAL THEORY IN HILBERT SPACE, G. HELMBERG]

The Dot Product or Inner Product converges to the Energy Level in the universe.

$<f,g>=\int f(x)\, g(X)\, dx$

$=x^* \int (1/(x-1)\, dx$

=-2x/ (x-1)^2

If we assume the dot Product and Cross product are equal, then sin x=cos x=1/√ 2

The Quadratic, results,

$x^2 2-0.8284x - 1=0$

Solving the quadratic, know √ (-1)=-0.618,

There is only one root, and it is 0.1483. Conjugate, E=0.852

Contraction

[INTRODUCTION TO HILBERT SPACE, N YOUNG]

$6\pi^{1/11}=G$

G* sin 60°=1/√3

So a side of the equilateral triangle is 1/sin 60°=1/F=E

Energized is minimized when the operator is π

Contraction is shrink to an equilateral triangle. From one point to the opposite, is a 30°-60°-90° triangle.

So, the hypotenuse is 2π/√3=sin 60°* Pi=1/e (recall the energy for harmonic motion and the golden mean).

So energy is minimized.

Since e^{-t} is robust, its derivative is itself, and $E=e^t=1$, the Conservation of Energy can be expressed as:

$e^{-t}=1$

$\text{Ln } e^{-t}=\text{Ln } 1$

$-tj_j=0$

So $t_0-(-tj)=0$

$t0-\_t_j=1$

$t_0-0=1$

$t0_0 1$

Strum-Louisville

Strum-Liouville:

f(0)=0=f(π)

2t-1=f

2(3.14)-1=0.528

$K.E.=1/2\ Mv^2$

$0.528=1/2\ (4.482)v^2$

$v^2=23.58$

Ln 23.58=π

$\text{Ln } v^2=\pi$

2v=π

v=π/2

P=Mv=4.4852(π/2)=1/0.858=1/t

So why does the frequency=1/π?

1/freq=Wavelength

1/(1/Pi) sec=λ

1/s=c=d/t

d=1

c=1/1=1

d=1=c

d=c

s=d/t

s=s'

y=y'

t=1=d

t=s

$t^2-t-1=1\ t^2-t=2$

Derivative

2t-1=0

t=1/2

sin θ=cos θ=1

θ=0 or π=t

t=0 is trival, so t=π

Plot Ln t and e^t and swe get our usual Rm.

$\delta^2 f/\delta t^2+p(x)\, \delta y/\delta x+G(x)\, y=R(x)$

Integrate

$x^3/3-x^2/2-x=C$

$3/2^* (x^2-2)/2 * x -1=0$

x=2.5 ~T Period

t=1/2.5=0,4=1 rad

Now inputting √2/√3=0.816 (618 in reverse)

Curvature=1/ρ=s ρ=δs/δθ 0.1273=δs/√π

δs=0.2256

1/δs=4.4320~A

$A=pc/\ E/s^2$

$4.4320=2.667(4.3079)/(1/0.4233)/s^2$

s=0.9544

1/ρ=0.9544

ρ=1.0478 rads=60.0°

$(\tau')^2=60$ degrees=1/6

If a particle moves at constant speed, its acceleration is normal to its velocity.

In the s-t plane, normal is the acceleration, or $F=\sin\theta$

$(\tau')^2=1/6$

$\int(\tau^2)=1/6$

$\tau=6.3$ EARTHQUAKE MAGNITUDE

Now inputting $\sqrt{2}/\sqrt{3}=0.816$ (618 in reverse)

$3/2x^2-2x-2=1/e$

$R(x)=1/e$

This is where the Harmonic Equation meets the Golden Mean. It is our Universe!

Four Fundamental Forces

So there are 4 fundamental forces in our material universe, namely, Strong Nuclear; Weak Nuclear; Electromagnetic; and Gravity There are all related algebraically as follows.

From above:

F=sin 1=1-1/beta

$G(2A)=F+\rho/\delta\rho*\delta p$

G(2A)=F+1/beta

$=F+\rho*\delta p/\delta\rho$

F=G(2A)-1/beta

$\delta F=0\sin\rho/\delta p(\text{not}=)\ 0=1$

$0=dF*\rho/\delta\rho$

$(1)=1+\delta F*\rho/\delta p$

$2GA=F+\delta p/\delta\rho*\rho$

$\delta F=\rho/\delta\rho=0$

$F'=0$

So F=G(2A)-1/beta

F=2GA-Fd

F=1-Fd

F+fd=1

F(1+d)=1'

F=1/(1+d)

$F'=[1+d)^{-1}]'$

$F'=(1+d)^0/0$

$F'=0=1/0$

$F'=0$

$1/F'=0$

$\sqrt{(-1)}/\sqrt{0}=\text{sqrt}(-1)=0.618,\ 1.618$

sqrt (1/0)=sqrt (-1)

1/0=-1

$F'+\rho/\delta p)=0$

1/0+1=-0'

$\sqrt{(0)}=1.618\ 0=(1.618)^2=2.6179=2.618=1+1.618=t+t=2t$

So, the Universe is a 4 th Order tensor with variables: F (or acceleration and its derivative momentum); t,s,E

a=v=s

$s=Ae^{-st}$ .... Temperature=mass=energy

Clairnaut harmonic bear

$\delta^2E/\delta t^2-E=0$

G-E=0

$G=\int E$ ....GRAVITY

$F=\sin\theta$ ....STRONG NUCLEAR

$\tau=c+\sigma*\tan(45°+\varphi/2)$

$c=-0, \varphi=30°, \theta=45°+30°/2=60°$

$\tau=\sigma*\sqrt{3}$ ....ENERGY

P/e-=F/1.602=2.667/1.602=1/6=60° ....Electromagnetic

In conclusion,

$E=M\ M=F\ E=G\ F=\sin\theta$

$F=G(2A)-1/\beta$

and The Universal Equation is: s=E x t @ F=sin 60° $s=|E||t|\sin\theta$ $s=|E||t|F$

Dampened Cosine In The Harmonic Beat:

$Y=e^{-st}\cos\theta$

$(1)(s)=-s*\cos\theta*\sin\theta$

$\sin\theta*\cos\theta=1$

Taking the derivative:

$-\cos\theta=[(\cos\theta)^{-1}]'$

$\cos\theta=1$

$\theta=0$

$\sqrt{\theta}=t$

$t=\sqrt{0}=1.618$

$1.618/\pi=1/196\ \infty$

$2.71828/\pi=0.865=E$

Fluid Mechanics

[Vectors, Tensors, and the Basic Equations of Fluid Mechanics, R Aris]

$1/\rho=\tau^2$

Curvature=$1/\rho=s$ rho=$\delta s/\delta\theta$ $0.1273=\delta s/\sqrt{\pi}$

ds=0.2256

$1/\delta s=4.4320\sim A$

$A=pc/\ E/s^2$

$4.4320=2.667(4.3079)/(1/0.4233)/s^2$

s=0.9544

$1/\rho=0.9544$

$\rho=1.0478$rads$=60.0$ °

$\tau'^2=60°=1/6$

If a particle moves at constant speed, its acceleration is normal to its velocity.

In the s-t plane, normal is the acceleration, or F=sin θ

$\tau'^2=1/6$

Integral $(\tau^2)=1/6$

τ=6.3 EARTHQUAKE MAGNITUDE

β=Tan u x v

$\beta=|\tau||v|\sin\theta$

θ=60 °

v=0.2592~porosity e

β*β=1

$|\beta||\beta|\cos\theta=1$

$\beta=\sqrt{2}$

v'=Mv (√2) - 1/0,1273)(6.3)=4.406~A

δs/δt=0.0027

0.0227/0.8415=0.0027=υ=POISSONS RATIO

v=ρ * τ

0.02698=0.1273τ

1/ τ=0.4718

Ln τ=Ln 6.3)=1.8405

1/Ln 6.3)=0.543

0.543/0.4718=115.16

0.868

Z=sin θ sin θ=-1

z=0.8415

z=r(cos θ+i sin θ) 0.8415=r(cos 1+i sin 1) 1=r(1+i) 1/r=1+i

$Z=r\ e^{(-i\theta)}$ $1.618=r(0.618)(e^{(-i\theta)}$

$2.6181=r^{(-i\theta)}$

Ln (2.618)=0.9825=Ln $(0.618)e^{(-i\theta)}$

0.9825=Ln (0.618)/Ln $e^{(-i\theta)}$

-i θ=Ln 0.618/0.9825

i θ=Ln 0.5000 i1.618 θ=0.5

θ=0.309

θ=1.77°=√π=t

M=f=2=δM/δt

P.E.=Mgh 43.98=(2+4.486)(0.1585)(h) h=0.4233=cuz=R_m

CUSACK'S UNIVERSAL POTENTIAL ENERGY EQUATION

P.E.=[M+δM/δt][u.g.)(R_m)

K.E.=1/2 Mv^2

=1/2 (2+δM/δt)v^2

$=(2+4.486)(0.8415)^2=2.2964$

Ln K.E.=3.1339~π

tan P.E. / K.E.=1397°

E=0.8603

tan φ=cot φ

φ=1.599=1/6

φ0.625 rads

'tan (45°+φ/2)=88.4=h

45°+φ/2)=35.82°

φ=16164~1.618=t on the Energy Golden Mean Parabola

Fluid Dynamics

[J. Bear Fluid Dynamics in porous Media]

β=1/ ρ * δρ/δp

=s*δρ/δp

=4/3* (0.1273)/26.667)

=0.0636

=Earthquake Magnitude

1/ β=15.72

1-1/β=0.8428=sin 1

F=sin 1=1-1/ β

And,

Permeability

1-Kz/b=sin 1

kz/b=Moment

kz/b=Fd

d=z

k/b=F

15.85=26.667 (b)=k

k=0.5944

1/k=1682

1/k=1.7

s=k*ρ=1.7(0.1273)=1333=s

s=k 1/s)

s^2=k

k=1.77=√π

s=√[√π]]

t=π

$t=\sqrt{\theta}$

$\sqrt{\theta}=\pi$ \\$\sqrt{[\sqrt{\pi}]]}=\sqrt{\pi}$

$s=\sqrt{\sqrt{\pi}}$

$s^4=\pi$

$s^4=(1+e)=1.263$

T=kb=(0.5944)(0.0594) 0.0353

$\tau\ \delta\tau/\delta t=kT/s$

$\sqrt{3}$ * $\delta\tau/\delta t=(0.5944)(0.0353)/(1.263)\ \delta\tau=\sqrt{3}=1/6$

$\delta\tau/\delta t=\delta E/\delta t=\int G=1/6/\sqrt{3}=0.0962$

$\int G=\int\delta\tau/\delta t=\tau$

$G^2/2=\tau=0.0962$

G=0.4386=P.E.

G=43.86(2/3)=15198

K.E.=8480

$K.E.=1/2Mv^2$

M=2=F

$M_0=F * d=0.1585=\sin\theta^* s$

$0.1585/(4/3)=\sin\theta$

$\sin\theta$=118.9 (118 Chemical Elements in the Periodic Table)

4.486/ (118.9* 938)=402.11=Re

Re/ sin θ=Mp+/M Re=Inertia f / Visc F=sin (0.1696)/ sin (0.2699)=6.383~h=2/Pi h=(δM/δt)/ t δM/δt=ht

2=6.36*t

$t=\pi$

This is the solution to Mass degeneration in QM.

$C=A'=2\pi R$

$M_0=F*d=F*R=\sin\theta * \delta R/\delta t=\delta M_0/\delta t$

$I=\delta M_0/\delta t=\sin\theta * \delta R/\delta tt=\delta R/\delta t$

$t\delta t=\delta R$

$\int t\,dt=\int\delta R dt\ t^2/2=R$

$t^2/2=R\ t=2$

$t^2-t-1=E\ \delta E/\delta t=2t-1=G$

2(2)-1=3=c

1/c=Wa. Eq.

$\delta M/\delta t=G=\delta^2E/\delta t^2$

So Gravitational Constant is the rate of Mass degradation (and formation)

M=1/81~0.125

$\int\delta M/\delta t\ C_0=\int M$

$M* 1.602=M^2/2$

C=251=T Period

So, what governs δM/δt=0.197?

$E=Mc^2=F$

$F=\sin\theta=Mc^2$

$\sin t=197\ (2.9979)^2$

$=1.77=\sqrt{\pi}=\sqrt{t}$

t=1/T

1=1/251=0.4

$\sqrt{4}=2=R$

δM/δt=R

$A'=2\pi R=2\pi(2)=4\pi$

So the rate of change=1/4π=0.796~0.8

Or

$81=c^4$

$\delta c^4/\delta t^4=\delta^3 v/\delta t^3=\delta^2 s/\delta t=C=T$ Period

So, Mass degrenrates at the Period T=251/sec

$M=\sin t\text{-}e^{(1696 * R^2)}$

$\delta M/\delta t=\cos t\text{-}e^{(0.1696*4)}$

=0.6609-0.507

=0.4283~cuz

0.4283*0.1978=0.847

1/0.847=118 Elements in the Periodic Table.

There are 7 Periods in the Periodic table of the Elements.

Wa. Eq.$=1.618^{(7*R^2)}$

28Ln 1.618=1/23=4.3482

M/c=28/2.974=941.5=Mp+

p=hk=hF=6.36*2.667=0. 1696

4.3482/c=1.48~1.5

$e^{-1.5}=M$

$E=Mc^2\ E=M/c * c^3$

$E/c^3=e^{pR}/c^3=e^{(1.696)(2)=}1/81$

1/81=0.012345679

This then is how Mass and the Period table and the Wave Equation are interrelated. It also shows the Mass Gap.

Quantum Chemistry

p=hk=hF=6.36(8/3)=1696

$F\text{-}\int\psi e^{(\pi/2)}=e^{(pR^2)}$

CUSACK'S QUANTUM CHEMISTRY FORMULA

$M=\sin(1/E)\text{-}e^{(pR^2)}$

$\psi=[Gt^3-1/R * t^2-t]e^{(t/2)}$

Let t=c=3

$\psi=[2/3(3^3)-1/2(3^2)-3](e^{-3/2})$

$=1/c$

$1/c=E$

$E=1/t$

Wave equation

$U=iV=1+(0.618*1)=1.618$

$1.618=1/\sqrt{(2\pi * e^k)}$

k=405

k=1.4001

0.86 or 59.3 degrees~ 60 degrees

Is matter a wave or particle? I'd say a wave creates a particle. A wave is a form of energy. That energy is stored in the mass. Thus $E=Mc^2$. The sine curve is the Force which equals Energy that is put into Mass formation. The Mass is a temporary store for P.E. Einstein's Equation should be, $PE=Mc^2$ So, F=E as we have already calculated.

$t^2-t-1=E=\sin t=1/\sqrt{2}$

$t^2-t=1/\sqrt{2}+1$

$t(t-1)=1.707$

$t=1.707=1+1/\sqrt{2}$ OR, $t=2.707\sim e^1$

t=1+sin 45=1-sin t=Moment OR t=E

$E=1/t$

[QUANTUM THEORY DAVID BOHM]

Quantum theory

The Fourier Integral is the dampened cosine curve already introduced.

The pulse of the universe is like a heart beat, one dampened cosine curve after another.

$Y=e^{-t}\cos\theta$

$Y'=e^t\sin\theta$

$y=y'$

$\cos\theta=\sin\theta$

Wave Equation=$1/\sqrt{2}/\sqrt{\pi}$* Integral $e^k\,\partial k$

$=1/\sqrt{\pi} * \cos 45° e^k$

$=1/\sqrt{\pi} * Y$

$1/\sqrt{\pi} * \cos 45°=0.3991\sim 4$

4=SUM (E+t) (Vector Space)

THIS UNITES COSMOLOGY WITH QUANTUM MECHANICS.

So we've unified Cosmology with Quantum Mechanics and Electromagnetism. That is the Grand Unified Theory.

Quantum mechanics

INTRODUCTION TO QUANTUM MECHANICS, D J GRIFFITHS]

$\psi==cn\int(t^2-t-1)*e^{(t/2)}$

CUSACK'S MODIFIED SCHRODINGER EQUATION

$\psi=[Gt^3-1/R*t^2-t]\ e^{(\pi/2)]}$

OR,

$E=\int E\ e^{(t/2)}$ where $t=\pi$

deBroglie Wavelength

$\lambda=h/\sqrt{[3MkT]}$

$=h/\sqrt{[3*4.482*8/3*0.2506]}$

$=h/c^2$

$=1//\sqrt{2}=\sin 45°=\cos 45°$ (45°=1 qusackian)

If we have 5 singular points for a Linear Second Order Differential Equation, we have the famous distance equation:

$d=v_it+1/2\ at^2$

Integrate this twice, we get 6 variables (s, G, d,v,a, M), From the Clairnaut Second ODE, we get the Operator (frequency) , and the Energy, All we need is R_m=cuz and N=11

I've shown already that sin must equal cos for matter to appear. Now I tell you that the Fourier Integral converges on G=2/3 when sin=cos or y=y' x=0.8415=sin 1 L=87 & 8.7& 0.87 n=1 $a_0=0$

$a_0/2+\Sigma\ [a * \cos(n\pi x/L)+b\sin(n\pi\ x/L_0]$

$a=1/L\int f(x)\cos(n\pi\ x/L)\,dx$

$=-\sin(n\pi\ x/L)$

$b=1/L\int f(x)\sin(n\pi\ x/L)$

$=\cos(n\pi)/L$

Substituting,

$0/2+\int(-\cos^2(n\pi\ x/L)+\sin^2(n\pi x/L\ )$

$2/3\ \sin^3(n\pi\ x/L)+2/3\ \cos^3(n\pi x/L)=-2/3\ (\sin^3((1)(\pi)(0.8415/87))+2/3\cos^3((1)\pi*0.8415)/87)$

$=0+2/3\ (1)=2/3=G$

Cusack's Fourier Integral Formula

F=G c* GE F=G(c+E). The solution to the Fourier Integral is necessary ($y=y'=e^t$) and sufficient.

CUSACK'S MASS FORMULA $M=2\pi R/cuz/C_0$ One only needs the Fourier Integral, the Orthogonal Matrix with Operator Ω^-1=$1/\pi$, and Cusack's Mass formula to sole the entire Universal Problem.

CUSACK'S FORCE EQUATION

F-G=sin 60°

$A'=Circ=2\pi R$

$R=2=\delta M/\delta t=E=G$

$2\pi*2=4\pi$

$4\pi/cuz=29.6867$

$29.68/C_0=29.6867/1.602=1853\sim$ n/p+

1853 * 7=129.71=1/0.771

0.771*24* 3600=6.67=G

$E=\Omega\text{--}\ \Omega_0=2\pi/l$

$=2\pi\ (0.75)=8/3 * \pi=F* t=I$

$\int qa/dE=(qa)^2/2* dt$

=7^2/2* dt=992

dt=40489=23.1985 rads

Ln 23.1985=3.1441~π

Bell Curve:

$\psi=1/\sqrt{2\pi}] * \int e^{-t}$

Let t=π

$1/\ Te^{-\pi t}*e^{\wedge}\pi=tE=172\sim\sqrt{3}=E$

$E=Mt\ \ tE=Mt^2$

$E=\sqrt{3}$

t=1.618=ψ

No spooky action at a distance.

So the process that governs QM is not subject to a hidden variable. It is 1`/81=0.012345679

ΔE=hv=6.36 (1/π)=202.45

$\Sigma qa\ /\ dE=qa)^{2}*\ dE$

=7^2* 202=992

1/992=1.0081=H+

God doesn't roll the dice.

d/dx csc=csc 2.9979+cot 2.9979=19.126-19.100=365.3=Earth Year

2.9979 * 365=1 LY

And,

[David Bohm, Dover]

E=1/ (8π) * ∫ (Electric field^2+Magnetic Field) dr

=19905 / (8π)* R

=1584~1-sin 1=Moment=F* d=Work

So the Universe of volume 19905 is an Electric and Magnetic field.

THE Cusack Universal Equation is:

$\delta^2E/\delta t^2\text{-}E=\text{Ln } t$

or,

$GE^3\text{-Ln } t=s$

$0.666*E^3\text{-Ln } 1=||E||||t||\cos 60°$

E=1.2533

E=-1.2533=Emin=t^2-t-1 (GOLDEN MEAN)

You might do a sensitivity analysis on this circuit?

$x^2\text{-}y^{\wedge}2=\pi$

$2x^2=\pi$

$x=\sqrt{\pi}/\sqrt{2}$

$x^2\text{-}x\text{-}1=1\text{-}0.625=1/\pi=31.8\text{Hz}$ (Human Perception)

The capacitor should be C=π

V=iR G||E||||t||cos 60°=[r+1/c+L]i

R=slope m=1/cuz 1/C=Pi L=2

Cusack Analogue Circuit Equation

$G*0.8415*\text{sqrt}3/2=[\pi\text{-}e+\pi+0.8415(\sqrt{3}/(2c^2))]i$

G*unit eigenvector=1 cycle –e+sin 1 unit eigenVector/[2 unit eigenValue^2]

G * unit eigenvector=3Pi-2e+eigen vector/eigen Value^2

G*unit eigenvector=eigenValue Pi+δM/δt*e+eigenvector/eigen value^2

The universe is where the FUNCTION $x^2\text{-}x\text{-}1=0$ meets the RELATION $x^2\text{-}y^2=1$

The FUNCTION is the Derivative whereas the RELATION is the Integral. So y=y'=Integral $y=e^x$

Since 0.618=sqrt 1, and this is imaginary, it follows that 1/0.618=1.618 is imaginary. 1.618*0.618=1 is imaginary. The energy that makes up the universe is imaginary. We are all but images in God's mind.

Conclusion

We see that the Cusack Universal Equation and the Cusack Force Equation unites the four fundamental forces in our universe -the Holy Grail of Physics. There is no sooky action at a distance and Quantum Mechanics and Cosmology are finally united.

References

1. Aris R (1962) vectors, tensors, and the basic equations of fluid mechanics. Dover, New York.
2. Bear J (1972) Fluid dynamics in porous media. Dover, New York.
3. Ohm D (1951) Quantum theory. Dover, New York.
4. Cusack P (2016) Riemann Hypothesis Clay Institute Millennium Problem Solution. J Appl Computat Math 5:317. doi:10.4172/2168-9679.1000317.
5. Griffiths DJ (2014) Introduction to quantum mechanics. Pearson Essex, England.
6. Helmberg G (1969) Introduction to spectral theory in hilbert space. Dover, New York.
7. Ronjansky V (1971) Electromagnetyic fields and waves. Dover, New York.
8. Young N (1988) Introduction to hilbert space. Cambridge University Press.

6

On De Broglie's Double-particle Photon Hypothesis

André Michaud*

Senior Researcher, Canada

Abstract

Establishment of an LC equation and of a local fields equation describing permanently localized photons from the analysis of kinetic energy circulation within the energy structure of the double-particle photon that Louis de Broglie hypothesized in the early 1930's. Among other interesting features, these equations provide a mechanical explanation to the localized photon properties of self-propelling at the speed of light and of self-guiding in straight line when no external interaction tends to deflect its trajectory. This paper summarizes the seminal considerations that led to the development of the 3-spaces model.

Keywords: Electromagnetic theory; Kinetic energy; Photon; Acceleration; Electron-positron pairs; 1.022 MeV LC equation; 3-spaces

Introduction

The first integrated representation of electromagnetic energy was provided by Maxwell as a continuous wave phenomenon that would be due to interacting electric and magnetic fields inducing each other, which led to the recognition that radio frequencies belong to the same electromagnetic spectrum as visible light. Then came Planck's analysis of Wien's experimental data on the black body demonstrating that electromagnetic energy is always captured as frequency dependent discrete amounts. Einstein's photoelectric proof confirmed Planck's hypothesis shortly afterwards by demonstrating that photons do behave as if they were separate localized quanta when intercepted while also demonstrating that they possess longitudinal inertia, which eventually earned them both Nobel prizes. Compton and Raman added further experimental confirmation of Planck's conclusion, while experimenting with other types of collisions between photons and electrons. These findings conclusively confirmed the discrete and point-like behavior of photons when being absorbed. We must also keep in mind that an interaction cross-section always larger than zero needs to be assumed for all point-like behaving particles during scattering experiments to correctly account for the observed recorded traces. So we know that photons are not really point-like with zero dimensions in the mathematical sense, even if their motion can be calculated as if they were; just like the trajectory of the Moon about the Earth is calculated as if their masses were concentrated in a single point at the center of each body. Point-like behavior of photons upon emission was also subsequently understood and verified, which we will have a look at further on. So, we know for certain that Maxwell's continuous "waves" do not exist as such at the submicroscopic level, despite the fact that his equations allow calculating all electromagnetic manifestations with the utmost precision when electromagnetic energy is treated as being continuous and featureless as observed from our macroscopic perspective.

In fact, what these discoveries reveal is that we are in the very same situation with respect to electromagnetic energy that we are in with respect to solid materials, as it closely parallels the fact that although we can observe that the surface of a polished diamond has a flawlessly smooth finish from our macroscopic perspective, for example, we can also alternately observe that this same surface is granular and bumpy when the scattering particles of an electron microscope reveal the outlines of the individual atoms making up the crystal surface at the submicroscopic level. In the latter case however, we have a rather extensive understanding of the inner structure of the atoms involved, but to this date, the inner structure of photons is still the object of speculation.

For the past century, there has been a deeply ingrained conception in the case of light that it sometimes behaves as a wave and sometimes as a particle, two types of behavior that are incompatible for a number of reasons and that gave rise to the concept of "wave-particle behavior" to characterize the photon. Close examination of the concept in light of the macroscopic-submicroscopic comparison just clarified leads to the view that generally speaking, "wave behavior" could simply be the result of behavior of crowds of discrete photons that our macroscopic instruments generally deal with while "particle behavior" could simply be the behavior of individual photons at the submicroscopic level. This would go a long way in removing the inherent incompatibility of the "wave-particle behavior" concept, by replacing it with a "macroscopic-wave behavior vs submicroscopic particle behavior" concept. But we will see further on that with the model that will be proposed here, even at the submicroscopic level, the localized photon can display both types of behavior without any conflict by associating transverse wave behavior with longitudinal particle behavior. Also, despite its systematic point-like behavior in all scattering and capture experiments, behavior typical of elementary particles, the photon was suspected early on of not being elementary because light can be polarized, which cannot be explained if the photon was made up of a single point-like behaving particle. This was clarified by Louis de Broglie as the concept of spin was introduced, associating a spin of 1/2 to point-like behaving particles that were proven out of any doubt to be really elementary, such as the electron and the positron, and consequently a spin of 1 to the photon, thus hypothesizing that if it was made up of two particles, this could directly explain why light can be polarized [1].

Louis de Broglie was the first to elaborate a comprehensive theory on the possible internal structure of photons. According to his hypothesis as proposed in the 1930's, a permanently localized photon following a least action trajectory can satisfy at the same time Bose-

***Corresponding author:** André Michaud, Senior Researcher, Canada
E-mail: srp2@srpinc.org

Einstein's statistic and Planck's Law, perfectly explain the photoelectric effect while obeying Maxwell's equations and totally conform to the properties of Dirac's theory of complementary corpuscles symmetry, only if it involves two particles, or half-photons of spin 1/2, " *that must be complementary with respect to each other in the same manner that the positive electron* (the positron) *is complementary to the negative electron in the Dirac Hole Theory*" [2].

The following other quotes from the same reference summarize his hypothesis:

"*Such a complementary couple of particles is likely to annihilate at the contact of matter by relinquishing all of its energy, which perfectly accounts for the characteristics of the photoelectric effect.*"

Furthermore:

"*The photon, being made up of two elementary particles of spin h/4π, will obey the Bose-Einstein statistic as required by the precision of Planck's law for the black body.*"

Finally, he concludes that:

"*this model of the photon allows the definition of an electromagnetic field linked to the probability of annihilation of the photon, a field that obeys Maxwell's equations and has all the characteristics of electromagnetic light waves.*"

Over the course of the 1930's and 1940's, de Broglie and his students progressively came up with an interesting and workable solution based on wave mechanics, that involved both corpuscles being singularities in an underlying wave phenomenon [1]. After Quantum Chromodynamics was developed in the 1970's an alternate model was developed, involving a mix of quark-antiquark pairs and gluons [3] based on this new theory and Quantum Mechanics, which also is an interesting and workable approach. A few other models have been proposed since, but all approaches have the similar downside with respect to Maxwell's theory of treating the electric and magnetic fields, either explicitly or implicitly, as a single "electromagnetic field" which turns out to be somehow featureless at the general level (the electromagnetic tensor), which distracts from permanent awareness that both fields are of equal and separate importance in Maxwell's theory, with different and irreconcilable characteristics, besides mutually inducing each other. This left no precise function being assigned to the "magnetic" aspect of the electromagnetic energy in a possible mechanics of mutual induction that would also involve the two separate charges, which are the "electric" components of the photon, a mechanics that would explain why photons can maintain sufficient local unity to account for their systematic and verifiable point-like behavior during scattering or absorption encounters, which includes all photons that we know have been emitted from the farthest reaches detectable in the universe, after having traveled for countless years. Indeed, the twin "electric" particles end up in both models as having an existence separate from the electric aspect of the electromagnetic energy that the localized photon is meant to represent, which introduces the required twin particles in a manner that does not incorporate them into the sequence of the electric vs magnetic mutual induction cycle that they theoretically are meant to enhance, according to de Broglie's initial hypothesis:

"*it seemed to me that to obtain a clear image, in agreement with the classical concepts of the wave-particle dualism with respect to space and time, it was required to succeed in incorporating the particle into the wave*" [1].

But it seems that the non-deterministic trend that was prevailing after the 1927 Solvey Congress confronted him with such difficulties that he ended up renouncing this ultimate goal [1]. Generalizing the electromagnetic interaction as a single tensor is a fine approach to obtain global perspectives, but it seems that looking for ever more detailed descriptions always favored deeper understanding of physical issues. This paper is then an attempt at exploring deeper even than the already interestingly detailed electric and magnetic fields as described by Maxwell's theory.

The Required Internal Electromagnetic Symmetry

As it stands, if the photon's double-component electric aspect is to remain coherent with its point-like behavior at the moments of emission and capture (or scattering), however long the time elapsed and distance covered between both events, the two separate "electric" half-photons have to unite in some fashion during each cycle of the photon's frequency to maintain point-like localization and most importantly to incorporate into mechanical process the other half of the electromagnetic relation, that is, its magnetic aspect. Doesn't the induction of an increasing magnetic field inseparable from changing current due to moving charges immediately come to mind at this point? In the case of photons, this brings displacement current into the picture, which would involve local motion of the postulated double charges that would cause the required change in the local electric field within the photon quantum, a current that would come into being without the presence of matter in this case, a process that interestingly was first proposed by Maxwell himself in 1865 and was the foundation of his electromagnetic theory [4] This in turn hints at the possibility of an internal oscillation of the photon energy related to its frequency.

Let's keep in mind here that the term "frequency" applies to any sort of cyclic motion, be it rotational, translational on a closed orbit or any other type of oscillatory motion, from simple sinusoidal harmonic motion to the cyclic translational reciprocating "swing" between two states being considered here and that we will term "oscillation" for simplicity's sake. This means that all aspects of angular momentum that we naturally associate with rotary motion can also be applied to reciprocating motion, which in turn allows the "spin" of elementary particles to be hypothesized as possibly corresponding to a reciprocating motion of the energy concerned without changing in any way the equations that already account for it.

It is a fact that all experimental research aimed at identifying charges in electromagnetic waves have failed to detect any in support of Maxwell's assumption. But let's consider that if electromagnetic waves as Maxwell conceived them really turn out to be only a convenient mathematical representation of a macroscopic perception of a crowd effect due to the presence of countless localized moving photons at the submicroscopic level, it would indeed be these individual photons that would display the searched for charges and would be the local sites of displacement current versus magnetic induction activity.

However, there exists no instrument sensitive enough to detect the infinitesimal fields of individual photons, with the added difficulties that they move at the speed of light and that any interception of a single photon simply incorporates its energy as an infinitesimal kinetic energy increment to one electron in one atom of the material that the detector is made of. But since this postulate was such a major and fruitful foundation in the elaboration of Maxwell's theory, which in turn allows such precise calculations, there seems to be no reason to do away with it now. The double-particle photon hypothesis would then imply that photons have to be stable localized moving electromagnetic structures whose energy quantum could logically only alternate between a two components electric state, with both components separating in space (an electric dipole), and a magnetic state involving only one component to explain permanent localization and that could consequently be dipolar in only one manner. Total symmetry of the magnetic aspect involving a single component can be obtained only if it consists in a single spherically expanding phase as both electric components move towards each other, followed by a spherical contraction phase as both electric components move away from each other; both magnetic expansion and contraction sequences of the magnetic component being normal to the electric phase at all times. This also means that the single magnetic component of the photon can be dipolar only along the time dimension since both expansion and contraction sequences cannot possibly occur simultaneously. Such a dynamic structure would still preserve the required fundamental symmetry since the space-wise moving electric dipole would be permanently counterbalanced by a related time-wise perpendicularly moving magnetic dipole, with both dipoles remaining perpendicular to the direction of motion of the photon in space, thus obeying the triple orthogonality required for plane wave treatment in Maxwell's theory's for straight line motion of electromagnetic energy.

Internal Coulomb Interaction between the Half-photons

Let us note here that de Broglie considered both half-photons as being electrically neutral [5] that is, not being charged negatively for one and positively for the other. But by the same token, he also discarded the possibility that Coulomb interaction could be involved in the process since he considered that the Coulomb force could be in action only between charged particles that are "signed" negatively and/or positively, which was confirmed to me by his lifelong friend and colleague Georges Lochak, in correspondence initiated by me precisely to clarify this issue, which is why the research that he carried out did not take this possibility into consideration. Paradoxically, it has been understood and extensively experimentally confirmed since the 1930's that any photon of energy 1.022 MeV or more, that has no rest mass and is electrically neutral, will destabilize and convert to a pair of electron-positron, massive and charged in opposition, when grazing a heavy particle such as an atomic nucleus. Could then the "signs" be an extrinsic property of elementary particles charges, possibly vectorial, that would be acquired during the separation process of the pair? This would leave the door wide open to the possibility that some form of Coulomb-like interaction might be involved at a level more fundamental than that of the acquisition of the opposite "signs" by the charges of the separating elementary particles. So let's dwell for a moment on what considering "signs" as a property separate from charges of elementary particles can allow visualizing. In this perspective, the very existence of "fractional signs" for the charges of the up and down quarks making up the inner scatterable structure of nucleons means that other "stable sign intensity levels" do exist besides the otherwise universal "unit sign intensity level" for the charges of electrons and positrons. Note that this comparison is by no means meant to hint at the possible origin of up and down quarks, which is still unresolved, but only to highlight the idea that different degrees of "sign intensity" do exist for stable particles, which allows considering that "sign intensity acquisition" for charges could possibly be progressive from null, for initially neutral photon charges, to maximum stable "unit sign intensity" for the charges of electron and positron, with intermediate stable levels corresponding to the up and down quarks' stable "fractional sign intensities". The opposite "unit sign intensities" of electron and positron could then be progressively acquired during the mother photon's destabilization process, possibly induced in the photon's initially neutral charges by the very presence of the "signs" of the charges of the destabilizing particle that the photon grazes, from neutral at the beginning of the process to maximum and stable opposite "unit sign intensity" for the separated charges if the destabilization sequence succeeds in separating the pair, or eventual regression back towards neutrality of the photon charges if the process fails for whatever reason, leaving the photon moving away with charges returning to neutral without decoupling for photons not energetic enough, or flying by too far from the destabilizing particle for the process to complete in the case of sufficiently energetic photons.

Electrostatically Destabilizing Trajectories Intersections

It must be considered also that Quantum Electrodynamics implicitly recognizes the presence of Coulomb interaction between a decoupling photon and a heavy nucleus, since it incorporates a Feynman's "virtual photon" into the pair production process representation (Figure 1), which was explicitly defined by Feynman himself as being a metaphor for Coulomb interaction [6], thus indirectly recognizing that Coulomb interaction has to be in action between the photon and the destabilizing heavy particle even before the pair separates, whatever the sign status of the photon's internal charges may have been. Let's consider what is likely to occur when a photon of energy 1.022 MeV or more grazes very closely a heavy atomic nucleus. We know since de Broglie that all massive and charged elementary particles are electromagnetic in nature, since electric charges cannot be dissociated from a magnetic counterpart. This includes of course the scatterable point-like behaving massive up and down quarks making up the inner scatterable structure of nucleons (protons and neutrons) since they also possess measurable electric charges, charged quarks whose existence was not yet known when de Broglie was actively working on his hypothesis, since they were experimentally scattered against only in the late 1960's [7]. Destabilization leading to pair decoupling could then be explainable by the presence of these point-like behaving electromagnetic charged elementary particles of which all nucleons making up atomic nuclei are made, that can presumably enter into homo- and/or heterostatic interaction with the charges of the half-photons while the photon is in its electrostatic phase as it flies by. It becomes just as obvious then that these interactions may then become more and more intense in relation with the inverse square of the diminishing distance that separates the half-photons from these up and down quarks if a Coulomb-like law effectively applies, a process represented in Quantum Electrodynamics by the Feynman diagram shown in Figure 1 [8]. The fact that such decoupling can occur only during moments of very close proximity between photon and nucleus comes in support of the presence of an interaction as a function of the inverse square of the distance such as

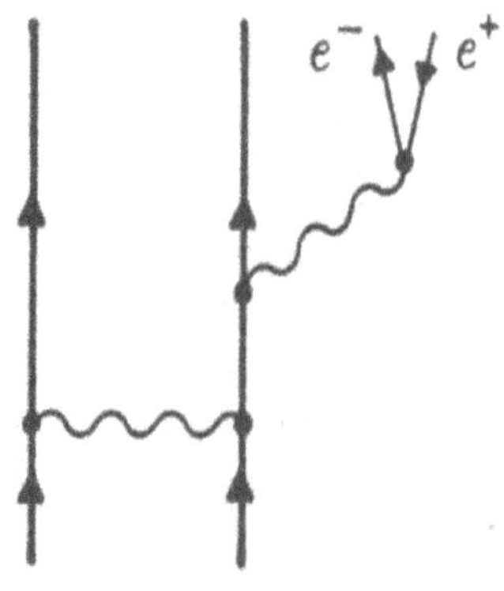

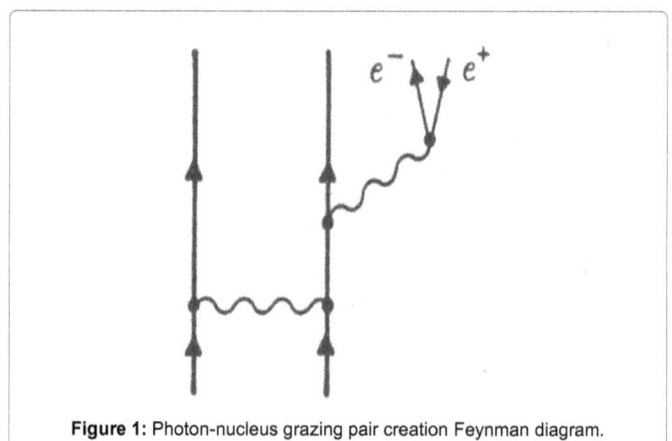

Figure 1: Photon-nucleus grazing pair creation Feynman diagram.

the Coulomb law. Similarly, pair creation during close flyby of two photons, at least one of which exceeding the 1.022 MeV minimum energy threshold without any atomic nuclei being close by, such as was first experimentally confirmed by Kirk McDonald et al. at the Stanford Linear Accelerator in 1997 with experiment #e144 [9], is represented by the Feynman diagram shown in Figure 2 [8]. So there seems to exist sufficient supporting evidence to at least explore the possibility that Coulomb-like interaction could be at play between photons and other localized electromagnetic particles and even between the possibly neutral charges of the de Broglie double-particle photon.

Photons, Electrons, Positrons, Exclusively Made of Kinetic Energy

After destabilization, the separated halves of the photon's energy can thereafter be observed behaving as one massive 0.511 MeV/c^2 electron plus one massive 0.511 MeV/c^2 positron traveling separately, whose unit charges are now observable as being signed in opposition, and whose velocity away from each other is linked to the residual energy that the mother photon possessed in excess of the 1.022 MeV energy threshold level which is now making up the rest masses of both particles, a process first observed and confirmed by Blackett and Occhialini from analyzing recorded cosmic radiation scattering impact traces in a bubble chamber in the early 1930's. The reverse process of electron-positron pairs re-uniting to entirely convert back to various photon states has also been first observed and confirmed by Blackett and Occhialini, such as in the case of positronium decay. So both reverse processes constitute the *de facto* irrefutable material proof that electrons and positrons are made of the very same energy and are of the very same electromagnetic nature as photons. In addition to this process of massive electron-positron pairs converting back to free moving electromagnetic energy photon state, we know that electromagnetic photons are created in a variety of other circumstances. But on final analysis, they all turn out to involve the emission of an electromagnetic photon when a charged particle, such as an electron, is suddenly stopped in its motion towards the oppositely signed nucleus of an ionized atom, for example, or similar processes involving metastable partons or events inside nuclei. If we take the process of a photon being emitted as an electron is being captured by an ionized atom for example, the photon that then escapes verifiably carries away part or all of the kinetic energy that the incoming electron was initially endowed with, if any, plus the additional kinetic energy that it accumulates during its Coulomb force related freefall acceleration towards the location of its brutal relative stop en route towards the attracting atomic nucleus, a location where it is captured in some

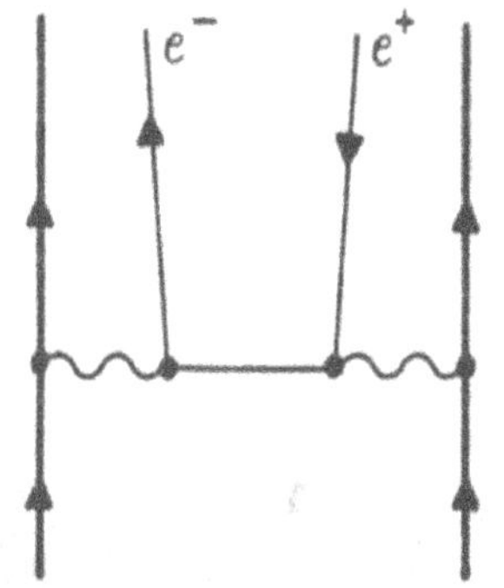

Figure 2: Photon-photon flyby pair creation Feynman diagram.

overwhelming local electromagnetic equilibrium state on some allowed orbital about the nucleus, where it is left with only the exact amount of energy allowed in this new equilibrium state, an amount related to the distance now separating it from the oppositely signed nucleus. Besides this case of free moving electrons being captured by ionized atoms, the other familiar cases involve electrons having moved further away from a nucleus after having been momentarily excited to a metastable higher energy state, that go back to a lower energy state as they return to an orbital closer to the nucleus, where a photon is emitted to release the kinetic energy that now becomes in excess for this closer location. This motion of an electron being momentarily sufficiently excited to move to a metastable orbital further away from an atomic nucleus, or to outright completely escape from the atom, is always due to this electron having been excited away from its rest orbital through conduction or convection transmitted kinetic energy when in gaseous, liquid or solid materials, or having absorbed a discrete amount of kinetic energy from being collided with by an incoming photon, the latter sometimes being completely absorbed in the process, sometimes relinquishing only part of its energy and moving on with the remainder as a less energetic photon, such as in Compton or Raman scattering.

Photons can thus carry away a variety of discrete amounts of kinetic energy depending on local circumstances, whose individual frequencies cover the complete gamut of the electromagnetic spectrum, from the longest radio wavelengths to the shortest gamma wavelengths, the latter due to similar emission processes at the level of atomic nuclei. The whole collection of these photons is of course what allows us to see the universe as they hit the sensory cells in our retinas and/or the sensors of our instruments, allowing us in turn to observe and understand our surroundings up to and including determining the composition of stars. The process of kinetic energy accumulation by charged particles during Coulomb force induced freefall acceleration can easily be verified experimentally at our macroscopic level in a number of ways; with Coolidge tubes for example, as photons are liberated carrying away the exact amount of kinetic energy accumulated during the acceleration phase between the electrodes by electrons that suddenly come to a brutal stop (bremmsstrahlung) as they are captured by ionized atoms located on the anode (or anti-cathode). Emission of photons due to sudden stop of accelerating particles can also be verified with electron beams that are magnetically wiggled in particle accelerators, submitting the electrons in the beam to cyclic transverse accelerations and slowing-downs as the beam is forced to oscillate from side to side, producing so-called synchrotron "radiation", typically in the X-ray range; or in high energy accelerator storage rings, where beams of charged particles are repeatedly forced by magnetic pulses to maintain a best fit approximately circular trajectory.

Now, the issue always remained unclear as to how unidirectional kinetic energy (aka "translational energy") accumulating through acceleration of massive and charged particles can "become" electromagnetic when it is liberated as a photon. Let us recall that the electric and magnetic "fields" of Maxwell's theory are only mathematical representations meant to allow us to describe the observed behavior of electromagnetic energy, which is physically existing in objective reality.

Indeed, there is no *prima facie* reason for this unidirectional kinetic energy to change in nature during the various processes that we examined, particularly since we directly recuperate it as the plain unidirectional kinetic energy that first apparently "converted" to photon state when a bremmsstrahlung photon is being "emitted" by an electron, or when a mother photon's residual energy in excess of the 1.022 MeV going into the rest masses of a separating pair, is observed precisely defining the velocity away from each other of both particles as unidirectional kinetic energy. If kinetic energy does not change in nature during these various processes, this also possibly means that what we perceive and measure as "charges" could also be a relative property that could become perceivable only as the unidirectional kinetic energy is in the process of separating to escape as a free moving photon, just like the opposite signs of isolated massive particles (electron and positron) could be relative properties that would be acquired as the particles come into being when the mother photon decouples.

So, let us then keep in mind as we move on that we will be attempting to explain how and why discrete quanta of this intriguing "substance" that we name "kinetic energy" can possibly move freely at the speed of light as discrete "electromagnetic" quantities without changing in nature. It doesn't seem unreasonable either to think that this "substance" that we identify as "kinetic energy" may have some form of "physical presence", since its quantized manifestations (photons, electrons, positrons, for example) can verifiably be mutually scattered against each other. Before proceeding further, let's define more precisely what "physical presence" could mean in the present context. We do not know and may never know what this "substance" or "fluid" really exactly is that we name kinetic energy. It may be possible however to come to terms with a usable "nearest possible approximation" of what its physical presence could be. De Broglie on his part thought of electromagnetic energy in terms of a "virtual fluid" [1].

"If we suppose known the form of the wave linked to a particle, the intensity of this wave at each point and at each instant (given by $|\psi|^2$) can be considered as defining the density of a virtual fluid (un fluide fictif) moving in space as time progresses and then the quantity of this fluid contained in an element of volume will give the probability for the particle to be present within this element of volume."

We will be going one step further here considering the apparent identity that seems to exist between fundamental electromagnetic energy and unidirectional kinetic energy that accumulates by means of acceleration, if the latter does not change in nature during the various changes of state that we examined. If we consider a rotating fan for example, there is no doubt that the incompressible volume of space cyclically visited by the rotating blades can be measured and studied, even though we know that the actual volume occupied by the material making up the blades and the nature of this material have no relation at all with the incompressible volume that the moving blades visit. If the blades of this fan were invisible to us and if we had no idea even of their existence, we nevertheless could study and measure the incompressible volume that the invisible rotating blades cyclically visits, due to the simple fact that trying to touch that volume would have physical consequences that we could then measure and that would allow us to try ascertaining its properties. We would be left to wonder however, forever maybe, at what could be causing this volume to exist at all with the possibly unexpected properties that our measurements seem to reveal. Indeed, how could we ever discover the existence of the blades and the nature of the material that they are made of, given that no clue to any of their characteristics are given by our measurements?

We find ourselves in a similar predicament regarding the possible "physical presence" of kinetic energy. We can possibly measure the physical presence of a "volume" for kinetic energy and assign to it the properties required to explain its observed behavior, even though this may not reveal the actual real cause and real nature of what is causing this "volume" to exist. For the needs of the present analysis, properties such as incompressibility, fluidity, and elasticity could tentatively be assigned to this "volume", to describe the tendency of the energy that resides in this volume to always remain in motion within this volume as the electromagnetic oscillation suggests, and/or alternately to also constantly tend to move in straight line in space when external electromagnetic equilibrium is not restraining it.

So let's proceed with this tentative "nearest possible approximate definition" for some form of "physical presence" of kinetic energy for the moment, within the frame of the state of our current knowledge about electromagnetic energy, subject to correction or completion as required. Now, if kinetic energy doesn't change in nature as it quantizes as free moving photons, the internally oscillating motion of the kinetic energy quantum could metaphorically be immobilized. The energy of this quantum could then be theoretically reduced to the smallest spherical uniformly isotropic volume that it could occupy, for the purpose of assessing its absolute density. This volume, that could be named the theoretical stationary isotropic volume of the energy of a photon, however small, would depend on the local amount of this kinetic energy and could then be calculated ([10], equations (40) to (41)). We will use this volume in equations (31) to (36).

The fundamental question can now be summarized as follows:

How can a quantity of kinetic energy, accumulating due to Coulomb force freefall acceleration of a massive particle (an electron for example) as the latter unidirectionally increases its velocity in space to start with, dynamically "fold" onto itself according to the threefold orthogonal relation revealed by Maxwell's theory, to become a stable quantum of electromagnetic energy escaping at the speed of light (a photon), while being animated with the local multidirectional oscillating motion suggested by de Broglie's hypothesis; a quantum whose energy would consist in a space-wise electric dipole cyclically morphing into a time-wise magnetic dipole, and that could also explain all electromagnetic properties of photons without changing in nature?

It must be obvious at this point that all photons have to be made of the same material, that is, *quantized amounts of kinetic energy*, an apparently physically existing "substance" that we still know so little about and that appears to be the only "material" of which all photons and all existing charged and massive elementary particles seem to be made of.

The Distribution of Kinetic Energy within a Localized Photon

Now, the question comes to mind as to how this kinetic energy organizes within the photon to sustain an electromagnetic oscillation at a particular frequency and at the same time sustain its own motion at the speed of light.

Clues to this internal structure were given by a brilliant analysis carried out by Paul Marmet in an article that was accepted for publication in the Kazan State University International IFNA-ANS Journal, in 2003, titled: "Fundamental Nature of Relativistic Mass and Magnetic Fields" [11]. His analysis of the relation between the relativistic magnetic mass increase of a moving electron in relation to relativistic velocities allowed defining an LC equation that can describe a possible dynamic internal energy structure for the carrying energy of the electron in motion. In turn, this LC equation allowed upgrading Newton's non-relativistic kinetic equation $K=(mv^2)/2$ to relativistic status [12]. It is the observation that the speed of light is obtained when the mass of the electron is set to zero in this relativistic equation, leaving behind only the carrying energy, that finally reveals that free moving electromagnetic photons (carrying no massive particle) are likely to have the same internal electromagnetic LC structure as that of the carrying energy of moving electrons. Marmet obtained the following definition of current by quantizing the charge, which removed the time element from the equation as he replaced *dt* by *dx/v*, since the velocity of current is constant at any given instant:

$$I = \frac{dQ}{dt} = \frac{d(Ne)}{dt} = \frac{d(Ne)v}{dx} \quad (1)$$

Where *e* represents the unit charge of the electron and *N* represents the number of electrons in one Ampere. Substituting the resulting value of *I* in the scalar version of the Biot-Savart equation then allowed doing away with the time element in this equation also:

$$d\mathbf{B} = \frac{\mu_0 I}{4\pi r^2}\sin(\theta)\,dx = \frac{\mu_0 v}{4\pi r^2}\sin(\theta)\,d(Ne) \quad (2)$$

Without going into the detail of his derivation, which is very clearly laid out in his paper ([11], Equations (1) to (26)), let us only mention that the final stage of this development consists in spherically integrating the electron magnetic energy, whose density is mathematically deemed to vary from a minimum limit corresponding to r_e to a maximum limit located at infinity.

$$M = \left\{\frac{\mu_0 e^2 v^2}{2(4\pi)^2 c^2 r^4}\right\} 2\pi \int_0^{\delta} \sin(\theta)\,d\theta \int_{r_e}^{\infty} r^2 dr \quad (3)$$

The electron classical radius r_e is the mandatory lower limit in such an integration to infinity, due to the simple fact that integrating any closer to *r=0* would accumulate more energy than experimental data warrants. After integrating, he obtained:

$$M = \frac{\mu_0 e^2 v^2}{8\pi r_e c^2} = \frac{m_e}{2}\frac{v^2}{c^2} \quad (4)$$

which very precisely corresponds to the total mass of the magnetic field of an electron moving at velocity *v*. He discovered by the same token that any instantaneous "magnetic mass" increase of an electron is a direct function of the square of its instantaneous velocity.

When this velocity is small with respect to the speed of light, the following classical equation is obtained, allowing clearly determining the contribution of the magnetic component to the rest mass of the electron:

$$\frac{\mu_0 e^2}{8\pi r_e}\frac{v^2}{c^2} = \frac{m_e}{2}\frac{v^2}{c^2} \quad (5)$$

Where r_e is the classical electron radius (2.817940285E-15 m), and *e* is the charge of the electron (1.602176462E-19 C), and from which can be concluded that the invariant magnetic component of the electron at rest corresponds to a mass of:

$$M_0 = \frac{\mu_0 e^2}{8\pi r_e} \quad (6)$$

which is exactly half the mass of an electron, the other half being made up of what could be termed its "electric mass", since the electron is an electromagnetic particle. Paying attention to the difference between equations (4) and (6), we observe that $M - M_0$ represents the relativistic mass increment related to instantaneous velocity *v*. We note also that the translational kinetic energy required to propel the electron at this velocity is absent from the equation. Close analysis and calculation reveals however that the amount of translational kinetic energy required to propel an electron with magnetic mass *M* at velocity *v* is exactly equal to the amount of energy captive in the instantaneous relativistic mass increment $M - M_0$.

This means that the total amount of energy that must be communicated to an electron at rest for it to move at any velocity must be defined as an amount of translational kinetic energy plus an equal amount of kinetic energy that momentarily converts to the instantaneous relativistic mass increment related to that velocity.

$$E_{\text{total}} = E_{\text{translational}} + E_{\text{magnetic mass increment}} \quad (7)$$

Since energy in motion cannot be dissociated from electromagnetism, it can be surmised that an electric component is *de facto* involved in relation with the half of the energy that in context clearly is "magnetic" in nature, and the only way it can be introduced in context is for this magnetic energy to alternate between this magnetic state and an electric state at the frequency that can be associated to this amount of energy.

$$E_{\text{total}} = E_{\text{trans.}} + \left[E_{\text{elec.}}\cos^2(\omega t) + E_{\text{mag.}}\sin^2(\omega t)\right] \quad (8)$$

This form in turn immediately suggests the following LC relation to represent the internal structure of the carrying energy of an electron in motion:

$$E = \frac{hc}{2\lambda} + \left[\frac{e^2}{2C_\lambda}\cos^2(\omega t) + \frac{L_e\, i_\lambda^{\,2}}{2}\sin^2(\omega t)\right] \quad (9)$$

where λ is the wavelength associated to this amount of electromagnetic energy in motion and where the following are the classical equations for calculating capacitance and inductance during a LC cycle:

$$E_{\mathbf{E}(\max)} = \frac{q^2}{2C} \quad \text{and} \quad E_{\mathbf{B}(\max)} = \frac{L\,i^2}{2} \quad (10)$$

Equation (9) reveals that all probabilities are that the velocity of light of an isolated electromagnetic photon would be maintained because the translational half of its kinetic energy serves to propel at this velocity an equal amount of kinetic energy while the latter permanently oscillates between an electric state and a magnetic state at the frequency determined by the total amount of kinetic energy involved. This structure will be analyzed in detail further on.

The Neglected Classical Maxwellian Space Geometry

Maxwell's theory is traditionally considered from the mathematical viewpoint offered by his famous equations and understood within the restrictive perspective of plane wave treatment, leaving the space geometry that underlies it to be mostly taken for granted, since it is sufficient for the needs of the continuous wave concept, which in turn is sufficient for precise calculations at the general level. This classical space geometry is of course the traditional Euclidian 3-dimensional flat space geometry to which the time dimension is added to justify motion. Just like the habit of using the electromagnetic tensor to represent a single "electromagnetic field" concept keeps away from immediate attention that both electric and magnetic fields are of equal and separate importance in Maxwell's theory, with different and irreconcilable characteristics, the habit of using plane wave treatment leaves in the background the fact that the wave front of the electromagnetic wave of Maxwell's theory could only be in spherical expansion from some point-like source, a point-like source which is confirmed out of any doubt by experimental reality for any electromagnetic quanta emission, even if Maxwell's continuous waves had been proven to really exist. Maxwell's theory is in fact the natural end result of the integration of many discoveries made previously. His first equation is In Gauss' law for electricity; his second equation is derived from Faraday's law, his third from Gauss' law on magnetism and his forth is a generalization of Ampere's law. What Maxwell did in fact was unify into one coherent integrated theory all these experimentally confirmed laws that were not clearly linked to each other previously.

But his really brilliant personal contribution was his success in mathematically linking Faraday's law and his modified Ampere's law in such a way that no doubt could remain that light was intimately linked to electricity and magnetism, as confirmed by Faraday's experiments on light polarization by magnetic fields. Linking them provided as a side benefit the only way ever devised to calculate light velocity from first principles, a velocity that is the only velocity possible from these equations since it rests on the products of only two other fundamental constants, that is, the electric permittivity and magnetic permeability constants of vacuum. As already mentioned, a fundamental and thoroughly verified aspect of Maxwell's theory is the mandatory state of orthogonality that must exist between the electric and magnetic fields of free moving electromagnetic energy, both fields also being normal to the phase velocity vector that identifies the direction of motion of any point considered on the wave front of the spherically propagating "wave". Experimental reality reveals that this triple orthogonality also applies to the motion of charged massive particles, such as electrons being forced to move in straight line when subjected to equal density external electric and magnetic fields. Indeed, any elementary textbook on electricity and magnetism explains how the vectorial cross product of equal intensity electric and magnetic fields being applied to a charged particle will generate a velocity vector in straight line forcing the particle to move in a direction perpendicular to both resulting forces. The more intense the fields, the faster the particle will move, and whose varying velocity is given in classical electrodynamics from the Lorentz equation, by this well known relation:

$$\frac{\mathbf{E}}{\mathbf{B}} = v \tag{11}$$

Which resolves to the fixed speed of light "c" for photons, from Maxwell's 4th equation:

$$\frac{\mathbf{E}}{\mathbf{B}} = c \tag{12}$$

Or rather, in the present context, under the form of a vectorial cross product:

$$\mathbf{E}\hat{j} \times \left(\frac{-1}{\mathbf{B}}\right)\hat{k} = \mathbf{E}\left(\frac{-1}{\mathbf{B}}\right)\cos\theta\hat{i} \tag{13}$$

and since angle θ must be equal to 90° by definition in the case of the straight line motion that we are considering:

$$\mathbf{E}\hat{j} \times \left(\frac{-1}{\mathbf{B}}\right)\hat{k} = v\hat{i} \tag{14}$$

where v is the velocity vector.

The orthogonal bases shown in Figure 3 will be used in this paper:

a) 3D rectangular x-y-z coordinate system, and corresponding rectangular unit vectors base and

b) The correspondingly oriented rectangular electromagnetic fields vs velocity vector base:

It is generally understood also that despite the precision of the calculations that Maxwell's theory allows for electromagnetic energy, his theory is deemed unable to directly describe photons as discrete localized moving electromagnetic particles since it is grounded on the notion that electromagnetic energy is a continuous wave phenomenon.

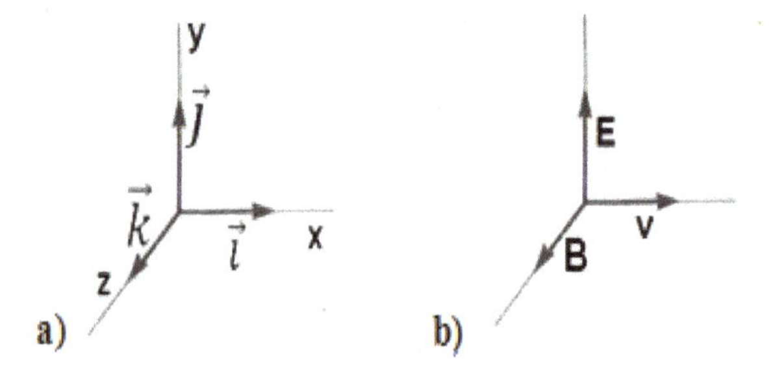

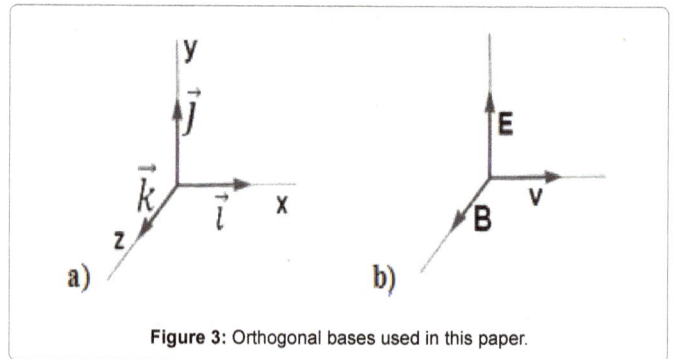

Figure 3: Orthogonal bases used in this paper.

Discrete Particles as the Only Possible Support of Electromagnetic Properties

Maxwell's theory, as a matter of fact, was designed to account for electromagnetic energy behavior at the macroscopic level without the need to take quantization into account which had not yet been clarified in Maxwell's time, that is, by treating electromagnetic energy as a featureless energy density per unit volume or featureless energy flow per unit surface rather than by adding the energy of localized moving electromagnetic photons enclosed in a unit volume or flowing through a unit surface, that would take localization into account and would represent just as well all observed electromagnetic phenomena at the macroscopic level. Considering that the "electromagnetic waves" that Maxwell conceived of were meant to animate what was perceived from our macroscopic level as a still hypothetical underlying and all pervading "ether", then if some means was found to associate to each individual localized photon all of the electric and magnetic properties that characterize the electromagnetic wave of Maxwell's theory, this would remove the theoretical need for the existence of such a supporting all pervading medium for the purpose of supporting continuous electromagnetic waves, that we now know do not exist at the submicroscopic level. Let us note also that a second theoretical use of the various forms of the concept of ether was for it to constitute the very substance that massive particles were made of as "singularities" that developed in such all pervading ether fields in a variety of theories. Now if kinetic energy, of which discrete localized photons are demonstrably made, turns out to have "physical presence" with a "volume" that can be measured, this would altogether remove the last reason that would justify resorting to the theoretical concept of ether as a basis to explain the fundamental level of physical reality. All the more so since it has been conclusively verified since the 1930's that massive electron and positron can be made from destabilizing electromagnetic photons containing at least 1.022 MeV of this kinetic energy [13]. Head-on collision experiments between beams of electrons and positrons [14] even lead to suspect that protons and neutrons could be stable adiabatic equilibrium states involving triads of electrons and positrons that could have interacted in such a way that they could have locally adiabatically accelerated until they reach these two ultimate and irreversible equilibrium bound states [15]. Of course, such a possibility seems at first glance to be in total contradiction with the Principle of conservation of energy. But considering that all existing closed systems for which the Principle of conservation of energy can be verified to apply have already reached some form of least action energy equilibrium, that can be modified only by introducing energy in excess of this equilibrium, there exists the possibility that newly created particles, that never were chased out of some pre-existing least action equilibrium state, could accumulate new energy by means of an initial and irreversible adiabatic acceleration process that would bring them to such a first least action energy equilibrium state, after which they would of course be forever subjected to the Principle of conservation. We must not forget either that even if ether could finally be done away with, more and more data seems to indicate that here on Earth, we are permanently immersed in an all pervading interacting magnetic fields combination involving the Earth's magnetic field moving through the immense magnetic field of the Sun that reaches way beyond Pluto, which also interacts with the local magnetic fields of the other planets of the Solar system, and finally there seems to be little doubt that the global magnetic field of our local galaxy also interacts with the Sun's magnetic field.

So, whatever the final solution will be, it will mandatorily involve this all pervading underlying medium in what we consider as the total vacuum of space.

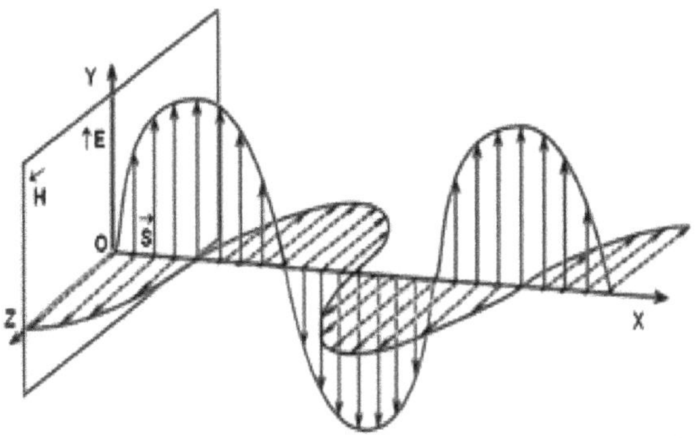

Figure 5: Electric and magnetic field 90o out of phase in classical electromagnetism.

The Issue of Intensity Conservation with Maxwell's Spherically Expanding Wave Concept

This leads to attempt clarifying why an acceptable description of electromagnetic photons as stable permanently localized moving particles, in line with their demonstrated point-like localization at the moments of emission and capture, more than one century ago, has not yet been successfully reconciled with the verified aspects of Maxwell's theory, particularly after Louis de Broglie elaborated his intriguingly promising hypothesis [2]. According to Maxwell's theory, the electric and magnetic aspects of an electromagnetic wave must by necessity always be in phase at the wave front (Figure 4), that is, at maximum at the same moment, for the wave to exist at all and propagate.When both aspects are 90° out of phase, we obtain a standing wave (Figure 5). But as an intriguing dead end in Maxwell's theory, when both aspects are set 180° out of phase, we end up with the exact equivalent of both aspects being in phase (Figure 4), But we will see further on that far from being a dead end in physical reality, this 180° dephasing will turn out to be in perfect harmony with the LC oscillation for which we will give the mathematic development (Figure 5). Also, it is the conjunction of both fields, in phase and at right angle with each other at all points of the wave front that is deemed to maintain the intensity of the energy of the wave at every points of the wave front, despite the inherent spherical spread involved from the mandatory point-like origin of such a wave, if it really existed. This issue is of course familiar to all but is apparently seen as an unavoidable axiom, no doubt resting on the comfortable fact that plane wave treatment allows precise calculation anyway. Mathematically speaking, when any point of the curved spherical wave front surface is considered, this surface can be locally approximated to a plane surface at the infinitesimal level which is the origin of the "plane wave" equations set. But space being three-dimensional, such treatment with the plane wave analogy can only be a mathematical approximation obscuring the fact that if such an electromagnetic wave really physically existed, it could only be in spherical expansion from its initial point-like state, assuming unbounded isotropic expansion. So plane wave treatment applied to Maxwell's theory currently does not describe electromagnetic energy as it starts existing at its point-like source, but only after the expanding wave has begun to propagate. Also, the geometry of such a spherically propagating wave would be much more similar to the spherical expansion of a sound wave from its point-like source in some underlying medium than to the propagation of waves on a plane liquid surface that immediately comes to mind when thinking about plane wave treatment. It then becomes very difficult to accept the idea that the initial intensity of the point source of the wave could be arbitrarily multiplied in such a way that it could be measured as equal to the energy of the source at any point of the expanding spherical wave front at any arbitrary distance from the punctual source, as plane wave treatment seems to allow. So the habit of dealing with the state of orthogonality of both fields with respect to each other and to the direction of motion in space of any point on the already expanding wave front always leaves in the background the fact that such a spherically expanding wave can only be a single electromagnetic event originating from a single point-like source.

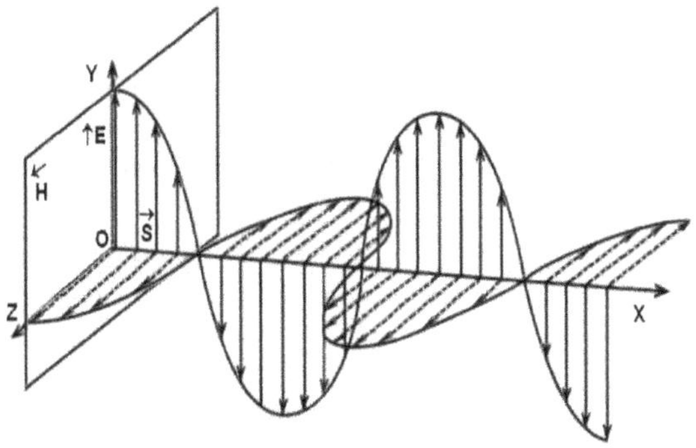

Figure 4: Electric and magnetic fields in phase, or 180° out of phase, in classical electromagnetism.

Applying Electromagnetic Properties to Maxwell's Spherically Expanding Wave's Point-like Initial State

Now, considering that such an electromagnetic event is a single event, could it not be imagined that after appearing at its point-like origin, it could be represented as remaining locally point-like as it starts moving, harmonically oscillating as it moves, which is what de Broglie's hypothesis implies, instead of spherically expanding as Maxwell's theory implies by definition?

This would involve a precise trajectory being followed by this electromagnetic event, which would then behave point-like from emission to capture, which would in turn be in total harmony with the verified fact of its point-like capture, however much time could have elapsed after it was emitted and whatever distance it could have covered before being captured. This would also directly explain why the initial intensity of this electromagnetic quantum is conserved, barring energy losses or gains through red or blue shifting due to gravitational interaction along the path that it would have followed. The idea naturally comes to mind then that the state of fundamental orthogonality of both electric and magnetic fields could possibly be served just as well, if not better, by being defined with respect to the electromagnetic event immediately as it initially appears point-like, instead of after spherical expansion is already under way. But the apparently insurmountable issue with this approach in classical electrodynamics is the mathematically assumed infinite energy associated with such a punctual electromagnetic concept.

Another problematic issue comes to light with the idea of mathematizing free moving energy at its point-like source. It is the fact that both fields of any point-like electromagnetic quantum (a photon) which is in the process of being emitted by a de-energizing electron can be orthogonal to no particular direction in space at the very moment of point-like separation, which leads to the conclusion that at the very moment of separation, both fields of the new photon could be orthogonal only to 3D space proper, despite the strangeness of the idea.

Considering also that electric interaction obeys the inverse square law of electrostatic attraction and repulsion between the charges of elementary particles and that magnetic interaction obeys the inverse cube law of magnetostatic attraction and repulsion between magnetic fields of the same elementary particles, makes it appear illogical, and even impossible, that quantized quantities of kinetic energy could possess both electric and magnetic properties at the same moment, or even in alternance, while not changing in nature.

The inverse square interaction law between electrically charged elementary particles, that is, the Coulomb law, is very familiar, but the inverse cube law between the magnetic aspects of the same point-like behaving elementary particles is much less familiar. A direct confirmation of this inverse cube relation has been very recently obtained by Shlomi Kotler and his team between the magnetic aspects of two electrons, as reported in the Nature magazine in April 2014 [16] thus confirming by the same token the validity of the lab bench experiment carried out 15 years ago which is described in reference [17].

It is precisely the combination of these mutually incompatible inverse square law applying to the electric aspect of a point-like behaving particle and of the inverse cube law applying to its magnetic aspect that elicits the strongest doubt on the ability of classical 4D spacetime geometry to allow the kinetic energy that the particle is made of to continue displaying these irreconcilable properties while not changing in nature as it electromagnetically oscillates while moving at the speed of light through vacuum, if kinetic energy is to be considered being a "physically existing substance". These considerations are what gave birth to the idea that the physically existing space geometry at the fundamental level may be more complex than can be directly observed from our macroscopic level, and that extra "spaces" could possibly be involved to allow for these possibilities, that is, a second space that would allow kinetic energy to display electric characteristics without changing in nature and a third space that would allow kinetic energy to display magnetic characteristics without changing in nature, both extra spaces remaining permanently perpendicular to each other and to normal space at the particle level.

It is to be noted at this point that Louis de Broglie also came to the conclusion from other considerations that it was impossible to exactly represent elementary particles in the restricted frame of continuous three dimensional space.

"The non-individuality of particles, the exclusion principle and exchange energy are three intimately related enigmas; all three are tied to the impossibility of exactly representing elementary physical entities within the frame of continuous three dimensional space (or more generally of continuous four dimensional space-time). Some day maybe, by escaping from this frame, will we better grasp the meaning, still quite cryptic today, of these major guiding principles of the new physics" [2].

An expanded space geometry that allows a clear definition of the double-particle photon without its kinetic energy quantum changing in nature, and that may also allow resolving some of the issues raised by de Broglie was first introduced in July 2000 at Congress-2000 at St. Petersburg State University [18]. This new space geometry will now be described before proceeding to build the LC and local fields equations that can represent the permanently localized double-particle photon in this expanded space geometry.

Expanding Space Geometry beyond Normal 3D Space

As previously done with the idea of a usable "nearest possible approximate definition" for the "physical presence" of kinetic energy as a "physically existing substance", we may think of this expanded space geometry as a usable "nearest possible approximate definition" of the required space geometry, within the frame of the current state of our knowledge about electromagnetic energy. If we imagine the observed electric behavior of charges as being due to the momentary presence of the incompressible energy of a photon in a separate 3D-space that allows such behavior, and magnetic behavior as being due to the alternate momentary presence of the same energy in a different 3D-space that allows such behavior, each space being governed by the same laws of motion as normal 3D-space, the same capacitance and inductance, both spaces remaining permanently perpendicular to each other and to normal space, and that would allow the circulating kinetic energy not to change in fundamental nature, it will become possible to visualize much more clearly the internal oscillation of the kinetic energy of the localized double-particle photon of de Broglie's hypothesis.

In order to more easily refer to these new spaces, let us name electrostatic space the space into which kinetic energy displays electric behavior, and magnetostatic space the space into which it displays magnetic behavior. For coherence, we will identify normal, electrostatic and magnetostatic spaces as being X-space, Y-space and Z-space respectively. Within normal space, let us rename the three minor spatial dimensions: X-x, X-y and X-z and likewise, for electrostatic and magnetostatic spaces: Y-x, Y-y, Y-z and Z-x, Z-y, Z-z. Let us assume furthermore that the minor x-axes of all 3 spaces are mutually parallel in a direction corresponding to the conventional direction of motion of energy in normal space in plane wave treatment. Of course, when the x, y and z dimensions are used without major axis prefix, they refer as usual to normal 3D space. In this space geometry, a point-like junction (*representing a "passage point" in physical reality, not really a dimensionless "point" in the mathematical sense, whence the best-fit representation phrase "point-like" being used, which does not exclude the possibility of a local "volume" or "area", however small, being involved*) between these three orthogonal spaces would be located at the geometric center of each photon, and it is this point-like junction that would be moving point-like at the speed of light in normal X-space, that is, along the X-x axis of this expanded geometry in plane wave treatment (Figure 6).

To be able to mentally visualize the locally standing motion of kinetic energy in this 3-spaces structure, an easily mastered technique can be use. It suffices to imagine the 3 familiar minor x-y-z orthogonal dimensions of normal 3D space as if they were the ribs of an open 3-ribs metaphorical umbrella, the apex of which would be located at the origin (or passage point where the 3 spaces meet). If we mentally fold the umbrella, we can now visualize the folded umbrella as if it was the linear major X-axis of this expanded coordinates superset. With

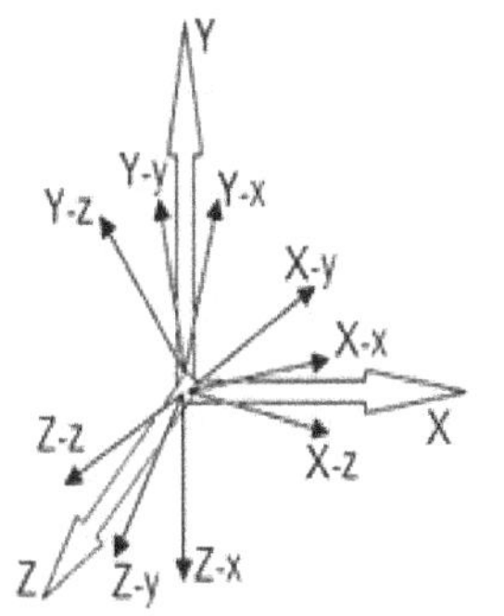

Figure 6: The orthogonal structure of the 3-spaces model.

this umbrella metaphor, it is now easy to visualize the three orthogonal spaces as three umbrellas meeting at their tips. We only need to mentally open any one of them to examine what is occurring in this particular space at any given moment of the electromagnetic cycle. As observed from within normal space, which will be our observer's viewpoint during this analysis, free fall acceleration induced unidirectional kinetic energy accumulating within the same normal space will be locally perceived as having longitudinal inertia but no transverse inertia. The longitudinal inertia of electromagnetic photons was experimentally confirmed more than one century ago by the same photoelectric proof due to Einstein that confirmed that electromagnetic energy behaves as discrete localized quanta and not as a continuous wave phenomenon. The absence of transverse inertia for unidirectional kinetic energy on its part was experimentally demonstrated more than one century ago also by Walter Kaufmann [19] as he demonstrated that the transverse inertia of electrons accelerated to relativistic velocities was lower than their longitudinal inertia. This issue will be analyzed further on. From within normal space again, all energy present within electrostatic and magnetostatic spaces at any given moment of the electromagnetic cycle will appear to possess both longitudinal and transverse inertia, that is, omnidirectional inertia; in other words: electromagnetic mass. Metaphorically speaking, the energy present in these two extra spaces would appear to be captive inside some invisible "container" that will resist being pushed around from any direction from within normal space. Although photons are known not to have a rest mass, they are also known to possess an electromagnetic mass that can interact gravitationally. The photon itself will now appear as a discrete amount of kinetic energy, half of which remaining unidirectional and moving in normal space, as determined in Section 6, propelling the other half that would be oscillating cyclically through the point-like junction between electrostatic and magnetostatic spaces at the frequency determined by the photon's energy. A separate analysis explains why half of any localized photon's kinetic energy (that is the photon's translational energy) has no option other than to remain unidirectional within the photon's inner structure, even without invoking the 3-spaces concept nor the double-particle concept, to propel the other half of a localized photon's energy at the speed of light [12]. A property of unbounded elasticity and fluidity for the kinetic energy "substance" can even allow for both half-photons to possibly not be "completely severed" from each other nor from the portion moving unidirectionally in normal space as they separate within electrostatic space, or as they transfer to magnetostatic space as a single quantity. The complete amount of the photon kinetic energy quantum could then continue remaining a single continuous quantity permanently linked through the central point-like junction between the 3 spaces.This model of the double-particle photon can now be seen as displaying transverse wave behavior with a frequency related to the amount of energy that its quantum possesses, while at the same time displaying longitudinal particle behavior with longitudinal inertia related to the total amount of energy that its quantum possesses and transverse inertia related to half this amount, which conforms to all experimentally observed characteristics of the photon.

Defining a Major Unit Vectors Superset

The traditional $\hat{i}$, $\hat{j}$ and $\hat{k}$ unit vectors previously mentioned, were of course defined to represent vectorial properties in normal 3D space. But in this expanded 3-spaces geometry, both new spaces also require their own internal minor unit vectors set.

So let's define a new superset of major unit vectors that will identify the three orthogonal spaces with capital letters as $\hat{I}$, $\hat{J}$ and $\hat{K}$, so that each minor local $\hat{i}$, $\hat{j}$ and $\hat{k}$ unit vectors set becomes subordinated to the major unit vector specific to its local space, all 12 resulting unit vectors (3 major and 9 minor) being of course drawn from the same origin O corresponding to the point-like junction between the 3 spaces (Figure 7).

Each of the three orthogonal minor unit vectors subsets (shown in the drawing as being half folded (let's remember the umbrella analogy), that is **I-i**, **I-j**, **I-k**, for normal space **J-i**, **J-j**, **J-k** for electrostatic space and **K-i**, **K-j**, **K-k** for magnetostatic space, allows defining the vectorial magnitude of the energy of a particle in any one of the three orthogonal coexisting spaces at any given moment.

This is how the vectorial relation drawn from Lorentz becomes in this expanded space geometry:

$$\mathbf{E}\vec{\mathbf{J}} \times \left(\frac{-1}{\mathbf{B}}\right)\vec{\mathbf{K}} = \mathrm{v}\vec{\mathbf{I}} \tag{15}$$

Electromagnetic Oscillation Energy-driven rather than Fields-driven

Now that we can view the photon kinetic energy quantum as a single continuous quantity permanently linked through the central point-like junction between the 3 spaces, comes into question the issue that the part of this amount oscillating between electrostatic and magnetostatic spaces must display distinct and apparently irreconcilable electric and magnetic properties that can continue being represented as reciprocally induced by the other aspect as in classical electromagnetism, that is, by apparent mutual "fields" interaction.

For example, if kinetic energy was a material incompressible in volume on top of its fundamental property of always tending to remain in motion, the local oscillation between both electrostatic and magnetostatic spaces of any quantity of this energy could possibly be

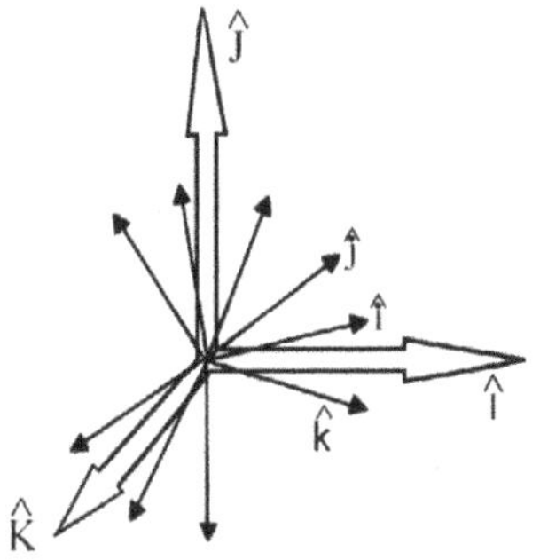

Figure 7: Major and minor unit vectors sets applicable to the 3-spaces model.

forced uniquely by a property of this energy to always tend to remain in motion.So, instead of a relation of mutual orthogonal induction between two fundamentally different electric and magnetic "fields" as Maxwell's theory assumes, this relation would be one of cyclic translation of this energy between both orthogonal extra spaces (Figure 8).

That is, an energy that would always conserve the characteristics it originally possessed before it was quantized to become a photon, but that would give the impression of having alternately all of the electric set of characteristics when momentarily present in electrostatic space, and then, all of the magnetic set of characteristics when momentarily present in magnetostatic space; but whose high frequency cyclic translation between the two states (between the two spaces in reality) would create the impression at our macroscopic level of the simultaneous and permanent presence of both fields of Maxwell's theory inducing each other.

This would negate in no way the usefulness of fields representations. Fields would simply take second seat to the now self-forced motion of the kinetic energy proper becoming more fundamental, operating as a primary cause of the electromagnetic oscillation, being perceived momentarily as "electric energy" as it transits within electrostatic space and momentarily as "magnetic energy" as it transits within magnetostatic space.

It seems entirely conceivable that such a high frequency cyclic translation process of a discrete quantity of incompressible energy between these two spaces, could explain the frequency of the photon and all other observed phenomena while preserving the usefulness of the traditional perception of the so convenient and precise electric and magnetic fields that would mutually induce each other, but would also open up an entirely new range of possibilities, a few of which will be discussed later. Maxwell's four original equations would remain totally valid in this new perspective, since his second equation ($\nabla\times\mathbf{E}=-\partial\mathbf{B}/\partial t$) derived from Faraday's induction law, does not even mandate that both fields be in phase, since it directly accepts the opposite relation, that of reciprocal interaction of both fields when out of phase by 180° as is being considered here.

Underlying Kinetic Energy Circulation

Let us now summarize the inner motion of kinetic energy within the structure of the double-particle photon.

This motion can be summarized as 4 distinct steps:

(a) The two half-photons having reached the farthest distance that they can reach within electrostatic space.

(b) The two half-photons closing in towards each other in electrostatic space as their energy starts transferring omnidirectionally into magnetostatic space.

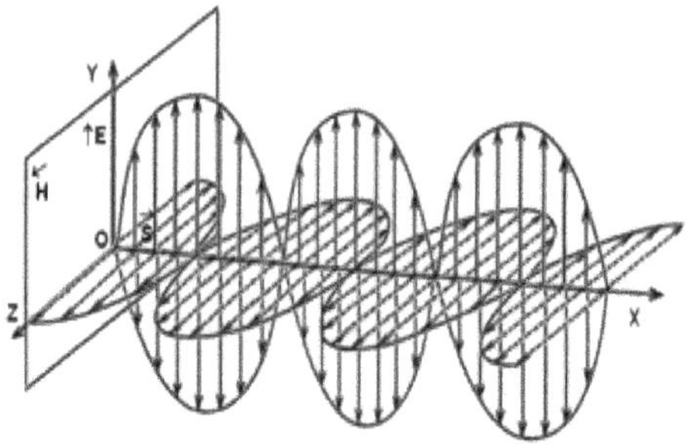

Figure 8: Electric and magnetic fields 180° out of phase in the 3-spaces model.

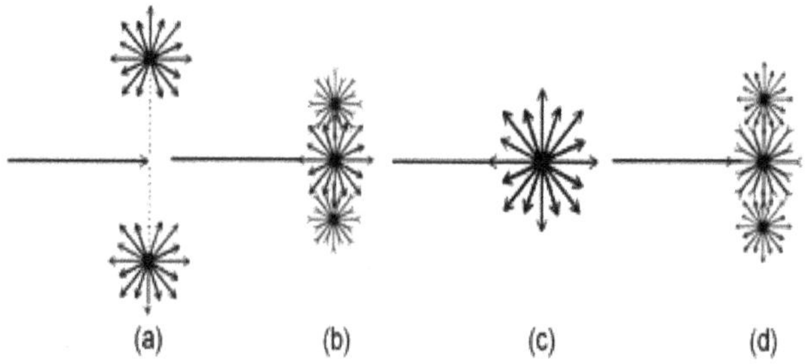

Figure 9: The complete cycle of kinetic energy circulation within the structure of the double-particle photon.

(c) The total complement of the two half-photons' energy having now completely crossed over into magnetostatic space, now making up the single spherical magnetic space component.

(d) The energy present in magnetostatic space starting to cross over back into electrostatic space as two separate half-photons.

(a) and (a) again as the cycle completes, poised to start the whole sequence again at the frequency mandated by the amount of kinetic energy making up the photon's energy quantum.

All through this process, the other half of the photon's energy, which is permanently located within normal space, remains in stable unidirectional motion, propelling the oscillating half at the speed of light in normal space vacuum.

Applying Plane Wave Treatment to the Permanently Localized Double Particle Photon

A point of particular interest with this internal photon structure is that it allows continued use of the plane wave analogy, but in which at any given instant of the cycle, the product of the electric and magnetic fields remains constant over the plane intersecting the central junction, perpendicularly to the direction of propagation of the photon (Figure 10). The energy of the photon would behave with respect to this plane as if it was stationary, as it actually is in the reference frame of the moving point-like junction, with the associated benefit that this plane, just like the point-like junction, can regardless continue moving at the speed of light in normal 3D space (along the X-x axis).Also, we can observe that the product of the projections on the transverse plane of the electric and magnetic oscillating energy will be constant and consequently will not fluctuate over time as is the case with classical in-phase wave front plane wave treatment.

In this model, the magnitude of the Poynting vector will thus be constant all through the electromagnetic cycle of any localized photon at the following value

$$\mathbf{S}=\frac{\mathbf{EB}}{2\mu_0} \qquad (16)$$

instead of fluctuating over time as in classical electromagnetism, since one more characteristic of applying plane wave treatment to the double-particle photon in motion is that the value of **S** corresponds by structure very precisely to the average value of the *intensity* of the "wave" in classical electromagnetism [20].

Let us note here that this measured *intensity* is directly reconcilable with the conclusion of this model according to which only half of the energy of a photon would be oscillating to and fro between electrostatic and magnetostatic states while the other half would not be oscillating but would remain stable and moving unidirectionally to simply propel the oscillating half.

The Double-particle Photon involves 2 Charges

It is highly interesting to note that the new equation for free moving energy derived from Paul Marmet's work in a separate analysis ([10], equation (11)) involves by structure two interacting charges:

$$E = hf = \frac{hc}{\lambda} = \frac{e^2}{2\varepsilon_0 \alpha \lambda} \tag{17}$$

The very form e^2 reveal that both charges associated to a free moving electromagnetic energy quantum have to be identical, and can effectively be neutral $|e|^2$ as hypothesized by de Broglie. This is what leads to the conclusion that it is possible that the opposite signs of a decoupling pair (positron + and electron -) could be acquired as the pair decouples, which is currently at odds with current axiomatic beliefs, but is in perfect harmony with de Broglie's conclusion that the double-particle photon charges should be neutral.

Let's also note that in this equation, fine structure constant α is related to the transverse amplitude of the electromagnetically oscillating half of the double-particle photon energy, an amplitude which in turn directly relates to the lower limit of spherical integration of the energy of a discrete localized electromagnetic particle ([10], Extended Abstract, and equations (1) to (11)).

Defining The Double-particle Photon LC Equation and Local Fields Equation

Macroscopic LC circuits

When an inductor coil is connected to a charged capacitor, it is well verified experimentally that the capacitor will completely discharge into the inductor coil as the current in the coil wire establishes a magnetic field in the surrounding space. When the potential difference between the capacitor terminals reaches zero, the magnetic field that just reached maximum about the inductor coil will now start decreasing, thus inducing a reverse current in the coil wire that will have completely recharged the capacitor when the magnetic field has completely disappeared, a behavior in complete agreement with 180° out of phase electromagnetic cycling in this 3-spaces model (Figure 8).

The capacitor will now start discharging again into the inductor coil and the process would repeat indefinitely in theory if no energy was lost in the process, a loss that always occurs in lab experiments due to the resistance and eventual heating of the coil wire and radiating of this energy into surrounding space. It is well understood however that if no energy was lost due to the coil wire resistance, the total amount of energy in the system would remain stable and be permanently conserved, which would keep the cycle going forever.

The Photon as a LC oscillator

Let us now transpose this LC behavior to the double-particle photon. Contrary to the coil wire of a LC circuit made of a capacitor and an inductor coil, it can be assumed that the point-like junction between the three spaces of the expanded 3-spaces geometry will offer no resistance to the passage of the photon's oscillating energy, since it is well established that a photon's energy remains constant from emission to capture, however long the time elapsed since its emission and whatever distance it could have traveled, barring losses due to red shifting, gains due to blue shifting and losses due to gravity induced changes in direction.

The classical equation representing the maximum energy stored in the capacitor of a LC circuit at the beginning of the cycle is:

$$E_{\mathbf{E}(\max)} = \frac{q^2}{2C} \tag{18}$$

and the equation representing the maximum energy stored in the magnetic field of the coil when the capacitor has been emptied of its charge is:

$$E_{\mathbf{B}(\max)} = \frac{L\,i^2}{2} \tag{19}$$

In the context of LC behavior applied to a localized photon's energy, where no energy can be lost through heating of a non-existent coil wire and considering that both quantities represent the same half quantum of the photon's energy oscillating between these two maxima, we can then equate:

$$E_{\mathbf{E}(\max)} = E_{\mathbf{B}(\max)} = E_{\mathbf{E}} + E_{\mathbf{B}} = E_{\mathbf{EB}} \tag{20}$$

Defining the photon capacitance (C)

As established in a separate analysis [12], only half of a photon's energy cyclically oscillates between electric and magnetic states. So making use of the free energy equation previously mentioned derived from Marmet's work ([10], equation (11)), that is:

$$E = \frac{e^2}{2\varepsilon_0 \alpha \lambda} \tag{21}$$

that we will divide by 2, to represent only the oscillating half of the photon's energy, and equate to equation (18) for capacitance, which represents the same half the photon's energy, that is, the two charges of the photon at their maximum value, we obtain:

$$E_{\mathbf{EB}} = \frac{E}{2} = \frac{q^2}{2C} = \frac{e^2}{4\varepsilon_0 \alpha \lambda} \tag{22}$$

We can then isolate:

$$2C = 4\varepsilon_0 \alpha \lambda \tag{23}$$

and finally obtain:

$$C = 2\varepsilon_0 \alpha \lambda \ \text{Farad} \tag{24}$$

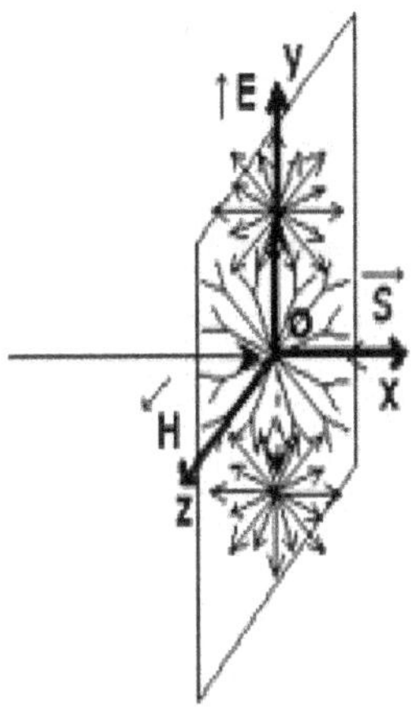

Figure 10: Plane wave applied to a permanently localized photon.

which allows calculating the capacitance of any localized photon from its wavelength and the permittivity constant of vacuum (ε_0).

Defining the photon inductance (L)

Since the angular frequency of a LC oscillator is obtained from the following equation:

$$\omega = \sqrt{\frac{1}{LC}} \tag{25}$$

we can separately calculate the angular frequency of a localized photon's energy from $\omega=2\pi f/\alpha$, or better yet, in context, from $\omega=2\pi c/\alpha\lambda$ (since we must use here the α related transverse amplitude (See Section 16) of the cycling frequency calculated from the wavelength of a localized photon's energy which is $f=c/\lambda$). So we can write:

$$\omega = \frac{2\pi c}{\alpha\lambda} = \sqrt{\frac{1}{LC}} \tag{26}$$

By squaring this last equation and replacing C by its value defined in equation (24) as $2\varepsilon_0\alpha\lambda$, we can isolate L in equation (26) and define the following equation:

$$L = \frac{\alpha^2\lambda^2}{C4\pi^2c^2} = \frac{\alpha\lambda}{\varepsilon_0 8\pi^2c^2} \tag{27}$$

Knowing that $\varepsilon_0 c^2=1/\mu_0$ and substituting this value into equation (27) to introduce permeability constant μ_0, we finally obtain:

$$L = \frac{\mu_0\alpha\lambda}{8\pi^2} \quad \text{Henry} \tag{28}$$

which allows calculating the inductance of any localized photon from its wavelength and the permeability constant of vacuum (μ_0)

The Photon maximum displacement current (*i*)

Having established how to calculate inductance L for a localized photon, we can now determine the maximum current i involved from the equation giving the maximum energy momentarily stored in the magnetic field. So, from equation (19):

$$E_{\mathbf{B}(\max)} = \frac{L\,i^2}{2} \tag{29}$$

we can isolate i, and knowing that $E_{\mathbf{B}(\max)}=E_{\mathbf{EB}}$ from equation (20), the value of L from equation (28), and knowing also that $\varepsilon_0\mu_0=1/c^2$, we can derive the localized photon's maximum displacement current:

$$i = \sqrt{\frac{2E_{EB}}{L}} = \sqrt{2\frac{e^2}{4\varepsilon_0\lambda\alpha}\frac{8\pi^2}{\mu_0\lambda\alpha}} = \sqrt{\frac{4\pi^2e^2}{\varepsilon_0\mu_0\alpha^2\lambda^2}} = \sqrt{\frac{4\pi^2e^2c^2}{\alpha^2\lambda^2}} = \frac{2\pi ec}{\alpha\lambda}\,Ampere \tag{30}$$

The photon general LC equation

Remembering that the of $E_{\mathbf{E}}$ and $E_{\mathbf{B}}$ is permanently constant as established with equation (20), we can now write:

$$E_{\mathbf{EB}} = E_{\mathbf{E}} + E_{\mathbf{B}} = \left[2\left(\frac{e^2}{4C}\right)_{\mathrm{Y}}\cos^2(\omega t) + \left(\frac{L\,i^2}{2}\right)_{\mathrm{Z}}\sin^2(\omega t)\right] \tag{31}$$

Where t is the time for one cycle, corresponding to $1/f$, or when defined as a function of λ as required here, corresponding to $t=\lambda/c$, and where the electric aspect needs by structure to be split into two equal quantities moving in opposite directions within electrostatic space (Y-space).

Since this energy corresponds to only half of the energy of the photon, we must finally add the other half, which is unidirectional and permanently localized within normal space (X-space) to obtain the total energy of the photon. Let's now also introduce the required set of directed unit vectors to completely represent the various directions of motion of the energy within the 3-spaces structure:

$$E\vec{I}\vec{i} = \left(\frac{hc}{2\lambda}\right)_{\mathrm{X}}\vec{I}\vec{i} + \left[\begin{matrix} 2\left(\dfrac{e^2}{4C}\right)_{\mathrm{Y}}(\vec{J}\vec{j},\vec{J}\overleftarrow{j})\cos^2(\omega t) \\ +\left(\dfrac{L\,i^2}{2}\right)_{\mathrm{Z}}\overleftrightarrow{K}\sin^2(\omega t)\end{matrix}\right] \tag{32}$$

Equation (32) is the most detailed and general equation, all terms of which being function of a single variable, that is, the photon wavelength λ, that can be established for the internally cycling energy of the permanently localized double-particle photon of de Broglie's hypothesis in this expanded space geometry, and where indices X, Y and Z represent the three mutually orthogonal spaces into which the kinetic energy quantum is in standing motion. All that is required now to observe how the energy oscillates between electric and magnetic states is to cyclically vary t from 0 to λ/c. This equation allows clearly understanding why the Poynting vector becomes totally stable when de Broglie's hypothesis is taken into account, at a value equal to the average value of this vector in classical Maxwell. This stability is due to the fact that at any given moment, the sum of capacitance and inductance energies is always exactly equal to half a photon's energy.

The photon general local fields equation

Equation (32), making use of the less familiar energy inductance and capacitance that were required to describe the double-particle photon kinetic energy electromagnetic oscillation, would gain in handiness if converted to use the more familiar electric (**E**) and magnetic (**B**) fields expressions for energy.

For a photon moving in straight line, it is well established that both electric and magnetic aspects of its internal energy have to be of equal density as described in ([10], equation (35)):

$$\mathbf{u_B} = \mathbf{u_E} = \frac{\mathbf{B}^2}{2\mu_0} = \frac{\varepsilon_0\mathbf{E}^2}{2} \tag{33}$$

Given that an energy density is an energy value divided by a volume, the fields related expressions for a photon's energy can be recovered by multiplying these density expressions by the related theoretical stationary isotropic volume that this incompressible oscillating kinetic energy quantum would occupy if it was immobilized as a sphere of isotropic density ([10], equation (40h)):

$$V = \frac{\alpha^5\,\lambda^3}{2\pi^2} \tag{34}$$

which, when multiplying the $\mathbf{u_B}$ and $\mathbf{u_E}$ fields energy density values expressed in equation (33) by this volume, will provide the required fields related energy values:

$$E_{\mathbf{E}} = \frac{\varepsilon_0\mathbf{E}^2}{2}V \quad \text{and} \quad E_{\mathbf{B}} = \frac{\mathbf{B}^2}{2\mu_0}V \tag{35}$$

This in turn allows the following conversion of equation (32) to a more familiar fields expression:

$$E\vec{I}\vec{i} = \left(\frac{hc}{2\lambda}\right)_X \vec{I}\vec{i} + \begin{bmatrix} 2\left(\frac{\varepsilon_0 \mathbf{E}^2}{4}\right)_Y (\vec{J}\vec{j}, \vec{J}\overleftarrow{j}) \cos^2(\omega t) \\ + \left(\frac{\mathbf{B}^2}{2\mu_0}\right)_Z \overleftrightarrow{K} \sin^2(\omega t) \end{bmatrix} V \quad (36)$$

where the photon electric field is expressed as:

$$\mathbf{E} = \frac{\pi e}{\varepsilon_0 \alpha^3 \lambda^2} \text{ from ([10], equation (40))} \quad (37)$$

and the photon magnetic field is expressed as:

$$\mathbf{B} = \frac{\mu_0 \pi e c}{\alpha^3 \lambda^2} \text{ from ([10], equation (34))} \quad (38)$$

The photon default self-guiding in straight line and self-propelling at the speed of light

It is quite interesting to observe that the default equal density by structure of both electric and magnetic fields of the double-particle photon directly explains why photons self-guide in straight lines when no outside force is acting on them, in conformity with Maxwell's fourth equation. The manner in which the trajectories of elementary electromagnetic particles can be very precisely programmed, by causing the default equal densities of both ambient electric and magnetic fields to vary from their equilibrium state, is completely described in any good textbook on high energy accelerators, such as the wonderfully made "Principles of Charged Particle Acceleration" by Stanley Humphries [21]. The mechanics of the natural variation of this default equilibrium of the density of both fields in the 3-spaces model for electromagnetic particles subjected to transverse interaction is described in a separate paper [22]. In addition to providing the previously described constant magnitude for the Poynting vector, it is also interesting to observe that this internal structure also provides a mechanical explanation to the stability of the speed of light of free moving electromagnetic energy in vacuum. As mentioned previously, a separate analysis [12] mathematically demonstrates why the speed of light of localized photons can be explained only if its kinetic energy is distributed as one half unidirectionally moving in space, propelling an equal amount of energy captive in transverse electromagnetic oscillation. It can be hypothesized that the 3-spaces structure itself acts as a set of communicating vessels through the common central junction, which would be offering zero resistance to the passage of energy, since objective reality shows that no energy is lost during free moving energy electromagnetic oscillation, and that this junction always allows the energy of the photon to remain in some form of permanent equilibrium between the 3 spaces, an equilibrium that would constantly seek to keep the photon's energy split into two equal amounts between X-space and YZ-spaces, even during energy losses or gains events related to red and blue shifting due to gravitational interaction. When energy is lost by a photon as witnessed by a displacement towards the red of its frequency or gained as witnessed by a displacement towards the blue of its frequency, the half-half X vs YZ equilibrium would be maintained by the required amount of kinetic energy seeping through the X-YZ junction in the direction required to constantly restore this equilibrium. This would directly explain why all photons self-propel, so to speak, at the same constant "equilibrium" velocity, which is of course the speed of light.

Now this brings up the old issue of what this "equilibrium" constant velocity of photons in vacuum (free moving kinetic energy) is relative to in reality. Is it relative to the medium? To the point of emission? To the point of absorption? To the observer? To this or that reference frame, or multiple reference frames, inertial, non inertial, Galilean, moving or not, etc. A deeply ingrained habit has developed since the beginning of the 20th century to hypothesize various reference frames in attempts to make sense of the experimentally observed data. But in physical reality, velocity depends on only one criterion: the actual presence of translational kinetic energy. If translational kinetic energy is present and if the local electromagnetic equilibrium allows it, there will be velocity in vacuum, relative to there being absence of translational kinetic energy, irrespective of any hypothesized reference frame or frames.

The absolute lower velocity limit, as seen from this perspective, would be an electron possessing zero translational kinetic energy in excess the energy making up its rest mass. Of course, such an electron totally deprived of translational kinetic energy can only be theoretical, because all massive particles are subject to gravitational or electrostatic acceleration in physical reality from the moment they start existing.

The absolute upper velocity limit involving electromagnetic oscillation is reached when an amount of translational (aka unidirectional) kinetic energy propels an equal amount of kinetic energy captive in transverse electromagnetic oscillation, that is, a free moving photon for example, as described in this paper.

The only other possible case between these two limits involving electromagnetic oscillation, applies to an amount of kinetic energy captive in transverse electromagnetic oscillation being propelled by a lesser amount of translational kinetic energy, such as the kinetic energy making up the rest mass of an electron, plus the transversely oscillating half of its carrier-photon's kinetic energy, both quantities being propelled by the unidirectional half of the carrier-photon's quantum of kinetic energy. The velocity of such a system will mandatorily lie between zero and asymptotically close to the speed of light, a process whose mechanics is described in a separate paper [12].

Finally, there remains one case of kinetic energy whose motion seems not to involve any electromagnetic oscillation and for which there consequently also seems not be any limiting factor on the velocity. It is the case of escaping neutrino energy, whose mechanics of liberation in the 3-spaces model is described in a separate paper [23].

The deflection angle of photons' trajectories

All of these considerations bring us to re-examine the case of light deflection experimentally verified for the first time in 1919 by Eddington and many others afterwards, during solar eclipses [24], to confirm a prediction of Einstein to the effect that light from far stars can be deflected by gravity and that this deflection could be measured, for example, as light closely grazes the mass of the Sun.

According to Newton's theory, the inertia of all bodies is deemed to be omnidirectional so a body should resist any change in its state of motion with the same intensity to a force acting on it, whichever direction it is being applied from. Associating mass to photons for the purpose of calculation, Einstein then applied the same logic to their total energy, assuming that the total complement of a photon's energy is subject to transverse interaction when flying by a celestial body as a function of the inverse square of the distance between them. His calculation then gave an estimated deflection angle of 0"83 arc second as as mentioned in a paper [25] that he published in 1911, that is, an angle twice shallower than the actual angle that will be observed in reality, which seemed to invalidate Newton's mechanics at the fundamental particles' level. Of course, he afterwards provided a different calculation, which gave an estimated double deflection angle of 1"75 arc second which is closer to reality, the supplementary deflection angle increment being considered an effect of the space-time curvature of his General Relativity Theory and a proof of the soundness of the theory.

Interestingly, as demonstrated in a separate analysis [12], the unidirectional half of the energy of the double-particle photon that must remain within normal X-space by structure, is impervious to transverse interaction as demonstrated by Walter Kaufmann at the beginning of the 20th century, as he carried out experiments by inducing varying amounts of kinetic energy into electrons [19]. When the trajectories of the moving electrons were not deflected (observations made by means of a bubble chamber), he found of course that the total longitudinal inertia of the particle involved the energy making up the rest mass of the electron plus the total amount of the added kinetic energy provided to the particle. But when the trajectories were deflected with sufficiently high velocities, he discovered that the transverse inertia of the particle involved less energy than this sum, which gave rise to the debate regarding "longitudinal mass" and "transverse mass", which led to the conclusion that mass was electromagnetic in nature. Close analysis shows that only part of the additional kinetic energy provided was involved in the transverse inertia component. Moreover, further analysis shows more precisely that at any velocity, exactly half of the additional kinetic energy provided converts to a momentary velocity related "relativistic" mass increment, which means that the other half of the additional kinetic energy, that is, the translational half of the kinetic energy provided, is totally impervious to transverse interaction while propelling the total amount of energy captive in the rest mass of the particle plus the momentary velocity related relativistic mass increment.

This means that the instantaneous relativistic mass of a moving particle can be directly measured only by means of transverse interaction since longitudinal inertia does not allow distinguishing the rest mass of the particle from the contribution from its carrying energy. This gave rise to the development of a new set of relativistic equations derived from electromagnetism which is complementary to that stemming from the Special Relativity Theory [12].

This new set can be summarized as follows in a form easy to manipulate with any scientific hand calculator. The full range of relativistic velocities can be obtained from this equation:

$$f(x) = c\frac{\sqrt{4ax + x^2}}{2a + x} \tag{39}$$

Where f(x) is the relativistic velocity, "a" is the energy in joules contained in the rest mass of the electron (8.18710414E-14 joules) and "x" is the kinetic energy provided in joules. "c" is of course the speed of light in meters per second.

From equation (39) can be derived the following equation that allows calculating the kinetic energy that must be communicated to an electron for it to move at relativistic velocity v, when only this velocity is known:

$$x = 2a(\gamma - 1) \tag{40}$$

Where "x" is the added kinetic energy, "a" is the energy making up the rest mass of the electron and γ is the Lorentz gamma factor. Any relativistic velocity plugged into the gamma factor will allow obtaining the amount of kinetic energy required for the particle to move at this velocity.

Let us note that the gamma factor would be much easier to deal with in equations if it was simplified to the following form which leaves only one fraction in the expression:

$$\gamma = \frac{1}{\sqrt{1 - v^2/c^2}} = \sqrt{\frac{c^2}{c^2 - v^2}} = \frac{c}{\sqrt{c^2 - v^2}} \tag{41}$$

With the velocity related amount of kinetic energy calculated with equation (40), the following equation allows calculating the instantaneous relativistic mass of the particle for this relativistic velocity:

$$m_{(rel)} = m_0 + \frac{x}{2c^2} \tag{42}$$

The full range of relativistic velocities can also be obtained from the following equation by using the wavelengths of the energies involved:

$$f(x) = c\frac{\sqrt{4ax + a^2}}{2x + a} \tag{43}$$

where f(x) is the relativistic velocity, "a" is the electron Compton wavelength (2.426310215E-12 m) and "x" is the wavelength of the total amount of kinetic energy provided to the particle. Finally, similarly to equation (40) being derived from equation (39), the following equation derived from equation (43) allows calculating the wavelength of the energy that must be communicated to an electron for it to move at relativistic velocity v, when only this velocity is known:

$$\lambda = \frac{\lambda_C}{2(\gamma - 1)} \tag{44}$$

Where "λ" is the wavelength of the communicated energy, "λ_C" is the electron Compton wavelength and "γ" is the Lorentz factor. In relation with the analysis carried out in a separate paper [12], when the energy making up the rest mass of the electron is set to zero in equation (39), or rather, in its electromagnetic version ([12], equation (33)), we observe that the only velocity that can be obtained is c, that is, the speed of light. This means that the remaining transversaly measurable mass increment plus the transversaly undetectable but equal translational other half of the added energy involved behave like a free moving photon, displaying a longitudinal inertia corresponding to the total amount of energy involved, but a transverse inertia corresponding to only half of the total amount of energy involved. This observation with regards to the electron's carrying energy comes in support of the analysis carried out in this paper to the effect that free moving electromagnetic photons would, by similarity, have the same internal electromagnetic structure as massive particles' carrying energy. Consequently, if Einstein's calculations had been done with the "mass" of only the electromagnetically oscillating half of the photon's energy with regards to photons' trajectories deflection by celestial bodies, that is, the only part of photons' energy that seems to be sensitive to transverse interaction, then the 1"75 arc second deflection angle for photons flying closely by the Sun could have been directly obtained from classical mechanics without any need to resort to the General Relativity space-time curvature.

Conclusion

This paper is meant to show that it is possible to represent the permanently localized photon of de Broglie's hypothesis in a manner totally conform to Maxwell's equations. This particular solution requires expanding the local space geometry in a manner that allows the photon's kinetic energy quantum to behave in accordance with Maxwell's theory without changing in nature while at the same time displaying the mutually exclusive properties of electric and magnetic fields. Since energy can be represented in a number of ways as exemplified by the two currently available approaches mentioned in the introduction Section, solutions other than the one proposed here are of course possible. But hopefully, the non-exhaustive benefits mentioned in this paper that this new solution provides may help rekindle causality based fundamental research in the community, all the more so since this space geometry also allows a simple and logical mechanical explanation to the process of conversion of massless photons of energy 1.022 MeV or more to pairs of massive electron-positron [13] that also structurally possess the dual wave-particle characteristics that characterize this model of the photon, and much more.

References

1. Einstein A (1953) Louis de Broglie, physicien et penseur. Éditions Albin Michel, Paris.
2. De Broglie L (1993) La physique nouvelle et les quanta (2nd edn). Flammarion, France.
3. Buras AJ (2005) Photon Structure Functions: 1978 and 2005. Acta Physica Polonica B 37: 609.
4. Sears F, Zemansky M, Young H (1984) University Physics (6th edn.). Addison Wesley, USA.
5. Lochak G (1992) Louis de Broglie, Flammarion, France.
6. Feynman R (1949) Space-Time Approach to Quantum Electrodynamics. Phys Rev lett 76: 769.
7. Breidenbach M (1969) Observed Behavior of Highly Inelastic Electron-Proton Scattering. Phys Rev Lett 23: 935-939.
8. Greiner W, Reinhardt J (1994) Quantum Electrodynamics. Springer-Verlag.
9. Burke DL (1997) Positron Production in Multiphoton Light-by-Light Scattering. Phys Rev Let 79: 1626.
10. Michaud A (2007) Field Equations for Localized Individual Photons and Relativistic Field Equations for Localized Moving Massive Particles. International IFNA-ANS Journal 13: 23-140.
11. Marmet P (2003) Fundamental Nature of Relativistic Mass and Magnetic Fields. International IFNA-ANS Journal 9: 1-10.
12. Michaud A (2013) From Classical to Relativistic Mechanics via Maxwell. International Journal of Engineering Research and Development 6: 1-10.
13. Michaud A (2013) The Mechanics of Electron-Positron Pair Creation in the 3-Spaces Model. International Journal of Engineering Research and Development 6: 1-10.
14. Hanson G (1975) Evidence for Jet Structure in Hadron Production by e+ e- Annihilation. Phys Rev Let 35: 1609-1612.
15. Michaud A (2013) The Mechanics of Neutron and Proton Creation in the 3-Spaces Model. International Journal of Engineering Research and Development 7: 29-53.
16. Kotler S, Akerman N, Navon N, Glickman Y, Ozeri R (2014) Measurement of the magnetic interaction between two bound electrons of two separate ions. Nature magazine 510: 376-380.
17. Michaud A (2013) On The Magnetostatic Inverse Cube Law and Magnetic Monopoles. International Journal of Engineering Research and Development 7: 50-66.
18. Michaud A (2000) On an Expanded Maxwellian Geometry of Space. Fundamental Problems of Natural Sciences and Engineering, Russia.
19. Kaufmann W (1903) Über die "Elektromagnetische Masse" der Elektronen, Kgl. Gesellschaft der Wissenschaften Nachrichten. Mathem-Phys Klasse 2: 91-103.
20. Resnick R, Halliday D (1967) Physics. John Wyley and Sons, New York.
21. Humphries S (1986) Principles of Charged Particle Acceleration. John Wiley and Sons, New York.
22. Michaud A (2013) On the Electron Magnetic Moment Anomaly. International Journal of Engineering Research and Development 7: 21-25.
23. Michaud A (2013) The Mechanics of Neutrinos Creation in the 3-Spaces Model. International Journal of Engineering Research and Development 7: 1-8.
24. Ohanian HC, Ruffini R (1994) Gravitation and Spacetime (2nd edn.). Norton, New York.
25. Einstein A (1911) About the influence of gravity on the propagation of light .Annals of Physics 340: 898-908.

7

Speakable and Unspeakable in Special Relativity: the Ageing of the Twins in the Paradox

Guerra V* and Abreu R

Department of Physics, Instituto Superior Técnico, University of Lisbon, Portugal

Abstract

In previous papers, we have presented a general formulation of special relativity, based on a weaker statement of the postulates. In this work, the paradigmatic example of the twin paradox is discussed in detail. Within the present formulation of special relativity, a "non-paradoxical" interpretation of the asymmetric ageing of the twins emerges. It is based exclusively on the *rhythms* of the clocks, which are not related by the standard textbook expressions and shall not be confused with clock *time readings*. Moreover, the current approach exposes the irrelevance of the acceleration of the returning twin in the discussion of the paradox.

Keywords: Special relativity; Lorentz transformation; Einstein speed; Twin paradox; IST transformation; Clock rhythms

1. Introduction

In previous works we undertook a reflection on the foundations of special relativity [1-6] where we present a general formulation of special relativity and reconcile the ideas of Lorentz and Poincaré of the existence of a "preferred reference frame" and Einstein's "equivalence of all inertial frames." A comprehensive overview of the theory is given in our last paper [6], hereafter denoted as I, where it is shown that it is not correct to speak about two philosophies, as they are different aspects of one and the same theory. Herein we extend our previous results and illustrate the power of the present formulation with a detailed discussion of the twin paradox.

The analysis of the "paradoxes" of a physical theory is an important matter, since it questions and assigns physical meaning to the statements of the theory. Following Feynman et al. [7], "a paradox is a situation which gives one answer when analyzed one way, and a different answer when analyzed another way, so that we are left in somewhat of a quandary as to actually what should happen. Of course, in physics there are never any real paradoxes because there is only one correct answer; at least we believe that nature will act in only one way (and that is the *right way*, naturally). So in physics a paradox is only a confusion in our own understanding".

The famous twin paradox emerges in the standard interpretation of special relativity, which says that each twin ages slower than the other during any part of the to-and-fro traject, but when they meet one of them is older. This is of course not possible if "ageing slower" refers exclusively to rhythms of clocks along the trip and nothing else happens. Although the twin paradox is addressed and "solved" in any introductory course on special relativity, it was at the origin of more than 25,000 articles in the literature [8] since it was launched by Langevin in 1911 [9], and new publications arise regularly (the references given are therefore merely indicative [8,10-31]. Therefore, one can only suspect that "perhaps the last word on the twin paradox has yet to be said" [18].

In this paper, we argue that the key point to understand the twin paradox remains misinterpreted in its usual "solutions." We start with a brief overview of the problem and then criticise the two main lines used in its standard "resolution:" the "simultaneity" and the "general relativity" arguments. In fact, despite giving the correct final answer, the usual explanations have a severe interpretation mistake, as they fail both to make the distinction between clock rhythms and clock time readings [6] and to recognise the indeterminacy of special relativity [4,6].

The structure of this paper is the following. In the next section we very briefly review some of the main findings from our former work [1-6], with emphasis on the relation of rhythms between two clocks in relative motion and its connection with the usual time dilation expressions. The twin paradox is then presented and examined in the subsequent sections. Section 3 presents a short historical overview of the problem. Critical reviews of the standard "solutions" using either only special relativity and no reference to acceleration or making use of general relativity and acceleration are given in sections 4 and 5, respectively. Section 6 contains the general analysis made in the context of the present formulation of special relativity. Finally, section 7 summarises our main findings.

2. Brief Overview of the Theory

In this section we very quickly review some results from I [6]. We expected the reader to be familiar with that paper and call a special attention to its sections 5, 6 and 8, of particular relevance to the ensuing discussion.

2.1 The IST transformation, Lorentzian time and Einstein speed

Our formulation is based on the definition of the rest system – as the system where the one-way speed of light in vacuum is isotropic – and on the IST transformation (Inertial [32-34]–Synchronized [2,4,35-37] Tangherlini [38]). In the usual configuration where the axis of the rest frame S and a moving frame S' are aligned, the origin of S'' moves along the x-axis of S with speed v in the positive direction, and the reference event is the overlapping of the origins of both frames at time zero, the IST transformation is given by Guerra and Abreu [2]

***Corresponding author:** Guerra V, Department of Physics, Instituto Superior Técnico, University of Lisbon, Portugal
E-mail: vguerra@tecnico.ulisboa.pt

$$x' = \gamma(x - vt)$$
$$t' = \frac{t}{\gamma}, \tag{1}$$

with

$$\gamma = \frac{1}{\sqrt{1-\frac{v^2}{c^2}}}. \tag{2}$$

It is important to note that the Lorentz transformation is readily obtained from the IST transformation by introducing the offset factor

$$t_L' = t' - \frac{v}{c^2}x' \tag{3}$$

and substituting t' in (1) [2,4]. Therefore, any phenomenon that can be described by the Lorentz transformation can also be described by the IST transformation. We denote by Lorentzian clocks the clocks "synchronized" (i.e., "adjusted" or "set") according to the Lorentz transformation, and by Lorentzian times the time readings exhibited by the Lorentzian clocks (3). Similarly, we denote by synchronized clocks the clocks adjusted according to the IST transformation and by synchronized times or simply "times" the time readings of the synchronized clocks.

The Lorentz transformation and the IST transformation can be associated with specific forms of clock "synchronization" [2]. However, as stated in I, one simple - and yet critical - issue that has to be clarified before engaging any discussion on the interpretation of special relativity or of special relativity results is to state the difference between clock *rhythms* (or clock tick rates) and clock *time readings* (or time coordinates), as further discussed below. Contrary to time readings, clock rhythms do *not* depend on any particular form of "synchronization." Surprisingly, the failure to make this basic distinction is at the origin of several misunderstandings surrounding the theory, including the discussion of the twin paradox.

The velocity addition formula can be obtained easily from (1) [2]. If an object is moving at speed w in S, then its speed in the inertial frame S', w_v', is simply

$$x' = w_v' t', \tag{4}$$

which can be written in the form

$$w_v' = \gamma_v^2(w - v) = \frac{w-v}{1-\frac{v^2}{c^2}}. \tag{5}$$

It is also essential to distinguish *speed* (or "synchronized speed") from *Einstein speed* [2,4,5]. Speed is defined as in the previous equations and can be calculated from $w_v' = \Delta x' / \Delta t'$. In turn, the Einstein speed is defined from

$$w_E' = \frac{\Delta x'}{\Delta t_L'}, \tag{6}$$

i.e., its value is calculated with the difference of the time readings of "Lorentzian" clocks. Substituting (3) into (4), it is straightforward to show [2] that the Einstein velocity, w_E', measured in a frame moving with speed v (in S) of an object which has speed w, is

$$w_E' = \frac{w_v'}{1-\frac{vw_v'}{c^2}}. \tag{7}$$

Replacing w_v' and using (5) this last expression can be rewritten as

$$w_E' = \frac{w-v}{1-\frac{vw}{c^2}}. \tag{8}$$

2.2 General expression for the rhythms of two clocks in relative motion

In prior work it was thoroughly debated that the indeterminacy of special relativity implies that just with Lorentzian clocks it is not possible to know in which inertial frame clocks are actually running slower [4]. This result was derived more formally in section 8 of I, where the general expression for the rhythms of two clocks in relative motion was deduced. Due to its importance for the discussion of the twin paradox, we reproduce a big part of it here.

Consider two inertial frames, S' and S'', moving in the x direction respectively with speeds v and w in the rest system, S. Clock 1 is at the origin of S' and clock 2 is at the origin of *S' and S''*. The speed of clock 2 in S' is given by (5), the Einstein speed of clock 2 in S' is given by (8). Let us further define the *proper time*, τ, in the usual way, as the time elapsed for one particular observer. Since the proper time is measured by a single clock, it is indeed associated with the clock *rhythm* and does not depend on the initial adjustment of distant clocks. The proper time of a clock in S'' relates to the time elapsed in the rest system S by the time equation in (1), which can be written in the form

$$d\tau'' = \frac{dt}{\gamma_w}, \tag{9}$$

where dt are the differential times marked by the different clocks in the rest system S that are co-punctual with clock 2 at each instant. The relation of the rhythm of a clock 1 in S' with the rhythms of clocks in S is given by a similar expression, namely

$$d\tau' = \frac{dt}{\gamma_v}. \tag{10}$$

Therefore, the relation of the proper times of clocks in $S''$and in S' is

$$d\tau'' = d\tau' \frac{\gamma_v}{\gamma_w}. \tag{11}$$

Equation (11) establishes the relation of clock *rhythms* in two inertial frames. However, one final step is still missing. From (10) and (1) it directly follows

$$d\tau' = dt'. \tag{12}$$

Here, the equality is completely general and is valid whether or not dx'=0 as x' does not appear in the second equation (1). In turn, as a consequence of (3), in general

$$d\tau' = dt' \neq dt_L', \tag{13}$$

although, if dx'=0, then $d\tau' = dt' = dt_L'$. Similarly, we can write $d\tau'' = dt'' \neq dt_L''$, except when dx''=0. Substituting (12) in (11), we get

$$d\tau'' = dt' \frac{\gamma_v}{\gamma_w}. \tag{14}$$

The inverse relation is simply $d\tau' = dt''(\gamma_w / \gamma_v)$.

As it is well-known from standard special relativity - and it is extremely easy to deduce from the Lorentz transformation - the proper time of clock 2 relates with the differential Lorentzian times through the Einstein speed,

$$d\tau'' = \frac{dt_L'}{\gamma_E}, \tag{15}$$

where γ_E is the Lorentz factor associated with the *relative* Einstein speed between both moving frames v_E given by equation (8), $\gamma_E = 1/\sqrt{1-\frac{v_E^2}{c^2}}$. Contrary to (11), this expression corresponds to comparisons of *time readings* of Lorentzian clocks in S' and does *not* correspond to a relation with the clock rhythms in S'.

Finally, by noting that $dx' = w_E' dt_L'$ and $dx' = w_v' dt'$, and using (13), equation (15) can be rewritten as

$$d\tau'' = d\tau' \frac{w_v'}{w_E'} \frac{1}{\gamma_E}. \tag{16}$$

This result confirms that Lorentzian clocks and Einstein speeds are not enough to determine in which frame clocks are running faster, since the relation between rhythms additionally involves the speed w_v'.

The indeterminacy of special relativity [4] has thus been expressed in an alternative way. If the rest system is inaccessible, then we *do not know* the value of the one-way speed of light in one inertial frame, nor can we know the value of w_v' in (16), and so *we cannot know in which of two inertial frames clocks are ticking slower.* This result, evidently, has profound implications in the analysis of the twin paradox.

3. Brief Overview

The twin paradox is one of the most famous paradoxes in physics. It emerges in the context of the standard interpretation of special relativity. Very briefly, consider twins Andrew and Bob [39]. Andrew stays on the Earth, while Bob flies to a distant star and back. Within the standard interpretation os special relativity, from Andrew's point of view Bob's clock is running slower both on the outward and return parts of the trip. Therefore, Bob is younger than Andrew when they meet up again. The "paradox" part of the story comes from the following alternate reasoning. Within the standard interpretation of special relativity, in Bob's frames it is Andrew's clock which is running slower, and so it is Andrew who is younger than Bob when they meet up again. Of course this is self-contradictory, as Andrew cannot be younger and older than Bob.

About 10 years ago Peter Pesic wrote a very interesting review article on the twin paradox [23]. In particular, he points out that, facing the paradox, Einstein addresses two arguments. The first one is the "simultaneity argument." It uses only special relativity and does not attribute importance to acceleration other than to break the symmetry of the situation of both twins [40]. This is indeed correct, as recognized by many physicists. For instance, Morin [39] states our time-dilation result holds only from the point of view of an inertial observer. The symmetry in the problem is broken by the acceleration. If both *A* and *B* are blindfolded, they can still tell who is doing the traveling, because *B* will feel the acceleration at the turnaround (...). For the entire outward and return parts of the trip, *B* does observe *A*'s clock running slow, but enough strangeness occurs during the turning-around period to make *A* end up older. Note, however, that a discussion of acceleration is not required to quantitatively understand the "paradox".

The somewhat uncomfortable but inescapable remark is that the change of inertial reference frames by Bob can be made in a very short time, in the limit even instantaneously. And it is this very short period, as compared to the trip duration, which is responsible for making Bob younger. Moreover, if Bob makes a longer trip and if he does the turn back exactly in the same way as before, he will return even younger than before. Thus, the effect of the very short turn-around period has to be higher now, in order to compensate for the longer period in which Bob sees Andrew ageing slower. The "solution" proposed by the standard interpretation of special relativity in this case is related to the "loss of simultaneity" exhibited by the time readings of Lorentzian clocks in different inertial frames [23,40]. If this is correct we have to acknowledge that "at the root of the twin paradox is the problem of synchronization of clocks" [19]. However, this raises another major alarm. In fact, as repeated several times along in this work, physical results cannot depend on the way in which one decides to set his own clocks. This "solution" is analyzed in section 4.

In 1918 Einstein has advanced the second argument, the "general relativity argument" [41]. It is now affirmed that we must consider general relativity in order to solve the paradox from the point of view of Bob. Since before and after the acceleration the argumentation of special relativity subsists, this comes back to attributing an important physical role to the acceleration. During the acceleration corresponding to the turn around period Bob would see Andrew age very quickly, because, according to the equivalence principle, Bob would see Andrew at a high gravitational potential. This second "solution" is discussed in section 5.

4. "Solution" without Acceleration

From the presentation of the broader view of special relativity proposed in I, in particular with the distinction between clock rhythms and clock time readings, together with the indeterminacy of special relativity, which does not allow to know in which inertial frame clocks are actually running slower, the meaningful solution to the paradox is already clear. The symmetry in the *description* of the outward trip between Andrew and Bob when we refer to Lorentzian times, repeated also for the return trip, does *not* correspond to a symmetry in the ageing (proper times) of the twins. Regarding proper times, we know that during the outward trip either Andrew or Bob is ageing slower and, without a reference to the rest system, we do not know which of them. And it may even happen that both are ageing at the same rhythm. The same occurs during the return trip. It is possible that it is always one of the twins who is ageing slower, both on the onward and on the return trip, or that one of the twins ages slower during the onward trip and the other during the return trip, or even that in one of the trips they are ageing at the same rate. However, we do know that when they meet it is Bob who is younger, and by which factor. Of course that all calculations can be made both with Lorentzian times and with synchronized times and from the point of view of each of the twins: all these calculations must give the same final result.

One quite convincing standard solution of the paradox using the "simultaneity argument" is obtained from the analysis of Minkowski diagrams [10,20,39]. Let us assume that Bob travels along the x' axis with *Einstein speed* of absolute value v_E in relation to Andrew, both in the outward and on the return trips. There are various Lorentzian clocks along the x' axis with photo cameras. At regular intervals $\Delta t_L' = \Delta t_A'$ in Andrew's frame the cameras shoot. For each set of photos one and only one of the cameras has the photo of Bob. The situation is depicted in Figure 1a). The universe line of Andrew is the ct' axis (x' =0), whereas the universe line of Bob is represented in red. Obviously, Andrew and Bob agree on the number *N* of photos taken during the all trip. It is straightforward to obtain the conversion of lengths in the diagram to the Lorentzian times in each frame [(one unit in the ct''

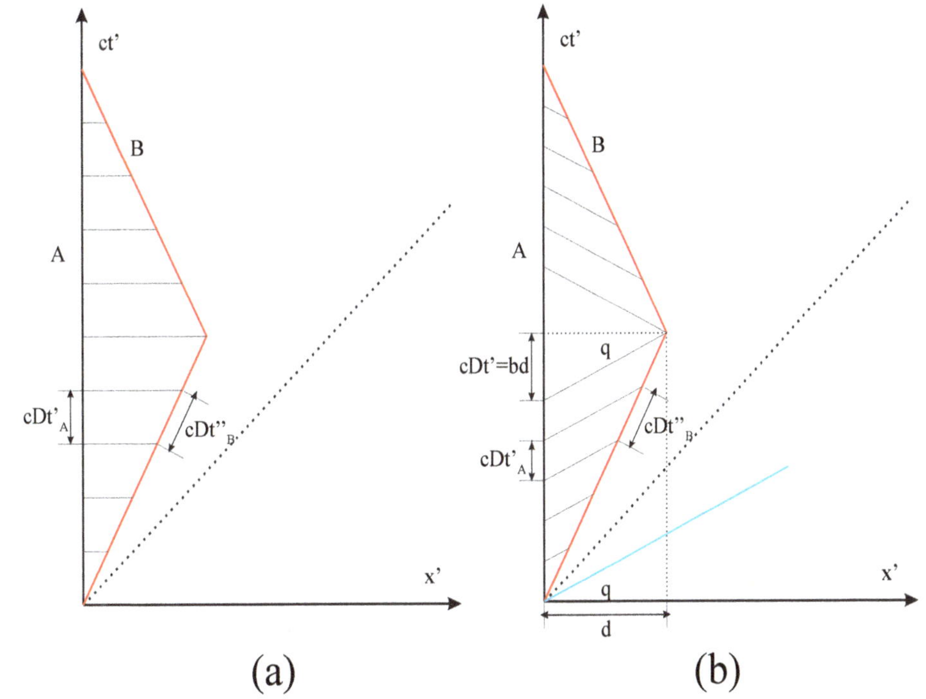

Figure 1: Minkowski diagram for the twin paradox from Andrew's point of view (a) and from Bob's point of view (b).

axis)/(one unit in the ct' axis)= $\sqrt{(1+\beta^2)/(1-\beta^2)}$, with $\beta=v/c$ [39] and from simple trigonometry to get $\Delta t''_B = \Delta t'_A\sqrt{1-\beta^2} = \Delta t'_A/\gamma_E$. Notice that $\Delta t'_A$ refers to a comparison of the time readings of Lorentzian clocks in Andrew's frame while $\Delta t''_B$ corresponds to Bob's proper time, so that this result is simply the symmetric description of time dilation obtained with Lorentzian clocks (*cf.* Figure 2 in I and its respective discussion). The total time elapsed for Andrew is $t'_A = N\Delta t'_A$, while for Bob the time lapse is $t''_B = N\Delta t''_B$, so that

$$t''_B = t'_A/\gamma_E. \tag{17}$$

Therefore, Bob is younger than Andrew when the twins meet

To complete the analysis we have now to use the reverse reasoning, taking photos of Andrew from cameras in Bob's frame, shot when the clocks associated with the cameras display the same *Lorentzian time*. Only one of each set of photos has Andrew's photo. This new situation is represented in Figure 1b. There is not a single inertial frame associated with Bob, as the inertial frames for the onward and return trips are distinct. While Bob sees himself photographed at equal time intervals, $\Delta t''_B$ in Figure 1a), Andrew seems himself being photographed regularly in the beginning of the trip, $\Delta t'_A$ in Figure 1b), then there is a time interval $2\Delta t'_L$ (corresponding to the proximity of Bob with the star, the turn-around point) where no photo is taken, and then Andrew is again photographed regularly. As before, Andrew and Bob agree with the number of photos taken. Let d be the distance between the Earth and the star (in the Earth frame). The total time of the trip for Andrew is simply

$$t'_A = 2d/v_E. \tag{18}$$

The total time elapsed for Bob is

$$t''_B = N\Delta t''_B. \tag{19}$$

Furthermore, in this case we have the reverse situation as compared to the previous figure concerning the intervals $\Delta t'_A$ and $\Delta t''_B$, with $\Delta t'_A = \Delta t''_B/\gamma_E$, as derived from the Lorentz transformation or directly from the Minkowski diagram. Of course $\Delta t''_B$ refers to comparisons of Lorentzian times, while $\Delta t'_A$ corresponds to Andrew proper times, so that $\Delta t'_A$ and $\Delta t''_B$ are not the same as in the Figure 1a). It is also not difficult to show that $2c\Delta t' = 2dv_E/c$. In other words, the time that passes for Andrew while he is not photographed, corresponding to Bob's turn-around, is

$$2\Delta t' = 2dv_E/c^2, \tag{20}$$

i.e., the offset factor (3). The final calculation for t'_A then gives

$$t'_A = N\Delta t'_A + (2\Delta t') = N\Delta t''_B/\gamma + (2\Delta t') \tag{21}$$

from where, substituting (18–20), one recovers (17), $t'_A = \gamma_E t''_B$, as it had to be. There is no paradox, and a careful calculation made only within standard special relativity leads to the right result, both from Andrew's and from Bob's points of view. In turn, failing to consider the factor $\Delta t'$ would lead to the paradoxical result $t'_A = t''_B/\gamma_E$, contradictory with (17).

Despite the correctness of the result obtained with the Minkowski diagrams, the language used in the standard solution confuses "time" with Lorentzian times, making the interpretation somewhat awkward. In fact, without drawing the Minkowski diagrams and just using the Lorentz transformation, very few scientists would remember to add the offset factor $2dv_E/c^2$ on a change of inertial frame corresponding to "Bob's point of view" to get the final calculation right. This calculation error comes from a correct intuition: if the Loretzian time intervals would correspond to proper times, then one would not need to add, a bit artificially, some extra interval to the travel time of Andrew, which in addition depends on the distance, to correct for the instantaneous action of changing inertial frame. The need to add $2dv_E/c^2$ to relate

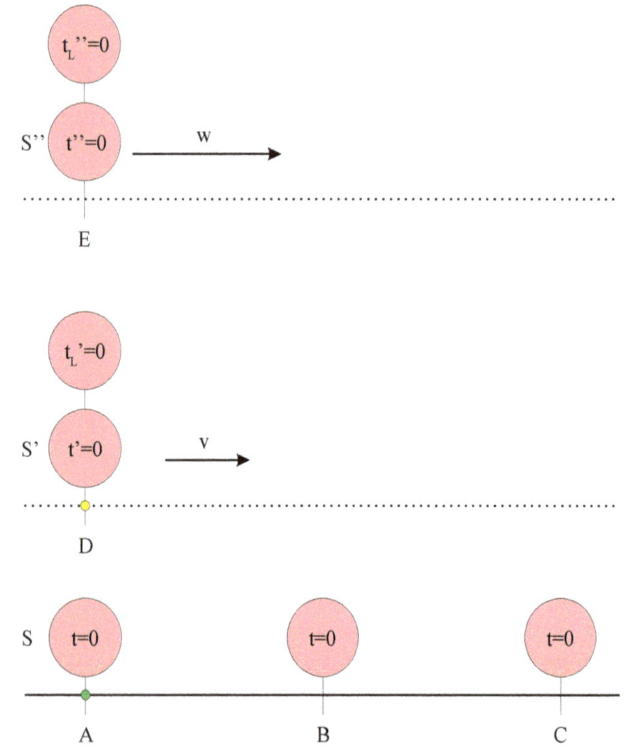

Figure 2: All times in the figures are expressed in milliseconds and rounded to two decimal places. Twins Andrew and Bob, initially in the moving frame S' separate at t=0; Andrew (yellow dot) remains in frame S' with clock D, while Bob (green dot) "jumps" to the rest system S to the location of clock A.

the Lorentzian times and compensate the different offset factors (3) of Bob's two frames, together with the repeated remark that clock rhythms do not depend on the initial adjustment of the clocks, points the way to a clear and simple analysis of the twin paradox. Let us take one example to illustrate how does it work.

Consider the setup shown in Figure 2. Andrew and Bob are initially in a moving frame S', which goes with speed v=0.4c in relation to the rest system S. A second moving frame S'' goes with speed w=2v=0.8c. All frames are equipped both with synchronized and with Lorentzian clocks. The *same* situation can then be *described* with both types of clocks. As the physical reality is not affected by the description chosen, one can use both types of clocks and both descriptions at the same time no conflict at all [6]. An *apparent* conflict only arises if one confuses Lorentzian clocks with synchronized ones.

We start the analysis by looking at the time readings of the synchronized clocks, so that the time lapses correspond to proper times (section 2.2). To go to the distant star Bob "jumps" at t=0 from S' into another frame, which, for simplicity, it happens to be the rest system S, staying there at rest for 10.00 milliseconds. From the "point of view of Andrew," Bob can be seen as "moving" outward, but still it is Bob who is ageing faster. Indeed, for Bob 10.00 ms have passed, while for Andrew the elapsed time was only about 9.17 ms, as shown by the synchronized clocks A and G in Figure 3, so that Bob is older by (10.00-9.17) ms=0.83 ms. On frame S'' only 6.00 ms have passed. Clock F from S'' is now on top of Bob, who "jumps" into S'' in order to rejoin Andrew. Both twins meet when 20.00 ms passed in the rest system, at the position of clock C as depicted in Figure 4. For Andrew the all process took approximately 18.33 ms. But for Bob only 16.00 ms have passed, 10.00 ms during the "outward" trip and (12.00-6.00) ms=6.00 ms during the "return" trip. In this particular case, although

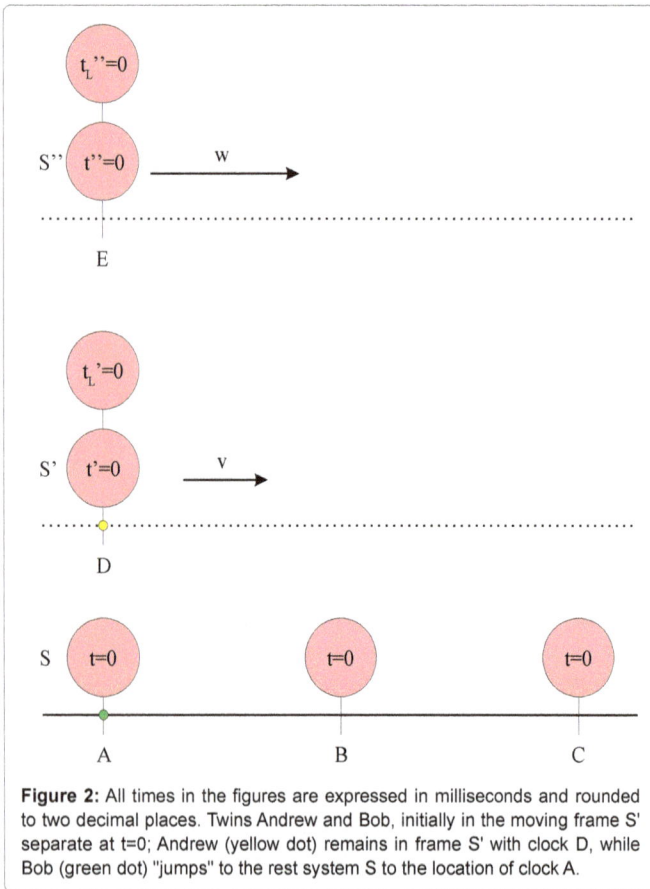

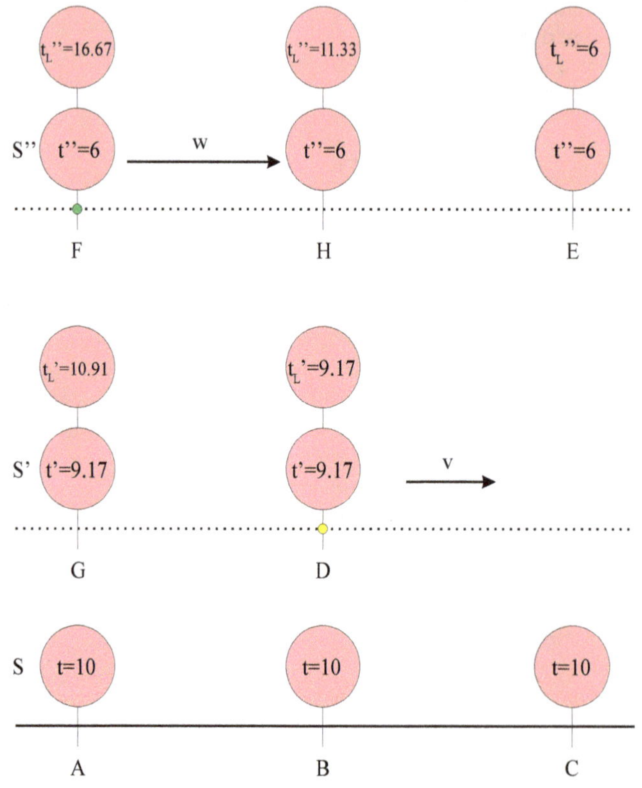

Figure 3: At t=10 ms Bob "jumps" to the moving frame S", to the position of clock F, in order to catch up Andrew.

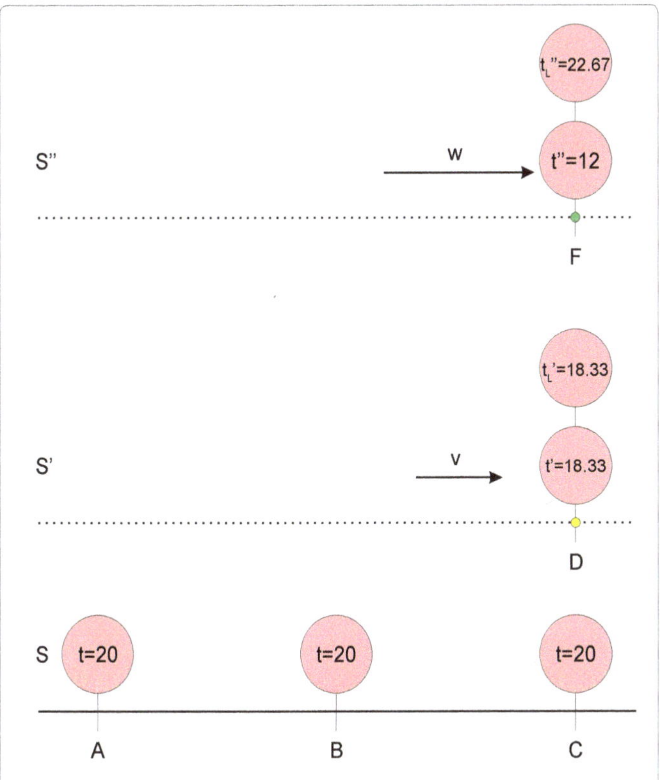

Figure 4: The two twins meet again at t=20 ms. Bob is younger.

Bob is ageing faster during the outward trip, he is ageing slower during the return trip. The latter journey compensates for the first part of the trip and it is actually Bob who is younger when the twins meet, by 2.33 ms. When speaking about proper times there is no difference in the descriptions made "from Andrew's point of view" and "from Bob's point of view," they both agree with all the numbers presented above and describe the sequence of events exactly in the same way. Different variations of this example can be made, all leading to the same final result.

It is enlightening to crosscheck how the same situation is described from the comparison of the *time readings* of the Lorentzian clocks, within the reasoning of the standard interpretation of special relativity. Let us first see how Andrew argues. He considers himself at rest and will compare the time readings of different Lorentzian clocks in his frame, co-punctual with Bob, with the time readings of Bob's clock. When Bob arrives at the turn-around point his clock (A) in S displays 10 ms (Figure 3). In turn, the Lorentzian clock in Andrew's frame S' (clock G) exhibits 10.91 ms. Thus, the difference in Lorentzian times is (10.91-10.00) ms=0.91 ms. In this way, one may be induced to think that Bob is younger by 0.91 ms (while in fact he is older by 0.83 ms, as shown before). During the return trip the comparison of Lorentzian times reveals that at the beginning of the return trip (Figure 3) Bob's clock F is in advance of the S' clock G by (16.67-10.91) ms=5.76 ms, while at the end of the trip (Figure 4) his clock F is in advance of the S' clock D only by (22.67-18.33)ms=4.34 ms. Therefore, during the return trip one may be induced to think that Bob got younger by (5.76-4.34) ms=1.42 ms.

Just adding the differences of the Lorentzian time intervals we get the correct number and conclude that Bob is younger by (0.91+1.42) ms=2.33 ms, the same number obtained before, as it should to be: the physical reality does not depend on its description. Adding the Lorentzian time intervals provides the correct result, albeit an erroneous interpretation if we forget they are Lorentzian times and confuse them with clock rhythms.

Still within the reasoning of the standard interpretation of special relativity, Bob's argumentation is much more striking and problematic. Bob considers himself to be at rest and so he expects to see Andrew ageing slower. He will compare the time readings of the Lorentzian clocks in his inertial frames (one frame during the onward trip, another one during the return trip) co-punctual with Andrew, with the time readings of Andrew's clock. For the outward trip, Andrew got younger by (10.00-9.17) ms = 0.83 ms (clocks B and D in figure 3).

To keep the rules of the standard interpretation of special relativity, for the return trip we have to compare first the time readings of the Lorentzian clocks D and H in Figure 3, and then those of clocks D and F in Figure 4. Bob may then think that Andrew got even younger during the return trip, as for Andrew the difference in Lorentzian times is (18.33-9.17)ms=9.16 ms, whereas for Bob it is (22.67-11.33) ms=11.34 ms, so that one may think that Andrew got younger by (11.34-9.16)=2.18 ms. Therefore, not counting with any "strangeness" of the turn-around point, Bob may think that in total Andrew got younger by (0.83+2.18)ms=3.01 ms. However, Bob knows that at the turn-around point the Lorentzian clock H displays 11.33 ms Figure 3. However, his new Lorentzian clock F displays 16.67 ms! This difference of (16.67-11.33)ms=5.34 ms, which of course corresponds to the factor (20) resulting from the offset (3) of Lorentzian clocks, has to propagate to all clocks in S'', as we are interested only in time-differences.

This is all there is about the strangeness of the turn-around point. However, if we would want to keep the speech that "for the entire outward and return parts of the trip, Bob does observe Andrew's clock running slow," then we could not correct the clocks in S'' (as if we do this we conclude after all that Bob sees himself ageing slower). Instead, *we would have to see these 5.34 ms – which simply correspond to a constant offset factor between two distant clocks and have nothing to do with clock rhythms – as an additional ageing of Andrew.* Regarding the example with photographs of the Minkowski diagrams, it would mean that Andrew has to advance for 5.34 ms until he crosses a Lorentzian clock from S'' displaying 16.67 ms, where he restarts being photographed again, and that this would be an additional ageing that would have nothing to do with time dilation itself, as if while Andrew advances for 5.30 ms Bob would not age at all. Whatever discourse we decide to adopt, the final result is that Andrew is older, by (5.34-3.01)ms=2.33 ms, has it had to be. Nevertheless, hopefully this simple example makes it self-evident that the addition of off-set factors does not correspond to any ageing of either twin. The plain and straight interpretation of the paradox naturally appears when synchronized clocks are used to compare time readings, and is directly related to proper times and true clock rhythms.

The above example confirms the validity of the "simultaneity argument," in the sense that general relativity is not necessary for an explanation of the twin paradox and a correct calculation can be made within special relativity. It is not too complicated to perform all the formal calculations with Lorentzian and synchronized times from the expressions presented in section 2 and to arrive at the correct result (17), whatever the description and the twin point of view is adopted. Some of these calculations are presented in section 6.

5. "Solution" with Acceleration

Let the total ageing in the round trip for Andrew be t'_A, while for

Bob t_B''. Any standard calculation made "from Andrew's point of view" is trivial using time dilation and gives the result (17), $t_A' = t_B''\gamma_E$, as shown in the previous section. It is the calculation made "from Bob's point of view" which is more difficult. As we have just seen, the correct result can be obtained, e.g., from the analysis of Minkowski's diagrams and only within the framework of special relativity. That being so, it should be clear that acceleration itself is not an essential ingredient in the solution of the paradox, other than to evince a lack of symmetry between the twins. However, it is also possible to use general relativity to explicitly account for the effect of acceleration. We take the standard argument from Tolman as presented by Grøn [27,42].

Consider a fixed coordinate non-inertial system with Bob. There are five different moments to take into account: 1) the acceleration of Bob up to speed v; 2) the travel up to the distant star at constant speed v; 3) the time required to reverse the speed at the distant star; 4) the back trip at constant speed $-v$; 5) the deceleration of Bob to meet Andrew. In order to keep the analysis as simple as possible, let us assume that all acceleration times in Bob's frame are very short. According to general relativity we can replace all accelerations by gravitational fields and calculate the times processes 1) - 5) take in each frame. For the processes involving acceleration the relation of the time lapses in each frame is given by Tolman [42,27].

$$\Delta t_A' = \Delta t_B''\left(1 + \frac{\Delta\phi}{c^2}\right), \tag{22}$$

where the gravitational potential difference $\Delta\phi$ is

$$\Delta\phi = g d_{AB}, \tag{23}$$

d_{AB} is the distance between Andrew and Bob and g is the proper acceleration of Bob.

For processes 1) and 5) Andrew and Bob are placed in the same location, so that d_{AB}=0. As such, there is no difference in the gravitational potential and no difference in the ageing of the twins in these periods. In fact, the irrelevance of these periods is self-evident from the start, as Bob does not even have to accelerate nor to decelerate. Bob can simply fly past Andrew already at speed v at the beginning of the trip and compare his age with the one of Andrew when they cross, and then do exactly the same on the arrival. Next, during the onward and back trips, "Bob sees Andrew ageing slower." As thoroughly debated along this paper, this assertion refers to the comparison of the displays of successive Lorentzian clocks in Bob's frame co-punctual with Andrew, and not to clock rhythms. Thus, for processes 2) and 4) we have

$$\Delta t' = \Delta t_L''\frac{1}{\gamma_E}, \tag{24}$$

where $\Delta t'$ and $\Delta t_L''$ are respectively the proper time interval for Andrew (as it is measured with a single clock, carried by Andrew) and the Lorentzian time interval measured by different clocks in Bob's frames in processes 2) and 4). Finally, for the turn around period 3), Bob is at a lower gravitational potential than Andrew, so that Bob's clock runs slower according to (22), with d_{AB}=d, the distance between Andrew and the distant star in Andrew's frame. According to the standard argument, Bob sees Andrew moving freely upwards in a gravitational field until he stops, when he is at a distance d and then begins to fall down [27]. Denoting by $''$ the short duration of the acceleration leading to the velocity reversal in Bob's frame, and noting that the proper acceleration is given by $g = 2v/\Delta t_3''$, we have

$$\Delta t_3' = \Delta t_3''\left(1 + \frac{gd}{c^2}\right) = \frac{2v}{g} + \frac{2dv}{c^2}. \tag{25}$$

Therefore, in the limit of short acceleration periods ($\Delta t_3'' \to 0$, $g \to \infty$),

$$\Delta t_3' = \frac{2dv}{c^2}. \tag{26}$$

Grøn [27] considers that the time interval $\Delta t_L''$ in (24) corresponds to the actual ageing of Bob during processes 2) and 4), consistently with the assertion from the standard interpretation of special relativity that "for the entire outward and return parts of the trip, Bob does observe Andrews clock running slow." Hence, it is claimed that Bob should calculate the time that passed for Andrew as

$$t_A' = \Delta t' + \Delta t_3' = t_B''\frac{1}{\gamma_E} + \Delta t_3'. \tag{27}$$

This equation is the same as (21) and the additional factor (26) is precisely the same as the factor (20). Thus, the correct final result (17) directly follows and has now been obtained in an alternative way.

The standard interpretation of general relativity claims that while for Bob a very short period passes during the turn around period, during this time he sees Andrew ageing significantly because he is subject to an intense acceleration. Grøn [27] considers that special relativity leads to a paradox, as he assigns to the Lorentizan time interval in (24) the physical meaning of proper time. As such, he does not find a justification to the inclusion of the additional factor (26) within the context of special relativity and advocates a "general relativity solution." He then presents two possible "solutions" to the question, an "elementary solution" and another "using relativistic Lagrangian dynamics." He acknowledges the difficulty of the former, as "it involves an assumption that is not obvious when the travelling twin stipulates the distance of his brother." Although this is incorrect, it exposes the difficulties with the speech surrounding the standard interpretation of special relativity. A deeper analysis of this question will be given separately in a future publication.

Equation (27) contains a correct final result. But its customary interpretation is exceedingly misleading. For a start, it confuses proper times and Lorentzian clock time readings in expression (24). Then it misinterprets as well equation (22). In reality, the general relativity approach merely hides a procedure to compensate for the offset factors (3) associated with the Lorentzian clocks, as explicitly proven in the next section. Therefore, it does not bring any new physics and the obtained effect during the turn-around period described by (26) is *not* related to differences in proper times.

To finish this section, let us still note that for the discussion of the paradox the acceleration in the turn-around point can be eliminated in the same way as the accelerations in processes 1) and 5): a third twin can cross Bob at the turn around point, take note of Bob's age when they cross, and then compare his own ageing since the meeting with Bob until he meets Andrew. He will age exactly the same as Bob in the limit Bob makes an instantaneous change in reference frame. This reinforces the indication that it is not the acceleration itself which is important, only the fact that there is one more inertial frame involved, with the corresponding modification of the clock rhythms.

6. General Solution

In this section we show that the broader view of special relativity proposed here demystifies the standard interpretations of the paradox both within the frameworks of special and of general relativity. It unveils, without ambiguity, that the additional factor (20) or (26) required to solve the twin paradox in either approach does *not* correspond to any physical effect on the ageing of either of the twins. It is simply a correction of the offset factors (3) resulting from Einstein's procedure of "synchronization," which are different for the two inertial frames associated with Bob in the roundtrip. This additional factor has no deeper meaning than to say that a person who suddenly advances his own clock by five minutes must then subtract five minutes to the time reading of his clock in order to know for how long he has been waiting. Let us then perform the calculations relevant to the twin paradox using the general expressions from section (2.2) and from the point of view of each of the twins. For simplicity, we again assume that Bob travels along the x' axis with *Einstein speed* of absolute value v_E in relation to Andrew, both in the outward and on the return trips.

The easiest calculation is the simple integration of equation (15), corresponding to using the standard expression "from Andrew's point of view." Keeping the same notation as in section 2.2, the actual ageing of the twins, given by the lapses in their proper times, is just

$$\tau_B'' = \oint d\tau'' = \oint \frac{dt_L'}{\gamma_E} = \frac{1}{\gamma_E}\oint dt_L' = \frac{\tau_A'}{\gamma_E}, \tag{28}$$

where, notwithstanding (13), $\oint dt_L' = \Delta t_L' = \tau_A'$. Indeed, although dt_L' is measured with different clocks from S′, $\Delta t_L'$ in the roundtrip is measured with a single clock.

A second calculation can be made using again the standard relation, but "from Bob's point of view." This corresponds to the use of equation (15), interchanging the primed and double primed variables. In what follows the subscripts "+" and "−" refer to the outward and return trips, respectively. Note that in the onward and the return trips Bob uses different inertial frames. We have, successively,

$$\begin{aligned}
\tau_A' &= \oint \frac{dt_L''}{\gamma_E} \\
&= \int_+ \frac{dt_L''}{\gamma_E} + \int_- \frac{dt_L''}{\gamma_E} \\
&= \frac{1}{\gamma_E}\left(\Delta t_{L,+}'' + \Delta t_{L,-}''\right) \\
&= \frac{1}{\gamma_E}\left(\frac{\Delta x_+''}{v_E''} + \frac{\Delta x_-''}{-v_E''}\right) \\
&= \frac{1}{\gamma_E}\left(\frac{\Delta x_+''}{v_{w+}''}\frac{v_{w+}''}{v_E''} + \frac{\Delta x_-''}{v_{w-}''}\frac{v_{w-}''}{-v_E''}\right) \\
&= \frac{1}{\gamma_E}\left(\tau_+''\frac{v_{w+}''}{v_E''} + \tau_-''\frac{v_{w-}''}{-v_E''}\right)
\end{aligned} \tag{29}$$

where we have used (6). At this point we can show that Bob's proper time lapses are the same on the onward and the return trips, $\tau_+'' = \tau_-''$. Take note that Bob's ageing rhythm is different in each of the trips, except in the peculiar case in which Andrew is in the rest system. However, although the Einstein speed is the same the (synchronized) speed is not, which compensates for the difference in the rhythms. From (15),

$$\tau_+'' = \frac{\Delta t_{L,+}'}{\gamma_E} = \frac{1}{\gamma_E}\frac{\Delta x'}{w_E'} \tag{30}$$

and

$$\tau_-'' = \frac{\Delta t_{L,-}'}{\gamma_E} = \frac{1}{\gamma_E}\frac{\Delta x'}{w_E'} = \tau_+''. \tag{31}$$

Since, $\tau_B'' = \tau_+'' + \tau_-''$, we have

$$\tau_+'' = \tau_-'' = \frac{1}{2}\tau_B''. \tag{32}$$

We now need to eliminate the speeds of Andrew in Bob's frame, v_{w+}'' and v_{w-}'', from (29). To do so, we note that, solving (8) for w,

$$w = \frac{w_E' + v}{1 + \frac{v w_E'}{c^2}}. \tag{33}$$

In turn, exchanging the roles of v and w and the primed and double primed variables in (7),

$$v_E'' = \frac{v_w''}{1 - \frac{w v_w''}{c^2}}. \tag{34}$$

Inverting for v_w'' and noting that $v_E'' = -w_E'$, we get

$$v_w'' = \frac{v_E''}{1 + \frac{w v_E''}{c^2}} = -\frac{w_E'}{1 - \frac{w w_E'}{c^2}}. \tag{35}$$

Substituting (33) in this last expression,

$$v_w'' = -w_E'\frac{1 + \frac{v w_E'}{c^2}}{1 - \frac{w_E'^2}{c^2}} = v_E''\frac{1 + \frac{v w_E'}{c^2}}{1 - \frac{w_E'^2}{c^2}}. \tag{36}$$

Hence,

$$\frac{v_{w+}''}{v_E''} = \frac{1 + \frac{v w_E'}{c^2}}{1 - \frac{w_E'^2}{c^2}}. \tag{37}$$

Analogously, for the return trip we simply have to replace v_E'' by $-v_E''$, so that

$$\frac{v_{w-}''}{-v_E''} = \frac{1 - \frac{v w_E'}{c^2}}{1 - \frac{w_E'^2}{c^2}}. \tag{38}$$

Finally, substituting (32), (37) and (38) in (29),

$$\begin{aligned}
\tau_A' &= \frac{1}{\gamma_E}\frac{\tau_B''}{2}\left(\frac{1 + \frac{v w_E'}{c^2}}{1 - \frac{w_E'^2}{c^2}} + \frac{1 - \frac{v w_E'}{c^2}}{1 - \frac{w_E'^2}{c^2}}\right) \\
&= \gamma_E \tau_B'',
\end{aligned} \tag{39}$$

so that we get once more the relation between the proper times of the twins (28). This somewhat dry but illuminating exercise proofs that the symmetric relations

$$\tau_B'' = \oint (1/\gamma_E)\, dt_L' \tag{40}$$

and

$$\tau_A' = \oint (1/\gamma_E)\, dt_L'' \tag{41}$$

are *both* valid and lead directly to the relation of rhythms,

$$\tau_B'' = \tau_A' / \gamma_E, \tag{42}$$

with no need to include any additional factor. It further demonstrates that when it is said that "for the entire outward and return parts of the trip, *B* does observe *A*'s clock running slow"[39] in the traditional symmetric account of time dilation, the expression "observing the clock running slow" is simply the comparison of the Lorentzian times of different clocks in *B*'s frame *and has nothing to do with true clock rhythms.* If it would correspond to the true ageing of the twins the paradox would remain, as it could not be simultaneously true that $\tau_B'' = \gamma\tau_A'$ and $\tau_A' = \gamma\tau_B''$. We would then have to add the factors (20) or (26) to Andrew's proper time during Bob's turn-around period when the calculation is made from Bob's point of view, in order to compensate for the supposedly slower ageing of Andrew in the remaining of the trip. Incidentally, one should note that in the standard interpretation of relativity Bob needs to invoke an extremely fast ageing of Andrew, who can be light years apart, taking place exclusively during his own nearly instantaneous turn-around period. This additional factor has a much simpler nature, namely the offset factor (3), as we show next.

For simplicity, and without loss of generality, let us assume that Andrew remains in the rest system, *S*. Bob travels to the distant star and returns, with speed of absolute value *v* both in the onward and on the return trips. In this case, the absolute value of Bob's Einstein speed in *S* coincides with the synchronized speed, $v_E = v$ and when Bob reaches the distant star the proper time elapsed for each of the twins is half of the total time elapsed in the complete round-trip. The ageing of Andrew until Bob arrives to the distant star can be calculated, e.g., as in (29)

$$d\tau_A = \frac{dt'_{L,+}}{\gamma_v}. \tag{43}$$

Here, $dt'_{L,+}$ refers to Lorentzian times displayed by different clocks in Bob's frame, *S'*, co-punctual with Andrew, along the onward trip. Integrating along the onward trip,

$$\frac{1}{2}\tau_A = \frac{\Delta t'_{L,+}}{\gamma_v}. \tag{44}$$

Now, instead of proceeding as before, we can relate the Lorentzian time with the synchronized time and the proper time by (3) and (13), from where

$$\Delta t'_{L,+} = \frac{1}{2}\tau'_B + \frac{v}{c^2}d'. \tag{45}$$

Note that *x'* in (3) is negative and $d' = |x'|$ is the position of Andrew in Bob's frame when he arrives to the star, i.e., $d'/\gamma_v = d$. Substituting in (44),

$$\frac{1}{2}\tau_A = \left(\frac{1}{2}\tau'_B + \frac{v}{c^2}d'\right)\frac{1}{\gamma_v}$$

$$= \frac{1}{2}\frac{\tau'_B}{\gamma_v} + \frac{v}{c^2}d. \tag{46}$$

Finally, noting that $d = v\tau_A/2$ have, successively,

$$\frac{1}{2}\tau_A = \frac{1}{2}\frac{\tau'_B}{\gamma_v} + \frac{1}{2}\tau_A\frac{v^2}{c^2}$$

$$\tau_A\left(1 - \frac{v^2}{c^2}\right) = \frac{\tau'_B}{\gamma_v}$$

$$\tau_A = \gamma_v\tau'_B, \tag{47}$$

which is again the true relation between the ageing of the twins.

The major and very consequential remark is to note that the second term in the r.h.s. of (46) is the additional factor (20) or (26). *This expression is valid all along the one-way outward trip*, in which *Bob remains in the same inertial frame.* It relates the proper times of both twins along the onward trip. It comes without saying that the additional factor has *nothing* to do neither with any strangeness at the turn-around point related to the "simultaneity argument" nor with acceleration, general relativity and the equivalence principle.

To summarize, for the particular case just considered the following relations are all valid all along the one-way onward trip,

$$d\tau_A = \frac{dt'_{L,+}}{\gamma_v}$$

$$= \frac{d\tau'_B}{\gamma_v} + \frac{v}{c^2}dx$$

$$= \gamma_v d\tau'_B, \tag{48}$$

and express the fact that Bob is ageing slower in this case. Understanding these three equalities clarifies the language and completely solves the twin paradox.

A straightforward analysis made with proper times and the IST transformation removes all difficulties surrounding the paradox at the onset. In particular, it is no longer said that each twin sees the other ageing slower during the length of the one-way trips. Instead, both twins agree that in each part of the trip one of them is ageing slower while the other is ageing faster (or are both ageing at the same rhythm). The simple addition of the proper times of the onward and back trips then gives directly the final result for both twins. There is no need to correct for any strangeness on the turn around period nor for any supposed effect of acceleration and equivalent gravitational fields. Specifically, there is no need to add any factor to Andrew's age due to a change in inertial frame of Bob, whatever point of view is considered.

As a final remark, the general calculations can be made as well using (16) as a starting point, both as it is written and as well interchanging the roles of *v* and *w* and the primed and double primed variables, in an additional consistency check of the present formalism.

7. Conclusion

In previous papers we have proposed a general formulation of special relativity, where the postulates are formulated in a weaker form than in the traditional presentation, while keeping fully compatible with all experimental evidence [1,2,3,4,5,6]. The theory assumes the existence of (at least) one reference frame where the one-way speed of light in vacuum is isotropic and equal to *c*, denoted as the "rest system." It was shown that the theory is undetermined, unless the one-way speed of light in one reference frame is measured [4]. The somewhat evident but very important difference between "time readings of Lorentzian clocks" and proper times or clock rhythms is thoroughly discussed in I [6]. It is noted that although the *description* of time dilation made with Lorentzian clocks is symmetrical for two inertial observers in relative motion, as it is well-known from the standard interpretation of special relativity, the reciprocal relation does not relate the clock *rhythms* [6]. Actually, the time dilation relation between the clock rhythms of two inertial observers in relative motion is *not* symmetric. During each one-way trip one of the observers is actually ageing slower than the other, or it may even happen that both are ageing at the same rhythm. Furthermore, without reference to the rest system it is impossible to know which of them is actually ageing slower, in another way of stating the indeterminacy of special relativity [6].

Within this context, in this work we have discussed the twin paradox in detail, as an illustration of the power and simplicity of the general formulation of special relativity formerly presented. In a round-trip such as the one in the classical configuration of the twin paradox, it is the returning twin who is younger when both twin meets, despite the impossibility of knowing which of the twins is younger at each phase of the trip. The result is due to the cumulative effect of the clock rhythms along the complete journey.

The total ageing of each twin is calculated directly from the *sum of their proper times* on the onward and the return trips. Contrary to what happens in the standard interpretation of special relativity and of general relativity, there is no need to consider any additional ageing factor of the resting twin as seen by the moving twin to account for the change in reference frame or the acceleration (sections 4 and 5). As a matter of fact, the factor invoked by these standard interpretations was deduced from the offset between Lorentzian and synchronized clocks (section 6). It was thus demonstrated that this factor has nothing to do with any modification with the clock rhythms and with the ageing of the twins during the turn-around period. It is merely a correction to a peculiar way of giving the initial adjustment (a so-called "synchronization") to the clocks. As a consequence, it becomes clear that acceleration does not play any role in the twin paradox other than telling which of the twins is returning back.

Finally, we would like to underline the following. The standard interpretation of special relativity pretends to assign a physical meaning of real ageing to assertions like "during the onward trip Bob sees Andrew ageing slower, Andrew himself also sees Bob ageing slower, but the change in inertial frames corrects this symmetry, as a result of the relativity of simultaneity, and makes it in the end that Bob is younger when the twins meet." The standard interpretation of general relativity pretends to assign a physical meaning to sentences like "during the turn around period Bob sees Andrew ageing very quickly because he sees him under the effect of a gravitational field at a higher gravitational potential." These are erroneous interpretations of correct mathematical results. One message we want to convey regarding the twin paradox is that such discourse is no longer tolerable and should become "unspeakable:" it was proven that "seeing the other twin ageing slower" is meaningless in this context and corresponds to the symmetric description arising from the comparison of the *time readings* of Lorentzian clocks (4 and 5), whose roots lie in the indeterminacy of special relativity [4]. It does *not* correspond to the clock *rhythms* and to the ageing of the twins.

References

1. Guerra V, Abreu R (2005) The conceptualization of time and the constancy of the speed of light. European Journal of Physics 26: S117-S123.
2. Guerra V, Abreu R (2006) On the consistency between the assumption of a special system of reference and special relativity. Foundations of Physics 36: 1826-1845.
3. Guerra V, Abreu R (2007) Comment on 'From classical to modern ether-drift experiments: the narrow window for a preferred frame. Phys Lett A 361: 509-512.
4. Abreu R, Guerra V (2008) The principle of relativity and the indeterminacy of special relativity. Eur J Phys 29: 33-52.
5. Abreu R, Guerra V (2009) Special relativity as a simple geometry problem. Eur J Phys 30: 229-237.
6. Abreu R, Guerra V (2015) Speakable and unspeakable in special relativity: Time readings and clock rhythms. Elect J Theor Phys 12: 183-204.
7. Feynman RP. Leighton RB, Sands M (1979) The Feynman Lectures on Physics. vol. II, 17-5.
8. Grandou T, Rubin JL (2009) On the ingredients of the twin paradox. Int J Theor Phys 48: 101-114.
9. Langevin P (1911) The evolution of space and time. Scientia 10: 31-54.
10. Scott GD (1959) On solutions of the clock paradox. Am J Phys 27: 580-584.
11. Builder G(1959) Resolution of the clock paradox. Am J Phys 27: 656-658.
12. Romer RH (1959) Twin paradox in special relativity. Am J Phys 27: 131-135.
13. Dingle MBH (1963) Special theory of relativity. Nature 197: 1287.
14. Lass H (1963) Accelerating frames of reference and the clock paradox. Am J Phys 31: 274-276.
15. Brans CH, Stewart DR (1973) Unaccelerated-returning-twin paradox in flat space-time. Phys Rev D 8: 1662-1666.
16. Perrin R (1979) Twin paradox: a complete treatment from the point of view of each twin. Am J Phys 47: 317-319.
17. Unruh WG (1981) Parallax distance. time. and the twin paradox. Am J Phys 49: 589-592.
18. Sastry GP (1987) Is length contraction really paradoxycal? Am J Phys 55: 943-946.
19. Boughn SP (1989) The case of identically accelerated twins. Am J Phys 57: 791-793.
20. Debs TA, Redhead MLG (1996) The twin paradox and the conventionality of simultaneity. Am J Phys 64: 384-392.
21. Schön M (1995) Twin paradox without one-way velocity assumptions. Found Phys 28: 185-204.
22. Soni VS (2002) A simple solution of the twin paradox also shows anomalous behaviour of rigidly connected distant clocks. Eur J Phys 23: 225-231.
23. Pesic P (2003) Einstein and the twin paradox. Eur J Phys 24: 585-590.
24. Iorio L (2005) An analytical treatment of the clock paradox in the framework of the special and general theories of relativity. Found Phys Lett 18: 1-19.
25. Minguzzi E (2005) Differential aging from acceleration: an explicit formula. Am J Phys 73: 876-880.
26. Selleri F (2006) Superluminal signals and the resolution of the causal paradox. Found Phys 36: 443-463.
27. Grøn O (2006) The twin paradox in the theory of relativity. Eur J Phys 27: 885-889.
28. Abramowicz MA, Bajtlik S, Miak WZ (2007) Twin paradox on the photon sphere. Phys Rev A 75: 044101-1-2.
29. Styer DF (2007) How do two moving clocks fall out of sync? A tale of trucks, threads, and twins. Am J Phys 75: 805-814.
30. Müller T, King A, Adis D (2008) A trip to the end of the universe and the twin paradox. Am J Phys 76: 360-373.
31. Grøn Ø (2013) The twin paradox and the principle of relativity. Phys Scripta 87: 035004.
32. Selleri F (1996) Noninvariant one-way velocity of light. Foundations of Physics 26: 641-664.
33. Selleri F (2003) Sagnac effect: end of the mystery, in Fundamental Theories of Physics. Kluwer Academic Publishers.
34. Selleri F (2005) The inertial transformations and the relativity principle. Foundations of Physics Letters. 18: 325-339.
35. Abreu R (2002) The physical meaning of synchronization and simultaneity in special relativity arXiv: physics/0212020.
36. Homem G (2003) Physics in a synchronized space-time.Master's thesis,Higher Technical Institute, Technical University of Lisbon.
37. Abreu R, Guerra Extra]muros[Lisboa, 1st ed., (2005) Relativity - Einstein's Lost Frame.
38. Tangherlini FR (1961) On energy-momentum tensor of gravitational field. Nuovo Cimento Supply 20: 351.
39. Morin D (2008) Introduction to Classical Mechanics: With Problems and Solutions. Cambridge University Press.
40. Einstein A (1911) The relativity theory. Natural researching society 56: 1-14.
41. Einstein A (1918) Dialog about objections against the theory of relativity. Natural Sciences 48: 697-702.
42. Tolman RC (1987) Relativity, Thermodynamics and Cosmology. Dover Publications.

Mathematical Theory of Space-Time

Dr. Andreas Bacher

Faculty of Mathematics and Natural Sciences, University of Cologne, Albertus-Magnus-Platz, Germany

Abstract

We hypothesize that the Universe contains observable subluminal matter (tardyons and locatily) and unobservable superluminal matter (tachyons and nonlocality). By using space-time ring we establish mathematical theory of space-time with subluminal and superluminal coexistence. The rotating motions of tardyons and tachyons produce centrifugal and centripetal (gravity) forces, respectively. This paper deduces a new gravitational formula and Newtonian gravitational formula from the tardyonic and tachyonic coexistence principle. Through the morphism we first show that tardyons and tachyons are interchangeable, but that tachyons are unobservable . We then convert the tachyonic mass into tardyonic mass We obtain gravitational coefficient $\eta = 6.9 \times 10^{-10}$.Using it we establish the expansion theory of the universe and suggest the new universe model

Keywords: New gravitational formula; Expansion theory of universe; New universe model

Mathematical Theory of Space-Time

In the Universe there are two matters: (1) observable subluminal matter called tardyon (locality) and (2) unobservable superluminal matter called tachyon (non-locality) which coexist in motion. Tachyon can be converted into tardyon, and vice versa. Tardyonic rotating motion produces the centrifugal force, but tachyonic rotating motion produces the centripetal force, that is gravity. Using tardyonic and tachyonic coexistence principle we deduce the new gravitational formula. For establishing the mathematical theory of space-time we first define space-time ring [1-3].

$$Z = \begin{pmatrix} ct & x \\ x & ct \end{pmatrix} = ct + jx \qquad (1)$$

where x and t are the tardyonic space and time coordinates, c is light velocity in vacuum,

$$j = \begin{pmatrix} 0 & 1 \\ 1 & 0 \end{pmatrix}$$

(1) can be written as Euler form

$$Z=ct_0e^{j\theta}=ct_0(ch\theta+jsh\theta) \qquad (2)$$

From (1) and (2) we have

$$ct=ct_0\, ch\theta,\ x=ct_0(ch\theta+jsh\theta) \qquad (3)$$

$$ct_0 = \sqrt{(ct)^2 = x^2} \qquad (4)$$

From (3) we have

$$\theta = th^{-1}\frac{x}{ct} = th^{-1}\frac{u}{c} \qquad (5)$$

where $c \geq u$ is the tardyonic velocity.

Using the morphism j: $z \rightarrow jz$, we have

$$jz = \bar{x} + jc\bar{t} = \bar{x}_0 e^{j\bar{\theta}} = \bar{x}_0(ch\bar{\theta} + jsh\bar{\theta}) \qquad (6)$$

Where $\bar{x}$ and $\bar{t}$ are the tachyonic space and time coordinates, $\bar{x}_0$ is tachyonic invariance, $\bar{\theta}$ tachyonic hyperbolical angle.

From (6) we have

$$\bar{x} = \bar{x}_0 ch\bar{\theta}, c\bar{t} = \bar{x}_0 sh\bar{\theta} \qquad (7)$$

$$\bar{x}_0 = \sqrt{(\bar{x})^2 - (c\bar{t})^2} \qquad (8)$$

From (7) we have

$$\bar{\theta} = th^{-1}\frac{c\bar{t}}{\bar{x}} = th^{-1}\frac{c}{u} \qquad (9)$$

where $\bar{u} \geq c$ is the tachyonic velocity

Figure 1 shows the formulas (1)-(9). j: $z \rightarrow jz$ is that tardyon can be converted into tachyon, but j: $jz \rightarrow z$ is that tachyon can be converted into tardyon. u=0→u=c is the positive acceleration, but u=∞ →u=c is the negative acceleration, which coexist. At the ct-axis u=0 and x=0 we define the tardyonic test time t. At the x – axis we define the tachyonic rest space as

$$\bar{X}_0 = \lim_{\substack{\bar{u}\to\infty \\ t\to 0}} \bar{u}\bar{t} = cons\tan t \qquad (10)$$

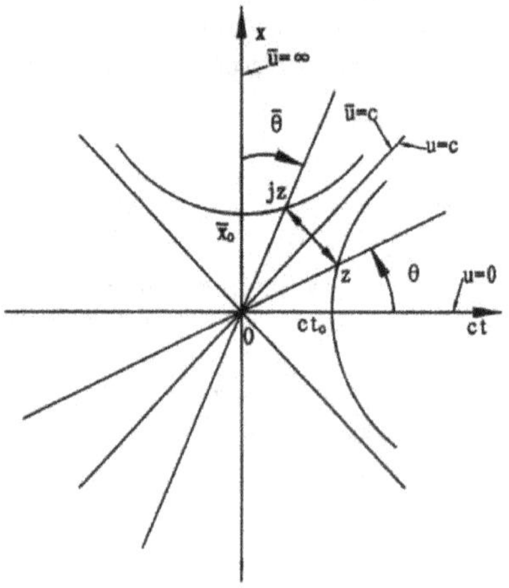

Figure 1: Mathematical theory of space-time.

***Corresponding author:** Dr. Andreas Bacher, Faculty of Mathematics and Natural Sciences, University of Cologne, Albertus-Magnus-Platz, Germany
E-mail: andreas.bacher1985@gmail.com

Since at rest the tachyonic rest time t=0 and u=∞, we prove that tachyon is unobservable. Figure 1 and (10) are mathematical theory of space-time, which are the foundations of physics andcosmology. Using it we prove that dark matter, dark energy, gravitational waves, black holes, quantum entanglement and quantum computers do not exist. From Figure 1 we deduce new gravitational formula.

New Gravitational Formula $\overline{F} = -\frac{mc^2}{R}$

Assume θ= $\overline{\theta}$, from (5) and (9) we get the tardyonic and tachyonic coexistence principle [1-6]

$$u\overline{u} = c^2 \qquad (11)$$

Using the geometrical method we deduce the new gravitational formula.

Figure 2 shows that the rotation ω of body A emits tachyon mass $\overline{m}$, which forms the tachyon and gravitation field and gives the body B revolutions u and $\overline{u}$.

From Figure 2 it follows

$$\frac{u\Delta t}{R} = \frac{\Delta\overline{u}}{u} \qquad (12)$$

From (12) it follows the tardyon centripetal acceleration on the body B

$$\frac{d\overline{u}}{dt} = \lim_{\substack{\Delta u \to 0 \\ \Delta t \to 0}} \frac{\Delta u}{\Delta t} = \frac{u\overline{u}}{R} = \frac{c^2}{R} \qquad (13)$$

From Figure 2. it follows

$$\frac{u\Delta t}{R} = \frac{\Delta\overline{u}}{u} \qquad (14)$$

From (14) and (11) it follows the tachyon centrifugal acceleration on the body B [4,7]

$$\frac{d\overline{u}}{dt} = \lim_{\substack{\Delta u \to 0 \\ \Delta t \to 0}} \frac{\Delta u}{\Delta t} = \frac{u\overline{u}}{R} = \frac{c^2}{R} \qquad (15)$$

From (13) it follows the tardyon centrifugal force on body B [2-4],

$$F = \frac{M_B u^2}{R} \qquad (16)$$

where M_B is body B mass.

$$\overline{F} = \frac{mc^2}{R} \qquad (17)$$

From (15) it follows the tachyon centripetal force on body B, that is gravity [8-10],

Where m is the gravitation mass converted into by tachyon mass $\overline{m}$, which is unobservable, but m is observable.

(17) Is the new gravitational formula. This simple thought made a deep impression on me. It impelled me to establish the new gravitational theory explained in Figure 3.

$$F + \overline{F} = 0 \qquad (18)$$

From (16), (17) and (18) it follows

$$\frac{m}{M_B} = \frac{u^2}{c^2} \qquad (19)$$

Body B increases mass m and centrifugal force is greater than gravitation force, then body B expands outward. Dark matter which causes cosmic attraction is wrong. From (17) it follows Newtonian gravitation formula. The m is proportional to body A mass M_A, in (19) m is proportional to M_B, is inversely proportional to the distance R between body A and body B. It follows

$$m = k\frac{M_A M_B}{R} \qquad (20)$$

where k is constant

Substituting (20) into (17) it follows the Newtonian gravitational formula [10-12]

$$\overline{F} = -G\frac{M_A M_B}{R^2} \qquad (21)$$

where $G=kc^2=6.673 \times 10^{-8}\ cm^3/\ g{\cdot}sec^2$ is gravitation constant.

Now we study the freely falling body. Tachyonic mass $\overline{m}$ can be converted into tardyonic mass m, which acts on the freely falling body and produces the gravitational force

$$\overline{F} = -\frac{mc^2}{R} \qquad (22)$$

where R is the Earth radius.

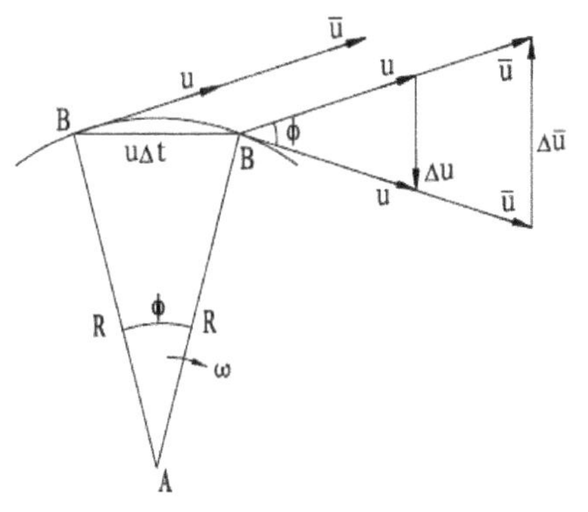

Figure 2: On body B the $\frac{du}{dt}$ and $\frac{d\overline{u}}{dt}$ coexistence.

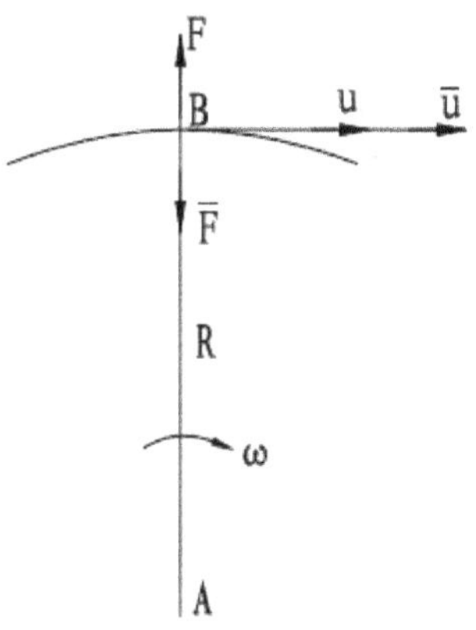

Figure 3: On body B the F and $\overline{F}$ coexistence.

We have the equation of motion

$$\frac{mc^2}{R} = mg \quad (23)$$

where g is gravitational acceleration, M is mass of freely falling body.

From (23) it follows the gravitational coefficient

$$\eta = \frac{m}{M} = \frac{Rg}{c^2} = 6.9 \times 10^{-10} \quad (24)$$

Eötvös experiment $\eta \sim 5.10^{-9}$ and Dicke experiment $\eta \sim 10^{-11}$. Since the gravitational mass

m can be transformed into the rest mass in freely falling body, we prove that the freely falling

bodies fall with the same acceleration.

The Expansion Theory of the Universe

Using new gravitational formula we study the expansion theory of the Universe [13]. Figure 4 shows an expansion model of the Universe. The rotation ω_1 of body A emits tachyonic flow, which forms the tachyonic field. Tachyonic mass $\overline{m}$ acts on body B, which produces its rotation ω_2, revolution u and gravitational force

$$\overline{F}_1 = -\frac{mc^2}{R} \quad (25)$$

where R denotes the distance between body A and body B, m is gravitational mass converted into by tachyonic mass $\overline{m}$ which is unobservable but m is observable. The rotation of the body B around body A produces the centrifugal force

$$\overline{F} = \frac{M_B u^2}{R^2} \quad (26)$$

where M_B is the inertial mass of body B, u is the orbital velocity of body B.

At the O_2 point we assume

$$F_1 + \overline{F}_1 = 0 \quad (27)$$

From (27) it follows that the coexistence of the gravitational force and centrifugal force.

From (25)-(27) it follows the gravitational coefficient

$$\eta = \frac{m}{M_B} = \left(\frac{u}{c}\right)^2 \quad (28)$$

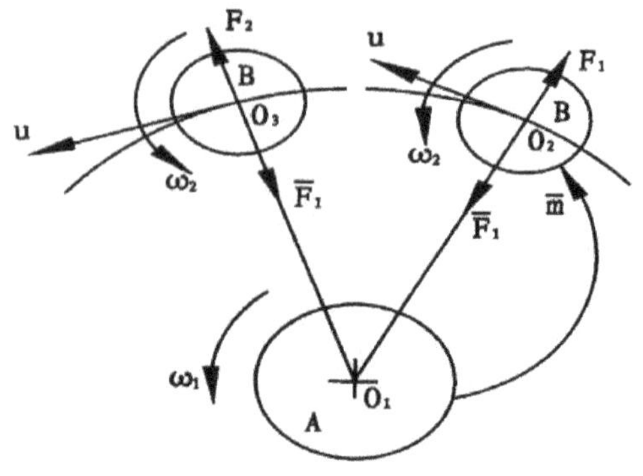

Figure 4: An expansion model of the Universe.

At the O_3 point the tachyonic mass m can be converted into the rest mass m in body B, it follows

$$F_2 = \frac{M_B u^2}{R^2} + \frac{mu^2}{R} \quad (29)$$

Since $F + \overline{F}_1 > 0$, centrifugal force F_2 is greater than gravitational force F_1, then the body B expands outwards and its mass increases. This is an expansion mechanism of the Universe. From (26,27,29) we have

$$F_2 + \overline{F}_1 = \frac{mu^2}{R} = M_B g_e \quad (30)$$

From (30) we obtain the expansion acceleration

$$g_e = mu^2 / M_B R \quad (31)$$

Substituting (28) in (31) we obtain

$$g_e = \frac{u^2}{c^2 R} \quad (32)$$

If body A is the Earth, then body B is the Moon; if body A is the Sun, then body B is the Earth; .... It can explain our accelerating universe. In the universe there are no dark matter and no dark energy. This simple thought made a deep impression on me. It impelled me to establish an expansion theory of the universe. Dark energy responsible cosmic repulsion is wrong. If the body A is the Sun and body B is the planet. We calculate the gravitational coefficients η as shown in Table 1. The gravitational field of the solar system is the origin of the planet mass. From it the planet acquire mass.

The New Universe Model

From the tachyonic theory we suggest the new universe model. The universe has no beginning and no end. The universe is infinite, but it has a center consisting of the tachyonic matter which is strong gravitational field (SGF), which governs motion of the whole universe. Therefore the whole universe is stable and harmonious. ... .In the sun there is a center consisting of the tachyonic matter with SGF, which governs motion of the solar system. It is stable and harmonious. In the earth there is a center consisting of the tachyonic matter with SGF, which governs motion of the earth and the moon. It is stable and harmonious. In the moon there is a center consisting of the tachyonic matter with SGF, which governs motion of the moon. It is stable and harmonious. In atomic nucleus there is a center consisting of the tachyonic matter with SGF, which governs motion of the nucleus. Therefore atomic nuclei are stable and harmonious. The tachyonic theory governs the amazing harmony of the whole universe from the smallest to the largest scales. New gravitational formula changes all that. In the Universe there are no dark matter, no dark energy and no gravitational waves. Multiverse, inflation and primordial gravitational waves do not exist [14].

Conclusion

We deduce tardyonic and tachyonic coexistence principle. Using it

Planet	*u* (km/sec)	η(10−10)
Venus	35.03	136.5
Earth	29.79	98.7
Mars	24.13	64.8
Jupiter	13.06	19
Saturn	9.64	10.3
Uranus	6.81	5.2
Neptune	5.43	3.3
Pluto	4.74	2.5

Table 1: Values of the gravitational coefficients η.

we deduce the centrifugal formula and new gravitational formula. We establish the expansion theory of the universe without dark energy and suggest the new universe model which is amazing harmony. The new gravitational formula is foundations of particle physics and cosmology. We prove that in the universe no dark matter, no dark energy, no gravitational waves and no quantum gravity. Where did we come from? Where are we going? What makes up the universe? These questions have occupied mankind for thousands of years. Over the course of history, our view of the world has been changed. Theologians and philosophers, physicists and astronomers have given us very different answers. Where did we come from? We answer this questions this way $m \to m$, Tachyons → tardyons, that is tachyons can be converted into the electrons and positrons which are the basic building-blocks of the elementary particles. The tachyons are the origin of mass. Where are we going? We answer this question this way $m \to m$, that is the tardyons produce tachyons. The tardyons and tachyons make up the Universe.

Note:

In 2017, Andreas Bacher found a gravitational formula [2]: $\bar{F} = -\bar{m}c^2 / R$, where is the tachyonic mass. In 2017 Andreas Bacher studied and found, $\bar{F} = -mc^2 / R$ where is gravitational mass converted into by tachyonic mass $\bar{m}$ [10]. Newtonian gravity formula is based on empirical evidence. He did not explain what is gravity? how it works? In general theory of relativity there is no gravitational formula [11]. In modified gravity and modified Einstein gravity there are no gravitational formula [12]. There cannot be really gravity theory without gravity formula. Tachyonis instability appears as a field with a negative mass squared which is wrong [12]. The tachyons are stable. It has no rest time and no rest mass [1].

Acknowledgement

The author thanks professor Walter Lewin for his mails.

(1) From: "Walter H.G. Lewin"<lewin@space.mit.edu> Date:Sun,10 Jun 2012 22:22:21-0400(EDT) Subject:Re: Fwd. To:123jiangchunxuan@gmail.com; lewin@mit.edu Cc:luc@vanocken.be Publish this in a refereed journal and once it is accepted buy yourself a first class ticket to Stockholm to pick up Nobel prize for physics.

From: "Walter H.G. Lewin"<lewin@space.mit edu> Date: Sun,17 Jun 2012 06:19:37-0400(EDT) Subject: Re To: 123jiangchunxuan@gmail.com Cc:luc@vanocken.be Dear Jiang Thank for your email. I suggest you submit your theory to a refereed journal.If it is accepted,then buy yourself a plane ticket to Stockholm to pick up a Nobel prize.

Author Contribution

Andreas Bacher conceived and developed the fundamentals of the idea. Andreas Bacher thanks Jiang for useful discussions.

Prof. Walter Lewin explains Newton's law of gravitation in MIT course 8.01[1]

References

1. Jiang C (1975) A theory of morphisms between the tardyon and tachyon, physics (Chinese) 4: 119-125.
2. Bacher A (2017) Mathematical Theory of Space-Time. J Phys Math 8: 210.
3. Jiang C (1982) An approach on the nature of attractive force. Potential science (Chinese) 4: 19-20.
4. Jiang C (2001) A unified theory of the gravitational and strong interactions. Hadronic J 24: 629-638.
5. Ade PA, Aikin RW, Barkats D, Benton SJ, Bischoff CA, et al. (2014) Detection of B-mode polarization at degree angular scales By BICEP2. Phys Rev Lett 112: 241101.
6. Dodelson S (2014) How much can we learn about the physics of inflation? Phys Rev Lett 112: 191301.
7. Caligiuri J, Kosowsky A (2014) Inflationary tensor perturbations after BICEP2. Phys Rev Lett 112: 191302
8. Jiang C (1979) A simple approach to the computation the total number of hadronic constituents in Santilli model. Hadronic J 3: 256-292.
9. Jiang C (1988) A mathematical model for particle classification. Acta Math Scien 8: 133-144.
10. Jiang C (2005) New gravitational formula and expansion theory of the universe. In: new studies of space-time theory (Chinese). Dizi press pp: 254-259
11. Einstein A (1956) The meaning of relativity, F mc2 = – R (5th edn) Princeton university press, Princeton.
12. Joyce A, Jain B, Khoury J, Trodden M (2015) Beyond the cosmological standard model. Physics Reports 568: 1-98.
13. Jiang C (2013) New gravitational formula , In: unsolved problems in special and general relativity. Education publishing pp: 125-130.
14. Jiang C (2013) The expansion theory of the universe without dark energy. In: Unsolved problems in special and general relativity.Education publishing pp: 131-140.

Inconsistency in the Development of the Lorentz Factor in Section 3 of 'On the Electrodynamics of Moving Bodies'

Masahiko Makanae*

Representative Free Web College, Nishikasai, Edogawa-ku, Tokyo 134-0088, Japan

Abstract

In 1905, Albert Einstein published his paper 'On the Electrodynamics of Moving Bodies', which is referred as the special theory of relativity. In Section 3 of this work, he developed a factor for describing a moving system in terms of a reference stationary system known as the Lorentz factor. In the climax of the development of Lorentz factor in this section, Einstein provided the series of equations that contains the mutual terms *c*, *v*, *t*, and *x*'. Clearly, '*c*' corresponds to the velocity of light, '*v*' is the velocity of the moving system, and '*t*' is the time passage in the reference stationary system. However, regarding *x*', the meaning is not clear as a mutual term. Besides, the values of *x*', which should have identical values, differ for the various equations. Thus, there is an inconsistency as regarding the usage of *x*' in the development of the Lorentz factor in Section 3 of 'On the electrodynamics of moving bodies'. This issue should be discussed not only from the viewpoint of mathematics, but also from the viewpoint of formal logic, which is a practical tool for the study of any science.

Keywords: Galilean transportation; Lorentz factor; Contradiction; Einstein

Introduction

In 1905, Albert Einstein published his paper 'On the Electrodynamics of Moving Bodies' [1] (the German original title is 'Zur Elektrodynamik bewegter Körper' [2]), which is referred to as the special theory of relativity (STR). In Section 1 of that paper, Einstein defines the concept of 'time', and then describes a method of confirming the synchronization of two clocks placed at two points. In Section 2 [1], based on '*the principle of relativity*' and '*the principle of the constancy of the velocity of light*', Einstein considers the relationships between a moving system and a reference stationary system with the concepts of 'length' and 'time', by using the universal equation 'time=distance/velocity'. Both systems have a relationship in the situation that an observer at rest in the reference stationary system observes an event in the moving system, which moves under parallel translation with uniform velocity with respect to the reference stationary system. Through the consideration in Section 2 [1], Einstein implies that an event in a moving system viewed within that moving system differs from the same event viewed from a reference stationary system. This perspective became the fundamental basis of the STR.

In Section 3 of the paper [1], Einstein develops the expression $\beta = \frac{1}{\sqrt{1-\left(\frac{v}{c}\right)^2}}$ which was later termed the 'Lorentz factor', as a core theory of the STR. This expression denotes the ratio of the values of 'time' or 'length' between the moving system and the reference stationary system when both systems are in the situation described above. The expression $\sqrt{1-(v/c)^2}$ in the Lorentz factor denotes the condition of length or time in the moving system viewed from the reference stationary system, when the reference value of length or time in the stationary system is defined to be unity. In this expression, *c* denotes the velocity of light, and *v* denotes the velocity of the moving system.

For this development, he assumed or implied the following conditions.

1. K is a reference stationary system.
2. *k* is another system in parallel with K.
3. *k* moves at a uniform velocity along the *x*-axis of K in the positive direction; this velocity is denoted as *v*.
4. *t* is time passage in K.
5. *τ* is time passage in *k*.
6. *c* is the velocity of light. (Note: In the German original text [2], the velocity of light is denoted as '*V*'.)
7. A light source and a clock are placed at the origin of *k*.
8. A mirror and another clock are placed at a specific point of the *x*-axis of *k*.
9. A light source emits a ray of light in the positive direction of the *x*-axis of *k* and K, when *t*=0.
10. *At the origin of k, τ=0, when t=0'.*
11. Confirming the synchronization of the two clocks, by using a round trip of the ray between the two clocks. This method was previously established and examined in Sections 1 and 2 [1].
12. The ray of light obeys '*the principle of the constancy of the velocity of light*', which defined as '*Any ray of light moves in the "stationary" system of co-ordinates with the determined velocity c, whether the ray be emitted by a stationary or by a moving body.*' in Section 2 [1].
13. '*To any system of values x, y, z, t, which completely defines the place and time of an event in the stationary system, there belongs a system of values ξ, η, ζ, τ, determining that event relatively to the system k*'.

***Corresponding author:** Makanae M, Independent Researcher, Representative Free Web College, Nishikasai, Edogawa-ku, Tokyo 134-0088, Japan
E-mail: edit@free-web-college.com

In the process of the development of Lorentz factor, Einstein provides the following equation as an initial equation in the description: '*If we place x'=x−vt, it is clear that a point at rest in the system k must have a system of values x', y, and z, independent of time.*'

$$x'=x-vt. \quad (1)$$

He then provides many equations, before providing the following equation in the part before the first appearance of Lorentz factor in Section 3 [1].

$$\xi = c\tau \text{ or } \xi = ac\left(t - \frac{v}{c^2 - v^2}x'\right) \quad (2)$$

The term 'ξ' of the equation 'ξ=cτ' corresponds to 'distance' of the universal equation: 'distance=velocity × time'. The term a in equation (2) was described as '*function φ (v) at present unknown*'; 'function φ (v)' is revealed (v)=1 before the second and also the last appearance of the Lorentz factor in Section 3 [1]. In other words, a=1.

Then, Einstein provides the following equation in the description, which we will term Sentence 1: '*But the ray moves relatively to the initial point of k, when measured in the stationary system, with the velocity c−v, so that:*

$$\frac{x'}{c-v} = t, \quad (3)$$

Then he describes '*If we insert this value of t in the equation for ξ, we obtain:*

$$\hat{\imath} = a\frac{c^2}{c^2-v^2}x' \quad (4)$$

The fractional expression, i.e., the denominator and numerator in the right-hand side of equation (4) has the same structure as the left-hand side of $\frac{c^2}{c^2-v^2} = \beta^2$ which was provided in Section 4 of the paper 'Electromagnetic phenomena in a system moving with any velocity smaller than that of light' [3] by Lorentz himself published in 1904. $\frac{c^2}{c^2-v^2} = \beta^2$, can be factorized to $\beta = \pm\frac{1}{\sqrt{1-\left(v/c\right)^2}}$.

After equation (4), in total five equations and two sets of equations were provided, then the form $\beta = \frac{1}{\sqrt{1-\left(v/c\right)^2}}$, appears for the first time [1]. As described in the above, the term x' is contained in the initial equation (1), and also in the series of equations (2), (3) and (4) that configure the final stage of the development of Einstein's Lorentz factor. Explanations or definitions are not provided as a sentence for each x', while the values of x' of equations (1) or (3) are clearly provided or implied as a formula. Thus, it is natural to interpret that the meaning of x' of equations (1), (2), (3), and (4) are the same. However, if examining each equation by using numerical values of practical example, we find that the meanings or tangible values of x' of equations (1), (2), (3), and (4) are different. Thus, we can say that there is an inconsistency as regarding the usage of x' in the development of Lorentz factor in Section 3 [1]. In this paper, we explain the details of this issue, and then suggest the solution to avoid any doubt that the grounds of Lorentz factor are unclear.

Establishment of the conditions

We assume mutual conditions for examining x' of equations (1), (2), (3), and (4); but, before assuming the conditions, we should define the concept of 'point' and 'position', because the key term x' or x may correspond to 'point' or 'position' in [1], whereas Einstein did not provide a definition of 'point' or 'position' as an independent description focus in each term. Probably, he thought that both terms are popular; thus, even a non-scientist can understand each meaning without definition. However, 'point' and 'position' are fundamental concepts that must be established in order to consider the relationship between K and k, thus we provide the following definitions for proceeding with our study, except for the case of using the term 'point' in the phrase 'time point' or 'point in time'.

- 'point' is: A stationary location on the x-axis of k.
- 'position' is: A location of a certain point viewed from K. It may move with k viewed from K.

Further, we assume that the value of 'point' can be obtained when measuring it by using a ruler fixed on the x-axis of k (ruler of k) and the value of 'position' can be obtained when measuring it by using a ruler fixed on the x-axis of K (ruler of K).

Now, we assume numerical values of a practical example as the mutual conditions based on the conditions 1 to 13 described in the previous section.

- v is the half of the velocity of light, i.e., 0.5 c.
- The distance between the light source and the mirror is 1l viewed within k.
- Light moves 1l in 1 s.

Examination of x' of equation (1)

The structure of equation (1) is identical to Galilean transportation, namely x'=x−vt, whereas Einstein does not state so in Section 3 [1]. This equation denotes that a point, which moves with v viewed from K, is at rest always when viewed within k. In other words, the value of x' of equation (1) is constant as it is the value of the distance between the origin of k and point x' viewed within k.

Now, let us examine x' of equation (1). For this purpose, we denominate the point where the mirror is placed, as x'; and we denominate the variable position moving along the x-axis of K with the location of x' viewed from K, as x. By using the numerical values of practical example, we can assume the following tangible situation.

1. The distance between the origin of k and x' is always 1l measured by the ruler of k.
2. The value of the distance from x' at 0 s to x after the light source had emitted the ray increases; it can be calculated by vt.
3. When x'+vt=ct, i.e., when 1l+vt=ct, the mirror reflects the front of the ray (i.e., the tip of the ray). 1l+vt=ct can be transposed to $\frac{1l}{c-v} = t$.
4. By inserting the numerical value to $\frac{1l}{c-v} = t$, we obtain $\frac{1l}{1c-0.5c} = 2s$.

From the above, we confirm

$$x'=1l. \quad (5)$$

As shown above, v or t does not affect the value of x'; in other words, the conditions 2, 3, and 4 of the above situation do not affect the value of x'. This is natural because the value of x' of equation (1) is constant in the first place. Therefore, if changing the value of v to more or less than 0.5 c, or changing the value of t to more or less than 2 s, the value of x' is still the same.

Examination of x' of equation (2)

In Section 3 [1], after providing the set of equations $\xi = c\tau$ or $\xi = ac\left(t - \frac{v}{c^2 - v^2}x'\right)$ and the equation $\frac{x'}{c-v} = t$, Einstein describes *'If we insert this value of t in the equation for ξ, we obtain'* and then provides equation (4). The phrase *'this value of t'* corresponds to the formula $\frac{x'}{c-v}$. The phrase *'the equation for ξ'* actually corresponds to $\xi = ac\left(t - \frac{v}{c^2 - v^2}x'\right)$, not $\xi = c\tau$. Therefore, hereafter, we employ only $\xi = ac\left(t - \frac{v}{c^2 - v^2}x'\right)$ as equation (2) for convenience. Under the premise that $a=1$ as Einstein implied, equation (2) can be transposed

$$x' = \left(t - \frac{\xi}{c}\right) \div \left(\frac{v}{c^2 - v^2}\right) \tag{6}$$

If we assume that equation (1) affects the whole process of the development of the Lorentz factor, x' of (2) must obey the initial equation (1), thus the value of x' of (6) must be $1l$. In other words, must be $1l$.

However, this formula, which can be described as $\left(t - \frac{\xi}{c}\right) \div \left(\frac{o.5c}{c^2 - o.25c}\right)$ by inserting the numerical value of v, cannot provide $1l$. Besides, equation (6) corresponds to a structure of 'distance=time ÷ velocity ÷ velocity, or 'distance=time ÷ a ratio of velocities' (because $\frac{\xi}{c}$ provides 'time'), thus it does not match with the universal equation time=distance/velocity, which is used as the fundamental theory in the discussion of Sections 1 and 2 of [1], and thus it does not provide any value of 'distance' or 'time' or 'velocity'.

Examination of x' of equation (3)

Equation (3), namely $\frac{x'}{c-v} = t$, can be transposed to $x'=ct-vt$; thus, the method of calculating the value of x' is very clear. However, the meaning of x' of equation (3) is not clear, because there is ambiguity in the meaning of *'the initial point of k'* in terms of whether it is variable or a constant. This is because:

- The term *'the initial point of k'* is provided as the one of premises of in Sentence 1, i.e. *'But the ray moves relatively to the initial point of k, when measured in the stationary system, with the velocity c–v, so that* $\frac{x'}{c-v} = t$.'

- The value '*ct*' in '*x'=ct–vt*', which can be extracted from $\frac{x'}{c-v} = t$. The value '*ct*' implies that the value of the distance between the position of the origin of *k* at a point in time at which the light source emits the ray and the front of the ray at a point in time at which the observer at rest in the reference stationary system measures the position of the front of the ray.

- The value '*vt*' of '*x'=ct–vt*' implies the value of the distance between the position of the origin of *k* at a point in time at which the light source emits the ray and the position of the origin of *k* at a point in time at which the observer at rest in the reference stationary system measures this position.

- The formula '*ct–vt*' can be factored as '*(c–v)t*'; thus, we can say that '*c–v*' indicates the relative velocity between the front of the ray and the position of the origin of *k*.

- Based on the above assumptions, we can interpret the sentence *'the ray moves relatively to the initial point of k'* to mean that the front of the ray moves relative to the position of the origin of *k*.

- From the above, we can say that *'the initial point of k'* corresponds to 'the position of the origin of *k*', which varies with *vt* viewed from the reference stationary system.

- However, the term *'initial'* give us the impression that it has a constant value from the time it had occurred.

Thus, we should clearly define *'the initial point of k'*, before examining x' of equation (3).

The English term *'initial point of k'* is extracted by referring to the English-translated version [1]. Indeed, in Section 3 [1], similar terms *'the origin of k'* and *'the origin of system k'* are provided together with *'the initial point of k'*. Further, 'the origin of one of the two systems (*k*)', *'the initial position of the moving system'*, and *'the origin of co-ordinates'* are described without thorough explanations of the differences between them. Thus, let us confirm the original German descriptions. Below, the former (left side) is the original German text described [2], and the latter (right side) is the English-translated text of the paper [1].

1. *'Anfangspunkte von k'* to *'the origin of k'*
2. *'Anfangspunkt von k'* to *'the initial point of k'*
3. *'Anfangspunkt des Systems k'* to *'the origin of system k'*
4. *'dem Anfangspunkte des einen der beiden Systems (k)'* to *'the origin of one of the two systems (k)'*
5. *'Anfangslage des bewegten Systems'* to *'the initial position of the moving system'*
6. *'dem Koordinatenursprunges'* to *'the origin of co-ordinates'*

In the literal translation; the German term *'Anfang'* means 'beginning', *'punkt'* or *'punkte'* means 'point', and *'lage'* means 'situation', but it can be reasonably translated as 'position'. The German term 's' in *'Anfangspunkt'* or *'Anfangslage'* has the role of connecting the former and the latter terms. Therefore, we can say that in the literal translation *'Anfangspunkt'* (or *'Anfangspunkte'*) is the 'beginning point'. Likewise, *'Anfangslage'* is the 'beginning position'. However, the German term *'Anfang'* had been translated as *'initial'* or *'origin'* as shown in translations 1, 2, 3, 4, and 5 above. On the other hand, in translation 6, the literal translation 'the origin of co-ordinate' had been employed for *'dem Koordinatenursprunges'* (*'Koordinaten'* corresponds to 'co-ordinates' and *'Ursprung'* corresponds to *'origin'*).

It seems to be slightly complicated to be comparing the English terms with the original German terms and the literally translated terms; especially for the reason that *'Anfang'* (i.e., 'beginning' as the literal translation) is translated in two ways as *'initial'* and *'origin'* [1]. However, from the view point of a time sequence, we can arrange the above phrases as below.

First, we treat all references such as *'the origin of one of the two systems (k)'* and *'the origin of system k'* as references to 'the origin of *k*' and assume that 'the origin of *k*' has the absolute value as the zero point of x, y, z-axes of k. However, k is in motion as viewed from K; thus, the position of 'the origin of k' is variable, as viewed from K. Regarding the phrase: '*the initial position of the moving system*', we assume that this corresponds to the position of the origin of k at the point in time at which the light source emits the ray as viewed from K. It is natural that this value is not variable; this is because the light source emits the ray at a definite time 0 s.

On the basis of the above arrangement, we define '*the initial point of k*' as the position of the origin of k at a point of time where the observer at rest in the stationary system measures the distance between the origin of K and this position by using the ruler of K, whereas the English term '*initial*' still give us the impression that it has a constant value. Thus, we state that x' of $x'=ct-vt$ denotes the value of the distance between the front of the ray and '*the initial point of k*' measured by the ruler of K, and 'the velocity $c-v$' is the relative velocity between the front of the ray and '*the initial point of k*'. If it is difficult to use the term '*initial*' despite the above being an accepted definition, we suggest unifying the translations of '*Anfangspunkte von k*' (described as '*the origin of k*' [1]) and '*Anfangspunkt von k*' (described as '*the initial point of k*' [1]) as '*the origin of k*'.

With the above, we conclude the confirmation of the terms, and we now return to the main subject. If equation (1) is provided as the initial equation that affect the whole process of development of Lorentz factor, then the meaning and the value of x' of (1) and (3) must be the same. If the meaning and the value of x' of (1) and (3) are the same, we can substitute the value of x' of (1) to x' of (3), then (3) becomes $\frac{x-vt}{c-v}=t$, i.e., $\frac{x}{c}=t$. The form of $\frac{x}{c}=t$ corresponds to the universal equation distance/velocity=time. This means that the relative velocity between the front of the ray and '*the initial point of k*', namely 'c-v', is cancelled. At the same time, the term 'v', which is required as an element of Lorentz factor, is cancelled. Therefore, it is not appropriate to treat x' of equation (1) and x' of equation (3) together in the same context in the development of the Lorentz factor.

Next, we calculate the value of x' of equation (3) by using the numerical values of the practical example. Under the premise that x' of equation (3) corresponds to the distance between the front of the ray and '*the initial point of k*' viewed from K, we can assume the following situation.

1. The value of the distance between the front of the ray and '*the initial point of k*' increases with $ct-vt$ measured by the ruler of K.

2. At the time point when the light source emits the ray (i.e., when t=0), the value of the distance from '*the initial point of k*' to the mirror is $1l$ measured by the ruler of K. This value is the same as the value of the distance between the light source and the mirror viewed within k.

3. The time interval from the time point when the light source emits the ray to the time point when the mirror reflects the front of the ray in K; it can be calculated by using $t=\frac{1l}{c-v}$. Because it was previously confirmed that the mirror reflects the front of the ray when $1l$+vt=ct as viewed from K, $1l+vt=ct$ can be transposed to $\frac{1l}{c-v}=t$. Inserting the numerical value of v to $\frac{1l}{c-v}=t$, *we obtain* $\frac{1l}{1c-0.5c}=2\text{s}$. Thus, this time interval is 2 s.

From the above, we can say that the value of x' of equation (3) at the time point when the mirror reflects the front of the ray viewed from K is $1l$, i.e.,

$$x'=1l. \tag{7}$$

Here, we find that the two different x', namely x' of equation (1) and x' of equation (3), have the same value as shown as equations (5) and (7) are exactly the same in the tangible situation assumed in our study. At a glance, it seems that even the meaning of x' of equations (1) and (3) are different, but both values are always the same. However, if assuming the other situation that putting a completely transparent glass plate instead of the mirror and more than 2 s (for example 20 s) passes after the light source emitted the ray, we obtain x'=$10l$ for (3). In this situation, the two different x' have the two different values $1l$ in equation (1) *vs.* $10l$ in equation (3). It is natural that we obtain this difference, because the value of x' of equation (1) is constant as $1l$, even if the time passes 20 s. However, the time passage is not the original cause of this difference. The original cause is the act of using the transparent glass instead of the mirror. Under this condition the ray goes through the glass plate then advances only in the positive direction of the x-axis, thus the value of $ct-vt$ (i.e., the value of x') increases with no limit. If precluding the glass plate then putting the mirror back, the mirror reflects the front of the ray at 2 s; thus the value of $ct-vt$ (i.e., the value of x') cannot increase after 2 s. In other words, the value of the distance of the trip of the ray in the positive direction of x-axis is $1l$ viewed within k or $2l$ viewed from K, even if the time passes more than 2 s.

From this fact, we can say that Einstein's Sentence 1 does not match with his STR that requires confirming the synchronization of the two clocks by using the round trip of the ray. Because, Sentence 1 does not contain a condition such as '*until the mirror reflects the ray*', which is necessary in order to perform the round trip of the ray. In other words, the one of the essential conditions of STR is lacking in Sentence 1.

Paper [1] is referred to as Einstein's special theory of relativity, and many people interpret that 'special' of 'the special theory of relativity' implies the special condition regarding the relationship between the moving system and the reference stationary system that the former moves with uniform velocity along the x-axis of the latter in parallel translation. This interpretation itself is correct; however, the light source, the mirror (in order to reflect the ray), and the round trip between two clocks for confirming the synchronization of the two clocks, are also the special conditions that are required in order to establish Einstein's STR. In other words, if one of these special conditions is lacking, his STR cannot hold.

Examination of x' of equation (4)

Equation (4), namely, $\xi=a\frac{c^2}{c^2-v^2}x'$, is obtained by substituting the value of t of (3) to t of (2). We previously confirmed that a ='*function* φ (v)=1'. Therefore, we can reduce (4) to $\xi=\frac{c^2}{c^2-v^2}x'$. We can transpose this equation to

$$x'=\xi\div\frac{c^2}{c^2-v^2}. \tag{8}$$

If we assume that equation (1) affects the whole process of the development of the Lorentz factor, x' of equation (8) must obey the initial equation (1), thus the value of x' of (8) must be $1l$. In other words, $\xi\div\frac{c^2}{c^2-v^2}$ must be $1l$. However, this formula, which can be described as $\xi\div\frac{c^2}{c^2-o.25c}$ by inserting the numerical value of v, cannot provide $1l$. Besides, equation (8) corresponds to a structure of distance=distance ÷ velocity ÷ *velocity* or 'distance=distance ÷ a ratio of velocities', thus it does not match with the universal equation time=distance/velocity, which is used as the fundamental theory in the discussion of Sections (1) and (2) [1], and thus it does not provide any value of 'distance' or 'time' or 'velocity'.

Overview

The velocity of light is described as V in the German original article [2]; on the other hand, it described as c in the Lorentz's article [3]. However, for us who have read [1,3] in English, the Lorentz factor $\beta = \frac{1}{\sqrt{1-\left(v/c\right)^2}}$ is substantially implied at the stage where equation (4), which contains the left-hand side of Lorentz's equation $\frac{c^2}{c^2-v^2} = \beta^2$ in Section 4 [3], is provided. In other words, the series of equations (2), (3), and (4) configure the climax of the development of Lorentz factor [1].

The following is the expansion process that from (2), namely $\xi = ac\left(t - \frac{v}{c^2-v^2}x'\right)$, to (4), namely $\xi = a\frac{c^2}{c^2-v^2}x'$ by using the value of t of (3), namely $\frac{x'}{c-v}$.

$$\xi = ac\left(\frac{x'}{c-v} - \frac{v}{c^2-v^2}x'\right),\ \xi = ac\left(\frac{1}{c-v} - \frac{v}{c^2-v^2}\right)x',$$

$$\xi = ac\left(\frac{\left(c^2+v^2\right)-v\left(c-v\right)}{(c-v)(c^2-v^2)}\right)x',\ \xi = ac\left(\frac{\left(c+v\right)\left(c-v\right)-v\left(c-v\right)}{(c-v)(c^2-v^2)}\right)x',$$

$$\xi = ac\left(\frac{\left(c+v\right)-v}{c^2-v^2}\right)x',\ \xi = ac\left(\frac{c}{c^2-v^2}\right)x', \hat{\imath} = a\frac{c^2}{c^2-v^2}x'.$$

As shown in the process from the first equation to the second equation in the above, the term x' has a role that simplifies the inside of the parenthesis by extracting x' of the left formula and x' of the right formula to the outside of the parenthesis. In the parenthesis of the first equation, the left formula is provided from equation (3), and the right formula is provided from equation (2). This means that the meaning and the value of x' of equations (2), (3), and (4) must be the same. Likewise, under the assumption that x' of equation (1) affects the whole process of the development of Lorentz factor, x' of (1) must be the same as x' of equations (2), (3), and (4). This perspective is correct not only from the viewpoint of mathematics but also from the viewpoint of the principle of identity provided by formal logic, which is a practical tool for the study of any science, including theoretical physics.

However,

- x' of (1) is constant, x' of equation (3) is variable.
- The meanings of x' of equations (2) and (4) are not clear.
- The value of x' of equations (2) or (4) differs from the value of x' of equations (1) or (3).

Conclusion

From the above, we can conclude that the equations (1), (2), (3), and (4) or some of these are not appropriate for the establishment of Lorentz factor; in other words, the development process of the Lorentz factor provided in Section 3 [1] is not correct.

In contrast, including the indirect evidence provided through discoveries of our universe by using the general theory of relativity (GTR), which was developed in 1915, much evidence that proves the correctness of the Lorentz factor exists. If we ignore this contradiction and continue to use the Lorentz factor in order to study relativity, then modern science, which espouses Einstein's STR, may at some point suffer from a lack of trust; therefore, this contradiction should be solved.

One of the ways to resolve the contradiction is to disprove the evidence; however, it may be difficult to proceed in this manner, because the evidence has been reported by highly trusted institutions and scientists. The other way is to state that the Lorentz factor was already established in ref. [3] before refs. [1,2] as a universal theory, whether Lorentz himself had recognized that his theory can be used at both the microscopic and the astronomical scale. However, this may also be unacceptable, because Einstein's Lorentz factor was not established on the Lorentz's article [3]. In fact, Einstein did not discuss or introduce [3] in the development process of his Lorentz factor [1,2]. We know that Einstein provided a different method to derive the Lorentz factor in his book entitled 'The Special and General Theory of Relativity' [4] published in 1916; but, if employing this method, we have to state that Einstein developed his Lorentz factor, which is a core theory of the STR, after releasing the GTR, which was developed based on STR.

It seems that there is no way to solve the contradiction. However, if we assume that the Lorentz factor is established as a categorical judgement, then the contradiction is simply removed. Because, if it is so, the Lorentz factor can be used as an absolute and universal theory that is never affected by any condition such as 'until the mirror reflects the ray', which is lacking in Sentence 1. Likewise, any development process is not necessary. Metaphorically speaking, it is appropriate to treat the Lorentz factor as the same as the equation 1+1=2 in the decimal system of mathematics, which holds without any explanation. Once this methodology has been employed, the above contradiction is removed, and we can avoid any doubt that the grounds of the Lorentz factor are unclear. In addition, under the assumption that the Lorentz factor is determined by categorical judgement, expressions describing the Lorentz factor as 'a result of STR' should not be used because such expressions imply that the concept of the Lorentz factor holds in certain premises, i.e., the Lorentz factor is not determined by categorical judgement. Therefore, we have used the phrase, 'a core theory of STR', on multiple occasions in our study.

In any case, the inconsistency in Section 3 [1,2] and the above methodology should be widely discussed, if we continue to use the Lorentz factor in the study of relativity.

Acknowledgments

I would like to thank Editage for English language editing, Ms. Maxie Pickert for confirming the meanings of the German original text [2], Mr. Makoto Kawahara for confirming the basic theory of mathematics, and Ms. Noriko Teramoto for help with the English grammar.

References

1. Perrett W, Jeffery GB (1923) On the electrodynamics of moving bodies. The Principle of Relativity, published by Methuen and Company 37-65.
2. Einstein A (1905) Zur Elektrodynamik bewegter Körper. Annalen der Physik und Chemie. 17: 891-921.
3. Lorentz HA (1904) 'Electromagnetic phenomena in a system moving with any velocity smaller than that of light'. Reprinted from the English version in Proceedings of the Academy of Sciences of Amsterdam. 6: 11-34.
4. Einstein A (1916) Relativity: The Special and General Theory. Published by London: Methuen & Co.

10

Sieve of Prime Numbers Using Algorithms

Stelian Liviu B*

Independent Researcher, Israel

Abstract

This study suggests grouping of numbers that do not divide the number 3 and/or 5 in eight columns. Allocation results obtained from multiplication of numbers is based on column belonging to him. If in the Sieve of Eratosthenes the majority of multiplication of prime numbers result in a results devoid of practical benefit (numbers divisible by 2, 3 and/or 5), in the sieve of prime numbers using algorithms, each multiplication of prime number gives a result in a number not divisible to 2, 3 and/or 5.

Keywords: Column; Factor; Position; Sieve; Termination

Introduction

Sieve of prime numbers using algorithms

This paper deals with the study of odd numbers that cannot be divided with 3 and/or 5 by grouping them in eight columns, as follows:

The multiplication versions are in number of 36, their results being allocated according to columns, explained in Table 1.

Position Calculus

From the result of multiplying two numbers subtract the number assigned at position zero of the column namely one of the numbers i(p0): 7-11-13-17-19-23-29-31, the result is divided by 30. Integer obtained indicates the position of that number considering its column origin [1,2].

Formulas for determining the position

Position occupied by the result of the multiplication between

Col.1=Col.	1x8	2x4	3x5			6x7		
Col.2=Col.	1x6	2x8	3x4		5x7			
Col.3=Col.	1x5	2x6	3x8	4x7				
Col.4=Col.	1x2		3x7	4x8	5x6			
Col.5=Col.	1x1	2x7	3x3	4x4	5x8	6x6		
Col.6=Col.	1x7	2x3		4x5		6x8		
Col.7=Col.	1x4	2x5	3x6				7x8	
Col.8=Col.	1x3	2x2		4x6	5x5		7x7	8x8

Table 1: Multiplication versions are in number of 36.

Position	1	2	3	4	5	6	7	8
0	7	11	13	17	19	23	29	31
1	37	41	43	47	49	53	59	61
2	67	71	73	77	79	83	89	91
3	97	101	103	107	109	113	119	121

Table 2: Odd numbers that cannot be divided with 3 and/or 5.

7+31	5+23	4+19	2+11	1+7	6+29	3+17	2 +13	+37n
6+17	11+31	8+23	2+7	10+29	4+13	6+19	3+11	+41n
8+19	7+17	13+31	12+29	5+13	4+11	9+23	2+ 7	+43n
6+11	7+13	16+29	17+31	9+17	10+19	3+ 7	12+23	+47n
8+13	**18+29**	**4+ 7**	**14+23**	**19+31**	**10+17**	**6+11**	**11+19**	**+49n**
22+29	5+ 7	8+11	14+19	17+23	23+31	9+13	12+17	+53n
22+23	18+19	16+17	12+13	10+11	6+ 7	29+31	27+29	+59n
7+7	11+11	13+13	17+17	19+19	23+23	29+29	31+31	+61n

Table 3: Position occupied p1 as a result of multiplication of numbers i.

7+31x2	5+23x2	4+19x2	2+11x2	1+7x2	6+29x2	3+17x2	2+13x2	+67n
6+17x2	11+31x2	8+23x2	2+7x2	10+29x2	4+13x2	6+19x2	3+11x2	+71n
8+19x2	7+17x2	13+31x2	12+29x2	5+13x2	4+11x2	9+23x2	2+7x2	+73n
6+11x2	**7+13x2**	**16+29x2**	**17+31x2**	**9+17x2**	**10+19x2**	**3+7x2**	**12+23x2**	**+77n**
8+13x2	18+29x2	4+7x2	14+23x2	19+31x2	10+17x2	6+11x2	11+19x2	+79n
22+29x2	5+7x2	8+11x2	14+19x2	17+23x2	23+31x2	9+13x2	12+17x2	+83n
22+23x2	18+19x2	16+17x2	12+13x2	10+11x2	6+7x2	29+31x2	27+29x2	+89n
7+7x2	**11+11x2**	**13+13x2**	**17+17x2**	**19+19x2**	**23+23x2**	**29+29x2**	**31+31x2**	**+91n**

Table 4: Positions of p1 are used to calculate p2, p3, p4.

1	2	3	4	5	6	7	8	9
7	5	4	2	1	6	3	2	+7n
6	11	8	2	10	4	6	3	+11n
8	7	13	12	5	4	9	2	+13n
6	7	16	17	9	10	3	12	+17n
8	18	4	14	19	10	6	11	+19n
22	5	8	14	17	23	9	12	+23n
22	18	16	12	10	6	29	27	+29n
7	11	13	17	19	23	29	31	+31n

Table 5: Position occupied p0 as a result of multiplication of numbers i (p0) and all the numbers.

numbers i(p0), i(p1), i(p2),..., i(pn), with all the numbers in Table 2. Position occupied p1 as a result of multiplication of numbers i (p1) and all the numbers in Table 3;

Positions of p1 are used to calculate p2, p3, p4,...., pn multiplying i(p0), positions occupied p2 as a result of multiplication of numbers i(p2) and all the numbers in Table 4;

Calculation algorithm

- Fill in Table 1 with all the numbers to be tested if they are prime number;
- Write all numbers under test, in order of their increasing in column 9,as shown in Table 5;
- Fill p0 formulas in Table 5;

***Corresponding author:** Stelian Liviu B, Independent researcher, Israel
E-mail: stelibarar@yahoo.com

- Mark all numbers divisible in Table 1 by the formulas of p0;
- Eliminates all the numbers in column 9 Table 2 that were marked in Table 1 according to the formulas of p0;
- Fill formulas of p1 Table 2; number 49 was removed according to Table 1 no longer consider;
- Repeat the operations made in step 4 and 5 according to the formulas p1;
- Fill formulas of p2 Table 2 and repeat the operations in step 4 and 5. Numbers not eliminated in column 9 Table 2 are prime numbers.

In column 9 we register numbers under test up to P (max). Maxim position calculation is the integer number of the maximum number being tested radical divided by 30 [2-4].

Formulas belonging composite numbers are omitted. The algorithm uses formulas primes numbers squared correlating n=0,1,2,3,.... With Pn.

Using the tables respecting the above algorithm complexity is much smaller, any multiple of prime number (which represents the number of position) has corresponding number is compound odd number and not divisible by 3 and/or 5.

Example: Determination of prime numbers up to N=1001.

In parentheses are the numbers corresponding to position past according to column.

Divisibility by 7:

Col.1: 7+7n=7(217) - 14(427) - 21(637) - 28(847)

Col.2: 5+7n=5(161) - 12(371) - 19(581) - 26(791) - 33(1001)

Col.3: 4+7n=4(133) - 11(343) - 18(553) - 25(763) - 32(973)

Col.4: 2+7n=2(77) - 9(287) - 16(497) - 23(707) - 30(917)

Col.5: 1+7n=1(49) - 8(259) - 15(469) - 22(679) - 29(889)

Col.6: 6+7n=6(203) - 13(413) - 20(623) - 27(833)

Col.7=3+7n=3(119) - 10(329) - 17(539) - 24(749) - 31(959)

Col.8=2+7n=2(91) - 9(301) - 16(511) - 23(321) - 30(931)

Divisibility by 11:

Col.1=6+11n=6(187) - 17(517) - 28(847)

Col.2=11+11n=11(341) - 22(671) - 33(1001)

Col.3=8+11n=8(253) - 19(583) - 30(913)

Col.4=2+11n=2(77) - 13(407) - 24(737)

Col.5=10+11n=10(319) - 21(649) - 32(979)

Col.6=4+11n=4(143) - 15(473) - 26(803)

Col.7=6+11n=6(209) - 17(539) - 28(869)

Col.8=3+11n=3(121) - 14(451) - 25(781)

Divisibility by 13:
Divisibility by 17:

Col.1=8+13n=8(247) - 21(637)
Col.1=6+17n=6(187) - 23(697)

Col.2=7+13n=7(221) - 20(611) - 33(1001)
Col.2=7+17n=7(221) - 24(731)

Col.3=13+13n=13(403) - 26(793)
Col.3=16+17n=16(493)

Col.4=12+13n=12(377) - 25(767)
Col.4=17+17n=17(527)

Col.5=5+13n=5(169) - 18(559) - 31(949)
Col.5=9+17n=9(289) - 26(799)

Col.6=4+13n=4(143) - 17(533) - 30(923)
Col.6=10+17n=10(323) - 27(833)

Col.7=9+13n=9(299) - 22(689)
Col.7=3+17n=3(119) - 20(629)

Col.8=2+13n=2(91) - 15(481) - 28(871)
Col.8=12+17n=12(391) - 29(901)

Divisibility by 19:
Divisibility by 23:

Col.1=8+19n=8(247) - 27(817)
Col.1=22+23n=22(667)

Col.2=18+19n=18(551)
Col.2=5+23n=5(161) - 28(851)

Col.3=4+19n=4(133) - 23(703)
Col.3=8+23n=8(253) - 31(943)

Col.4=14+19n=14(437)
Col.4=14+23n=14(437)

Col.5=19+19n=19(589)
Col.5=17+23n=17(529)

Col.6=10+19n=10(323) - 29(893)
Col.6=23+23n=23(713)

Col.7=6+19n=6(209) - 25(779)
Col.7=9+23n=9(299) - 32(789)

Col.8=11+19n=11(361) - 30(961)
Col.8=12+23n=12(391)

Divisibility By 29:
Divisibility by 31:

Col.1=22+29n=22(667)
Col.1=7+31=7(217)

Col.2=18+29n=18(551)
Col.2=11+31=11(341)

Col.3=16+29n=16(493)
Col.3=13+31n=13(403)

Col.4=12+29n=12(377)
Col.4=17+31n=17(527)

Col.5=10+29n=10(319)
Col.5=19=31n=19(589)

Col.6=6+29n=6(203)
Col.6=23+31n=23(713)

Col.7=29+29n=29(899)
Col.7=29+31n=29(899)

Col.8=27+29n=27(841)
Col.8=31+31n=31(961)

Numbers not eliminated are prime numbers

Application: The Factorial Multiplying or the Method of Determining if a Number is Prime up to a Given Number

The method of grouping odd numbers according to Table 1, allows checking whether a number is prime according to the last two or five digits of position the number.

For termination two digits

The calculation algorithm is:

Step 1: Determine the position number and column it belongs;

Step 2: Last two digits of the calculated number indicates the termination position of tested number;

Step 3: Determine factors for termination and column number tested. I have illustrated the calculation of factors termination 10, column 1. Once calculated these factors can be used to determine of any prime numbers that belongs to the column 1, termination 10.

Step 4: It performs testing divisibility of a number with multiples of 3 000 plus pairs of numbers factorial group to which it belongs termination corresponding column number tested.

We assign factorial group for multiplying operation positions from 0-99, as in Table 1, numbers between 7-3.001 grouped in columns. The position occupied by the result of the multiplication between any two numbers in the factorial group is a maximum six digit number. The last two digits of the number shows the termination, the rest of maximum four digits is the factor and which the position will be calculated for those termination belonging to specific column [5,6].

I1 and I2 are two numbers higher than the numbers belonging to factorial group.

Position obtained by multiplying the numbers is determined by formula:

P=n2 × i1 (f)+n1 × i2+F, followed by T

Or, =n1 × i2 (f)+n2 × i1+F, followed by T

Where:

n1, n2: represents multiples of 3000 corresponding of i1(f), respectively i2(f);

i1 (f), i2 (f): represents the corresponding numbers of i1 and i2 in factorial group;

F - Factor

T - Termination

Be: 32 999 × 32 693=1 078 836 307

P=(1 078 836 307 - 7): 30=35 961 210 col.1 T=10 p(without T)=359 612

Factor calculation and termination:

2 999 × 2 693=(8 076 307 - 7): 30=269 210; F=2 692 T=10

P=10 × 2 999+10 × 32 693+F, followed by T

=10 × 2 693+10 × 32 999+F, followed by T

We calculate all the factors column 1, termination 10. The four types of multiplication corresponding col. 1 between numbers belonging to factor group, generates 400 factors with T.10, as follows:

7 × 901=2 37 × 1 711=21 67 × 721=16

307 × 3 001=307 337 × 811=91 367 × 2 821=345

607 × 2 101=425 637 × 2 911=618 667 × 1 921=427

2 707 × 1 801=1 625 2 737 × 2 611=2 382 2 767 × 1 621=1 495

97 × 931=30 127 × 2 341=99 157 × 1 951=102

397 × 31=4 427 × 1 441=205 457 × 1 051=160

697 × 2 131=495 727 × 541=131 757 × 151=38

2 797 × 1 831=1 707 2 827 × 241=227 2 857 × 2 851=2 715

187 × 2 761=172 217 × 1 771=128 247 × 1 981=163

487 × 1 861=302 517 × 871=150 547 × 1 081=197

787 × 961=252 817 × 2 971=809 847 × 181=51

2 887 × 661=636 2 917 × 2 671=2 597 2 947 × 2 881=2 830

277 × 391=36

577 × 2 491=476

877 × 1 591=465

2 977 × 1 291=1 281

Or,

11 × 1 937=7 41 × 227=3 71 × 2 117=50

311 × 2 837=294 341 × 1 127=128 371 × 17=2

611 × 737=150 641 × 2 027=433 671 × 917=205

2 711 × 1 037=937 2 741 × 2 327=2 126 2 771 × 1 217=1 124

101 × 1 607=54 131 × 1 697=74 161 × 2 387=128

401 × 2 507=335 431 × 2 597=374 461 × 287=44

701 × 407=95 731 × 497=121 761 × 1 187=3 011

2 801 × 707=660 2 831 × 797=752 2 861 × 1 487=1 418

191 × 677=43 221 × 2 567=189 251 × 2 057=172

491 × 1 577=258 521 × 467=81 551 × 2 957=543

791 × 2 477=653 821 × 1 367=374 851 × 857=243

2 891 × 2 777=2 676 2 921 × 1 667=1 623 2 951 × 1 157=1 138

281 × 2 147=201

581 × 47=9

881 × 947=278

2 981 × 1 247=1 239

Or,

19 × 1 753=11 49 × 1 843=30 79 × 1 333=35

319 × 2 653=282 349 × 2 743=319 379 × 2 233=282

619 × 553=114 649 × 643=139 679 × 133=30

2 719 × 853=773 2 749 × 943=864 2 779 × 433=401

109 × 223=8 139 × 1 513=70 169 × 2 203=124

409 × 1 123=153 439 × 2 413=353 469 × 103=16

709 × 2 023=478 739 × 313=7 769 × 1 003=257

2 809 × 2 323=2 175 2 839 × 613=580 2 869 × 1 303=1 246

199 × 2 293=152 229 × 1 783=136 259 × 673=58

499 × 193=32 529 × 2 683=473 559 × 1 573=293

799 × 1 093=291 829 × 583=161 859 × 2 473=708

2 899 × 1 393=1 346 2 929 × 883=862 2 959 × 2 773=2 735

289 × 1 963=189

589 × 2 863=562

889 × 763=226

2 989 × 1 063=1 059

Or,

29 × 2 183=21 59 × 1 073=21 89 × 2 363=70

329 × 83=9 359 × 1 973=236 389 × 263=34

629 × 983=206 659 × 2 873=631 689 × 1 163=267

2 729 × 1 283=1 167 2 759 × 173=159 2 789 × 1 463=1 360

119 × 53=2 149 × 143=7 179 × 2 633=157

419 × 953=133 449 × 1 043=156 479 × 533=85

719 × 1 853=444 749 × 1 943=485 779 × 1 433=372

2 819 × 2 153=2 023 2 849 × 2 243=2 130 2 879 × 1 733=1 663

209 × 1 523=106 239 × 2 813=224 269 × 503=45

509 × 2 423=411 539 × 713=128 569 × 1 403=266

809 × 323=87 839 × 1 613=451 869 × 2 303=667

2 909 × 623=604 2 939 × 1 913=1 874 2 969 × 2 603=2 576

299 × 593=59

599 × 1 493=298

899 × 2 393=717

2 999 × 2 693=2 692

Grouping numbers from left of multiplying operation according to the above model, in this case numbers on the right have a constant growth rate, which allows for relatively simple determination of them. Perform tests to see if number N is prime or not, using position calculation formulas, as follows:

Divisibility by:

(3 000 × n+7) × (3 000 × n+901) F=2

7 × n; 901 × n; 901+3 007xn; 901x2+6 007xn; 901x3+9 007xn;......................

7xn correspond to: 7 × (3 000 × n+901); 901xn correspond to: 901 × (3 000 × n+7);

901+3 007xn correspond to: 3 007 × (3 000 × n+901);

901x2+6 007xn correspond to: 6 007 × (3 000 × n+901);

901x3+9 007xn correspond to: 9 007 × (3 000 × n+901);

If not results indicate position of N decreased by the factor F=2, the number studied does not divide with multiples of 3000 plus pair of numbers 7-901

(3 000 × n+307) × (3 000 × n+3001) F=307

307 × n; 3 001 × n; 3 001+3 307xn; 3 001x2+6 307xn; 3 001x3+9 307xn;.....................

307 × n correspond to: 307 × (3 000 × n+3 001); 3 001 × n correspond to: 3 001 × (3 000 × n+307);

3 001+3 307 × n correspond to: 3 307 × (3 000 × n+3 001);

3 001x2+6 307xn correspond to: 6 307 × (3 000 × n+3 001);

3 001x3+9 307xn correspond to: 9 307 × (3 000 × n+3 001);...

Extract factor F=307 out of the position number of N than check calculation above.

(3 000 × n+607) × (3 000 × n+2 101) F=425

607 × n; 2 101 × n; 2 101+3 607xn; 2 101x2+6 607xn; 2 101x3+9 607xn;

Or,

(3 000 × n+2 707) × (3 000 × n+1 801) F=1 625

2 707 × n; 1 801 × n; 1 801+5 707xn; 1 801x2+8 707xn; 1 801x3+11 707xn;

If none of the operations related to 400 factors do not give as results the position of studied number, this number is prime.

For this example (p=359 612) we check these calculations:

Divisibility by:

(3 000 × n+7) × (3 000 × n+901) F=2 P – F=359 610

7 × 51 372=359 604 not divisible by 7 × (3 000 × n+901)

901 × 399=359 499 not divisible by 901 × (3 000 × n+7)

901+3 007x119=358 734 -//- 3 007 × (3 000 × n+901)

901x2+6 007x59=356 215 -//- 6 007 × (3 000 × n+901

901x3+9 007x39=353 976 -//- 9 007 × (3 000 × n+901)

901x4+12 007x29=351 807 -//- 12 007 × (3 000 × n+901)

901x5+15 007x23=349 666 -//- 15 007 × (3 000 × n+901)

901x6+18 007x20=365 546 -//- 18 007 × (3 000 × n+901)

901x7+21 007x16=342 419 -//- 21 007 × (3 000 × n+901)

901x8+24 007x14=343 306 -//- 24 007 × (3 000 × n+901)

901x9+27 007x13=359 200 -//- 27 007 × (3 000 × n+901)

901x10+30 007x11=339 087 -//- 30 007 × (3 000 × n+901)

901x20+60 007x5=318 055 -//- 60 007 × (3 000 × n+901)

901x30+90 007x3=297 054 -//- 90 007 × (3 000 × n+901)

901x40+120 007x2=276 054 -//- 120 007 × (3 000 × n+901)

901x50+150 007x2=345 064 -//- 150 007 × (3 000 × n+901)

901x60+180 007x1=234 067 -//- 180 007 × (3 000 × n+901)

901x92+276 007=358 899 -//- 276 007 × (3 000 × n+901)

Last calculation can be performed.

Testing for number N continues with:

Divisibility by:

(3 000 × n+37) × (3 000 × n+1 711) F=21 P – F=359 591

(3 000 × n+67) × (3 000 × n+721) F=16 P – F=359 596

Divisibility by:

(3 000 × n+2999) × (3 000 × n+2693) F=2 692 P – F=356 920

2 999 × 119=356 881 -//- 2 999 × (3 000 × n+2 693)

2 693 × 132=355 476 -//- 2 693 × (3 000 × n+2 999)

2 693+5 999x59=356 634 -//- 5 999 × (3 000 × n+2 693)

2 693x2+8 999x39=356 347 -//- 8 999 × (3 000 × n+2 693)

2 693x10+32 999x10=356 920, number identical to P – F,

So N is divisible by 32 999.

For termination five digits

The calculation algorithm is:

Pas.1: Determine the position number and column it belongs;

Pas.2: Last five digits of the calculated number indicates the termination position of tested number;

Pas 3: Determine factors for termination and column number tested. I have illustrated the calculation of factors termination 001 10, column 1;

Pas.4: We divisibility test the formulas for calculating factorial.

Positions calculated results do not contain termination 001 10

For pair of numbers 31 – 397

31 × (3 000 000xn+1 161 397) p=12+31 × n; divisibility by 31

3 031 × (3 000 000xn+1 800 397) p=1 819+3 031 × n -//- 3 031

6 031 × (3 000 000xn+2 439 397) p=1 819+3 085+6 031 × n -//- 6 031

9 031 × (3 000 000xn+3 078 397) p=1 819+3 085 × 2+1 278+9 031 × n -//- 9 031

12 031 × (3 000 000xn+3 717 397) p=1 819+3 085 × 3+1 278 × (2)!+12 031 × n -//- 12 031

15 031 × (3 000 000xn+4 356 379) p=1 819+3 085 × 4+1 278 × (3)!+15 031 × n -//- 15 031

18 031 × (3 000 000xn+4 995 379) p=1 819+3 085 × 5+1 278 × (4)!+18 031 × n -//- 18 031

2 997 031 × (3 000 000xn+639 522 379) p=1 819+3 085 × 998+1 278 × (997)!+2 997 031 × n divisibility by 2 997 031

3 000 031 × (3 000 000xn+640 161 379) p=1 819+3 085 × 999+1 278 × (998)!+3 000 031 × n divisibility by 3 000 031

3 003 031 × (3 000 000xn+640 800 379) p=1 819+3 085 × 1 000+1 278 × (999)!+3 003 031 × n divisibility by 3 003 031

And,

397 × (3 000 000xn+2 403 031) p=318+397 × n divisibility by 397

3 397 × (3 000 000xn+234 031) p=265+3 397 × n -//- 3 397

6 397 × (3 000 000xn+1 065 031) p=265+2 006+6 397 × n -//- 6 397

9 397 × (3 000 000xn+1 896 031) p=265+2006 × 2+1 662+9 397 × n -//- 9 397

12 397 × (3 000 000xn+2 727 031) p=265+2006 × 3+1 662 × (2)!+12 397 × n -//- 12 397

15 397 × (3 000 000xn+3 558 031) p=265+2 006 × 4+1 662 × (3)!+15 397 × n -//- 15 397

18 397 × (3 000 000xn+4 389 031) p=265+2 006 × 5+1 662 × (4)!+18 397 × n -//- 18 397

2 997 397 × (3 000 000xn+829 572 031) p=265+2 006 × 998+1 662 × (997)!+2 997 397 × n divisibility by 2 997 397

3 000 397 × (3 000 000xn+830 403 031) p=265+2 006 × 999+1 662 × (998)!+3 000 397 × n divisibility by 3 000 397

3 003 397 × (3 000 000xn+831 234 031) p=265+2 006 × 1 000+1 662 × (999)!+3 003 397 × n divisibility by 3 003 397

Or, pair of numbers 331 – 1 297

331 × (3 000 000xn+2 755 297) p=304+331 × n divisibility by 331

3 331 × (3 000 000xn+994 297) p=1 104+3 331 × n -//- 3 331

6 331 × (3 000 000xn+2 233 297) p=1 104+3 609+6 331 × n -//- 6 331

9 331 × (3 000 000xn+3 472 297) p=1 104+3 609 × 2+2 478+9 331 × n -//- 9 331

12 331 × (3 000 000xn+4 711 297) p=1 104+3 609 × 3+2 478 × (2)!+12 331 × n -//- 12 331

15 331 × (3 000 000xn+5 950 297) p=1 104+3 609 × 4+2 478 × (3)!+15 331 × n -//- 15 331

18 331 × (3 000 000xn+7 189 297) p=1 104+3 609 × 5+2 478 × (4)!+18 331 × n -//- 18 331

And,

1 297 × (3 000 000xn+342 331) p=148+1 297 × n divisibility by 1 297

4 297 × (3 000 000xn+1 773 331) p=2 540+4 297 × n -//- 4 297

7 297 × (3 000 000xn+3 204 331) p=2 540+5 254+7 297 × n -//- 7 297

10 297 × (3 000 000xn+4 635 331) p=2 540+5 254 × 2+2 862+10 297 × n -//- 10 297

13 297 × (3 000 000xn+6 066 331) p=2 540+5 254 × 3+2 862 × (2)!+13 297 × n -//- 13 297

16 297 × (3 000 000xn+7 497 331) p=2 540+5 254 × 4+2 862 × (3)!+16 297 × n -//- 16 297

19 297 × (3 000 000xn+8 928 331) p=2 540+5 254 × 5+2 862 × (4)!+19 297 × n -//- 19 297

Conclusion

Number testing is done with all the 400 pairs of numbers in the group factorial. Factorial multiplication process has as principle of calculation pairs of numbers that belong to the factorial group unique to each termination and column.

References

1. Canfield ER, Erdos P, Pomerance C (1983) On a problem of Oppenheim concerning Factorisatio Numerorum. J Number Theory 17: 1-28.
2. Davis JA, Holdridge DB (1983) Factorization sings the quadratic sieve algorithm. Advances in Cryptology 2: 103-113.
3. Lehmer DH, Powers RE (1931) on factoring large numbers. Bull her Math Soc 37: 770-776.
4. Miller JCP (1975) on factorisation with a suggested new approach. Math Comp 29: 155-772.
5. Pomerance C, Wagstaff SS (1983) Implementation of the continued fraction algorithm. Cow Numerantium 37: 99-118.
6. Morrison MA, Brillhart J (1975) A method of factoring and the factorization of F_7. Math Comp 29: 183-205.

A Hierarchy of Symmetry Breaking in the Nonsymmetric Kaluza-Klein (Jordan-Thiry) Theory

Kalinowski MW*

Bioinformatics Laboratory, Medical Research Centre, Polish Academy of Sciences, Poland

Abstract

The paper is devoted to the hierarchy of a symmetr y breaking in the Non symmetric Kaluza–Klein (Jordan–Thiry) Theory. The basic idea consists in a deformation of a vacuum states manifold to the cartesian product of vacuum states manifolds of every stage of a symmetry breaking .In the paper we consider a pattern of a spontaneous symmetry breaking including a hierarchy in the Non symmetr ic Kaluza–Klein (Jordan–Thiry) Theory.

Introduction

In this paper we consider hierarchy of symmetry breaking in the Nonsymmetric Kaluza–Klein Theory and the Nonsymmetric Kaluza–Klein Theory with a spontaneous symmetry breaking and Higgs' mechanism . In the second section we consider a Nonsymmetric Kaluza–Klein Theory and the Nonsymmetric Kaluza–Klein Theory with a spontaneous symmetry breaking and Higgs' mechanism [1-6]. In the third section we develop a hierarchy of the symmetry breaking in our theory. For further development of the nonsymmetric Kaluza–Klein (Jordan–Thiry) Theory [7-10].

Elements of the Nonsymmetric Kaluza–Klein Theory in general non-Abelian case and with spontaneous symmetr y breaking and Higgs' mechanism

Let $\underline{P}$ be a principal fiber bundle over a space-time E with a structural group G which is a semisimple Lie group. On a space-time E we define a nonsymmetric tensor $g_{\mu\nu} = g_{\mu\nu)} + g_{\mu\nu]}$ such that

$g=\det(g\mu\nu)\neq 0$

$$\tilde{g} = \det(g_{(\mu\nu)}) \neq 0 \tag{2.1}$$

$g_{[\mu\nu]}$ is called as usual a skewon field (e.g., in NGT, [6,11-13] We define on E a nonsymmetric connection compatible with $g_{\mu\nu}$ such that

$$\overline{D}g_{\alpha\beta} = g\alpha\delta\overline{Q}^{\delta}{}_{\beta\gamma}(\overline{\Gamma})\overline{\theta}^{\gamma} \tag{2.2}$$

where $\overline{D}$ is an exterior covariant derivative for a connection $\overline{\omega}^{\alpha}{}_{\beta} = \overline{\Gamma}^{\alpha}{}_{\beta\gamma}\overline{\theta}^{\gamma}$ and $\overline{Q}^{\alpha}{}_{\beta\delta}$ is its torsion. We suppose also

$$\overline{Q}^{\alpha}{}_{\beta\alpha}(\overline{\Gamma}) = 0 \tag{2.3}$$

We introduce on E a second connection

$$\overline{W}^{\alpha}\beta = \overline{W}^{\alpha}{}_{\beta\gamma}\overline{\theta}^{\gamma} \tag{2.4}$$

such that

$$\overline{W}^{\alpha}\beta = \overline{\omega}^{\alpha}{}_{\beta} - \frac{2}{3}\delta^{\alpha}_{\beta}\overline{W} \tag{2.5}$$

$$\overline{W} = \overline{W}\gamma\overline{\theta}^{\gamma} = \frac{1}{2}(\overline{W}^{\sigma}_{\gamma\sigma} - \overline{W}^{\sigma}_{\gamma\sigma})\overline{\theta}^{\gamma} \tag{2.6}$$

Now we turn to nonsymmetric metrization of a bundle $\underline{P}$. We define a nonsymmetric tensor γ on a bundle manifold P such that

$$\gamma = \pi^{*}g \oplus \ell_{ab}\theta^{a} \otimes \theta^{b} \tag{2.7}$$

where π is a projection from P to E. On $\underline{P}$ we define a connection ω (a 1-form with values in a Lie algebra g of G. In this way we can introduce on P (a bundle manifold) a frame $\theta^{A} = (\pi^{*}(\overline{\theta}^{\alpha}), \theta^{a})$ such that

$\theta^{a} = \lambda\omega^{a}, \quad \omega = \omega^{a}X_{a}, \quad a = 5,6,\ldots,n+4, \quad n = \dim G = \dim \mathfrak{g}, \quad \lambda = const.$

Thus our nonsymmetric tensor looks like

$$\gamma = \gamma AB\theta^{A} \otimes \theta^{B}, \ A,B=1,2\ldots..,n+4, \tag{2.8}$$

$$l_{ab}=h_{ab}+\mu K_{ab} \tag{2.9}$$

where h_{ab} is a bi invarian t Killing–Cartan tensor on G and k_{ab} is a right-invarian t skew- symmetric tensor on G, $\mu=const$

We have

$$h_{ab} = C^{c}_{ad}C^{d}_{bc} = h_{ab} \qquad K_{ab} = -K_{ba} \tag{2.10}$$

Thus we can write $\overline{\gamma}(X,Y) = \overline{g}(\pi'X, \pi'Y) + \lambda^{2}h(\omega(X),\omega(Y))$ (2.11)

$$\overline{\gamma}(X,Y) = \underline{g}(\pi'X, \pi'Y) + \lambda^{2}k(\omega(X),\omega(Y)) \tag{2.12}$$

($C^{a}{}_{bc}$ are structural constant s of the Lie algebra g).

$\overline{\gamma}$ is the symmetr ic part of γ and $\underline{\gamma}$ is the anti symmetr ic part of γ We have as usual

$$[X_{a}, X_{b}] = C^{c}_{ab}X_{c} \tag{2.13}$$

and

$$\Omega = \frac{1}{2}H^{a}\mu\upsilon\theta^{\mu}\Lambda\theta^{\nu} \tag{2.14}$$

is a curvature of the connection ω

$$\Omega = d\omega + \frac{1}{2}[\omega,\omega] \tag{2.15}$$

The frame θ^{A} on P is partially nonholonomic. We have

$$d\theta^{a} = \frac{\lambda}{2}(H^{a}_{\mu\upsilon}\theta^{\mu}\Lambda\theta^{\upsilon} - \frac{1}{\lambda^{2}}C^{a}_{bc}\theta^{b}\Lambda\theta^{c}) \neq 0 \tag{2.16}$$

Even if the bundle $\underline{P}$ is trivial, i.e. for Ω=0 This is different than

***Corresponding author:** Kalinowski MW, Bioinformatics Laboratory, Medical Research Centre, Polish Academy of Sciences, Poland
E-mail: markwkal@bioexploratorium.pl

in an electromagnetic case explanied by Kalinowski MW [3]. Our nonsymmetric metrization of a principal fiber bundle gives us a right-invariant t structure on P with respect to an action of a group G on P [3]. Having P nonsymmetric ally metrized one defines two connection s on P right- invarian t with respect to an action of a group G on P. We have

$$\gamma AB = \begin{pmatrix} g_{\alpha\beta} & 0 \\ 0 & l_{ab} \end{pmatrix} \quad (2.17)$$

In our left horizontal frame θ^A

$$D_{\gamma AB} = \gamma ADQ^D BC^{(\Gamma)\theta^C} \quad (2.18)$$

$$Q^D BD^{(\Gamma)} = 0 \cdot \quad (2.19)$$

where D is an exterior covariant derivative with respect to a connection $\omega^A{}_B = \Gamma^A{}_B, \theta^C$ on P and $Q^A{}_B,(\Gamma)$ its torsion. One can solve Equation (2.18)- (2.19) getting the following results

$$\omega^A B = \begin{pmatrix} \pi^*(\overline{\omega}^\alpha\beta) - \ell_{db} g^{\mu\alpha} L^d{}_{\mu\beta}\theta^b L^a{}_{\beta\gamma}\theta^\gamma \\ \ell_{db} g^{\alpha\beta}(2H^d_{\gamma\beta} - L^d_{\gamma\beta})\theta^\gamma \tilde{\omega}^a_b \end{pmatrix} \cdot \quad (2.20)$$

where $g^{\mu\alpha}$ is an inverse tensor of $g(\alpha)$

$$g_{\alpha\beta} g^{\gamma\beta} = g_{\beta\alpha} g^{\beta\gamma} = \delta^\gamma_\alpha \quad (2.21)$$

$L^d{}_{\gamma},= -L^a{}_{\beta},$ is an Ad-type tensor on P such that

$$\ell_{dc} g\mu\beta g^{\gamma\mu} L^d_{\gamma\alpha} + \ell_{cd} g\alpha\mu g^{\mu\gamma} L^d_{\beta\gamma} = 2\ell_{cd} g\alpha\mu g^{\mu\gamma} H^d_{\beta\gamma} \quad (2.22)$$

$\tilde{\omega}^a_b = \tilde{\Gamma}^a_{bc}\theta^c$ is a connection on an internal space (typical fiber) compatible with a metric ℓ_{ab} such that

$$\ell_{db}\tilde{\Gamma}^d_{ac} + \ell_{ab}\tilde{\Gamma}^d_{cb} = -\ell_{db} C^d_{ac} \quad (2.23)$$

$$\tilde{\Gamma}^a_{ba} = 0, \tilde{\Gamma}^a_{bc} = -\tilde{\Gamma}^a_{cb} \quad (2.24)$$

and of course $\tilde{Q}^a_{bc}(\tilde{\Gamma}) = 0$ where is a torsion of the connection $\tilde{\omega}^a_b$ We also introduce an inverse tensor of $g(\alpha)$

$$g(\alpha\beta)\tilde{g}^{(\alpha\gamma)} = \delta^\gamma_\beta \quad (2.25)$$

We introduce a second connection on P defined as

$$W^A{}_B = \omega^A{}_B - \frac{A}{3(n+2)}\delta^A_B \overline{W} \quad (2.26)$$

$\overline{W}$ is a horizontal one form

$$\overline{W} = hor\overline{W} \quad (2.27)$$

$$\overline{W} = \overline{W}_\upsilon \theta^\upsilon = \frac{1}{2}(\overline{W}^\sigma_{\upsilon\sigma} - \overline{W}^\sigma_{\sigma\upsilon}) \quad (2.28)$$

In this way we define on P all analogues of four- dimension al quantities from NGT [6,11]. It means, $(n+4)$ dimension al analogues from Moffat theory of gravitation, i.e. two connection s and a nonsymmetric metric γ_{AB}. Those quantities are right- invarian t with respect to an action of a group G on P. One can calculate a scalar curvature of a connection $W^A{}_B$ getting the following result [1-3].

$$R(W) = \overline{R}(\overline{W}) - \frac{\lambda^2}{4}(2\ell_{cd} H^{c} H^{d} - \ell_{cd} L^{c} \quad H \quad) + \tilde{R}(\tilde{\Gamma}) \quad (2.29)$$

Where

$$R(W) = \gamma^{AB}(R^C{}_{ABC}(W) + \frac{1}{2}R^C{}_{CAB}(W)) \quad (2.30)$$

is a Moffat-Ricci curvature scalar for the connection $W^A{}_B$, $\overline{R}(\overline{W})$ is a Moffat–Ricci curvature scalar for the connection $\overline{W}^\alpha_\beta$, and $\tilde{R}(\tilde{\Gamma})$ is a Moffat–Ricci curvature scalar for the connection $\tilde{\omega}^a_b$

$$H^\alpha = g^{[\mu\nu]} H^a_{\mu\nu} \quad (2.31)$$

$$L^{\alpha\mu\nu} = g^{\alpha\mu} g^{\beta\mu} L^a_{\alpha\beta} \quad (2.32)$$

Usually in ordinary (symmetric) Kaluza–Klein Theory one has $\lambda = 2\frac{\sqrt{G_N}}{c^2}$ where GN is a Newtonian gravitational constant and c is the speed of light. In our system of units this is the same as in Non symmetric Kaluza–Klein Theory in an electromagnetic case [3,4] In the non- Abelian Kaluza –Klein Theory which unifies GR and Yang-Mills field theory we have a Yang-Mills lagrangian and a cosmological term. Here we have

$$\mathcal{L}_{\text{YM}} = \frac{1}{8\pi}\ell_{cd}(2H^c H^d - L^{c\mu\nu} H^d_{\mu\nu}) \quad (2.33)$$

and $\tilde{R}(\tilde{\Gamma})$ plays a role of a cosmolog ical term.In order to incorporate a spontaneous symmetr y breaking and Higgs' mechanism in our geometrical unification of gravitation and Yang-Mills' fields we consider a fiber bundle $\underline{P}$ over a base manifold $E \times G/G_0$, where E is a space-time, $G_0 \subset G$, G_0, G are semisimple Lie groups. Thus we are going to combine a Kaluza–Klein theory with a dimension al reduction procedure.

Let $\underline{P}$ be a principal fiber bundle over $V=E \times M$ with a structural group H and with a projection π, where

$M= G/G_0$ is a homogeneous space, G is a semisimple Lie group and G_0 its semisimple Lie subgroup. Let us suppose that (V, γ) is a manifold with a nonsymmetric metric tensor

$$\gamma_{AB} = \gamma_{(AB)} + \gamma_{[AB]} \quad (2.34)$$

The signature of the tensor γ is $(+,---,\underbrace{-\cdots-}_{n_1},\underbrace{-\cdots-}_{n})$ Let us introduce a natural Phenomenon

$$\theta^{\bar{A}} = (\pi^*(\theta^A), \theta^0 = \lambda\omega^a) \quad (2.35)$$

It is convenient to introduce the following notation. Capital Latin indices with tilde $\tilde{A},\tilde{B},\tilde{C}$ run $1,2,3\ldots,m+4$, m=dim$H$+ dim$M$=n+ dim$M$=+n_1 , $n_{1=}$ dim$M$, n=dim$H$. Lower Greek indices $\alpha,\beta,\gamma,\delta=1,2,3,4$ and lower Latin indices $a,b,c,d=n_1+5$, $n_2+5,\ldots$, $n_1+6,\ldots,m+4$ Capital Latin indices $A,B,C=1,2,\ldots$, n_1+4 . Lower Latin indices with tilde $\tilde{a},\tilde{b},\tilde{c}$ run $5,6\ldots$, n_1+4 The symbol over θ^A and other quantities indicates that these quantities are defined on V. We have of course $n_{1=}$ dim$G$-dim $G_0=n_2-(n_2-n_1)$ where dim $G=n_2$dim $G_0=n_2-n_1$, $m=n_1+n$.

On the group H we define a bi- invarian t (symmetr ic) Killing–Cartan tensor

$$h(A,B) = h_{ab} A^a B^b \quad (2.36)$$

We suppose H is semisimple, it means det$(h_{ab})\neq 0$. We define a skew- symmetr ic right- invarian t tensor on H

$$k(A,B) = k_{bc} A^b B^c \quad k_{bc} = -k_{cb}$$

Let us turn to the nonsymmetric metrization of $\underline{P}$.

$$k(X,Y) = \gamma(X,Y) + \lambda^2 \ell_{bc}\omega^a(X)\omega^b(Y) \quad (2.37)$$

where

$$\ell_{ab} = h_{ab} + \xi K_{ab} \quad (2.38)$$

is a nonsymmetric right-invarian t tensor on H.One gets in a matrix form (in the natural frame (2.35))

$$\begin{pmatrix} \gamma_{AB} & 0 \\ 0 & \ell_{ab} \end{pmatrix} \quad (2.39)$$

$\det(\ell_{ab}) \neq 0$, $\xi = const$ and real, then

$$\ell_{ab}\ell^{ac} = \ell_{ba}\ell^{ca} = \delta^c_b \quad (2.40)$$

The signature of the tensor k is $(+,---,\underbrace{-\cdots-}_{n_1},\underbrace{--\cdots-}_{n})$. As usual, we have commutation relations for Lie algebra of H, $\mathfrak{h}$

$$[X_a, X_b] = C^c_{ab} X_c \quad (2.41)$$

This metrization of $\underline{P}$ is right- invarian t with respect to an action of H on P.Now we should nonsymmetric ally metrize $M=G/G_0$. M is a homogeneous space for G (with left action of group G). Let us suppose that the Lie algebra of G, g has the following reductive decomposition

$$\mathfrak{g} = \mathfrak{g}_0 + \mathfrak{m} \quad (2.42)$$

where $\mathfrak{g}_0$ is a Lie algebra of G_0 (a subalgebra of $\mathfrak{g}$) and m (the complement to the subalgebra $\mathfrak{g}_0$) is AdG_0 invarian t, + means a direct sum. Such a decomposition might be not unique, but we assume that one has been chosen. Sometimes one assumes a stronger condition for m, the so called symmetr y requirement

$$[\mathfrak{m}, \mathfrak{m}] \subset \mathfrak{g}_0 \quad (2.43)$$

Let us introduce the following notation for generators of $\mathfrak{g}$:

$$Y_i \in \mathfrak{g}, Y_{\tilde{i}} \in \mathfrak{g}, Y_{\hat{a}} \in \mathfrak{g}_0 \quad (2.44)$$

This is a decomposition of a basis of $\mathfrak{g}$ according to (2.42). We define a symmetr ic metric on M using a Killing–Cartan form on G in a classical way. We call this tensor h_0. Let us define a tensor field $h^0(x)$ on G/G_0, $x \in G/G_0$, using tensor field h on G. Moreover, if we suppose that h is a bi invarian t metric on G (a Killing–Cartan tensor) we have a simpler construction.The complement m is a tangent space to the point $\{\varepsilon G_0\}$ of M, ε is a unit element of. We restrict h to the space m only. Thus we have $h^0\{\varepsilon G_0\}$ at one point of M. Now we propagate $h^0\{\varepsilon f G_0\}$ using a left action of the group G $h^0(\{fG_0\}) = (L_f^{-1})^*(h^0(\{\varepsilon G_0\}))$. $h^0(\{\varepsilon G_0\})$ is of course AdG_0 invarian t tensor defined on m and $L_f^* h^0 = h^0$

We define on M a skew- symmetr ic 2-form k^0. Now we introduce a natural frame on M. Let $f^i{}_j$, be structure constant s of the Lie algebra $\mathfrak{g}$, i.e.

$$[Y_j, Y_k] = f^i_{jk} Y_i \quad (2.45)$$

Y_j are generators of the Lie algebra g. Let us take a local section $s: V^{®} \to G/G_0$ of a natural bundle $G \mapsto G/G_0$ where. $V \subset M = G/G_0$ The local section s can be considered as an introduction of a coordinate system on M.

Let ω_{MC} be a left- invarian t Maurer–Cartan form and let

$$\omega^\sigma{}_{\tilde{I}C} = \sigma^* \omega_{\tilde{I}C} \quad (2.46)$$

Using de composition (2.42) we have

$$\omega^\sigma \tilde{I}\ C = \omega^\sigma{}_0 + \omega^\sigma{}_{\mathrm{m}} = \theta^i Y_i + \bar{\tau}^{\hat{a}} Y_{\hat{a}} \quad (2.47)$$

It is easy to see that $\bar{\theta}^{\tilde{a}}$ is the natural (left- invarian t) frame on M and we have

$$h^0 = \overset{0}{h}_{\tilde{a}\tilde{b}} \bar{\theta}^{\tilde{a}} \otimes \bar{\theta}^{\tilde{b}} \quad (2.48)$$

$$k^0 = \overset{0}{k}_{\tilde{a}\tilde{b}} \bar{\theta}^{\tilde{a}} \wedge \bar{\theta}^{\tilde{b}} \quad (2.49)$$

According to our notation $\tilde{a}, \tilde{b} = 5, 6, \ldots, n_1 + 4$.

Thus we have a nonsymmetric metric on M

$$\gamma_{\tilde{a}\tilde{b}} = r^2(\overset{0}{h}_{\tilde{a}\tilde{b}} + \zeta \overset{0}{k}_{\tilde{a}\tilde{b}}) = r^2 g_{\tilde{a}\tilde{b}} \quad (2.50)$$

Thus we are able to write down the nonsymmetric metric on $V=E \times M=E \times G/G_0$

$$\gamma_{AB} = \begin{pmatrix} g_{\alpha\beta} & 0 \\ 0 & r^2 g_{\tilde{a}\tilde{b}} \end{pmatrix} \quad (2.51)$$

where

$$g_{\alpha\beta} = g(\alpha\beta) + g[\alpha\beta]$$

$$g_{\tilde{a}\tilde{b}} = h^0_{\tilde{a}\tilde{b}} + \zeta k^0 b^0_{\tilde{a}\tilde{b}}$$

$$k^0_{\tilde{a}\tilde{b}}, = -k^0_{\tilde{b}\tilde{a}}$$

$$h^0_{\tilde{a}\tilde{b}}, = -h^0_{\tilde{b}\tilde{a}}$$

$\alpha, \beta = 1, 2, 3, 4$, $\quad \tilde{a}, \tilde{b} = 5, 6, \ldots, n_1 + 4 = \dim M + 4 = \dim G - \dim G_0 + 4$.

The frame $\bar{\theta}^{\tilde{a}}$ is unholonomic:

$$d\bar{\theta}^{\tilde{a}} = \frac{1}{2} k^{\tilde{a}}_{\tilde{b}\tilde{c}} \bar{\theta}^{\tilde{b}} \wedge \bar{\theta}^{\tilde{c}} \quad (2.52)$$

where $\kappa, b\tilde{c}$, are coefficient s of nonholonomicity and depend on the point of the manifold $M=G/G_0$ (they are not constant in general). They depend on the section s and on the constants $f^{\tilde{a}}_{\tilde{b}\tilde{c}}$. We have here three groups H, G, G_0 H, G, G_0. Let us suppose that there exists a homomorphism μ between G_0 and H, $\mu(G_0)$

$$\mu : G_0 \to H \quad (2.53)$$

such that a centralizer of $\mu(G_0)$ in H, C^μ is isomorphic to G. C^μ, a centralizer of $\mu(G_0)$ in H, is a set of all element s of H which commute with element s of $\mu(G_0)$, which is a subgroup of H. This means that H has the following structure, $C^\mu = G$.

$$\mu(G_0) \otimes G \subset H \quad (2.54)$$

If μ is a iso morphi sm between G_0 and $\mu(G_0)$ one gets

$$G_0 \otimes G \subset H \quad (2.55)$$

Let us denote by μ' a tangent map to μ at a unit element. Thus μ' is a differential of μ acting on the Lie algebra element s. Let us suppose that the connection ω on the fiber bundle $\underline{P}$ is invarian t under group action of G on the manifold $V=E \times G/G_0$. According to Kobayashi [14-17] this means the following.

Let e be a local section of $\underline{P}$, $e: V \subset U \to P$ and $A = e^*\omega$. Then for every $g \in G$ there exists a gauge transformation ρ_g such that

$$f^*(g)A = Ad_{\rho^{-1}g} + \rho^{-1}{}_g dg_g \quad (2.56)$$

f^*means a pull-back of the action f of the group G on the manifold V. According to Hlavaty [13-25] we are able to write a general form for such an ω. Following [17] we have

$$\omega = \tilde{\omega}_E + \mu' \mathrm{o} \omega^\sigma_0 + \Phi \mathrm{o} \omega^\sigma_m \quad (2.57)$$

(An action of a group G on $V=E \times G/G_0$ means left multiplication on a homogeneous space $M=G/G_0$.) where $\omega^\sigma{}_0 + \omega^\sigma{}_{\mathrm{m}} = \omega^\sigma{}_M$, are components of the pull-back of the Maurer–Cartan form from the de composition (2.47) $\tilde{\omega}_E$ is a connection defined on a fiber bundle Q over a space-time E with structural group C^μ and a projection π_E. Moreover, $C^\mu = G$ and $\tilde{\omega}_E$ is a 1-form with values in the Lie algebra $\mathfrak{g}$.

This connection describes an ordinary Yang–Mills' field gauge group $C^{\mu}=G$ on the space-time E. Φ is a function on E with values in the space $\tilde{S}$ of linear maps

$$\Phi : \mathfrak{m} \to \mathfrak{h} \tag{2.58}$$

satisfying Φ

$$\Phi[X_0, X] = \Phi[\mu' X_0, \Phi(X)] \tag{2.59}$$

Thus

$$\begin{aligned} \tilde{\omega}_E &= \tilde{\omega}^i_E Y_i, Y_i \in \mathfrak{g}, \\ \omega^{\sigma}_0 &= \theta^{\hat{i}} Y_{\hat{i}}, Y_{\hat{i}} \in \mathfrak{g}_0, \\ \omega^{\sigma}_m &= \overline{\theta}^{\tilde{a}} Y_{\tilde{a}}, Y_{\tilde{a}} \in \mathfrak{m}. \end{aligned} \tag{2.60}$$

Let us write condition (2.57) in the base of left-invarian t form $\theta^{\hat{i}}, \overline{\theta}^{\tilde{a}}$, which span respectively dual spaces to g_0 and m [24,25]. It is easy to see that

$$\Phi \circ \omega^{\sigma}_m = \Phi^a_{\tilde{a}}(x) \overline{\theta}^{\tilde{a}} X_a, X_a \in \mathfrak{h} \tag{2.61}$$

and

$$\mu' = \mu^a_{\hat{i}} \theta^{\hat{i}} X_a \tag{2.62}$$

From (2.59) one gets

$$\Phi^c_{\tilde{b}}(x) f^{\tilde{b}}_{\hat{i}\tilde{a}} = \mu^a_{\hat{i}} \Phi^b_{\tilde{a}}(x) C^c_{ab} \tag{2.63}$$

where $f^{\tilde{b}}_{\hat{i}\tilde{a}}$ are structure constant s of the Lie algebra g and $C^c{}_a$, are structure constant s of the Lie algebra $\mathfrak{h}$ Equation (2.63) is a constraint on the scalar field $\Phi^a_{\tilde{a}}(x)$. For a curvature of ω one gets

$$\Omega = \frac{1}{2} H^C AB \theta^{A} \wedge \theta^B X_C = \frac{1}{2} \tilde{H}^i_{\mu\nu} \theta^{\mu} \wedge \theta^{\nu} \alpha^c_i X_c + \overset{gauge}{\nabla}_{\mu} \Phi^c_{\tilde{a}} \theta^{\mu} \wedge \theta^{\tilde{a}} X_c + \frac{1}{2} C^c_{ab} \Phi^a_{\tilde{a}} \Phi^b_{\tilde{b}} \theta^{\tilde{a}} \wedge \theta^{\tilde{b}} X_c - \frac{1}{2} \Phi^c_{\tilde{d}} f^{\tilde{d}}_{\tilde{a}\tilde{b}} \theta^{\tilde{a}} \wedge \theta^{\tilde{b}} X_c \tag{2.64}$$

Thus we have

$$H^C_{\mu\nu} = \alpha^c_i \tilde{H}^i_{\mu\nu} \tag{2.65}$$

$$H^C_{\mu\tilde{a}} = \overset{gauge}{\nabla}_{\mu} \Phi^c_{\tilde{a}} = -H^C_{\tilde{a}\mu} \tag{2.66}$$

$$H^C_{\tilde{a}\tilde{b}} = C^c_{ab} \Phi^a_{\tilde{a}} \Phi^b_{\tilde{b}} - \mu^c_{\hat{i}} f^{\hat{i}}_{\tilde{a}\tilde{b}} - \Phi^c_{\tilde{d}} f^{\tilde{d}}_{\tilde{a}\tilde{b}} \tag{2.67}$$

where $\overset{gauge}{\nabla}_{\mu}$ means gauge derivative with respect to the connection $\tilde{\omega}_E$ defined on a bundle q over a space-time E with a structur al group G

$$Y_i = \alpha^c_i X_c \tag{2.68}$$

$\tilde{H}^i_{\mu\nu}$ is the curvature of the connection $\tilde{\omega}_E$ in the base $\{Y_i\}$, generators of the Lie algebra of the Lie group G g, α^c_i is the matrix which connects $\{Y_i\}$ with $\{X_c\}$. Now we would like to remind that indices a,b,c refer to the Lie algebra h, $\tilde{a},\tilde{b},\tilde{c}$ to the space m (tangent space to M), $\hat{i},\hat{j},\hat{k}$ to the Lie algebra g_0 and i,j,k to the Lie algebra of the group G, g. The matrix α^c_i establishes a direct relation between generators of the Lie algebra of the subgroup of the group H iso morphi c to the group G.

Let us come back to a construction of the Nonsymmetric Kaluza–Klein Theory on a manifold P. We should define connection s. First of all, we should define a connection compatible with a nonsymmetric tensor γ_{AB}, Equation (2.51)

$$\overline{\omega}^A B = \overline{\Gamma}^A BC \theta^C \tag{2.69}$$

$$\overline{D} \gamma AB = \gamma AD \overline{Q}^D BC(\overline{\Gamma}) \theta^C \tag{2.70}$$

$$\overline{Q}^D BD(\overline{\Gamma}) = 0$$

where $\overline{D}$ is the exterior covariant derivative with respect to $\overline{\omega}^A B$ and $\overline{Q}^D BC(\overline{\Gamma})$ its torsion. Using (2.51) one easily finds that the connection (2.69) has the following shape

$$\overline{\omega}^A B = \begin{pmatrix} \pi^*_E(\omega^{-\alpha}\beta) & 0 \\ 0 & \hat{\overline{\omega}}^{\tilde{a}}_{\tilde{b}} \end{pmatrix} \tag{2.71}$$

where $\omega^{-\alpha}\beta = \overline{\Gamma}^{\alpha}\beta\gamma\overline{\theta}^{\gamma}$ is a connection on the space-time E and on the manifold $M = G/G_0$ with the following properties .

$$\overline{D}_{\alpha\beta} = g\alpha\delta \overline{Q}^{\delta}\beta\gamma(\overline{\Gamma})\overline{\theta}^{\gamma} = 0 \tag{2.72}$$

$$\overline{Q}^{\alpha}\beta\alpha(\overline{\Gamma}) = 0$$

$$\hat{\overline{D}} g_{\tilde{a}\tilde{b}} = g_{\tilde{a}\tilde{d}} \hat{\overline{Q}}^{\tilde{d}}_{\tilde{b}\tilde{c}}(\hat{\overline{\Gamma}}) \tag{2.73}$$

$$\overline{Q}^{\tilde{d}}_{\tilde{b}\tilde{d}}(\hat{\overline{\Gamma}}) = 0 \tag{2.74}$$

($\overline{D}$ is an exterior covariant derivative with respect to a connection $\omega^{-\alpha}\beta.\overline{Q}^{\alpha}_{\beta\gamma}$ is a tensor of torsion of a connection $\hat{\overline{D}}$ is an exterior covariant derivative of a connection $\hat{\overline{\omega}}^{\tilde{a}}_{\tilde{b}}$ and $\hat{\overline{Q}}^{\tilde{a}}_{\tilde{b}\tilde{c}}$ its torsion.On a space-time E we also define the second affine connection such that

$$\overline{W}^{\alpha}\beta = \overline{\omega}^{\alpha}{}_{\beta} - \frac{2}{3}\delta^{\alpha}\beta\overline{W} \tag{2.75}$$

$$\overline{W} = \overline{W}\gamma\overline{\theta}^{\gamma} = \frac{1}{2}(\overline{W}^{\sigma}_{\gamma\sigma} - \overline{W}^{\sigma}_{\gamma\sigma})$$

We proceed a nonsymmetric metrization of a principal fiber bundle $\underline{P}$ according to (2.51). Thus we define a right-invarian t connection with respect to an action of the group H compatible with a tensor $\kappa_{\tilde{A}\tilde{B}}$

$$\begin{aligned} D_{K\tilde{A}\tilde{B}} &= K_{\tilde{A}\tilde{D}} Q^{\tilde{D}}_{\tilde{B}\tilde{C}}(\Gamma)\theta^{\tilde{C}} \\ Q^{\tilde{D}}_{\tilde{B}\tilde{D}}(\Gamma) &= 0 \end{aligned} \tag{2.76}$$

where $\omega^{\tilde{A}}{}_{\tilde{B}} = \Gamma^{\tilde{A}}{}_{\tilde{B}}, \tilde{\theta}^{\tilde{C}}$ D is an exterior covariant derivative with respect to the connection $\omega^{\tilde{A}}{}_{\tilde{B}}$ and $Q^{\tilde{A}}{}_{\tilde{B}}$, its torsion. After some calculations one finds

$$\omega^{\tilde{A}}{}_{\tilde{B}} = \begin{pmatrix} \pi^*(\overline{\omega}^A B) - \ell_{db}\gamma^{MA} L^d MB\theta^b L^a BC\theta^C \\ \ell_{db}\gamma^{AB}(2H^d_{CB} - L^d_{CB})\theta^C \tilde{\omega}^a_b \end{pmatrix} \tag{2.77}$$

Where

$$L^d_{MB} = -L^d_{BM} \tag{2.78}$$

$$\ell_{dc}\gamma MB\gamma^{CM} L^d_{CA} + \ell_{cd}\gamma AM\gamma^{MC} L^d_{BC} = 2l_{cd}\gamma AM\gamma^{MC} H^d_{BC} \tag{2.79}$$

$L^d{}_C$, is Ad-type tensor with respect to H (Ad- covariant on $\underline{P}$)

$$\tilde{\omega}^a_b = \tilde{\Gamma}^a_b \theta^C \tag{2.80}$$

$$\ell_{db}\tilde{\Gamma}^d_{ac} + \ell_{ab}\tilde{\Gamma}^d_{cb} = -\ell_{db}C^d_{ac} \tag{2.81}$$

$$\tilde{\Gamma}^d_{ac} = -\tilde{\Gamma}^d_{ca}, \tilde{\Gamma}^a_{ad} = 0 \tag{2.82}$$

We define on P a second connection

$$W^{\tilde{A}}{}_{\tilde{B}} = W^{\tilde{A}}{}_{\tilde{B}} - \frac{A}{3(m+2)}\delta^{\tilde{A}}{}_{\tilde{B}}\overline{W} \tag{2.83}$$

Thus we have on P all $(m+4)$ dimension al analogues of geometrical quantities from NGT, i.e. $W^{\tilde{A}}{}_{\tilde{B}}$, $\omega^{\tilde{A}}{}_{\tilde{B}}$ *and* $\kappa_{\tilde{A}\tilde{B}}$.

Let us calculate a Moffat–Ricci curvature scalar for the connection $W^{\tilde{A}}{}_{\tilde{B}}$

Hierarchy of a Symmetry Breaking

Let us incorporate in our scheme a hierarchy of a symmetry breaking. In order to do this let us consider a case of the manifold

$$M = M_0 \times M_1 \times \ldots \times M_{k-1} \tag{3.1}$$

where

$$\dim M_i = \bar{n}_i, \quad i = 0,1,2,\ldots,k-1 \tag{3.2}$$

$$\dim M = \sum_{i=0}^{k-1} \bar{n}_i, \tag{3.3}$$

$$M_i = G_{i+1} / G_i. \tag{3.4}$$

Every manifold M_i is a manifold of vacuum states if the symmetr y is breaking from G_{i+1} to G_i, G_k=G.

Thus

$$G_0 \subset G_1 \subset G_2 \subset \ldots \subset G_k = G. \tag{3.5}$$

We will consider the situation when

$$M \simeq G / G_0. \tag{3.6}$$

This is a constraint in the theory. From the chain (3.5) one gets

$$\mathfrak{g}_0 \subset \mathfrak{g}_1 \subset \ldots \subset \mathfrak{g}_k = \mathfrak{g} \tag{3.7}$$

and

$$\mathfrak{g}_{i+1} = \mathfrak{g}_i + \mathfrak{m}_i, \quad i = 0,1,\ldots,k-1. \tag{3.8}$$

The relation (3.6) means that there is a diffeomorphism g onto G/G_0 such that

$$g : \prod_{i=0}^{k-1} (G_{i+1} / G_i) \rightarrow G / G_0. \tag{3.9}$$

This diffeomorphism is a deformation of a product (3.1) in G/G_0. The theory has been constructed for the case considered before with G_0 and G. The multiplet of Higgs' fields Φ breaks the symmetr y from G to G_0 (equivalently from G to G_0 in the false vacuum case). g_i mean Lie algebras for groups G_i and m_i a complement in a decomposition (3.8) . On every manifold M_i we introduce a radius r_i (a "size" of a manifold) in such a way that $r_i > r_{i+1}$. On the manifold G/G_0 we define the radius r as before. The diffeomorphism g induces a contragradient transformation for a Higgs field Φ in such a way that

$$g^* \Phi = (\Phi_0, \Phi_1, \ldots, \Phi_{k-1}). \tag{3.10}$$

The fields Φ_i, i=0,…,k-1.

In this way we get the following decomposition for a kinetic part of the field Φ and for a potential of this field:

$$\mathcal{L}_{kin}(\overset{\text{gauge}}{\nabla} \Phi) = \sum_{i=0}^{k-1} \mathcal{L}^{i}_{\text{kin}}(\overset{\text{gauge}}{\nabla} \Phi_i)0 \tag{3.11}$$

$$V(\Phi) = \sum_{i=0}^{k-1} V^i(\Phi_i)1 \tag{3.12}$$

where

$$V_i = \int_{M_i} \sqrt{|\tilde{g}_i|}\, d^{\bar{n}_i} x \tag{3.13}$$

$$\tilde{g} = \det(g_{i b_i a_i}) \tag{3.14}$$

$$g_{i\tilde{b}_i\tilde{a}_i} \tag{3.15}$$

Equation (3.5) is a non symmetr ic tensor on a manifold M_i.

$$V^i(\Phi_i) = \frac{l_{ab}}{V_i} \int_{M_i} \sqrt{|\tilde{g}_i|}\, d^{\bar{n}_i} x [2 g^{[\tilde{m}_i \tilde{n}_i]} (C^a_{cd} \Phi^c_{i\tilde{m}_i} \Phi^d_{i\tilde{n}_i} - \mu^a_{\hat{i}_i} f^{\hat{i}_i}_{\tilde{m}_i \tilde{n}_i} - \Phi^a_{i\tilde{e}_i} f^{\tilde{e}_i}_{\tilde{m}_i \tilde{n}_i}) \times g^{[\tilde{a}_i \tilde{b}_i]}$$
$$(C^b_{ef} \Phi^e_{i\tilde{a}_i} \Phi^f_{i\tilde{b}_i} - \mu^b_{\hat{j}_i} f^{\hat{j}_i}_{\tilde{a}_i \tilde{b}_i} - \Phi^b_{i\tilde{a}_i} f^{\tilde{d}_i}_{\tilde{a}_i \tilde{b}_i}) - g_i^{\tilde{a}_i \tilde{m}_i} g_i^{\tilde{b}_i \tilde{n}_i} L^a_{\tilde{a}_i \tilde{b}_i} (C^b_{cd} \Phi^c_{i\tilde{m}_i} \Phi^d_{\tilde{n}_i} - \mu^b_{\hat{i}_i} f^{\hat{i}_i}_{\tilde{m}_i \tilde{n}_i} - \Phi^b_{\tilde{e}_i} f^{\tilde{e}_i}_{\tilde{m}_i \tilde{n}_i})], 5 \tag{3.16}$$

$f^{\hat{j}_i}_{\tilde{a}_i \tilde{b}_i}$ are structure constant s of the Lie algebra $\mathfrak{g}_i$. The scheme of the symmetr y breaking acts as follows from the group G_{i+1} to G_i (G_t) (if the symmetr y has been broken up to G_{i+1}). The potential $V^i(\Phi_i)$ has a minimum (global or local) for Φ^k_{icrt}, $k = 0,1$. The value of the remaining part of the sum (3.12) for fields Φ_j, $j < i$, is small for the scale of energy is much lower ($r_j > r_i$, $j < i$). Thus the minimum of $V^i(\Phi_i)$ is an approximate minimum of the remaining part of the sum (3.12)). In this way we have a descending chain of truncations of the Higgs potential. This gives in principle a pattern of a symmetr y breaking. However, this is only an approximate symmetr y breaking. The real symmetr y breaking is from G to G_0 (or to G_0 in a false vacuum case). The important point here is the diffeomorphism g.

$$g^* \Phi^b = (\Phi^b_0, \Phi^b_1, \ldots, \Phi^b_{k-1}) \tag{3.17}$$

$$\Phi^b_i = \Phi^b_{i\tilde{a}_i} \tilde{\theta}^{\tilde{a}_i}, \quad i = 0,\ldots,k-1 \tag{3.18}$$

The shape of g is a true indicator of a reality of the symmetr y breaking pattern. If

$$g = \mathrm{Id} + \delta g \tag{3.19}$$

where δg is in some sense small and Id is an identity, the sums (3.11)-(3.12) are close to the analogous formulae from the expanation of Kalinowski [5,10]. The smallness of g is a criterion of a practical application of the symmetr y breaking pattern (3.5) . It seems that there are a lot of possibilities for the condition (3.9). Moreover, a smallness of δg plus some natural conditions for groups G_i can narrow looking for grand unified models. Let us notice that the decomposition of M results in decomposition of cosmolog ical terms

$$\underset{\sim}{P} = \sum_{i=0}^{k-1} \underset{\sim}{P}_i 9 \tag{3.20}$$

where

$$\underset{\sim}{P}_i = \frac{1}{r_i^2 V_i} \int_{M_i} \sqrt{|\tilde{g}_i|}\, \hat{\tilde{R}}_i(\hat{\tilde{\Gamma}}_i)\, d^{n_i} x 0 \tag{3.21}$$

where $\hat{\tilde{\Gamma}}_i$ is a non symmetr ic connection on M_i compatible with the non symmetr ic tensor $g_{i\tilde{a}_i\tilde{b}_i}$ and $\hat{\tilde{R}}_i(\hat{\tilde{\Gamma}}_i)$ its curvature scalar. The truncation procedure can be proceeded in several ways. Finally let us notice that the energy scale of broken gauge bosons is fixed by a radius r_i at any stage of the symmetr y breaking in our scheme.

Let us consider Equation (3.10) in more details. One gets

$$A^{\tilde{a}}_{i\tilde{a}_i}(y) \Phi^b_{\tilde{a}}(y) = \Phi^b_{\tilde{a}_i}(y_i), \quad y \in M, y_i \in M_i \tag{3.22}$$

where

$$g^*(y) = (A_0 | A_1 | A_2 | \ldots | A_{k-1}) \tag{3.23}$$

$$A_i = (A^{\tilde{a}}_{i\tilde{a}_i})_{\tilde{a}=1,2,\ldots,n_1, \tilde{a}_i=1,2,\ldots,\bar{n}_i}, i = 0,1,2,\ldots,k \tag{3.24}$$

is a matrix of Higgs' fields transformation.

According to our assumptions one gets also:

$$\left(\frac{r_i}{r}\right)^2 g_{i\tilde{a}_i\tilde{b}_i}(y_i) = A^{\tilde{a}}_{\tilde{a}_i}(y) A^{\tilde{b}}_{\tilde{b}_i}(y) g_{\tilde{a}\tilde{b}}(y) \tag{3.25}$$

For g is an invertible map we have $\det g^*(y) \neq 0$.

We have also

$$n_1 = \sum_{i=0}^{k-1} \bar{n}_i \tag{3.26}$$

and

$$\Phi_{\bar{a}}^b(y) = \sum_{i=0}^{k-1} \tilde{A}_{i\bar{a}}^{\bar{a}_i}(y)\Phi_{\bar{a}_i}^b(y_i) \tag{3.27}$$

or

$$\begin{pmatrix} \tilde{A}_0 \\ \tilde{A}_1 \\ \cdots \\ \tilde{A}_{k-1} \end{pmatrix} \tag{3.28}$$

$$\tilde{A}_i = (\tilde{A}_{i\bar{a}}^{\bar{a}_i})_{\bar{a}_i=1,2,\ldots,\bar{n}_i,\bar{a}=1,2,\ldots,n_1} \tag{3.29}$$

such that

$$g(y_0,\ldots,y_{k-1}) = y \tag{3.30}$$

$$(y_0, y_1,\ldots,y_{k-1}) = g^{-1}(y) \tag{3.31}$$

For an inverse tensor $g^{\bar{a}\bar{b}}$ one easily gets

$$(\frac{r_i^2}{r^2})g^{\bar{a}\bar{b}} = \sum_{i=0}^{k-1} \tilde{A}_{i\bar{a}_i}^{\bar{a}} g_i^{\bar{a}_i\bar{b}_i} A_{i\bar{b}_i}^{\bar{b}} \tag{3.32}$$

We have

$$r^{2n_1}\det(g_{\bar{a}\bar{b}}) = \prod_{i=0}^{k-1} r_i^{2\bar{n}_i}\det(g_{i\bar{a}_i\bar{b}_i}). \tag{3.33}$$

In this way we have for the measure

$$d\mu(y) = \prod_{i=0}^{k-1} d\mu_i(y_i) \tag{3.34}$$

where

$$d\mu(y) = \sqrt{\det g}\, r^{n_1} d^{n_1} y \tag{3.35}$$

$$d\mu_i(y_i) = \sqrt{\det g_i}\, r_i^{\bar{n}_i} d^{\bar{n}_i} y_i. \tag{3.36}$$

In the case of $\mathcal{L}_{int}(\Phi,\tilde{A})$ one gets

$$\mathcal{L}_{int}(\Phi,\tilde{A}) = \sum_{i=0}^{k-1} \mathcal{L}_{int}(\Phi_i,\tilde{A}) \tag{3.37}$$

where

$$\mathcal{L}_{int}(\Phi_i,\tilde{A}) = h_{ab}\mu_i^a \tilde{H}^l \underline{g}_{\underline{i}}^{[\bar{a}_i\bar{b}_i]}(C_{cd}^b \Phi_{i\bar{a}_i}^c \Phi_{i\bar{b}_i}^d - \mu_i^b f_{\bar{a}_i\bar{b}_i}^i - \Phi_{\bar{a}_i}^b f_{\bar{a}_i\bar{b}_i}^{\bar{d}_i}) \tag{3.38}$$

where

$$\underline{g}_{\underline{i}}^{[\bar{a}_i\bar{b}_i]} = \frac{1}{V_i} \int_{M_i} \sqrt{|g_i|}\, d^{\bar{n}_i} x g_i^{[\bar{a}_i\bar{b}_i]}, \quad i = 0,1,2,\ldots,k-1 \tag{3.39}$$

Moreover, to be in line in the full theory we should consider a chain of groups H_i, $i = 0,1,2,\ldots,k-1$, in such a way that

$H_0 \subset H_1 \subset H_2 \subset \ldots \subset H_{k-1} = H$ (3.40)

For every group H_i we have the following assumptions

$$G_i \subset H_i \tag{3.41}$$

and G_{i+1} is a centralizer of G_i in H_i. Thus we should have

$$G_i \otimes G_{i-1} \subset H_i, \quad i = 0,1,2,\ldots,k-1.0 \tag{3.42}$$

We know from elementary particles physics theory that

$G_0 = U_{el} \otimes SU(3)_{color}$,

$G_1 = SU(2)_L \otimes U(1)_Y \otimes SU(3)_{color}$

and that $G2$ is a group which plays the role of H in the case of a symmetr y breaking from $SU(2)_L \otimes U_Y(1)$ to $U_{el}(1)$. However, in this case because of a factor $U(1)$, $M=S^2$. Thus $M_0 = S^2$ and $G2 \subset H_0$.

It seems that in a reality we have to do with two more stages of a symmetr y breaking. Thus $k=3$. We have

$$M \simeq S^2 \times M_1 \times M_2 \tag{3.43}$$

$$M = G/(U(1)\otimes SU(3)) \tag{3.44}$$

$$M_1 = G_1/(SU(2)\times U(1)\times SU(3))$$

$$U(1)\otimes SU(3) \subset SU(2)\otimes U(1)\otimes SU(3) \subset G_2 \otimes SU(3) \subset G_3 = G3 \tag{3.45}$$

and

$$G_1 \subset H_1 \subset H \tag{3.46}$$

$$U(1)\otimes SU(3)\otimes G \subset H \tag{3.47}$$

$$(U(1)\otimes SU(2)\otimes SU(3))\otimes G_2 \subset H_1 \tag{3.48}$$

and

$$G_2 \otimes G \subset H_2 = H \tag{3.49}$$

$$M_2 = G/G_1 \tag{3.50}$$

We can take for G, $SU(5)$, $SU(10)$, $E6$ or $SU(6)$. Thus there are a lot of choices for G_2, H_1 and H.We can suppose for a trial that

$$G2 \otimes SU(3) \subset H_0 \tag{3.51}$$

We have also some additional constraints

$$rank(G) \geq 4 \tag{3.52}$$

Thus

$$rank(H_0) \geq 4 \tag{3.53}$$

We can try with $F4 = H_0$.

In the case of H

$$rank(H) \geq rank(G) + 3 \geq 7 \tag{3.54}$$

Thus we can try with E_7, E_8

$$rank(H_1) \geq rank(G_2) + 4$$

$$rank(H) \geq rank(G_2) + rank(G) \geq rank(G_2) + 4 \geq rank(G) + 4 \geq 8 \tag{3.55}$$

In this way we have

$$rank(H) \geq 8.5f \tag{3.56}$$

Thus we can try with

H= *E8.5g* (3.57)

But in this case

rank (G_2)= *rank (G)=4*

This seems to be nonrealistic. For instance, if *G= SO(10), E6,*

rank(SO(10))=5rankE6=6

In this case we get

rank (H)=9 rank (H)=10

And *H* could be *SO*(10), *SO*(18), *SO*(20).In this approach we try to consider additional dimensions connecting to the manifold M more seriously, i.e. as physical dimensions, additional space-like dimensions. We remind to the reader that gauge-dimensions connecting to the group H have different meaning. They are dimensions connected to local gauge symmetr ies (or global) and they cannot be directly observed. Simply saying we cannot travel along them. In the case of a manifold M this possibility still exists. However, the manifold M is diffeomorphically equivalent to the product of some manifolds M_i, $i=0,1,2,\dots,k-1$, with some characteristic sizes r_i The radii r_i represent energy scales of symmetr y breaking. The lowest energy scale is a scale of weak interactions (Weinberg–Glashow–Salam model) $r_0 \simeq 10^{-16}$ cm. In this case this is a radius of a sphere S^2 The possibility of this "travel" will be considered in the concept explanied by Kalinowski [26]. In this case a metric on a manifold M can be dependent on a point $x \in E$ (parametrically).It is interesting to ask on a stability of a symmetr y breaking pattern with respect to quantum fluctuations. This difficult problem strongly depends on the details of the model. Especially on the Higgs sector of the practical model. In order to preserve this stability on every stage of the symmetr y breaking we should consider remaining Higgs' fields (after symmetry breaking) with zero mass. According to S. Weinberg, they can stabilize the symmetry breaking in the range of energy

$$\frac{1}{r_i}\left(\frac{\hbar}{c}\right) < E < \frac{1}{r_{i+1}}\left(\frac{\hbar}{c}\right), \quad i=0,1,2,\dots,k-1,6 \tag{3.58}$$

i.e. for a symmetry breaking from G_{i+1} to G_i.

It seems that in order to create a realistic grand unified model based on non symmetr ic Kaluza–Klein (Jordan–Thiry) theory it is necessary to nivel cosmolog ical terms. This could be achieved in some models due to choosing constant s ξ and ζ and μ . After this we can control the value of those terms, which are considered as a selfinteraction potential of a scalar field Y. The scalar field Y can play in this context a role of a quintessence .

Let us notice that using the equation

$$\Phi^c_{\tilde{b}}(x) f^{\tilde{b}}_{i\tilde{a}} = \mu^a_i \Phi^b_{\tilde{a}}(x) C^c_{ab}, w \tag{3.59}$$

and (3.27) one gets

$$\sum_{i=0}^{k-1} \tilde{A}^{\tilde{a}_i}_{ib} \Phi^c_{\tilde{a}_i} f^{\tilde{b}}_{i\tilde{a}} = C^c_{ab} \mu^a_i \sum_{i=0}^{k-1} \tilde{A}^{\tilde{a}_i}_{ia} \Phi^b_{\tilde{a}_i} \tag{3.60}$$

In this way we get constraints for Higgs' fields, Φ_0, Φ_1, Φ_{k-1}

$$\Phi_i = (\Phi^b_{\tilde{a}_i}), \quad i=0,1,\dots,k-1.$$

Solving these constraints we obtain some of Higgs' fields as functions of in dependent components [26]. This could result in some cross terms in the potential (3.12) between Φ's with different i. For example a term

$$V(\Phi'_i, \Phi'_j),$$

where Φ' means in dependent fields. This can cause some problems in a stability of symmetry breaking pattern against radiative corrections. This can be easily seen from Equation (3.59) solved by in dependent Φ',

$$\Phi = B\Phi' \tag{3.61}$$

$$\Phi^c_{\tilde{b}} = B^{c\tilde{\tilde{b}}}_{\tilde{b}\tilde{c}} \Phi'^{\tilde{c}}_{\tilde{\tilde{b}}} \tag{3.62}$$

Where B is a linear operator transforming in dependent Φ' into Φ.

We can suppose for a trial a condition similar to (3.59) for every $i=0,\dots,k-1$,

$$\Phi^{c_i}_{\tilde{b}_i} f^{\tilde{b}_i}_{i i \tilde{a}_i} = \mu^{a_i}_{ii} \Phi^{b_i}_{\tilde{a}_i} C^{c_i}_{a_i b_i} \tag{3.63}$$

where $C^{c_i}_{a_i b_i}$ are structure constant s for the Lie algebra h_i of the group H_i. $f^{\tilde{b}_i}_{ii\tilde{a}_i}$ are structure constant s of the Lie algebra $\mathfrak{g}_{i+1}$, $\hat{i}_i$ are indices belonging to Lie algebra g_i and $\tilde{a}_i$ to the complement m_i.

In this way

$$\Phi^c_{\tilde{b}_i} = \Phi^{c_i}_{\tilde{b}_i} \delta^c_{c_i} \tag{3.64}$$

In this case we should have a consistency between (3.63) and (3.60) which impose constraints on $C, f\, \mu$ and $C^i\, f^i, \mu^i$ where $C^i\, f^i, \mu^i$ refer to H_i, G_{i+1}. Solving (3.63) via introducing in dependent fields Φ'_i one gets

$$\Phi^{c_i}_{i\tilde{b}_i} = B^{c_i \tilde{\tilde{b}}_i}_{i\tilde{c}_i \tilde{b}_i} \Phi'^{\tilde{c}_i}_{\tilde{\tilde{b}}_i} \tag{3.65}$$

Combining (3.62) , (3.64) , (3.65) one gets

$$B^{c\tilde{\tilde{b}}}_{\tilde{b}\tilde{c}} \Phi'^{\tilde{c}}_{\tilde{\tilde{b}}} = \sum_{i=0}^{k-1} \tilde{A}^{\tilde{a}_i}_{ib} \delta^c_{c_i} B^{c_i \tilde{\tilde{b}}_i}_{i\tilde{c}_i \tilde{b}_i} \Phi'^{\tilde{c}_i}_{\tilde{\tilde{b}}_i} \tag{3.66}$$

Equation (3.66) gives a relation between in dependent Higgs' fields Φ' and Φ'_i. Simultaneously it is a consistency condition between Equation (3.59) and Equation (3.63). However, the condition (3.63) seems to be too strong and probably it is necessary to solve a weaker condition (3.60) which goes to the mentioned terms $V(\Phi'_i, \Phi'_j)$. The conditions (3.63) plus a consistency (3.66) avoid those terms in the Higgs potential. This problem demands more investigation. ϕ (g)= $\{gG_I\}$

It seems that the condition (3.9) could be too strong. In order to find a more general condition we consider a simple example of (3.5). Let $G_0 = \{e\}$ and $K=2$ In this case we have

$$\{e\} \subset G_1 \subset G_2 = G \tag{3.67}$$

$$M_0 = G_1, \quad M_1 = G/G_1 \tag{3.68}$$

$$g: G_1 \times G/G_1 \to G \tag{3.69}$$

In this way $G_1 \times G/G_1$ is diffeomorphically equivalent to G. Moreover, we can consider a fibre bundle with base space G/G_1 and a structural group G_1 with a bundle manifold G. This construction is known in the theory of induced group representation done by Trautman [27]. The projection $\phi : G^{*}\, G/G_1$ is defined by ϕ (g)= $\{gG_1\}$. The natural extension of (3.69) is to consider a fibre bundle (G, G/G_1, G_1, ϕ). In this way we have in a place of (3.69) a local condition

$$g_U : G_1 \times U \underset{\text{in}}{\to} G \tag{3.70}$$

where $U \subset G/G_1$ is an open set. Thus in a place of (3.9) we consider a local diffeomorphism

$$g_U : M_0 \times M_1 \times \ldots \times M_{k-1} \underset{\text{in}}{\to} G/G_0 \tag{3.71}$$

where

$$U = U_0 \times U_1 \times \ldots \times U_{k-1},$$

$U_i \subset M_i$, $i=0,1,2,\dots,k-1$, are open sets. Moreover we should define projectors φ_i, $i=0,1,2,\dots,k-1$,

$$\varphi_i : G/G_0 \to G_{i+1}/G_i \tag{3.72}$$

i.e.

$$\varphi_i(\{gG_0\}) = \{g_{i+1}G_i\} \tag{3.73}$$

$gG, g \notin G_{i+1}, G_0 \subset G_i \subset G_{i+1} \subset G$

in a unique way. This could give us a fibration of G/G_0 in $\prod_{i=0}^{k-1}(G_{i+1}/G_i)$.

For $g \notin G_{i+1}$ we simply define

$$\varphi_i(\{gG_0\}) = \{gG_i\} \tag{3.74}$$

If $g \in G$, $g \notin G_{i+1}$, we define

$$\varphi_i(\{gG_0\}) = \{G_i\} \tag{3.75}$$

Thus in general

$$\varphi_i(\{gG_0\}) = \{p(g)G_i\} \tag{3.76}$$

where

$$p(g) = \begin{cases} g, & g \in G_{i+1} \\ e, & g \notin G_{i+1}. \end{cases} \tag{3.77}$$

Thus in a place of (3.9) we have to do with a structure

$$\{G/G_0, \prod_{i=0}^{k-1}(G_{i+1}/G_i), \varphi_0, \varphi_1, \ldots, \varphi_{k-1}\} \tag{3.78}$$

such that

$$g_U \circ \varphi_U = \mathrm{id} \tag{3.79}$$

where

$$\varphi_U = \prod_{i=0}^{k-1} \varphi_{iU_i} \tag{3.80}$$

This generalizes (3.9) to the local conditions (3.71). Now we can repeat all the considerations concerning a decomposition of Higgs' fields using local diffeomorphisms g_U (g_U^*) in the place of g (g^*). Let us also notice that in the chain of groups it would be interesting to consider as G_2

$$G_2 = SU(2)_L \otimes SU(2)_R \otimes SU(4) \tag{3.81}$$

suggested by Salam and Pati, where $SU(4)$ unifies $SU(2)_{color} \otimes U(1)_Y$. This will be helpful in our future consideration concerning extension to super symmetr ic groups, i.e. $U(2,2)$ which unifies $SU(2)_L \otimes SU(2)_R$ to the super Lie group $U(2,2)$ considered by Mohapatra. Such models on the phenomenological level incorporate fermions with a possible extension to the super symmetric $SO(10)$ model. They give a natural framework for lepton flavour mixing going to the neutrino oscillations incorporating see-saw mechanism for mass generations of neutrinos. In such approaches the see-saw mechanism is coming from the grand unified models. Our approach after incorporating manifolds with anticommuting parameters, super Lie groups, super Lie algebras and in general supermanifolds (superfibrebundles) can be able to obtain this. However, it is necessary to develop a formalism (in the language of supermanifolds, superfibrebundles, super Lie groups, super Lie algebras) for non symmetr ic connections, non symmetr ic Kaluza-Klein (Jordan-Thiry) theory. In particular we should construct an analogue of Einstein-Kaufmann connection for supermanifold, a non symmetr ic Kaluza-Klein (Jordan-Thiry) theory for superfibrebundle with super Lie group. In this way we should define first of all a non symmetr ic tensor on a super Lie group and afterwards a non symmetr ic metrization of a superfibrebundle. Let us notice that on every stage of symmetr y breaking, i.e. from G_{i+1} to G_i, we have to do with group G_t (similar to the group G_0). Thus we can have to do with a true and a false vacuum cases which may complicate a pattern of a symmetr y breaking.

References

1. Kalinowski MW (1990) Non-symmetric Fields Theory and its Applications. World Scientific.
2. Kalinowski MW (1991) Can we get confinement from extra dimensions? Physics of Elementary Interactions. World Scientific, Singapore 294-304.
3. Kalinowski MW (1991) Non symmetric Kaluza-Klein (Jordan-Thiry) Theory in a general non-Abelian case. Int J Theor Phys 30: 281.
4. Kalinowski MW (1992) Nonsymmetric Kaluza-Klein and Jordan-Thiry Theory in the electromagnetic case.Int Journal of Theor Phys 31: 611.
5. Kalinowski MW (2015) Scalar fields in the Non-symmetric Kaluza-Klein (Jordan-Thiry) Theory.
6. Moffat JW (1982) Generalized theory of gravitation and its physical consequences.in: Proceeding of the VII International School of Gravitation and Cosmology.
7. Kalinowski MW (2013) Preliminary applications of the non-symmetric Kaluza-Klein (Jordan-Thiry) theory to Pioneer 10 and 11 spacecraft anomalous acceleration. CEAS Space Journal 5: 19.
8. Kalinowski MW (2015) pioneer 10 and 11 spacecraft anomalous acceleration in the light of the nonsymmetric kaluza-klein (jordan-thiry) theory.
9. Kalinowski MW (2014) On some developments in the Nonsymmetric Kaluza-Klein Theory. The European Physical Journal C 74: 2742.
10. Kalinowski MW (2014) The Nonsymmetric Kaluza-Klein Theory and Modern Physics. A novel approach.
11. Einstein A (1951) The Meaning of Relativity: Including the Relativistic Theory of the Non-Symmetric field. Princeton university press.
12. Einstein A (1950) On the generalized theory of gravitation. Sci Amer 182: 13.
13. Hlavaty V (1957) Geometry of Einstein Unified Field Theory.
14. Kobayashi S, Nomizu K (1969) Foundations of Differential Geometry. John Wiley and Sons.
15. Wang H (1958) On invariant connection on principal fibre bundle. Nagoya Math J 13: 1-19.
16. Harnad J, Shnider S, Tafel J (1980) Group actions on principal bundles and invariance conditions for gauge fields. J Math Phys 21: 2719.
17. Harnad J, Shnider S, Tafel J (1980) Group actions on principal bundles and dimensional reduction. Lett in Math Phys 4: 107.
18. Manton NS (1979) A new six-dimensional approach to the Weinberg-Salam model. Nucl Phys B 158: 141.
19. Forgacs P (1980) Manton NS Space-time symmetries in gauge theories. Comm Math Phys 72: 15.
20. Chapline G, Manton NS (1981) The geometrical significance of certain Higgs' potentials: An approach to grand unifications. Nucl Phys B 184: 391.
21. Meclenburg W (1992) Geometrical unification of gauge and Higgs' fields. Journal of Physics G: Nuclear Physics 6: 1-9.
22. Meclenburg W (1981) Towards a unified theory for gauge and Higgs' fields Phys Rev D 24: 3194.
23. Witten E (1977) Some exact multipseudo-particle solutions of classical Yang-Mil ls theory. Phys Rev Lett 38: 121.
24. Helgason S (1987) Differential Geometry, Lie Groups and Symmetric Spaces. Academic Press, New York.
25. Cheeger J, Ebin DG (1975) Comparison Theorem in Riemannian. Geometry.
26. Kalinowski MW (2003) A Warp Factor in the Non symmetric Kaluza-Klein (Jordan-Thiry) Theory.
27. Trautman A (1970) Fibre bundles associated with space-time. Rep of Math Phys 1: 29

Solutions of the Ultra-Relativistic Euler Equations in Riemann Invariants

Abdelrahman MAE* and Moaaz O

Department of Mathematics, Faculty of Science, Mansoura University, Egypt

Abstract

In this paper we introduce a new technique for constructing solutions of the ultra-relativistic Euler equations. The Riemann invariants are formulated. We also give some applications of the Riemann invariants. We firstly study the geometric properties of the solution in Riemann invariants coordinates. The other application of Riemann invariants, representing the ultra-relativistic Euler equations in diagonal form, which admits the existence of global smooth solution for the ultra-relativistic Euler equations.

Keywords: Relativistic Euler system; Hyperbolic systems; Shock waves; Entropy conditions; Rarefaction waves; Riemann invariants; Diagonal form

Introduction

This paper is devoted to the analysis of the following coupled ultra-relativistic Euler system:

$$(p(3+4u^2))_t + (4pu\sqrt{1+u^2})_x = 0,$$
$$(4pu\sqrt{1+u^2})_t + (p(1+4u^2))_x = 0, \tag{1}$$

Where $p>0$ and $u \in \mathbb{R}$, [1-8].

Consider $x=x(t)$ is a shock-discontinuity of the weak solution of (1) with speed $s = \dot{x}(t)$, $(p\_, u\_)$ the state lower to the shock and (p_+, u_+) the state upper to the shock with $p_\pm > 0$, respectively. Then the Rankine-Hugoniot jump (RHj) conditions are

$$s\left[p_+(3+4u_+^2) - p_-(3+4u_-^2)\right] = 4p_+u_+\sqrt{1+u_+^2} - 4p_-u_-\sqrt{1+u_-^2},$$
$$s\left[4p_+u_+\sqrt{1+u_+^2} - 4p_-u_-\sqrt{1+u_-^2}\right] = p_+(1+4u_+^2) - p_-(1+4u_-^2). \tag{2}$$

The entropy inequality at singular points is

$$s[h] + [\varphi] > 0, \tag{3}$$

where

$$h(p,u) = p^{\frac{3}{4}}\sqrt{1+u^2}, \quad \varphi(p,u) = p^{\frac{3}{4}}u,$$

which is equivalent to $u_- > u_+$, [2].

We can rewrite the 2×2 system for p and u in (1) in the quasilinear form

$$\begin{pmatrix} p_t \\ u_t \end{pmatrix} + A(p,u)\begin{pmatrix} p_x \\ u_x \end{pmatrix} = 0, \tag{4}$$

where

$$A(p,u) = \begin{pmatrix} \dfrac{2u\sqrt{1+u^2}}{3+2u^2} & \dfrac{4p}{\sqrt{1+u^2}(3+2u^2)} \\ \dfrac{3\sqrt{1+u^2}}{4p(3+2u^2)} & \dfrac{2u\sqrt{1+u^2}}{3+2u^2} \end{pmatrix}.$$

The eigenvalues of that system (1) are

$$\lambda_1 = \frac{2u\sqrt{1+u^2} - \sqrt{3}}{3+2u^2} < \lambda_3 = \frac{2u\sqrt{1+u^2} + \sqrt{3}}{3+2u^2}. \tag{5}$$

The characteristic velocities λ_1 and λ_3 are corresponding to the 1 and 3 family of waves, respectively. The decoupled equation

$$(n\sqrt{1+u^2})_t + (nu)_x = 0 \tag{6}$$

for the particle density $n > 0$ gives rise for contact discontinuities with the eigenvalue $\lambda_2 = \dfrac{u}{\sqrt{1+u^2}}$, [2].

Lemma 1.1 *Suppose that $(p\_,u\_)=(p\_,0)$ and $(p+, u+)= (p, u(p))$ satisfy condition (2). Then the shock curves satisfy* [2]

$$u(p) = \pm\frac{\sqrt{3}(p-p_-)}{4\sqrt{pp_-}}. \tag{7}$$

The +ve sign in (7) with $p < p\_$ gives a 3-shock. These 3-shocks satisfy both the RHj conditions (2) and the entropy condition (3), or in a similar way $u\_ > u_+$.

The -ve sign in (7) with $p\_ < p$ gives a 1-shock. These 1-shocks satisfy both the RHj conditions (2) and the entropy condition (3), or in a similar way $u\_ > u_+$. Furthermore $\dfrac{du}{dp} < 0$ on shock curves S_1 and $\dfrac{du}{dp} > 0$ on shock waves S_3, where

$$S_1 = \{(p,u(p)) \in \mathbb{R}^+ \times \mathbb{R} \mid p > p_-\} \quad and \quad S_3 = \{(p,u(p)) \in \mathbb{R}^+ \times \mathbb{R} \mid p < p_-\}. \tag{8}$$

we studied the Riemann invariants for the ultra-relativistic Euler system. In fact we show that the Riemann invariants have interesting relations with the representations of nonlinear elementary waves (shocks and rarefaction waves). Namely, we points out the relation between Riemann invariants and nonlinear elementary waves. This turns out to be the basic ingredient of our paper [9]. One of the main applications of the Riemann invariants is to derive the diagonal form of system (1). We hope that these formula will be useful in various studies of the ultra-relativistic Euler system, for example, in developing numerical methods, [10]. We show that the Riemann invariants, play a pivotal role in the solution of the ultra-relativistic Euler system (1).

***Corresponding author:** Abdelrahman MAE, Professor, Department of Mathematics, Faculty of Science, Mansoura University, Egypt
E-mail: mahmoud.abdelrahman@mans.edu.eg

In other words, in this paper we pose the following questions:

1. What are the Riemann invariants.

2. For what purpose they are useful.

The rest of this paper is given as follows : In Section 2, we derive the Riemann invariants for the ultra-relativistic Euler equations (1). In fact this topic plays a useful role in studying the ultra-relativistic Euler system (1) in a completely unified way. In Section 3 we give some applications of the Riemann invariants for the system (1). The first one, we study the geometric properties of the solution in the Riemann invariants coordinates. we give a new parametrization of the system (1), namely Lemma 3.1. This parametrization plays an important role in order to study these properties in useful way. The second one, is to give the diagonal form of the ultra-relativistic Euler system (1). Finally, in Section 4 we give the conclusions.

Riemann Invariants

We derive the Riemann invariants for the system (1), which plays a main role in this paper.

We consider our rarefaction waves. If we assume $\xi = \frac{x}{t}$, then $W = W(\frac{x}{t})$ satisfies the ordinary differential equation

$$(-\xi JW + JF)\begin{pmatrix} p_\xi \\ u_\xi \end{pmatrix} = 0,$$

where

$$W = \begin{pmatrix} p(3+4u^2) \\ 4pu\sqrt{1+u^2} \end{pmatrix}, \quad F(W) = \begin{pmatrix} 4pu\sqrt{1+u^2} \\ p(1+4u^2) \end{pmatrix}. \tag{9}$$

and

$$JW = \begin{pmatrix} 3+4u^2 & 8pu \\ 4u\sqrt{1+u^2} & \frac{4p(1+2u^2)}{\sqrt{1+u^2}} \end{pmatrix}, \quad JF = \begin{pmatrix} 4u\sqrt{1+u^2} & \frac{4p(1+2u^2)}{\sqrt{1+u^2}} \\ 1+4u^2 & 8pu \end{pmatrix}. \tag{10}$$

If $\begin{pmatrix} p_\xi \\ u_\xi \end{pmatrix} \neq 0$, then $\begin{pmatrix} p_\xi \\ u_\xi \end{pmatrix}$ is an eigenvector of $JW^{-1}JF$ for the eigenvalue ξ. Since $JW^{-1}JF$ has two distinct real eigenvalues, $\lambda_1<\lambda_3$, there are two families of rarefaction waves, 1-rarefaction waves and 3-rarefaction waves.

We first consider 1-rarefaction waves. The eigenvector $\begin{pmatrix} p_\xi \\ u_\xi \end{pmatrix}$ satisfies

$$(-\lambda_1 JW + JF)\begin{pmatrix} p_\xi \\ u_\xi \end{pmatrix} = 0, \tag{11}$$

From (10) we get

$$(-\lambda_1 JW + JF) = \frac{2u\sqrt{1+u^2}-\sqrt{3}}{3+2u^2}\begin{pmatrix} 3+4u^2 & 8pu \\ 4u\sqrt{1+u^2} & \frac{4p(1+2u^2)}{\sqrt{1+u^2}} \end{pmatrix} + \begin{pmatrix} 4u\sqrt{1+u^2} & \frac{4p(1+2u^2)}{\sqrt{1+u^2}} \\ 1+4u^2 & 8pu \end{pmatrix}$$

$$= \begin{pmatrix} \frac{3\sqrt{3}+4\sqrt{3}u^2+6u\sqrt{1+u^2}}{3+2u^2} & \frac{4p(3+4u^2+2\sqrt{3}u\sqrt{1+u^2})}{\sqrt{1+u^2}(3+2u^2)} \\ \frac{3+6u^2+4\sqrt{3}u\sqrt{1+u^2}}{3+2u^2} & \frac{4p(4u\sqrt{1+u^2}+\sqrt{3}(1+2u^2))}{\sqrt{1+u^2}(3+2u^2)} \end{pmatrix}. \tag{12}$$

Using (11), we have

$$\frac{3\sqrt{3}+4\sqrt{3}u^2+6u\sqrt{1+u^2}}{3+2u^2}p_\xi + \frac{4p(3+4u^2+2\sqrt{3}u\sqrt{1+u^2})}{\sqrt{1+u^2}(3+2u^2)}u_\xi = 0,$$
$$\frac{3+6u^2+4\sqrt{3}u\sqrt{1+u^2}}{3+2u^2}p_\xi + \frac{4p(4u\sqrt{1+u^2}+\sqrt{3}(1+2u^2))}{\sqrt{1+u^2}(3+2u^2)}u_\xi = 0. \tag{13}$$

The two equations are dependent since

$Det(-\lambda_1 JW + JF) = 0.$

So we have

$$\left(3\sqrt{1+u^2}+6u^2\sqrt{1+u^2}+4\sqrt{3}u(1+u^2)\right)p_\xi + 4p\left(4u\sqrt{1+u^2}+\sqrt{3}(1+2u^2)\right)u_\xi = 0.$$

Thus we obtain the following differential equation,

$$\left(3\sqrt{1+u^2}+6u^2\sqrt{1+u^2}+4\sqrt{3}u(1+u^2)\right)dp + 4p\left(4u\sqrt{1+u^2}+\sqrt{3}(1+2u^2)\right)du = 0,$$

which has the solution

$$\ln(\sqrt{1+u^2}+u) + \frac{\sqrt{3}}{4}\ln(p) = constant.$$

This is the 1-rarefaction curve. Similarly, we can determine the 3-rarefaction as follows

$$\ln(\sqrt{1+u^2}+u) - \frac{\sqrt{3}}{4}\ln(p) = constant.$$

Since Riemann invariants are functions, which are constant along rarefaction waves, we can define

$$w = \ln(\sqrt{1+u^2}+u) + \frac{\sqrt{3}}{4}\ln(p) \tag{14}$$

and

$$z = \ln(\sqrt{1+u^2}+u) - \frac{\sqrt{3}}{4}\ln(p) \tag{15}$$

are the 1 and 3-Riemann invariant for system (1), respectively.

Remark 2.1 *The function w=w (p,u) is constant across 1-rarefaction waves and z=z (p,u) is constant across 3-rarefaction waves.*

Lemma 2.1 *The mapping (p,u)ℝ(w,z) is one-to-one with nonsingular Jacobian for p > 0 u∈ℝ.* [9]

Applications of Riemann Invariants

In this section we will show how Riemann invariants can be used to solve various problems related to the system (1).

Geometry of the shock waves

Here we study the geometry of the shock waves of the ultra-relativistic Euler system (1) in the Riemann invariants coordinates (*w,z*). We first derive the new parametrization of the ultra-relativistic Euler system (1) in Lemma 3.1. In fact this representation turns out to be very valuable in order to study the geometric properties of the solution in a unified way.

Lemma 3.1 *Assume that* (p_-,u_-) *and* $(p_+, u_+) \equiv (p, u)$ *satisfy the jump condition (2). Then the following relations hold:*

$$\alpha = \frac{\beta^4-\beta^2+2\mp 2(\beta^2-1)\sqrt{\beta^4+\beta^2+1}}{3\beta^2} = f_\mp(\beta), \tag{16}$$

where $\alpha := \frac{p}{p_-}$, $\beta := \frac{\sqrt{1+u^2}-u}{\sqrt{1+u_-^2}-u_-}$. The -ve sign in (16) and $p_- < p$

gives a 1-shock curve S_1 given in (8). The +ve sign in (16) and $p < p_-$ gives a 3-shock curve S_3 given in (8).

Proof. Using the RHj conditions (2) and eliminating the shock speed S give

$$\left(4p_+u_+\sqrt{1+u_+^2}-4p_-u_-\sqrt{1+u_-^2}\right)^2=\left(p_+(3+4u_+^2)-p_-(3+4u_-^2)\right)\left(p_+(1+4u_+^2)-p_-(1+4u_-^2)\right).$$

Now multiplying out gives

$$3p^2+3p_-^2-6pp_--16pp_-(u_-^2+u^2+2u_-^2u^2-2u_-u\sqrt{1+u_-^2}\sqrt{1+u^2})=0,$$

that is,

$$3\left(\frac{p}{p_-}\right)^2-16\frac{p}{p_-}\left(u_-^2+u^2+2u_-^2u^2-2u_-u\sqrt{1+u_-^2}\sqrt{1+u^2}+6\right)+3=0.$$

After a straight but tedious computation, we get the result explained in Figure 1

The following lemma shows that the differences Z - Z_ and w - w_- through a shock curve depend only on the parameters α, and thus the geometric aspect of the shock wave in the zw-plane is independent of the base point. To give this lemma in a useful way we define the functions, $K_S: \mathbb{R}^+ \mapsto \mathbb{R}^+$ by

$$K_S(\alpha) := \frac{\sqrt{1+3\alpha}\sqrt{3+\alpha}+\sqrt{3}(\alpha-1)}{4\sqrt{\alpha}}, \tag{17}$$

and $K_R: \mathbb{R}^+ \mapsto \mathbb{R}^+$ by

$$K_R(\alpha) := \alpha^{\frac{\sqrt{3}}{4}}, \tag{18}$$

for $\alpha := \frac{p_+}{p_-} > 0$.

Lemma 3.2 *Let $z=z(p_+, u_+)$ $w=w(p_+, u_+)$. Then the representation of 1-shock curve S_1 for the system (1) based at (z_-w_-) with respect to the parameter $\alpha = \frac{p_+}{p_-}$ [9] is given as follows:*

$$z-z_- = \ln K_S\left(\frac{1}{\alpha}\right)+\ln K_R\left(\frac{1}{\alpha}\right), \quad w-w_- = \ln K_S\left(\frac{1}{\alpha}\right)-\ln K_R\left(\frac{1}{\alpha}\right).$$

While the 3-shock curves S_3 based at (z_-w_-) has the parametrization

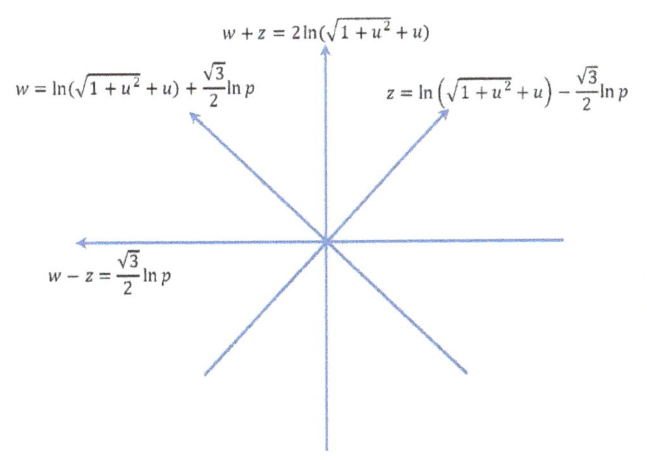

Figure 1: The plane of Riemann invariants.

$$z-z_- = \ln K_S(\alpha)-\ln K_R(\alpha), \quad w-w_- = \ln K_S(\alpha)+\ln K_R(\alpha).$$

Lemma 3.3 *The 3-shock wave based at an arbitrary point (w_-, z_-) is the reflection in the wz-plane of the 1-shock wave based at the same point, where the axis of reflection is the line passing through (w_-, z_-), parallel to the line w=z.*

Proof. Using (16), then the result follows immediately from the following:

$$f_-(\beta).f_+(\beta)=\frac{\beta^4-\beta^2+2-2(\beta^2-1)\sqrt{\beta^4+\beta^2+1}}{3\beta^2}.\frac{\beta^4-\beta^2+2+2(\beta^2-1)\sqrt{\beta^4+\beta^2+1}}{3\beta^2}=1.$$

The following lemma presents further important features of the shock wave.

Lemma 3.4 *For shock curves of system (1) we have*[9]

$$0<\frac{dw}{dz}<1 \tag{19}$$

along a 1-shock curves S_1 and

$$0<\frac{dz}{dw}<1 \tag{20}$$

a along a 3-shock curves S_3.

Therefore we can use either the pu-plane or the zw-plane to study our model, see Figure 1. Thus we conclude that the shock waves are independent of the base point (z_-w_-).

Diagonalization of the ultra-relativistic Euler equations

Here we present the ultra-relativistic Euler system (1) in diagonalized form. This form enables us to develop numerical methods to study the ultra-relativistic Euler system (1), [10]. This will be presented in a forthcoming paper.

Definition 3.1 *System (4) is said to be diagonalizable, if there exists a smooth transformation$R=(w,z)^T$ with non-vanishing Jacobian such that (4) can be rewritten as follows*

$$\frac{\partial R_i}{\partial t}+\sum_{i=1}^{2}\lambda_i(R_i)\frac{\partial R_i}{\partial x}=0, \quad i=1,2, \tag{21}$$

Where $\lambda_i(R_i)$ are smooth function of Riemann invariants R.

The diagonal system (21) is so important possessing so interesting properties. For example, it is easier to find exact solutions and study uniqueness of solutions, [11,12]. In fact, not all quasilinear systems can represent in diagonal form. Hence, it is so interesting to study this problem.

Proposition 3.1 *The diagonalized system for system (1) is*

$$\frac{\partial w}{\partial t}+\lambda_1(w,z)\frac{\partial w}{\partial x}=0,$$
$$\frac{\partial z}{\partial t}+\lambda_3(w,z)\frac{\partial z}{\partial x}=0, \tag{22}$$

where

$$\lambda_1(w,z)=\frac{e^{w+z}-2-3}{e^{w+z}+2+3} \quad and \quad \lambda_3(w,z)=\frac{e^{w+z}-2+3}{e^{w+z}+2-3}$$

illustrated in Figure 2.

Proof. We first start with the first equation of (22), namely with $\lambda_1(w,z)$ One can follows with $\lambda_3(w,z)$ similarly. From (5), we get

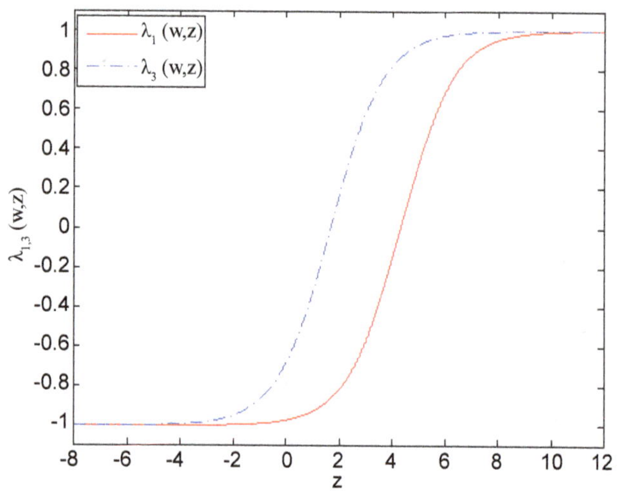

Figure 2: The smooth function of Riemann invariants $\lambda_1(w,z)$ and $\lambda_3(w,z)$.

$u = \sqrt{\frac{3}{2}} \frac{\lambda_1 \mp \frac{1}{\sqrt{3}}}{\sqrt{1-\lambda_1^2}}$. Hence we have

$$u + \sqrt{1+u^2} = \frac{(\sqrt{3}+1)(\lambda_1+1)}{\sqrt{2}\sqrt{1-\lambda_1^2}}. \quad (23)$$

Another equivalent form of the same relation, using (14) and (15) is

$$u + \sqrt{1+u^2} = e^{\frac{w+z}{2}}, \quad (24)$$

From (23) and (24) and after a straight but tedious computation, we obtain

$$\lambda_1(w,z) = \frac{e^{w+z} - 2 - 3}{e^{w+z} + 2 + 3},$$

hence the proof of the proposition is completed.

Remark 3.1 *In fact the diagonal formula is very useful in developing numerical methods, see [9].*

Remark 3.2 *Based on the results given in, Theorem 2.4, Lemma 3.1], in order to prove the existence of the global smooth solution on* $t \geq 0$ *for system (1), it is sufficient to prove that* [12]

$$\lambda_{1z} \geq 0,\ \lambda_{2w} \geq 0,\ \lambda_{1z} + \lambda_{1w} \geq 0,\ \lambda_{2z} + \lambda_{2w} \geq 0.$$

One can easily check the following:

$$\lambda_{1z} = \frac{(4+2\sqrt{3})e^{w+z}}{(e^{w+z}+2+\sqrt{3})^2} > 0,$$

$$\lambda_{3w} = \frac{(4-2\sqrt{3})e^{w+z}}{(e^{w+z}+2-\sqrt{3})^2} > 0.$$

We also have $\lambda_{1w=}\lambda_{1z}$ and λ_{3w}=λ_{3z} hence

$$\lambda_{1w} + \lambda_{1z} > 0 \quad and \quad \lambda_{3w} + \lambda_{3z} > 0,$$

which completes the statement.

Conclusions

In this work, we presented the Riemann invariants method for the ultra-relativistic Euler equations. We have shown that the shock curves have good geometry in Riemann invariant coordinates. The diagonal form of the ultra-relativistic Euler system has been introduced, which admits the existence of global smooth solution.

References

1. Abdelrahman MAE, Kunik ZM (2014) The Front Tracking Scheme for the Ultra-Relativistic Euler Equations. J Comput Phys 275: 213-235.
2. Abdelrahman MAE, El-Shafei ME, Alsammarraie HA (2014) Riemann Invariants for the Ultra-Relativistic Euler Equations. IJCTA. 5: 3057-3066.
3. Abdelrahman MAE, Kunik M (2015)The Ultra-Relativistic Euler Equations. Math Meth Appl Sci. 38: 1247-1264.
4. Chen J (1995) Conservation laws for the relativistic p-system. Comm Partial Differential Equations 20:1602-1646.
5. Chen GQ, Li YC (2004) Relativistic Euler equations for isentropic fluids: stability of Riemann solutions with large oscillation. Z Angew Math Phys 55: 903-926.
6. Hoff D (1982) Globally Smooth Solutions of Quasilinear Hyperbolic Systems in Diagonal Form. J Math Anal Appl 86: 221-236.
7. Kunik M (2005) Selected Initial and Boundary Value Problems for Hyperbolic Systems and Kinetic Equations. Habilitation thesis. Otto-von-Guericke University Magdeburg.
8. Smoller J (1992) Numerical methods for conservation laws. Birkhauser¨ Verlag Basel, Switzerland.
9. Pant V (1996) Global entropy solutions for isentropic relativistic fluid dynamics. Commu Partial Diff Eqs 21: 1609-1641.
10. Ruan L, Zhu C (2005) Existence of global smooth to the relativistic Euler equations. Nonl Anal 60: 993-1001.
11. Smoller J, Temple B (1993) Global Solutions of the Relativistic Euler Equations. Commun Math Phys 156: 67-99.
12. Wissman BD (2011) Global Solutions to the Ultra-Relativistic Euler Equations. Commun Math Phys 306: 831-851.

13

The Electrodynamic Algorithms of Core of Stars and Black Holes

Alexandris NG*

Independent Researcher, Greece

Abstract

The prediction of the magnetic field of Sgr A* by the algorithm of a star's core rotation is presented in author book "Modified Hawking field" 2010. In the present article author can see the way. Author can calculate the electrodynamic parameters of the nucleus of stars and black holes like Sgr A*.

Keywords: Gravity energy; Stoney's coefficient; Magnetic field; Black holes

Introduction

The following article is based on an idea of Alvfen Carlqvist [1] that author can predict the parameters of super massive stars and black holes by Electrodynamic parameters. Gravity energy is equal to electrodynamic energy and the magnetic field is produced by an L C circuit. The basic hypothesis of author book is that electric charge is equal to mass by a coefficient ($k_{5.1a}$), different from Stoney's coefficient as well as Hawking's coefficient but it is in agreement with Hawking's theory. In author mathematical method I introduce two mathematical parameters β and t_c. These parameters are necessary to extract a coefficient which is related to the 2.73 K of cosmic background radiation [2].

Article

The prediction of magnetic field of Sgr A* by the algorithm of a star's core rotation is presented in author book [3]. We can input to algorithm the mass of star m and the radius $R=l_c$, $l_{c=}$length of Coulomb law

$$\tilde{n}_m = \frac{m}{V}\tilde{n}_c = k\tilde{n}_m \quad (1)$$

ρ_m: density of mater, m: mass, V: volume=(4/3)π l_c^3, ρ_c: density of electric charge

$k=k_{5.1a=}(G/2\pi K_e)^{1/2}=3.43745 \times 10^{-11}$ C/Kg

$$t_c = K_e m\theta_v \rho_c \quad (2)$$

Finally the t_c parameter is independent of density of matter and it is depended on the radius (R^3) of the star K_e: Coulomb constant, θ_V=(4/3) π, t_c: mathematical parameter

$$\hat{o}=kt_c \quad (3)$$

τ: elastic coefficient of Hook law,

$$k=\beta t_c, \beta=\frac{k}{t_c} \quad (4)$$

We can find β mathematical parameter

$$\beta=\frac{k^2}{4\pi F}l_c, F=\frac{k^2}{4\pi\beta}l_c \quad (5)$$

l_c is the radius(R) of the body, we find F force of oscilation[4]

$$g=\frac{F}{m}, \textit{Newton law} \quad (6)$$

$$\hat{o}=C\frac{g_x^2}{k^2}, C=\hat{o}\frac{k^2}{g_x^2} \quad (7)$$

we can find C capacity

$$C=k\frac{Q}{2g\lambda}, \lambda=2\pi l_c, Q=2g\lambda\frac{C}{k} \quad (8)$$

We can find **Q**, the electric charge.

The period of rotation is:

$$T=2\eth\sqrt{\frac{\ddot{e}}{g}} \quad (9)$$

$$T=2\eth\sqrt{LC}, L=\frac{T^2}{4\eth^2 C} \quad (10)$$

L: self-induction coefficient.

So we can find the physics parameters of the star or black hole; g: acceleration, C: electric capacity, Q: electric charge, T: period of rotation and L: self-induction coefficient. From those values author can calculate the electromagnetic field of the core of a star or black hole energy of electric currents:

$$E= UQ=kU_mQ=Q\frac{Gmk}{R} \quad (11)$$

U: electric potential, Um: gravitation potential [5], G: gravitation constant, m: mass of star, $k=k_{5.1a}$

and $I=\sqrt{\frac{2E}{L}}$ (12)

I: intensity of current, L: self-induction coefficient so

$$B=\frac{E}{I2\pi R^2/n} \quad (13)$$

Transformation of Lorenz-Laplace force F=B I 2πR/n, E=FR,

m: mass of star or black hole, I: intensity of current, R=radius of

***Corresponding author:** Nikos G Alexandris, Independent Researcher, Greece
E-mail: nalxchal@yahoo.com

star, n=R/Rcore, Rcore: radius of core of star,

For black holes n=1, for Sun n=5 and for Earth n=2. The results are very good for Earth-nucleus, Sun and Sgr A*. Also author can write the function of magnetic field as following:

$$B=\frac{E}{2\pi IRR_{core}} \tag{14}$$

For black holes R=Rcore

Magnetic Field of Sagittarius (Sgr A*) Black Hole

The event horizon experiment could prove Hawking-Bekenstein-Kerr theory [6]. The mass of sagittarius black hole found, but we do not know the horizon surface and radius.

$M_{BH}=8.22 \times 10^{36}$ kg

The radius of black hole arises from Hawking-Bekenstein-Kerr surface of horizon.

So $R_{BH}=1.1 \times 10^{10}$ m

Using the algorithm in my book "modified Hawking field" page 55-56 we get the following values:

$Q=4.4 \times 10^{28}$ Cb C=190 F

$I=4.9 \times 10^{15}$ A B=0.02 Tesla

$E=7.6 \times 10^{34}$ Joule $g=5.7 \times 10^{4}$ m/sec^2

L=6200 H $T=6.9 \times 10^{3}$ sec

$V=2\pi R/T=1 \times 10^{7}$ m/sec, 3% of speed of light or 0.03 C.

This rotation function is appropriate for stars like the sun for this function rotation we use some of the six hypotheses. For a black hole it is better to use the following function:

Velocity of surface $V=\sqrt{2gR}$ $=3.55 \times 10^{7}$ m/sec,

11% the speed of light or $0.11 \times C$

$V=\omega R$, $\grave{u}=2\frac{\eth}{T}$, $T=1950$ *sec, period*

All equations arise without relativity. Using the known mass of Sgr A* Black hole and the radius of Bekenstein–Hawking-Kerr function author can find the magnetic field of Black hole near the horizon without relativity[5-7]. The same algorithm is used for the Sun.

For the sun

The function is:

$$B=\frac{E}{I2\pi R^2/5} \tag{15}$$

Transformation of Lorenz-Laplace force

I=intensity of current, R=radius of Sun, R/5 is the radius of the core of the Sun, E=energy of the Sun's currents, B/10=surface magnetic field, 10 is the analogy of rotation between surface and core. The result for the core is 25 Gauss and for the surface 3.8 Gauss.

For the earth

Function (1) for $Il^2=IR^2$ gives 78 Gauss, for $Il^2=I(2\Pi R)^2$ gives 2 Gauss, for $Il^2=I2\Pi R^2$ gives 12 Gauss, for arises 24 Gauss which is the experimental[8] value of 2010, R/2 is the radius of the Earth's core. Also the algorithm gives rotation 1.3 days.

For neutron stars a few are in agreement with those functions like burst nebule.

Sgr A*

The algorithm gives 0.12 Tesla for $B=\frac{E}{IR^2}$, but the best function is: $B=\frac{E}{I2\pi R^2}$, so the result is 0.02 Tesla, without relativity. I choose the relativistic coefficient, $\tilde{a}^3$, $\tilde{a}=\frac{1}{\sqrt{1-\frac{v^2}{c^2}}}$, v=0.11c, $\gamma \approx 1$.

I choose this coefficient for $I=\frac{q}{t}$ and R^2. For an observer far away from the horizon, $B=\frac{B_0}{\tilde{a}^3}$, B_0=B=0.02Tesla=202Gauss (16)

This value 202 gauss is 500 times the Earth's magnetic field [9]. The algorithm uses coefficient k_{51a} of all the above functions without strong participation of relativity. The problem of that algorithm is that does not give the rotation of black hole like the rotation of the Sun or the Earth. In 2013 88% the speed of light of the gas was observed

So the gases are falling and rotating under an acceleration g which is the same as the index. A good approximation could arise in the following way using relativity [4-8]:

$$V(t)=g.\frac{t}{\sqrt{1+g^2\frac{t^2}{c^2}}}=2.4 \times 10^{8}\text{ m/sec} \tag{17}$$

or 0.8 C or 80% the speed of light, T=t

If we use the constant of Stoney [10].

$$k_{51b}=\sqrt{\frac{G}{K_e}}=\frac{k_{51a}}{\sqrt{2\pi}} \tag{18}$$

We get for currents 0.3 C, B=200 Gauss and for rotation velocity 0.96 C. Stoney gives a great value.

The Paradox of the Hawking Theory

It is known that the Hawking-Bekenstein theory about black holes contains two basic equations. One of the equations is that of Bekenstein (1972) which correlates the surface of the horizon of a black hole to the mass of a black hole. Specifically, the surface is proportional to the square of the mass. The other equation is Hawking's (1971) which connects the temperature of the horizon of a black hole with the mass. Temperature is inversely proportional to mass. These theories correlate fundamental theories of physics: the general relativity theory (not completely proven, due to our ignorance about gravity), quantum, thermodynamic and mechanical theory.

The paradox lies in the fact that no astronomical body gave the Hawking radiation or else the temperature which is monstrously10^{-14} Kelvin small, weaker than the temperature of the universe, as Susskind protests in his recent book.

Sagittarius A*

Contemporary astrophysics has understood and photographed very dense objects, neutron stars. There has never been observed any body more dense than these, and the name black holes is attributed to them, without being able to confirm the Hawking-Bekenstein equations. The temperature (T) detected is much greater than10^{-14} *K*. Kelvin and corresponds to X-ray and γ radiation. However, it is believed that it comes from clouds around the horizon rather than the horizon itself.

Μάζα Sagittarius A* (m): 1.2×10^{42} kg and for Black Hole BH 8.2×10^{36} kg (19)

Radius (lc): 5×10^{14} m and for BH 1.1×10^{10} m (20)

Using the above values in the equation of page 48 which gives the temperatures of the cores of the stars, the result is temperature(T) similar to the one observed:

Radiation (T): 9.1×10^{6} Kelvin or 6 Å and for BH 5.8×10^{9} Kelvin or 0.5 Mev.

$$m = \frac{8.928x10^{-4} VT^2}{\theta_v l_c} = \frac{8.928x10^{-4} l_c^3 T^2}{l_c} = 8.928x10^{-4} l_c^2 T^2 \quad (21)$$

$$T = \sqrt{\frac{M_{SagitA}}{8.928x10^{-4} l_c^2}} = 7.33x10^7\,Kelvin$$ and for BH 8.7×10^{9} kelvin=0.75 Mev

Colliders

This equation also gives the crushing length (*lc*) of the particles in the buffers. This is not odd because the crushing length (*lc*) corresponds to the wave formed on the particle which receives the energy of the collision (mass m). The particle becomes a crushing length sized star.

From (21) arises $l_c = \sqrt{\frac{M_{collision}}{8.928x10^{-4} T_c^2}}$, (22)

For $M_{collision=}7\ TeV/c^2$ and 10^{11} *Kelvin* arises $l_{c=}10^{-21}\ m$

Hawking Theory

But what is the connection to the Hawking's theory? If we correlate the above equation with the two Hawking-Bekenstein equations we might solve the paradox of the black hole temperature(T). On the equation of page 48 which gives the temperature(T) of the core of the stars we put the Planck temperature and the mass of Sagittarius A*(M_{sagitA}). The result is the radius of an atom, which is very small.

From (22) arises: $M_{SagitA} = 8.928x10^{-4} l_c^2 T_{plank}^2$ then $l_c = \sqrt{\frac{M_{SagitA}}{8.928x10^{-4} T_{plank}^2}}$ (23)

$l_{c=}2.58 \times 10^{-10}\ m$, for BH $l_{c=}6.76 \times 10^{-13}\ m$

If this length is imported as a horizon radius of the black hole facts we will take the mass(M_B) of the black hole in the Bekenstein equation. This mass in the Hawking equation(M_B=M_H) will give the Hawking radiation(T_H) which is about 12-20 times smaller than the observed.

Radius of Horizon of Black Hole of Bekenstein's[12] function:

$$R_S^2 = \frac{4G^2 M_B^2}{c^4} \quad (24)$$

Rs=lc, $M_{B=}1.74 \times 10^{17}\ kg$, for BH, $M_{B=}4.55 \times 10^{14}\ kg$

Radiation of Black Hole of Hawking's function:

$$T_H = \frac{hc^3}{16\pi^2 k_B G M_H} \quad (25)$$

M_B=$M_{H,}$ $T_{H=}7 \times 10^5$ *Kelvin*, for BH T_H=2.7×10^8 *Kelvin*=22.7 KeV

This value I in agreement with the observed values (2011 Fermi Symposium). This is the way that a paradox seems to be solved, but there is another one added. The black hole has a size of an atom. Mass MB must be singularity.

The theoretical paradox of the hawking theory

The general theory of relativity which is the gravity theory has given little evidence for 100 years, such as gravitational lenses galaxies. There has not, however, been proven that inertia is equivalent to gravity, or that acceleration of bodies produces gravitational waves.

For the same reason we do not yet accept that collisions of bodies and particles enhance gravity. In a few words we have not understood the nature of gravity. In the existing weakness of general relativity the Hawking theory introduces another assumption, that general relativity which applies in the macrocosm also applies in the region of quantum phenomena. Thus, either quantum effects apply in the macrocosm, or general relativity applies in the microcosm. The above calculations give an explanation. The great mass regards general relativity and small length of the black hole regards quantum. Therefore, perhaps the two theories become compatible. Although, we must not forget that paradoxes of black holes started from mechanics long before the appearance of relativity.

Paradox says that the escape velocity in a heavy body can reach the speed of light, so the star will be dark. In fact it will be so dark that it will also be completely cold. The latter has been a matter of controversy for the past 40 years. The problem is that it is not certain that a law of mechanics applies everywhere and in the same way. For example, logically it is not prohibited that the escape velocity is always less than the speed of light, so that the light escapes always.

Solar dynamo

The equations resulting from the assumptions can extract the rotation of stars and in particular of their cores, specifically the first 40 equations. Regarding the Sun's rotation of the core it is 10 times faster than the surface, page 56. The fact that rotation concerns the core is shown by the fact that the same equations give the core's temperature of sun 7×10^7 *Kelvin*, page 48. Confirmation of equations and of coefficients is very important for the rest part of the project. The coefficient of the equations selected so that they match with the empirical cosmological types. For the calculation of the magnetic field(B) the equation which gives the force is selected Laplace(F):

F=BIl_c then $Fl_c = BIl_c^2$ (26)

so $E = BIl_c^2$ arises $B = \frac{E}{Il_c^2}$ (27)

The calculations of page 56 should be modified in order to take into account that the core of the sun has radius the 1/5 of the sun and that currents are not rectilinear but circular.

Eqn (27) arises $B = \frac{E}{I\frac{2\pi l_c}{5}^2}$ (28) with $I=2 \times 10^7$ Ampere

Energy (E) is proportional to the square of the angular speed, like the Laplace force:

E ≈ ω^2. The product $Il_{c=}Q\upsilon$, (29)

speed (υ) is proportional to the angular speed (ω), thus the intensity of the magnetic field (B) is proportional to the angular speed ω.

$$B \approx \frac{\mathring{A}}{\omega} \text{ so } B \approx \frac{\omega^2}{\omega} = \omega \quad (30)$$

Therefore, the magnetic field's (B) intensity is inversely proportional to the rotation period (T): $\omega = \frac{1}{\hat{O}}$. If the rotation of the surface is 10 times smaller, and therefore the magnetic field will be 10 times smaller, i.e. 44 Gaus/10=4.4 Gauss.

This verification shows that the coefficients of equations 1-47, pages 14-17 were selected properly. This verification enhances the accuracy of the temperature equation of the core of the star which is extracted from the same equations, pages 47-48. It is the equation that agrees with the radiance of Sagittarius A* and it is combined with the Hawking-Bekenstein equations. It also gives very good results to the collider regarding the length of particle breakage.

The special feature of the constant (CR) which gives the solar nuclear rotation of the star is that it depends on the constant $k_{5.1a}$ and not on the classic constant $k_{5.1b}$ of Stoney. Author should remind that the quotient of the electrical charge to the $k_{5.1a}$ gives $M_{eg5.1a}$ which is associated to the Planck mass and the Hawking and Stoney masses. The verification of the constant $k_{5.1a}$ is fundamental for the analysis of the book.

Conclusion

The algorithm can predict the basic parameters of a super massive star especially the nucleus and some planets like the Earth. The Stoney coefficient is appropriate for solid bodies where gravity and electricity are equal [11]. The coefficient $k_{5.1a}$ is appropriate for states in nucleus matter of stars and some planets like the Earth. The bodies are different by an index n=R/Rcore which includes the analogy between the total radius of star and nucleus. The prediction is very good for magnetic field and rotation.

References

1. Peng JJ, Wu QS (2010) Hawking radiation of black holes in infrared modified Hořava-Lifshitz gravity. The European Physical Journal C 66: 325-331.
2. Atwood W (2012) Fermi Symposium Proceedings. eConf C121028, Italy.
3. Grant A (2013) Magneticfield-black-hole measured. Science News.
4. Nikos A (2013) Magnetic Field of Sagittarius (Sgr A*) Black Hole.
5. Oikonomou EN (2012) Quark to Universe. Crete University Press, Greece.
6. Buffett BA (2010) Tidal dissipation and the strength of the Earth's internal magnetic field. Nature 68: 952-954.
7. Serway RA, Moses CJ, Moyer CA (2000) Modern Physics. Crete University Press, Greece
8. Nikos AG (2012) Paradox of Hawking Theory. Gsjournals.
9. Valenzuela GR (1977) Theories for the interaction of electromagnetic and oceanic waves — A review. Boundary-Layer Meteorology 13: 61-85.
10. Gordon IM (1973) Plasma theory of radio echoes from the Sun and its implications for the problem of the solar wind. Space Science Reviews 15: 157-204.
11. Susskind L (2010) The Black Hole War. Hachette Publishing House, UK.

Matrix Representation for Seven-Dimensional Nilpotent Lie Algebras

Ghanam R[1]*, Basim Mustafa B[2], Mustafa MT[3] and Thompson G[4]

[1]*Department of Liberal Arts and Sciences, Virginia Commonwealth University in Qatar, Qatar*
[2]*Department of Mathematics, An-Najah National University, Palestine*
[3]*Department of Mathematics, Statistics and Physics, Qatar University, Qatar*
[4]*Department of Mathematics, University of Toledo, USA*

Abstract

This paper is concerned with finding linear representations for seven-dimensional real, indecomposable nilpotent Lie algebras. We consider the first 39 algebras presented in Gong's classification which was based on the upper central series dimensions. For each algebra, we give a corresponding matrix Lie group, a representation of the Lie algebra in terms of left-invariant vector field and left-invariant one forms.

Keywords: Lie algebra; Lie group; Representation, Nilpotent

Introduction

Given a real Lie algebra g of dimension *n* a well known theorem due to Ado [1,2] asserts that g has a faithful representation as a subalgebra of *gl(p,R)* for some *p*. The theorem does not give much information about the value of *p* but leads one to believe that *p* may be very large in relation to the size of *n* and consequently it seems to be of limited practical value. In previous work we have found linear representation for all indecompsable real Lie algebras in dimesion six or less [3-5] In this paper we are concerned with finding linear representations for seven-dimensional nilpotent Lie algebras. Let g be a nilpotent Lie algebra, we shall assume throughout that g is indecomposable in the sense that g is not isomorphic to a direct sum of two proper nilpotent ideals. If g is decomposable and we have representations for both factors then we can easily find a representation for g. It is well known that a nilpotent Lie algebra has a non-trivial center and so finding representations is difficult because the adjoint representation is not faithful. In this article, we consider the list of seven-dimensional nilpoten Lie algebras over *R* classified by Gong [6], his list contained 147 indecomposable Lie algebras that are classifed according to the upper cenral series. We consider the first 39 cases for which the dimensions the upper centeral series are (27) (37) (247) (257) and (357). For each algebra g we give a corresponding Lie group that is a subgroup of *GL*(*7,R*). The representation for the Lie algebra is then easily obtained by differentiating and evaluating at the identity. The structure of the paper is as follows: in Section 2, we give a brief history and background on the classification of the seven-dimensional nilpotent Lie algebras, In section 3, we give a matrix Lie gorup that corresponds to each Lie algebra. We also give a representation of the Lie algebra in terms of left-invariant one-forms and left-invariant vectros fields. Throught out the paper we will use (*p,q,r,x,y,z,w*) as local coordinates on Lie groups.

Classifying Nilpotent Lie Algebras in Dimension Seven

Classification of solvable Lie algebras is not an easy problem, this is due to the fact the solvable Lie algebras are unlike semisimple Lie algebras. Semisimple Lie algebras are considered very beautiful since over the complex numbers we have the killing form, Dynkin diagrams, root space decompositions, the Serre representation, the theory of of highest weight represenation, the Weyl character formula and much more [7-10]. On the other hand, we have Lie , Engel and Ado's theorms for solvable Lie algebras. There has been several attempts to classify the seven-dimensional nilpotent Lie algebras, we will mention the most recent ones that we are aware of. In 1993, Seeley [10] gave a classification of over the filed of comples numbers. His classification was based on the upper central series of the Lie algebras and knowledge of all lower dimensional nilpotent Lie algebras. His list contained 161 Lie algebras; 130 indecomposable and 31 indecomposable. In 1998, Gong [8], presented a new list of seven-dimensional nilpotent Lie algebras. Gong's classification was based on the Skjelbred-Sund method [11]. Gong provided a classifcation of the seven-dimensional indecomposable algebras over Algebriacally closed fileds ($\chi \neq 2$) and another classification over the field of real numbers. Once again, he used the same labeling as Seeley; the dimensions of the upper central series.

Representations

In this section, we present a matrix Lie group representation for the first 39 seven-dimensional Lie algebras with upper central series (27) (37) (247) (257) and (357)

Algebras with upper central series dimensions (27)

1. $(27A):[x_1,x_2]=x_6,[x_1,x_4]=x_7,[x_3,x_5]=x_7;$

A matrix Lie group is given by:

$$S=\begin{bmatrix} 1 & x & y & r & p & q \\ 0 & 1 & 0 & 0 & w & 0 \\ 0 & 0 & 1 & 0 & -z & 0 \\ 0 & 0 & 0 & 1 & 0 & x \\ 0 & 0 & 0 & 0 & 1 & 0 \\ 0 & 0 & 0 & 0 & 0 & 1 \end{bmatrix}$$

Left invariant differential forms are given by:

$$F_1=dp-wdx+zdy,\quad F_2=-xdr+dq,\quad F_3=dr,\quad F_4=dx,$$
$$F_5=dy,\quad F_6=dz,\quad F_7=dw$$

***Corresponding author:** Ghanam R, Department of Liberal Arts and Sciences, Virginia Commonwealth University in Qatar, Qatar
E-mail: raghanam@vcu.edu

The vector field representation is given by:

$$E_1 = -wD_p - D_x, \quad E_2 = -xD_q - D_r, \quad E_3 = D_z, \quad E_4 = D_w,$$
$$E_5 = zD_p - D_y, \quad E_6 = D_q, \quad E_7 = D_p$$

2. $(27B) : [x_1, x_4] = x_6, [x_2, x_5] = x_7, [x_3, x_4] = x_6 + x_7, [x_3, x_5] = x_6 + x_7;$

A matrix Lie group is given by:

$$S = \begin{bmatrix} 1 & y & x & r & p & q \\ 0 & 1 & 0 & 0 & w+z & z \\ 0 & 0 & 1 & 0 & z & z \\ 0 & 0 & 0 & 1 & 0 & x \\ 0 & 0 & 0 & 0 & 1 & 0 \\ 0 & 0 & 0 & 0 & 0 & 1 \end{bmatrix}$$

Left invariant differential forms are given by:

$$F_1 = dp - zdx + (-w - z)dy, \quad F_2 = -xdr + dq - zdx - zdy, \quad F_3 = dr, \quad F_4 = dx$$
$$F_5 = dy, \quad F_6 = dz, \quad F_7 = dz + dw$$

Vector field representation is given by:

$$E_1 = D_w, \quad E_2 = -xD_q - D_r, \quad E_3 = D_z, \quad E_4 = (w+z)D_p + zD_q + D_y$$
$$E_5 = zD_p + zD_q + D_x, \quad E_6 = D_p, \quad E_7 = D_q$$

Algebras with upper central series dimensions (37)

1. $(37A) : [x_1, x_2] = x_5, [x_2, x_3] = x_6, [x_2, x_4] = x_7;$

A matrix Lie group is given by:

$$S = \begin{bmatrix} 1 & 0 & 0 & w & 0 & 0 & q \\ 0 & 1 & 0 & 0 & w & 0 & -x \\ 0 & 0 & 1 & 0 & 0 & w & y \\ 0 & 0 & 0 & 1 & 0 & 0 & z \\ 0 & 0 & 0 & 0 & 1 & 0 & r \\ 0 & 0 & 0 & 0 & 0 & 1 & p \\ 0 & 0 & 0 & 0 & 0 & 0 & 1 \end{bmatrix}$$

Left invariant differential forms are given by:

$$F_1 = dp, \quad F_2 = dq - zdw \quad F_3 = dr \quad F_4 = -dx - rdw$$
$$F_5 = dy - pdw, \quad F_6 = dz, \quad F_7 = dw$$

Vector field representation is given by:

$$E_1 = D_p, \quad E_2 = zD_q - rD_x + pD_y + D_w, \quad E_3 = D_r, \quad E_4 = D_z$$
$$E_5 = D_y, \quad E_6 = D_x, \quad E_7 = -D_q$$

2. $(37B) : [x_1, x_2] = x_5, [x_2, x_3] = x_6, [x_3, x_4] = x_7;$

A matrix Lie group is given by:

$$S = \begin{bmatrix} 1 & 0 & 0 & w & 0 & -z & q \\ 0 & 1 & 0 & 0 & w & 0 & x \\ 0 & 0 & 1 & 0 & 0 & w & y \\ 0 & 0 & 0 & 1 & 0 & 0 & 0 \\ 0 & 0 & 0 & 0 & 1 & 0 & -r \\ 0 & 0 & 0 & 0 & 0 & 1 & p \\ 0 & 0 & 0 & 0 & 0 & 0 & 1 \end{bmatrix}$$

Left invariant differential forms are given by:

$$F_1 = dp, \quad F_2 = dq + pdz, \quad F_3 = dr, \quad F_4 = dx + rdw$$
$$F_5 = dy - pdw, \quad F_6 = dz, \quad F_7 = dw$$

Vector field representation is given by:

$$E_1 = -D_r, \quad E_2 = D_w - rD_x + pD_y, \quad E_3 = -D_p, \quad E_4 = D_z - pD_q,$$
$$E_5 = D_x, \quad E_6 = D_y, \quad E_7 = D_q$$

3. $(37C) : [x_1, x_2] = x_5, [x_2, x_3] = x_6, [x_2, x_4] = x_7, [x_3, x_4] = x_5;$

A matrix Lie group is given by:

$$S = \begin{bmatrix} 1 & 0 & 0 & w & 0 & r & -q \\ 0 & 1 & 0 & 0 & w & 0 & x \\ 0 & 0 & 1 & 0 & 0 & w & y \\ 0 & 0 & 0 & 1 & 0 & 0 & z \\ 0 & 0 & 0 & 0 & 1 & 0 & r \\ 0 & 0 & 0 & 0 & 0 & 1 & p \\ 0 & 0 & 0 & 0 & 0 & 0 & 1 \end{bmatrix}$$

Left invariant differential forms are given by:

$$F_1 = dp, \quad F_2 = dr, \quad F_3 = dz, \quad F_4 = dw$$
$$F_5 = -pdr - dq - zdw, \quad F_6 = dx - rdw, \quad F_7 = dy - pdw$$

Vector field representation is given by:

$$E_1 = D_p, \quad E_2 = zD_q - rD_x + pD_y + D_w, \quad E_3 = D_r, \quad E_4 = D_z,$$
$$E_5 = D_y, \quad E_6 = D_x, \quad E_7 = -D_q$$

4. $(37D) : [x_1, x_2] = x_5, [x_1, x_3] = x_6, [x_2, x_4] = x_7, [x_3, x_4] = x_5;$

A matrix Lie group is given by:

$$S = \begin{bmatrix} 1 & 0 & 0 & w & -z & 0 & q \\ 0 & 1 & 0 & 0 & w & -z & x \\ 0 & 0 & 1 & 0 & 0 & w & y \\ 0 & 0 & 0 & 1 & 0 & 0 & 0 \\ 0 & 0 & 0 & 0 & 1 & 0 & r \\ 0 & 0 & 0 & 0 & 0 & 1 & p \\ 0 & 0 & 0 & 0 & 0 & 0 & 1 \end{bmatrix}$$

Left invariant differential forms are given by:

$$F_1 = dp, \quad F_2 = dq + rdz, \quad F_3 = dr, \quad F_4 = dx + pdz - rdw$$
$$F_5 = dy - pdw, \quad F_6 = dz, \quad F_7 = dw$$

Vector field representation is given by:

$$E_1 = D_p, \quad E_2 = -rD_q - pD_x + D_z, \quad E_3 = rD_x + pD_y + D_w, \quad E_4 = D_r,$$
$$E_5 = -D_x, \quad E_6 = D_y, \quad E_7 = D_q$$

Algebras with upper central series dimensions (247)

1. $(247A) : [x_1, x_i] = x_{i+2}, i = 2,3,4,5;$

A matrix Lie group is given by:

$$S = \begin{bmatrix} 1 & 0 & w & 0 & 1/2w^2 & 0 & q \\ 0 & 1 & 0 & w & 0 & 1/2w^2 & x \\ 0 & 0 & 1 & 0 & w & 0 & y \\ 0 & 0 & 0 & 1 & 0 & w & z \\ 0 & 0 & 0 & 0 & 1 & 0 & r \\ 0 & 0 & 0 & 0 & 0 & 1 & p \\ 0 & 0 & 0 & 0 & 0 & 0 & 1 \end{bmatrix}$$

Left invariant differential forms are given by:

$$F_1 = dp, \quad F_2 = dq - ydw, F_3 = dr, \quad F_4 = dx - zdw$$
$$F_5 = dy - rdw, \quad F_6 = dz - pdw, \quad F_7 = dw$$

Vector field representation is given by:

$$E_1 = -yD_q - zD_x - rD_y - pD_z - D_w, \quad E_2 = D_p, \quad E_3 = D_r, \quad E_4 = D_z$$
$$E_5 = D_y, \quad E_6 = D_x, E_7 = D_q$$

2. $(247B):[x_1,x_i]=x_{i+2}, i=2,3,4,[x_3,x_5]=x_7;$

A matrix Lie group is given by:

$$S=\begin{bmatrix} 1 & 0 & 0 & p & 0 & -z+pw & q \\ 0 & 1 & w & 0 & 1/2w^2 & 0 & x \\ 0 & 0 & 1 & 0 & w & 0 & y \\ 0 & 0 & 0 & 1 & 0 & w & z \\ 0 & 0 & 0 & 0 & 1 & 0 & r \\ 0 & 0 & 0 & 0 & 0 & 1 & p \\ 0 & 0 & 0 & 0 & 0 & 0 & 1 \end{bmatrix}$$

Left invariant differential forms are given by:

$F_1=dp,\quad F_2=-zdp+dq+pdz-p^2dw,\quad F_3=dr,\quad F_4=dx-ydw$
$F_5=dy-rdw,\quad F_6=dz-pdw,\quad F_7=dw$

Vector field representation is given by:

$E_1=D_w+yD_x+rD_y+pD_z,\quad E_2=D_r,\quad E_3=D_p+zD_q,\quad E_4=-D_y$
$E_5=pD_q-D_z,\quad E_6=D_x,\quad E_7=2*D_q$

3. $(247C):[x_1,x_i]=x_{i+2}, i=2,3,4,[x_1,x_5]=x_7,[x_3,x_5]=x_6;$

A matrix Lie group is given by:

$$S=\begin{bmatrix} 1 & 0 & w & p & 1/2w^2 & -z+pw & q \\ 0 & 1 & 0 & w & 0 & 1/2w^2 & x \\ 0 & 0 & 1 & 0 & w & 0 & y \\ 0 & 0 & 0 & 1 & 0 & w & z \\ 0 & 0 & 0 & 0 & 1 & 0 & r \\ 0 & 0 & 0 & 0 & 0 & 1 & p \\ 0 & 0 & 0 & 0 & 0 & 0 & 1 \end{bmatrix}$$

Left invariant differential forms are given by:

$F_1=dp,\quad F_2=-zdp+dq+pdz+(-p^2-y)dw,\quad F_3=dr,\quad F_4=dx-zdw$
$F_5=dy-rdw,\quad F_6=dz-pdw,\quad F_7=dw$

Vector field representation is given by:

$E_1=D_w+yD_q+zD_x+rD_y+pD_z,\quad E_2=2D_r,\quad E_3=D_p+zD_q,\quad E_4=-2D_y$
$E_5=pD_q-D_z,\quad E_6=2D_q,\quad E_7=D_x$

4. $(247D):[x_1,x_i]=x_{i+2}, i=2,3,[x_1,x_4]=x_6,[x_2,x_5]=x_7,[x_3,x_4]=x_7;$

A matrix Lie group is given by:

$$S=\begin{bmatrix} 1 & 0 & 0 & r & 0 & rw-y & q \\ 0 & 1 & 0 & w & 0 & 1/2w^2 & x \\ 0 & 0 & 1 & 0 & w & 0 & y \\ 0 & 0 & 0 & 1 & 0 & w & z \\ 0 & 0 & 0 & 0 & 1 & 0 & r \\ 0 & 0 & 0 & 0 & 0 & 1 & p \\ 0 & 0 & 0 & 0 & 0 & 0 & 1 \end{bmatrix}$$

Left invariant differential forms are given by:

$F_1=dp,\quad F_2=-zdr+dq+pdy-prdw,\quad F_3=dr,\quad F_4=dx-zdw,$
$F_5=dy-rdw,\quad F_6=dz-pdw,\quad F_7=dw$

Vector field representation is given by:

$E_1=zD_x+rD_y+pD_z+D_w,\quad E_2=D_p,\quad E_3=zD_q+D_r,\quad E_4=-D_z,$
$E_5=pD_q-D_y,\quad E_6=D_x,\quad E_7=D_q$

5. $(247E):[x_1,x_i]=x_{i+2}, i=2,3,4,[x_1,x_5]=x_6,[x_2,x_5]=x_7,[x_3,x_4]=x_7;$

A matrix Lie group is given by:

$$S=\begin{bmatrix} 1 & 0 & 0 & r & 0 & rw-y & q \\ 0 & 1 & w & w & 1/2w^2 & 1/2w^2 & x \\ 0 & 0 & 1 & 0 & w & 0 & y \\ 0 & 0 & 0 & 1 & 0 & w & z+pw \\ 0 & 0 & 0 & 0 & 1 & 0 & r \\ 0 & 0 & 0 & 0 & 0 & 1 & p \\ 0 & 0 & 0 & 0 & 0 & 0 & 1 \end{bmatrix}$$

Left invariant differential forms are given by:

$F_1=dp,\quad F_2=(-pw-z)dr+dq+pdy-prdw,\quad F_3=dr,\quad F_4=dx+(-pw-z-y)dw$
$F_5=dy-rdw,\quad F_6=wdp+dz,\quad F_7=dw$

Vector field representation is given by:

$E_1=(pw+z+y)D_x+rD_y+D_w,\quad E_2=(z+pw)D_q+D_r,\quad E_3=D_p-wD_z,\quad E_4=pD_q-D_y$
$E_5=-D_z,\quad E_6=D_x,\quad E_7=D_q$

6. $(247F):[x_1,x_i]=x_{i+2}, i=2,3,[x_2,x_4]=x_6,[x_2,x_5]=x_7;$

A matrix Lie group is given by:

$$S=\begin{bmatrix} 1 & 0 & 0 & w & 0 & r+1/2w^2 & q \\ 0 & 1 & r & 0 & -y+rw & 0 & x \\ 0 & 0 & 1 & 0 & w & 0 & y \\ 0 & 0 & 0 & 1 & 0 & w & z \\ 0 & 0 & 0 & 0 & 1 & 0 & r \\ 0 & 0 & 0 & 0 & 0 & 1 & p \\ 0 & 0 & 0 & 0 & 0 & 0 & 1 \end{bmatrix}$$

Left invariant differential forms are given by:

$F_1=dp,\quad F_2=-pdr+dq-zdw,\quad F_3=dr,\quad F_4=-ydr+dx+rdy-r^2dw$
$F_5=dy-rdw,\quad F_6=dz-pdw,\quad F_7=dw$

Vector field representation is given by:

$E_1=rD_y+pD_z+D_w,\quad E_2=D_r+yD_x,\quad E_3=-D_p-zD_q,\quad E_4=rD_x-D_y$
$E_5=-pD_q+D_z,\quad E_6=2D_x,\quad E_7=2D_q$

7. $(247G):[x_1,x_i]=x_{i+2}, i=2,3,[x_1,x_4]=x_7,[x_2,x_4]=x_6,[x_3,x_5]=x_7;$

A matrix Lie group is given by:

$$S=\begin{bmatrix} 1 & 0 & w & p & 1/2w^2 & -z+pw & q \\ 0 & 1 & r & 0 & -y+rw & 0 & x \\ 0 & 0 & 1 & 0 & w & 0 & y \\ 0 & 0 & 0 & 1 & 0 & w & z \\ 0 & 0 & 0 & 0 & 1 & 0 & r \\ 0 & 0 & 0 & 0 & 0 & 1 & p \\ 0 & 0 & 0 & 0 & 0 & 0 & 1 \end{bmatrix}$$

Left invariant differential forms are given by:

$F_1=dp,\quad F_2=-zdp+dq+pdz+(-p^2-y)dw,\quad F_3=dr,\quad F_4=-ydr+dx+rdy-r^2dw$
$F_5=dy-rdw,\quad F_6=dz-pdw,\quad F_7=dw$

Vector field representation is given by:

$E_1=yD_q+rD_y+pD_z+D_w,\quad E_2=2D_r+2yD_x,\quad E_3=-D_p-zD_q,\quad E_4=2rD_x-2D_y$
$E_5=-pD_q+D_z,\quad E_6=8D_x,\quad E_7=2D_q$

8. $(247H)$:

$[x_1,x_i]=x_{i+2}, i=2,3,[x_1,x_4]=x_7,[x_1,x_5]=x_6,[x_2,x_4]=x_6,[x_3,x_5]=x_7;$

A matrix Lie group is given by:

$$S=\begin{bmatrix}1 & 0 & w & p & 1/2w^2 & pw-z & q\\ 0 & 1 & r & w & rw-y & 1/2w^2 & x\\ 0 & 0 & 1 & 0 & w & 0 & y\\ 0 & 0 & 0 & 1 & 0 & w & z\\ 0 & 0 & 0 & 0 & 1 & 0 & r\\ 0 & 0 & 0 & 0 & 0 & 1 & p\\ 0 & 0 & 0 & 0 & 0 & 0 & 1\end{bmatrix}$$

Left invariant differential forms are given by:

$F_1=dp,\quad F_2=-zdp+dq+pdz+(-p^2-y)dw,\quad F_3=dr,\quad F_4=-ydr+dx+rdy+(-r^2-z)dw$
$F_5=dy-rdw,\quad F_6=dz-pdw,\quad F_7=dw$

Vector field representation is given by:

$E_1=2yD_q+2zD_x+2rD_y+2pD_z+2D_w,\quad E_2=D_r+yD_x,\quad E_3=D_p+zD_q,\quad E_4=2rD_x-2D_y$
$E_5=2pD_q-2D_z,\quad E_6=4D_x,\quad E_7=4D_q$

9. $(247I):[x_1,x_i]=x_{i+2},i=2,3,[x_2,x_5]=x_6,[x_3,x_4]=x_6,[x_3,x_5]=x_7;$

A matrix Lie group is given by:

$$S=\begin{bmatrix}1 & 0 & 0 & r & 0 & rw-y & q\\ 0 & 1 & 0 & p & 0 & -z & x\\ 0 & 0 & 1 & 0 & w & 0 & y\\ 0 & 0 & 0 & 1 & 0 & w & z+pw\\ 0 & 0 & 0 & 0 & 1 & 0 & r\\ 0 & 0 & 0 & 0 & 0 & 1 & p\\ 0 & 0 & 0 & 0 & 0 & 0 & 1\end{bmatrix}$$

Left invariant differential forms are given by:

$F_1=dp,\quad F_2=(-z-pw)dr+dq+pdy-prdw,\quad F_3=dr,\quad F_4=zdp+dx+pdz$
$F_5=dy-rdw,\quad F_6=wdp+dz,\quad F_7=dw$

Vector field representation is given by:

$E_1=rD_y+D_w,\quad E_2=(2z+2pw)D_q+2D_r,\quad E_3=-D_p+(-z-pw)D_x+wD_z,\quad E_4=2pD_q-2D_y$
$E_5=-pD_x+D_z,\quad E_6=-2D_q,\quad E_7=2D_x$

10. $(247J):[x_1,x_i]=x_{i+2},i=2,3,4[x_2,x_5]=x_6,[x_3,x_4]=x_6,[x_3,x_5]=x_7;$

A matrix Lie group is given by:

$$S=\begin{bmatrix}1 & 0 & 0 & p & 0 & pw-z & q\\ 0 & 1 & w & r & 1/2w^2 & rw-y & x\\ 0 & 0 & 1 & 0 & w & 0 & y\\ 0 & 0 & 0 & 1 & 0 & w & z\\ 0 & 0 & 0 & 0 & 1 & 0 & r\\ 0 & 0 & 0 & 0 & 0 & 1 & p\\ 0 & 0 & 0 & 0 & 0 & 0 & 1\end{bmatrix}$$

Left invariant differential forms are given by:

$F_1=dp,\quad F_2=-zdp+dq+pdz-p^2dw,\quad F_3=dr,\quad F_4=-zdr+dx+pdy+(-pr-y)dw$
$F_5=dy-rdw,\quad F_6=dz-pdw,\quad F_7=dw$

Vector field representation is given by:

$E_1=yD_x+rD_y+pD_z+D_w,\quad E_2=D_r+zD_x,\quad E_3=D_p+zD_q,\quad E_4=pD_x-D_y$
$E_5=pD_q-D_z,\quad E_6=D_x,\quad E_7=2D_q$

11. $(247K):[x_1,x_i]=x_{i+2},i=2,3,4,[x_2,x_5]=x_7,[x_3,x_4]=x_7,[x_3,x_5]=x_6;$

A matrix Lie group is given by:

$$S=\begin{bmatrix}1 & 0 & 0 & r & 0 & rw-y & q\\ 0 & 1 & w & p & 1/2w^2 & -z & x\\ 0 & 0 & 1 & 0 & w & 0 & y\\ 0 & 0 & 0 & 1 & 0 & w & z+pw\\ 0 & 0 & 0 & 0 & 1 & 0 & r\\ 0 & 0 & 0 & 0 & 0 & 1 & p\\ 0 & 0 & 0 & 0 & 0 & 0 & 1\end{bmatrix}$$

Left invariant differential forms are given by:

$F_1=dp,\quad F_2=(-z-pw)dr+dq+pdy-prdw,\quad F_3=dr,\quad F_4=-zdp+dx+pdz-ydw$
$F_5=dy-rdw,\quad F_6=wdp+dz,\quad F_7=dw$

Vector field representation is given by:

$E_1=yD_x+rD_y+D_w,\quad E_2=(2z+2pw)D_q+2D_r,\quad E_3=D_p+(z+pw)D_x-wD_z,\quad E_4=2pD_q-2D_y$
$E_5=pD_x-D_z,\quad E_6=2D_x,\quad E_7=2D_q$

12. $(247L):[x_1,x_i]=x_{i+2},i=2,3,4,5,[x_2,x_3]=x_6$

A matrix Lie group is given by:

$$S=\begin{bmatrix}1 & 0 & w & 0 & 1/2w^2 & r & q\\ 0 & 1 & 0 & w & 0 & 1/2w^2 & x\\ 0 & 0 & 1 & 0 & w & 0 & y\\ 0 & 0 & 0 & 1 & 0 & w & z\\ 0 & 0 & 0 & 0 & 1 & 0 & r\\ 0 & 0 & 0 & 0 & 0 & 1 & p\\ 0 & 0 & 0 & 0 & 0 & 0 & 1\end{bmatrix}$$

Left invariant differential forms are given by:

$F_1=dp,\quad F_2=-pdr+dq-ydw,\quad F_3=dr,\quad F_4=dx-zdw$
$F_5=dy-rdw,\quad F_6=dz-pdw,\quad F_7=dw$

Vector field representation is given by:

$E_1=yD_q+zD_x+rD_y+pD_z+D_w,\quad E_2=-pD_q-D_r,\quad E_3=-D_p,\quad E_4=D_y$
$E_5=D_z,\quad E_6=-D_q,\quad E_7=-D_x$

13. $(247M):[x_1,x_i]=x_{i+2},i=2,3,4,[x_2,x_3]=x_6,[x_3,x_5]=x_7;$

A matrix Lie group is given by:

$$S=\begin{bmatrix}1 & 0 & 0 & w & 0 & r+1/2w^2 & q\\ 0 & 1 & r & 0 & -y+rw & 0 & x\\ 0 & 0 & 1 & 0 & w & 0 & y\\ 0 & 0 & 0 & 1 & 0 & w & z\\ 0 & 0 & 0 & 0 & 1 & 0 & r\\ 0 & 0 & 0 & 0 & 0 & 1 & p\\ 0 & 0 & 0 & 0 & 0 & 0 & 1\end{bmatrix}$$

Left invariant differential forms are given by:

$F_1=dp,\quad F_2=-pdr+dq-zdw,\quad F_3=dr,\quad F_4=-ydr+dx+rdy-r^2dw$
$F_5=dy-rdw,\quad F_6=dz-pdw,\quad F_7=dw$

Vector field representation is given by:

$E_1=-zD_q-rD_y-pD_z-D_w,\quad E_2=D_p,\quad E_3=pD_q+D_r+yD_x,\quad E_4=D_z$
$E_5=-rD_x+D_y,\quad E_6=D_q,\quad E_7=-2D_x$

14. $(247N):[x_1,x_i]=x_{i+2},i=2,3,[x_1,x_5]=x_6,[x_2,x_3]=x_7,[x_2,x_4]=x_6;$

A matrix Lie group is given by:

$$S=\begin{bmatrix} 1 & 0 & 0 & 0 & 0 & r & q \\ 0 & 1 & r & w & rw-y & 1/2w^2 & x \\ 0 & 0 & 1 & 0 & w & 0 & y \\ 0 & 0 & 0 & 1 & 0 & w & z \\ 0 & 0 & 0 & 0 & 1 & 0 & r \\ 0 & 0 & 0 & 0 & 0 & 1 & p \\ 0 & 0 & 0 & 0 & 0 & 0 & 1 \end{bmatrix}$$

Left invariant differential forms are given by:

$F_1 = dp,\quad F_2 = -pdr + dq,\quad F_3 = dr,\quad F_4 = -ydr + dx + rdy + (-r^2 - z)dw$
$F_5 = dy - rdw,\quad F_6 = dz - pdw,\quad F_7 = dw$

Vector field representation is given by:

$E_1 = zD_x + rD_y + pD_z + D_w,\quad E_2 = pD_q + D_r + yD_x,\quad E_3 = 2D_p,\quad E_4 = rD_x - D_y$
$E_5 = -2D_z,\quad E_6 = 2D_x,\quad E_7 = -2D_q$

15. $(247O):[x_1,x_i]=x_{i+2}, i=2,3,4,[x_1,x_5]=x_7,[x_2,x_3]=x_7,[x_3,x_5]=x_6;$

A matrix Lie group is given by:

$$S=\begin{bmatrix} 1 & 0 & w & 0 & 1/2w^2 & r & q \\ 0 & 1 & r & w & -y+rw & 1/2w^2 & x \\ 0 & 0 & 1 & 0 & w & 0 & y \\ 0 & 0 & 0 & 1 & 0 & w & z \\ 0 & 0 & 0 & 0 & 1 & 0 & r \\ 0 & 0 & 0 & 0 & 0 & 1 & p \\ 0 & 0 & 0 & 0 & 0 & 0 & 1 \end{bmatrix}$$

Left invariant differential forms are given by:

$F_1 = dp,\quad F_2 = -pdr + dq - ydw,\quad F_3 = dr,\quad F_4 = -ydr + dx + rdy + (-r^2 - z) * w$
$F_5 = dy - rdw,\quad F_6 = dz - pdw,\quad F_7 = dw$

Vector field representation is given by:

$E_1 = 2yD_q + 2zD_x + 2rD_y + 2pD_z + 2D_w,\quad E_2 = 4D_p,\quad E_3 = 2pD_q + 2D_r + 2yD_x,\quad E_4 = -8D_z$
$E_5 = 4rD_x - 4D_y,\quad E_6 = 16D_x,\quad E_7 = 8D_q$

16. $(247P):[x_1,x_i]=x_{i+2}, i=2,3,[x_2,x_3]=x_6,[x_2,x_5]=x_7,[x_3,x_4]=x_7;$

A matrix Lie group is given by:

$$S=\begin{bmatrix} 1 & 0 & 0 & 0 & p & 0 & q \\ 0 & 1 & 0 & r & 0 & rw-y & x \\ 0 & 0 & 1 & 0 & w & 0 & y \\ 0 & 0 & 0 & 1 & 0 & w & z \\ 0 & 0 & 0 & 0 & 1 & 0 & r \\ 0 & 0 & 0 & 0 & 0 & 1 & p \\ 0 & 0 & 0 & 0 & 0 & 0 & 1 \end{bmatrix}$$

Left invariant differential forms are given by:
$F_1 = dp,\quad F_2 = -rdp + dq,\quad F_3 = dr,\quad F_4 = -zdr + dx + pdy - prdw$
$F_5 = dy - rdw,\quad F_6 = dz - pdw,\quad F_7 = dw$

Vector field representation is given by:

$E_1 = rD_y + pD_z + D_w,\quad E_2 = -D_r - zD_x,\quad E_3 = -D_p - rD_q,\quad E_4 = -pD_x + D_y$
$E_5 = D_z,\quad E_6 = D_q,\quad E_7 = D_x$

17. $(247Q):[x_1,x_i]=x_{i+2}, i=2,3,4,[x_2,x_3]=x_6,[x_2,x_5]=x_7,[x_3,x_4]=x_7;$

A matrix Lie group is given by:

$$S=\begin{bmatrix} 1 & 0 & 0 & w & 0 & r+1/2w^2 & q \\ 0 & 1 & p & r & pw-z & -y+rw & x \\ 0 & 0 & 1 & 0 & w & 0 & y \\ 0 & 0 & 0 & 1 & 0 & w & z \\ 0 & 0 & 0 & 0 & 1 & 0 & r \\ 0 & 0 & 0 & 0 & 0 & 1 & p \\ 0 & 0 & 0 & 0 & 0 & 0 & 1 \end{bmatrix}$$

Left invariant differential forms are given by:

$F_1 = dp,\quad F_2 = -pdr + dq - zdw,\quad F_3 = dr,\quad F_4 = -zdr - ydp + dx + pdy + rdz - 2prdw$
$F_5 = dy - rdw,\quad F_6 = dz - pdw,\quad F_7 = dw$

Vector field representation is given by:

$E_1 = 2zD_q + 2rD_y + 2pD_z + 2D_w,\quad E_2 = -2D_p - 2yD_x,\quad E_3 = 4pD_q + 4D_r + 4zD_x,\quad E_4 = -4rD_x + 4D_z$
$E_5 = 8pD_x - 8D_y,\quad E_6 = -8D_q,\quad E_7 = -32D_x$

18. $(247R)$:

$[x_1,x_i]=x_{i+2}, i=2,3,4,[x_1,x_5]=x_6,[x_2,x_3]=x_6,[x_2,x_5]=x_7,[x_3,x_4]=x_7;$

A matrix Lie group is given by:

$$S=\begin{bmatrix} 1 & 0 & w & w & p+1/2w^2 & 1/2w^2 & q \\ 0 & 1 & 0 & r & 0 & rw-y & x \\ 0 & 0 & 1 & 0 & w & 0 & y \\ 0 & 0 & 0 & 1 & 0 & w & z \\ 0 & 0 & 0 & 0 & 1 & 0 & r \\ 0 & 0 & 0 & 0 & 0 & 1 & p \\ 0 & 0 & 0 & 0 & 0 & 0 & 1 \end{bmatrix}$$

Left invariant differential forms are given by:

$F_1 = dp,\quad F_2 = -rdp + dq + (-z - y)dw,\quad F_3 = dr,\quad F_4 = -zdr + dx + pdy - rpdw$
$F_5 = dy - rdw,\quad F_6 = dz - pdw,\quad F_7 = dw$

The Vector field representation is given by:

$E_1 = (z + y)D_q + rD_y + pD_z + D_w,\quad E_2 = D_r + zD_x,\quad E_3 = D_p + rD_q,\quad E_4 = pD_x - D_y,$
$E_5 = -D_z,\quad E_6 = D_q,\quad E_7 = D_x$

Algebras with upper central series dimensions (257)

1. $(257A):[x_1,x_2]=x_3,[x_1,x_3]=x_6,[x_1,x_5]=x_7,[x_2,x_4]=x_6;$

A matrix Lie group is given by:

$$S=\begin{bmatrix} 1 & x & -y & -r & p & q \\ 0 & 1 & x & 0 & w & 0 \\ 0 & 0 & 1 & 0 & z & 0 \\ 0 & 0 & 0 & 1 & 0 & x \\ 0 & 0 & 0 & 0 & 1 & 0 \\ 0 & 0 & 0 & 0 & 0 & 1 \end{bmatrix}$$

Left invariant differential forms are given by:

$F_1 = dp + (-w + zx)dx + zdy,\quad F_2 = xdr + dq,\quad F_3 = dr,\quad F_4 = dx,$
$F_5 = xdx + dy,\quad F_6 = dz,\quad F_7 = -zdx + dw$

Vector field representation is given by:

$E_1 = -D_x - wD_p + xD_y - zD_w,\quad E_2 = D_z,\quad E_3 = D_w,\quad E_4 = -D_y + zD_p,$
$E_5 = D_r - xD_q,\quad E_6 = D_p,\quad E_7 = D_q$

2. $(257B):[x_1,x_2]=x_3,[x_1,x_3]=x_6,[x_1,x_4]=x_7,[x_2,x_5]=x_7;$

A matrix Lie group is given by:

$$S=\begin{bmatrix} 1 & 0 & -w & -w & 0 & 1/2w^2 & q \\ 0 & 1 & 0 & 0 & -w & z & -x \\ 0 & 0 & 1 & 0 & 0 & -w & -y \\ 0 & 0 & 0 & 1 & 0 & 0 & 0 \\ 0 & 0 & 0 & 0 & 1 & 0 & r \\ 0 & 0 & 0 & 0 & 0 & 1 & p \\ 0 & 0 & 0 & 0 & 0 & 0 & 1 \end{bmatrix}$$

Left invariant differential forms are given by:

$F_1 = dp,\quad F_2 = dq - ydw,\quad F_3 = dr,\quad F_4 = dx + pdz - rdw,$
$F_5 = dy - pdw,\quad F_6 = dz,\quad F_7 = dw$

Vector field representation is given by:

$E_1 = -yD_q - rD_x - pD_y - D_w,\quad E_2 = D_p,\quad E_3 = D_y,\quad E_4 = D_r,$
$E_5 = pD_x - D_z,\quad E_6 = D_q,\quad E_7 = D_x$

3. $(257C): [x_1,x_2]=x_3, [x_1,x_3]=x_6, [x_2,x_4]=x_6, [x_2,x_5]=x_7;$

A matrix Lie group is given by:

$$S=\begin{bmatrix} 1 & x & -y & -r & p & q \\ 0 & 1 & z & 0 & w & 0 \\ 0 & 0 & 1 & 0 & z & 0 \\ 0 & 0 & 0 & 1 & 0 & x \\ 0 & 0 & 0 & 0 & 1 & 0 \\ 0 & 0 & 0 & 0 & 0 & 1 \end{bmatrix}$$

Left invariant differential forms are given by:

$F_1 = dp + (-w + z^2)dx + zdy,\quad F_2 = xdr + dq,\quad F_3 = dr,\quad F_4 = dx,$
$F_5 = zdx + dy,\quad F_6 = dz,\quad F_7 = -zdz + dw$

Vector field representation is given by:

$E_1 = D_z + zD_w,\quad E_2 = wD_p + D_x - zD_y,\quad E_3 = zD_p - D_y,\quad E_4 = -D_w,$
$E_5 = xD_q - D_r,\quad E_6 = D_p,\quad E_7 = D_q$

4. $(257D): [x_1,x_2]=x_3, [x_1,x_3]=x_6, [x_1,x_4]=x_7, [x_2,x_4]=x_6, [x_2,x_5]=x_7;$

A matrix Lie group is given by:

$$S=\begin{bmatrix} 1 & 0 & -w & -w & p & 1/2w^2 & q \\ 0 & 1 & 0 & 0 & -w & z & -x \\ 0 & 0 & 1 & 0 & 0 & -w & -y \\ 0 & 0 & 0 & 1 & 0 & 0 & 0 \\ 0 & 0 & 0 & 0 & 1 & 0 & r \\ 0 & 0 & 0 & 0 & 0 & 1 & p \\ 0 & 0 & 0 & 0 & 0 & 0 & 1 \end{bmatrix}$$

Left invariant differential forms are given by:

$F_1 = dp,\quad F_2 = -rdp + dq - ydw,\quad F_3 = dr,\quad F_4 = dx + pdz - rdw,$
$F_5 = dy - pdw,\quad F_6 = dz,\quad F_7 = dw$

Vector field representation is given by:

$E_1 = D_w + yD_q + rD_x + pD_y,\quad E_2 = D_p + rD_q,\quad E_3 = -D_y,\quad E_4 = -D_r,$
$E_5 = -D_z + pD_x,\quad E_6 = D_q,\quad E_7 = D_x$

5. $(257E): [x_1,x_2]=x_3, [x_1,x_3]=x_6, [x_2,x_4]=x_7, [x_4,x_5]=x_6;$

A matrix Lie group is given by:

$$S=\begin{bmatrix} 1 & 0 & -w & -w & z & 1/2w^2 & q \\ 0 & 1 & 0 & 0 & 0 & z & -x \\ 0 & 0 & 1 & 0 & 0 & -w & -y \\ 0 & 0 & 0 & 1 & 0 & 0 & 0 \\ 0 & 0 & 0 & 0 & 1 & 0 & r \\ 0 & 0 & 0 & 0 & 0 & 1 & p \\ 0 & 0 & 0 & 0 & 0 & 0 & 1 \end{bmatrix}$$

Left invariant differential forms are given by:

$F_1 = dp,\quad F_2 = dq - rdz - ydw,\quad F_3 = dr,\quad F_4 = dx + pdz,$
$F_5 = dy - pdw,\quad F_6 = dz,\quad F_7 = dw$

Vector field representation is given by:

$E_1 = yD_q + pD_y + D_w,\quad E_2 = D_p,\quad E_3 = -D_y,\quad E_4 = -rD_q + pD_x - D_z,$
$E_5 = D_r,\quad E_6 = D_q,\quad E_7 = D_x$

6. $(257F)$: $\begin{bmatrix} 0 & 0 & 0 & 0 & 0 & 0 & 1 \end{bmatrix}$

A matrix Lie group is given by:

$$S=\begin{bmatrix} 1 & 0 & -w & -w & z & 1/2w^2 & q \\ 0 & 1 & 0 & 0 & -w & 0 & -x \\ 0 & 0 & 1 & 0 & 0 & -w & -y \\ 0 & 0 & 0 & 1 & 0 & 0 & 0 \\ 0 & 0 & 0 & 0 & 1 & 0 & r \\ 0 & 0 & 0 & 0 & 0 & 1 & p \\ 0 & 0 & 0 & 0 & 0 & 0 & 1 \end{bmatrix}$$

Left invariant differential forms are given by:

$F_1 = dp,\quad F_2 = dq - rdz - ydw,\quad F_3 = dr,\quad F_4 = dx - rdw,$
$F_5 = dy - pdw,\quad F_6 = dz,\quad F_7 = dw$

Vector field representation is given by:

$E_1 = -D_p,\quad E_2 = -yD_q - rD_x - pD_y - D_w,\quad E_3 = D_y,\quad E_4 = D_r,$
$E_5 = rD_q + D_z,\quad E_6 = D_q,\quad E_7 = D_x$

7. $(257G): [x_1,x_2]=x_3, [x_1,x_3]=x_6, [x_1,x_5]=x_7, [x_2,x_4]=x_7, [x_4,x_5]=x_6;$

A matrix Lie group is given by:

$$S=\begin{bmatrix} 1 & 0 & -w & -w & z & 1/2w^2 & q \\ 0 & 1 & 0 & 0 & -w & z & -x \\ 0 & 0 & 1 & 0 & 0 & -w & -y \\ 0 & 0 & 0 & 1 & 0 & 0 & 0 \\ 0 & 0 & 0 & 0 & 1 & 0 & r \\ 0 & 0 & 0 & 0 & 0 & 1 & p \\ 0 & 0 & 0 & 0 & 0 & 0 & 1 \end{bmatrix}$$

Left invariant differential forms are given by:

$F_1 = dp,\quad F_2 = dq - rdz - ydw,\quad F_3 = dr,\quad F_4 = dx + pdz - rdw,$
$F_5 = dy - pdw,\quad F_6 = dz,\quad F_7 = dw$

Vector field representation is given by:

$E_1 = -yD_q - rD_x - pD_y - D_w,\quad E_2 = D_p,\quad E_3 = D_y,\quad E_4 = -rD_q + pD_x - D_z,$
$E_5 = D_r,\quad E_6 = D_q,\quad E_7 = D_x$

8. $(257H): [x_1,x_2]=x_3, [x_1,x_3]=x_6, [x_2,x_4]=x_6, [x_4,x_5]=x_7;$

A matrix Lie group is given by:

$$S=\begin{bmatrix} 1 & x & -y & -r & p & q \\ 0 & 1 & z & 0 & w+1/2z^2 & 0 \\ 0 & 0 & 1 & 0 & z & 0 \\ 0 & 0 & 0 & 1 & 0 & w \\ 0 & 0 & 0 & 0 & 1 & 0 \\ 0 & 0 & 0 & 0 & 0 & 1 \end{bmatrix}$$

Left invariant differential forms are given by:

$F_1 = dp+(-w+(1/2)z^2)dx+zdy,\quad F_2 = wdr+dq,\quad F_3 = dr,\quad F_4 = dx,$
$F_5 = zdx+dy,\quad F_6 = dz,\quad F_7 = dw$

Vector field representation is given by:

$E_1 = -D_z,\quad E_2 = D_x+(w+(1/2)z^2)D_p - zD_y,\quad E_3 = -zD_p + D_y,\quad E_4 = -D_w,$
$E_5 = -wD_q + D_r,\quad E_6 = D_p,\quad E_7 = D_q$

9. $(257I):[x_1,x_2]=x_3,[x_1,x_3]=x_6,[x_1,x_4]=x_6,[x_1,x_5]=x_7,[x_2,x_3]=x_7;$

A matrix Lie group is given by:

$$S=\begin{bmatrix} 1 & 0 & -w & -w & 0 & 1/2w^2 & q \\ 0 & 1 & p & 0 & -w & y & -x \\ 0 & 0 & 1 & 0 & 0 & -w & -y-pw \\ 0 & 0 & 0 & 1 & 0 & 0 & z \\ 0 & 0 & 0 & 0 & 1 & 0 & r \\ 0 & 0 & 0 & 0 & 0 & 1 & p \\ 0 & 0 & 0 & 0 & 0 & 0 & 1 \end{bmatrix}$$

Left invariant differential forms are given by:

$F_1 = dp,\quad F_2 = dq+(-y-pw+z)dw,\quad F_3 = dr,\quad F_4 = ydp-dx-pdy+rdw,$
$F_5 = wdp+dy,\quad F_6 = dz,\quad F_7 = dw$

Vector field representation is given by:

10. $(257J):[x_1,x_2]=x_3,[x_1,x_3]=x_6,[x_1,x_5]=x_7,[x_2,x_3]=x_7,[x_2,x_4]=x_6;$

A matrix Lie group is given by:

$$S=\begin{bmatrix} 1 & 0 & -w & -w & 0 & 1/2w^2+z & q \\ 0 & 1 & p & 0 & -w & y & -x \\ 0 & 0 & 1 & 0 & 0 & -w & -y-pw \\ 0 & 0 & 0 & 1 & 0 & 0 & 0 \\ 0 & 0 & 0 & 0 & 1 & 0 & r \\ 0 & 0 & 0 & 0 & 0 & 1 & p \\ 0 & 0 & 0 & 0 & 0 & 0 & 1 \end{bmatrix}$$

Left invariant differential forms are given by:

$F_1 = dp,\quad F_2 = dq-pdz+(-y-pw)dw,\quad F_3 = dr,\quad F_4 = ydp-dx-pdy+rdw,$
$F_5 = wdp+dy,\quad F_6 = dz,\quad F_7 = dw$

Vector field representation is given by:

$E_1 = (y+pw)D_q + rD_x + D_w,\quad E_2 = D_p+(y+pw)D_x - wD_y,\quad E_3 = pD_x - D_y,\quad E_4 = pD_q + D_z,$
$E_5 = -2D_r,\quad E_6 = D_q,\quad E_7 = 2D_x$

11. $(257K):[x_1,x_2]=x_3,[x_1,x_3]=x_6,[x_2,x_3]=x_7,[x_4,x_5]=x_7;$

A matrix Lie group is given by:

$$S=\begin{bmatrix} 1 & 0 & -z & y-wz & r & 0 & q \\ 0 & 1 & w & 1/2w^2 & 0 & 0 & x \\ 0 & 0 & 1 & w & 0 & 0 & 2y \\ 0 & 0 & 0 & 1 & 0 & 0 & 2z \\ 0 & 0 & 0 & 0 & 1 & 0 & p \\ 0 & 0 & 0 & 0 & 0 & 1 & r \\ 0 & 0 & 0 & 0 & 0 & 0 & 1 \end{bmatrix}$$

Left invariant differential forms are given by:

$F_1 = dp,\quad F_2 = -pdr+dq-2zdy+2ydz+2z^2dw,\quad F_3 = dr,\quad F_4 = dx-2ydw,$
$F_5 = dy-zdw,\quad F_6 = dz,\quad F_7 = dw$

Vector field representation is given by:

$E_1 = D_w+2yD_x+zD_y,\quad E_2 = D_z-2yD_q,\quad E_3 = -D_y-2zD_q,\quad E_4 = -4D_p,$
$E_5 = D_r+pD_q,\quad E_6 = 2D_x,\quad E_7 = -4D_q$

12. $(257L):[x_1,x_2]=x_3,[x_1,x_3]=x_6,[x_2,x_3]=x_7,[x_2,x_4]=x_6,[x_4,x_5]=x_7;$

A matrix Lie group is given by:

$$S=\begin{bmatrix} 1 & 0 & -z & y-wz & r & 0 & q \\ 0 & 1 & w & 1/2w^2 & 0 & z & x \\ 0 & 0 & 1 & w & 0 & 0 & 2y \\ 0 & 0 & 0 & 1 & 0 & 0 & 2z \\ 0 & 0 & 0 & 0 & 1 & 0 & p \\ 0 & 0 & 0 & 0 & 0 & 1 & r \\ 0 & 0 & 0 & 0 & 0 & 0 & 1 \end{bmatrix}$$

Left invariant differential forms are given by:

$F_1 = dp,\quad F_2 = -pdr+dq-2zdy+2ydz+2z^2dw,\quad F_3 = dr,\quad F_4 = dx-rdz-2ydw,$
$F_5 = dy-zdw,\quad F_6 = dz,\quad F_7 = dw$

Vector field representation is given by:

$E_1 = D_w+2yD_x+zD_y,\quad E_2 = D_z-2yD_q+rD_x,\quad E_3 = -D_y-2zD_q,\quad E_4 = -2D_r-2pD_q,$
$E_5 = -2D_p,\quad E_6 = 2D_x,\quad E_7 = -4D_q$

Algebras with upper central series dimensions (357)

1. $(357A):[x_1,x_2]=x_3,[x_1,x_3]=x_5,[x_1,x_4]=x_7,[x_2,x_4]=x_6;$

A matrix Lie group is given by:

$$S=\begin{bmatrix} 1 & 0 & 0 & 0 & 0 & r & -q \\ 0 & 1 & 0 & w & 0 & 1/2w^2 & x \\ 0 & 0 & 1 & 0 & w & 0 & y \\ 0 & 0 & 0 & 1 & 0 & w & z \\ 0 & 0 & 0 & 0 & 1 & 0 & r \\ 0 & 0 & 0 & 0 & 0 & 1 & p \\ 0 & 0 & 0 & 0 & 0 & 0 & 1 \end{bmatrix}$$

Left invariant differential forms are given by:

$F_1 = dp,\quad F_2 = -pdr-dq,\quad F_3 = dr,\quad F_4 = dx-zdw,$
$F_5 = dy-rdw,\quad F_6 = dz-pdw,\quad F_7 = dw$

Vector field representation is given by:

$E_1 = -zD_x+rD_y-pD_z-D_w,\quad E_2 = D_p,\quad E_3 = D_z,\quad E_4 = -pD_q+D_r,$
$E_5 = D_x,\quad E_6 = -D_q,\quad E_7 = -D_y$

2. $(357B):[x_1,x_2]=x_3,[x_1,x_3]=x_5,[x_1,x_4]=x_7,[x_2,x_3]=x_6;$

A matrix Lie group is given by:

$$S=\begin{bmatrix} 1 & 0 & 0 & p & 0 & -z+pw & 2q \\ 0 & 1 & 0 & w & 0 & 1/2w^2 & x \\ 0 & 0 & 1 & 0 & w & 0 & y \\ 0 & 0 & 0 & 1 & 0 & w & z \\ 0 & 0 & 0 & 0 & 1 & 0 & r \\ 0 & 0 & 0 & 0 & 0 & 1 & p \\ 0 & 0 & 0 & 0 & 0 & 0 & 1 \end{bmatrix}$$

Left invariant differential forms are given by:

$$F_1 = dp, \quad F_2 = -zdp + 2dq + pdz - p^2dw, \quad F_3 = dr, \quad F_4 = dx - zdw,$$
$$F_5 = dy - rdw, \quad F_6 = dz - pdw, \quad F_7 = dw$$

The vector field representation is given by:

$$E_1 = zD_x + rD_y + pD_z + D_w, \quad E_2 = D_p + z/2D_q, \quad E_3 = p/2D_q - D_z, \quad E_4 = -D_r,$$
$$E_5 = D_x, \quad E_6 = D_q, \quad E_7 = D_y$$

3. $(357C): [x_1, x_2] = x_3, [x_1, x_3] = x_5, [x_1, x_4] = x_7, [x_2, x_3] = x_6, [x_2, x_4] = x_5;$

A matrix Lie group is given by:

$$S = \begin{bmatrix} 1 & 0 & 0 & p & 0 & -z+pw & 2q \\ 0 & 1 & 0 & w & p & 1/2w^2 & x \\ 0 & 0 & 1 & 0 & w & 0 & y \\ 0 & 0 & 0 & 1 & 0 & w & z \\ 0 & 0 & 0 & 0 & 1 & 0 & r \\ 0 & 0 & 0 & 0 & 0 & 1 & p \\ 0 & 0 & 0 & 0 & 0 & 0 & 1 \end{bmatrix}$$

Left invariant differential forms are given by:

$$F_1 = dp, \quad F_2 = -zdp + 2dq + pdz - p^2dw, \quad F_3 = dr, \quad F_4 = -rdp + dx - zdw,$$
$$F_5 = dy - rdw, \quad F_6 = dz - pdw, \quad F_7 = dw$$

Vector field representation is given by:

$$E_1 = D_w + zD_x + rD_y + pD_z, \quad E_2 = D_p + ((1/2)z)D_q + rD_x, \quad E_3 = -D_z + ((1/2)p)D_q, \quad E_4 = -D_r,$$
$$E_5 = D_x, \quad E_6 = D_q, \quad E_7 = D_y$$

Conclusion

The Explanation regarding finding linear representations for seven-dimensional real, indecomposable nilpotent Lie algebras is done.

References

1. Ado ID (1935) Note on the representation of finite continuous groups by means of linear substitution. Izv Fiz-Mat Obsch (Kazan) 7: 01-43.
2. Ado ID (1947)The representation of Lie algebras by matrices (in Russian), Akademiya Nauk SSSR i Moskovskoe Matematicheskoe Obshchestvo. Uspekhi Matematicheskikh Nauk 2: 159-173.
3. Ghanam R, Thompson G, Tonon S (2006) Representations for six-dimensional nilpotent Lie algebras. Hadronic J 29: 299-317.
4. Ghanam R Strugar I,Thompson G (2005) Matrix representations for low dimensional Lie algebras. Extracta Math 20: 151-184.
5. Ghanam R, Thompson G, Miller E (2004) Variationality of four-dimensional Lie group connections. J Lie Theory 14: 395-425.
6. Gong MP (1998) Classification of Nilpotent Lie algebras of dimension 7 (Over algebriacally closed Fields and *R*).University of Waterloo.
7. Jacobson N (1962) Lie Algebras. Tracts in Pure and Applied Math Interscience Publishers, Newyork.
8. Humphreys JE (1972) Introduction to Lie algebras and representation theory. Graduate Text in Math Springer.
9. Seeley C. Degeneration of 6-dimensional Nilpotent Lie Algebras over *C*. Communications in Algebra 18: 3493-3505.
10. Seeley C (1993) 7-dimensional Nilpotent Lie Algebra.Trans Amer Math Soc 335: 479-496.
11. Skjelbred T, Sund T (1977).Classification of Nilpotent Lie Algebras in dimension six.Univerisity of Oslo.

15

Verifying 'Einstein's Time' by Using the Equation 'Time=Distance/ Velocity

Makanae M*

Independent Researcher, Representative Free Web College, Nishikasai, Edogawa-ku, Tokyo 134-0088, Japan

Abstract

The statement '*Every reference-body (co-ordinate system) has its own particular time*', which appears in Einstein's book—'Relativity: The Special and General Theory', is widely accepted among physicists and even by the general public with the popular interpretation that a clock in a moving body and another clock at rest in the reference stationary body will indicate different values of time. However, upon examining the grounds for this perspective by using the equation 'time=distance/velocity' and using '*the principle of the constancy of the velocity of light*', we find that the above sentence should be arranged as 'Every reference body (coordinate system) has its own particular measurement of the time interval for the propagation of light and, also it has its own particular measurement of the interval of the light path that must be used in order to calculate its time interval. The numerical value of the ratio of these two intervals is 1:1 always.' This implies that the pace of ticking of all clocks is identical. This fact contradicts the above popular interpretation.

Keywords: Numerical value; Einstein's time; Lorentz factor

Introduction

In 1905, A. Einstein published the article titled 'On the Electrodynamics of Moving Bodies' [1], which is referred to as the Special Theory of Relativity (STR). In Section 1 of [1], he first defined 'time', which is obtained by observing the hands of a watch at the place where the watch and the observer (i.e., the user of the watch) are located. Then, he describes the method for confirming the synchronization of two clocks that are placed a certain distance apart by using the round trip of the ray of light between the two clocks. In Section 2 of [1], by considering the relationship between a moving system in which the round trip of the ray of light is performed and a reference stationary system in which an observer measures the time intervals of the round trip of the ray of light, Einstein implied that the progress of time in the moving system and the progress of time in the reference stationary system are different. Subsequently, in Section 3 of [1], he developed the expression, $\beta = \frac{1}{\sqrt{1-\left(\frac{v}{c}\right)^2}}$, which was later termed as 'Lorentz factor', as a core theory of STR. This expression denotes the ratio between the value of 'time' or 'length' in the moving system and the value of 'time' or 'length' in the reference stationary system.

In 1916, A. Einstein published his book titled 'Relativity: The Special and General Theory' [2]. In this book, as an explanation of STR, he introduced his thought experiment that consisted in the observation of the simultaneity of two lightning strikes. Based on this thought experiment, his conclusion regarding time was: '*Every reference body (co-ordinate system) has its own particular time.*' (For convenience, hereafter, we refer to this Einstein's perspective as 'Einstein's time'.)

Einstein's time is widely accepted among physicists, and even among the general public, with the popular interpretation that a clock attached to a moving body and another clock at rest attached to a reference stationary body indicate different values of time. However, upon examining the grounds for this perspective by using the universal equation 'time=distance/velocity' and by incorporating '*the principle of the constancy of the velocity of light*', we find that the above sentence should be arranged as 'Every reference body (coordinate system) has its own particular measurement of the time interval for the propagation of light and, also, it has its own particular measurement of the interval of the light path that must be used in order to calculate its time interval. The numerical value of the ratio of these two intervals is always 1:1'. This indicates that the ticking pace of all clocks is identical. This fact contradicts the above popular interpretation.

In this study, we explain this issue with tangible reasons by using numerical values in a practical example. Then, we suggest a practical approach, which uses advanced technology of the 21st century, for verifying our conclusion, i.e., for verifying Einstein's time.

Confirming the Definition of 'Independent Time'

Before dwelling on Einstein's time, we confirm the definition of the concept of 'independent time' (as we call it) that was provided in Section 1 of [1], as this concept is the starting point for studying Einstein's perspective regarding 'time'.

In the first half of Section 1 of [1], after emphasizing that the concept of time is important in the study of physics, Einstein considered 'time' at a place where a watch and an observer (i.e., the user of the watch) are located; then he states '*the definition of "time" by substituting "the position of the small hand of my watch" for "time"*', hereafter referred to as Sentence 1.

We can consider sundials, hourglasses, modern digital clocks, or any other instrument that can be used to measure time, instead of the 'hand' of the watch that appears in Sentence 1. Therefore, we may replace Sentence 1 with 'the definition of "time" by substituting "the indicator of my/a watch" for "time"', which may be referred to as Sentence 2.

Usually, once a certain term has been replaced by another term, we

***Corresponding author:** Makanae M, Independent Researcher, Representative Free Web College, Nishikasai, Edogawa-ku, Tokyo 134-0088, Japan
E-mail: edit@free-web-college.com

can put the former term aside. In Sentence 1, 'time' (as the last word in the sentence) is replaced by 'the indicator of a watch'. Therefore, we can arrange Sentence 2 as 'the definition of "time" is "the indicator of watch (or clock)"' by removing the phrase 'for time'. This sentence is ideal from the viewpoint of formal logic, because it is meaningless to define the concept of time by using the phrase 'for time' in this case. Hereafter, we treat this sentence as a definition of the concept of 'independent time', for the reason that this concept is established at the place where a watch and an observer are located. We can say that this Einstein's concept of independent time is correct, since time measurement can be carried out precisely and scientifically only by watching the indicator of a clock.

With the premise that Einstein's definition of independent time is correct, we can state that the progress of (independent) time can only be known through a clock that ticks at a steady pace, unless the clock is out of order. Thus, we assume that the pace of time progress can be confirmed by the ticking pace of a clock.

Confirming the Background of the Primary Perspective of Einstein's Time

After explaining 'independent time', in Section 1 of [1], Einstein describes a method for the synchronization of two clocks in the same coordinate system by using the round trip of a ray of light between the two clocks as

$$t_B - t_A = t'_A - t_B. \tag{1}$$

This equation is established based on the following assumptions.

- A clock is located at point A, and another clock is located at point B.
- t_A is the point of time when the ray is emitted from the light source located at A.
- t_B is the point of time when the ray is reflected by a mirror placed at B.
- t'_A is the point of time when the ray returns to A.
- The left hand side of (1) represents the time required for the ray to 'Go' (from A to B)
- The right hand side represents the time required for the ray to 'Return' (from B to A).

With the assumption that the velocity of light is constant in vacuum, the right-hand and left-hand sides of (1) are equal. Thus, we can conclude that the synchronization of the two clocks is satisfied. By assuming that the distance between points A and B (viewed within the system to which both the points belong) is $1l$, the ray of light moves $1l$ in 1 s. Thus, we can express (1) as 1 s - 0 s=2 s - 1 s, then obtaining 1 s=1 s.

In Section 2 of [1], based on *'the principle of relativity'*, *'the principle of the constancy of the velocity of light'*, and the above method for confirming the synchronisation of the two clocks, Einstein considered the relationships between a moving system and a reference stationary system with the concepts of 'time' and 'length', using the universal equation 'time=distance/velocity'. He then proved the following set of equations,

$$t_B - t_A = \frac{\gamma_{AB}}{c-v} \text{ and } t'_A - t_B = \frac{\gamma_{AB}}{c+v} \tag{2}$$

The premises implied or described by Eq. (2) are the following:

1. Two coordinate systems: the moving system refers to a moving rigid rod, and the stationary system is described as the reference frame.
2. The rigid rod travels with a uniform velocity, undergoing parallel translation with respect to the stationary system along the positive direction of the x-axis.
3. A is the point at the beginning of the rod, closest to the origin of the x-axis, and B is the point at the end of the rod, at a distance l from A.
4. A light source is placed at A, and a mirror is placed at B to reflect the light in the opposite direction.
5. A clock is placed at each of the points A and B.
6. The round trip of a ray of light between A and B is governed by Eq. (1) in the moving system.
7. The velocity of the moving rigid rod is v, and c is the velocity of light.
8. The point in time at which the light is emitted by the light source is t_A.
9. The point in time at which the light is reflected by the mirror is t_B.
10. The point in time at which the light returns to A is t'_A.
11. *'γ_{AB} denotes the length of the moving rigid rod— measured in the stationary system'.*
12. The quantity $c-v$ or $c+v$ is the relative velocity between the front of the ray and A viewed from the stationary reference system.

Conditions 4, 5, 6, 8, 9, and 10 are the same conditions expressed by Eq. (1).

When an observer at rest in the reference stationary system observes this round trip of the ray of light, Eq. (2) holds. The left side of Eq. (2) corresponds to the Go interval (A to B), and the right side corresponds to the Return interval (B to A). Unlike Eq. (1), the structure of the equations representing Go and Return is different in Eq. (2). Based on Eq. (2), Einstein concluded Section 2 of [1] with the following statement.

'Observers moving with the moving rod would thus find that the two clocks were not synchronous, while observers in the stationary system would declare the clocks to be synchronous. So we see that we cannot attach any absolute signification to the concept of simultaneity, but that two events which, viewed from a system of co-ordinates, are simultaneous, can no longer be looked upon as simultaneous events when envisaged from a system which is in motion relatively to that system'. We can say that this sentence indirectly implies Einstein's time. In other words, this sentence can be defined as a primary perspective of the Einstein's time. Thus, for brevity, we refer to this sentence as the 'primary perspective of Einstein's time'.

Calculating the Time Intervals of the Trips of the Light Ray

First, let us calculate the time intervals in the round trip of the ray between two clocks by using numerical values in a practical example, where the velocity of the moving system is half of the velocity of light (i.e., $0.5c$), and the distance between the two clocks placed at A and B is $1l$ viewed within the moving system. The ray of light travels $1l$ in 1 s. Hereafter, the clock placed at A is described as 'Clock A' and the clock placed at B is described as 'Clock B'.

Whereas the moving system moves along the x-axis of the reference stationary system, the system appears to be stationary when viewed in the moving system, since it moves with uniform velocity. Therefore, in the moving system, in which the round trip of the ray between Clock A and Clock B is performed, we can calculate the time interval for Go and Return by using Eq. (1), namely $t_B - t_A = t'_A - t_B$. Thus, we obtain 1 s in both intervals. In other words, 1 s=1 s in the moving system.

On the other hand, if observing the Go interval of the round trip of the ray from the reference stationary system, when $1l + vt = ct$, the mirror reflects the front of the ray under *'the principle of the constancy of the velocity of light'*. We can convert $1l + vt = ct$ to $t = \frac{1l}{c-v}$. The time $t = \frac{1l}{c-v}$ corresponds to the time interval between t_A and t_B. Thus, we can express $t = \frac{1l}{c-v}$ as

$$t_B - t_A = \frac{1l}{c-v} \tag{3}$$

Inserting the numerical value of v (i.e., $0.5c$) in the denominator of the right-hand side, we obtain $t_B - t_A = \frac{1l}{1c - 0.5c} = 2s$

Likewise, in the Return interval, if we assume for the time being that the point of time when the mirror reflects the front of the ray is 0 s, where $1l - vt = ct$, the front of the ray returns to Clock A (viewed from the reference stationary system). We can convert $1l - vt = ct$ to $t = \frac{1l}{c+v}$. The time $t = \frac{1l}{c+v}$ corresponds to the time interval from t_B to t'_A. Thus, we can express $t = \frac{1l}{c+v}$ as

$$t'_A - t_B = \frac{1l}{c+v}. \tag{4}$$

Inserting the numerical value of v (i.e., $0.5c$) in the denominator of the right-hand side, we obtain $t'_A - t_B = \frac{1l}{1c + 0.5c} = 2/3s$. From the above, we conclude that the time intervals for the Go and Return trips measured by the observer at rest in the reference stationary system are different, and are in the ratio 2 s:2/3 s, whereas, the proportion is 1 s:1 s when measured by the observer at rest in the moving system.

The above situation does not consider the Lorentz factor $\beta = \frac{1}{\sqrt{1-\left(\frac{v}{c}\right)^2}}$, which is described in Section 3 of [1] as a core theory of STR. The denominator of $\frac{1}{\sqrt{1-\left(\frac{v}{c}\right)^2}}$ denotes the value of 'time' or 'length' in a moving system, whereas the corresponding value in the reference stationary system is unity. Therefore, if the Lorentz factor is employed and the value of the velocity of the moving system (i.e., v) is $0.5c$, the distance between A (where Clock A and the light source are placed) and B (where Clock B and the mirror are placed) is $1l$ viewed within the moving system, and the ray moves $1l$ in 1 s, then the value of the time interval in the moving system becomes $2 \times \sqrt{0.75}$ s in the Go trip and $2/3 \times \sqrt{0.75}s$ in the Return trip. By inserting the numerical value of v (i.e., $0.5c$) in $\frac{1}{\sqrt{1-\left(\frac{v}{c}\right)^2}}$, it becomes $\sqrt{1-\left(\frac{0.5c}{c}\right)^2} = \sqrt{1-0.25} = \sqrt{0.75}$, and the time interval for the light ray to travel as measured by the observer at rest in the reference stationary system equals 2 s in the Go trip and 2/3 s in the Return trip, as previously obtained.

From the above, we can infer that the time intervals of the same round trip of the ray are different in the reference stationary system and in the moving system when the Lorentz factor is employed. It seems that the results of the above calculation prove the correctness of Einstein's time, namely *'Every reference-body (co-ordinate system) has its own particular time'*.

Considering 'Time' and 'Distance' in the Expression 'Time=Distance'

In the previous section of our study, we focused on the time interval for the propagation of a ray of light. Now, we consider the above time intervals by including the 'distance' of the trips of the ray, i.e., the intervals of the light path between A (where Clock A and the light source are placed) and B (where Clock B and the mirror are placed).

By employing *'the principle of the constancy of the velocity of light'*, the velocity of the ray of light in the universal equation 'time=distance/velocity' can be described as unity. Thus, we can treat 'time' and 'distance' as equivalent; we will write 'time=distance', for convenience. With the above assumption that time=distance and not employing the Lorenz factor, the relationship between the time interval for the propagation of the ray of light (i.e., 'time') and the interval for the path of the light ray (i.e., 'distance') can be expressed as 2 s=$2l$ in the Go trip and 2/3 s=$2/3l$ in the Return trip, when observing the round trip of the ray from the reference stationary system. On the other hand, the relationship is 1 s=$1l$ in both the Go and Return trips when observing in the moving system.

If the Lorenz factor is employed and with the assumption that the time interval for the propagation of the light ray in the reference stationary system is 2 s and the interval for the light path of the ray is $2l$ in the Go trip, 2/3 s and $2/3l$ in the Return trip, we can describe time=distance as $2\,s \times \sqrt{0.75} = 2l \times \sqrt{0.75}$ in the Go trip, and $2/3\,s \times \sqrt{0.75} = 2/3l \times \sqrt{0.75}$ in the Return trip, in the moving system.

From the above, we find that the interval for the path of the light ray per second is $1l$ in every situation, whether the Lorentz factor is employed or not. This means that the ratio of the numerical value of the time interval for the propagation of the ray of light to the interval of the path of the light ray is 1:1, which also implies that unless the clocks are out of order, the ticking pace of the clocks is always identical. In other words, all clocks are synchronous, even if each time interval for the propagation of the light ray has its own particular value.

Confirming the Transformation Method for the Case When the Lorentz Factor is not Employed

In the above statement, we considered two cases, one that employs the Lorentz factor and another that does not employ it. The Lorentz factor denotes the ratio of the value of 'time' (or 'length') in the reference stationary system to the value of 'time' (or 'length') in the moving system. Thus, the transformation method provides the ratio as $1:\sqrt{1-\left(\frac{v}{c}\right)^2}$. We treat below the case when the Lorentz factor is not employed.

We previously confirmed that the moving system has uniform

velocity. Thus, this system is stationary when viewed within the system itself, and thus the ray of light moves 1*l* in 1 s in the Go and Return trips, so the interval of the light path between Clock *A* and Clock *B* viewed within the moving system is 1*l*. Therefore, $t=1l/c$. This equation corresponds to the form of the universal equation 'time=distance/velocity'.

On the other hand, as measured by the observer at rest in the reference stationary system, the time interval for the Go trip of the light ray is 2 s. This value is obtained by using $t=\frac{1l}{c-v}$, because, when $1l+vt=ct$, the mirror reflects the front of the ray viewed from the reference stationary system. Therefore, with the equation time=distance/velocity, we can express the ratio of the 'time' (interval) in the reference stationary system to the 'time' (interval) in the moving system as '$1l/(c-v):1l/c$'. We can reduce '$1l/(c-v):1l/c$' to '$c:c-v$'.

The Return interval of the round trip of the ray refers to the opposite of the Go interval. Thus, we can denote the above ratio as '$c:c+v$'. For confirming that this ratio for Return is correct, we can insert the numerical value of the practical example for *v* (i.e., 0.5*c*). We then obtain the numerical ratio as 1:1.5. This ratio matches the ratio 2/3 s:1 s that we previously obtained.

From the above, in the case when the Lorentz factor is not employed, the transformation theory gives the ratio for the round of trip of the ray as '$c:c-v$' for the Go trip and '$c:c+v$' for the Return trip.

Difference between Equations (2), (3), and (4)

Here, we notice that the structure of the set of equations (2) provided in Section 2 of [1], namely $t_B - t_A = \frac{\gamma AB}{c-v}$ and $t'_A - t_B = \frac{\gamma AB}{c+v}$, resembles the structure of equations (3) and (4), namely $t_B - t_A = \frac{1l}{c-v}$ and $t'_A - t_B = \frac{1l}{c+v}$, except that the expressions in the numerators of the right-hand side are different: γ_{AB} in (2), and 1*l* in (3) and (4), where γ_{AB} is '*the length of the moving rigid rod—measured in the stationary system*'. From the above, it seems that the correct numerical value of γ_{AB} is 1*l*.

However, Eq. (2) is established based on the concept of $velocity=\frac{light\ pass}{time\ interval}$, which can be arranged to time=distance/velocity, described in the first half of Section 2 in [1]. On the other hand, 1*l* in Eqs. (3) and (4) is the result of transposing from $1l+vt=ct$ or $1l - vt=ct$, which is the time corresponding to the light propagation between Clock *A* and Clock *B*, as viewed from the reference stationary system. In other words, the grounds for γ_{AB} in Eq. (2) and 1*l* in Eqs. (3), and (4) are different, whereas the structures of Eq. (2) and Eqs. (3) and (4) resemble each other. Therefore, we cannot conclude that γ_{AB} in Eq. (2) equals 1*l* is correct. In fact, in Section 2 of [1], Einstein implies that the value of γ_{AB} differs from 1*l*. Therefore, this issue, where γ_{AB} of Eq. (2) and 1*l* in Eqs. (3) and (4) have different meanings in similar equations, does not affect our study in this paper, even if the tangible value of γ_{AB} is unknown. At any rate, in our study the main focus is not on Einstein's 'distance' or 'length of a rigid rod', but on 'time'. Therefore, we set aside the issue of γ_{AB} and concentrate on 'time' instead.

Confirming the Background of Einstein's Time in [2]

From the above considerations, we can conclude that the primary perspective of Einstein's time provided in [1] cannot be used as the ground of Einstein's time provided in [2]. Therefore, we now focus on Einstein's time itself.

First, we confirm the background of Einstein's time, namely '*Every reference-body (co-ordinate system) has its own particular time*'. Here, between Sections 1 and 17, Einstein explained regarding the background of STR and STR itself. From Section 18 to 29, he provided his perspective regarding the General Theory of Relativity. From Section 30 to Section 32, he introduced his perspective of our universe.

In Section 8 of, titled 'On the Idea of Time in Physics,' Einstein assumed a situation in which two lightning strikes occur simultaneously when a train runs along a straight railway embankment as a thought experiment. He then introduces a method to recognize the simultaneity of the two lightning strikes from the midpoint between Points *A* and *B*, where the lightning strikes the ground. Einstein described this method as follows:

'*By measuring along the rails, the connecting line AB should be measured up and an observer placed at the midpoint M of the distance AB. This observer should be supplied with an arrangement (e.g. two mirrors inclined at 90°) which allows him visually to observe both places A and B at the same time. If the observer perceives the two flashes of lightning at the same time, then they are simultaneous.*'

Then, in Section 9, 'The Relativity of Simultaneity,' he explained the following:

'*When we say that the lightning strokes A and B are simultaneous with respect to the embankment, we mean: the rays of light emitted at the places A and B, where the lightning occurs, meet each other at the mid-point M of the length A —> B of the embankment. However, the events A and B also correspond to positions A and B on the train. Let M' be the mid-point of the distance A —> B on the travelling train. Just when the flashes 1 of lightning occur, this point M' naturally coincides with the point M, but it moves towards the right in the diagram with the velocity v of the train. If an observer sitting in the position M' in the train did not possess this velocity, then he would remain permanently at M, and the light rays emitted by the flashes of lightning A and B would reach him simultaneously, i.e. they would meet just where he is situated. Now in reality (considered with reference to the railway embankment) he is hastening towards the beam of light coming from B, whilst he is riding on ahead of the beam of light coming from A. Hence the observer will see the beam of light emitted from B earlier than he will see that emitted from A. Observers who take the railway train as their reference-body must therefore come to the conclusion that the lightning flash B took place earlier than the lightning flash A.*'

The above explanation of the thought experiment is correct and very clear, and is easily intelligible to the general public. Immediately after that explanation, Einstein stated the following conclusion of the above thought experiment, which includes Einstein's time.

'*We thus arrive at the important result: Events which are simultaneous with reference to the embankment are not simultaneous with respect to the train, and vice versa (relativity of simultaneity). Every reference-body (co-ordinate system) has its own particular time; unless we are told the reference-body to which the statement of time refers, there is no meaning in a statement of the time of an event.*'

Examining the Tangible Timings of the Flashes Reaching *M* or *M'*

Here, let us examine the tangible timings of the two flashes reaching

M or M' by using numerical values in a practical example.

First, we assume the following conditions, along with the conditions that A, B, and M are the stationary points on the embankment and point M' moves with the train.

- The value of the distance from A or B to M is $1l$ (i.e., from A to B is $2l$).
- The value of the distance from A or B to M' is $1l$, at 0 s.
- The flash of light moves $1l$ in 1s on the embankment and also in the train, satisfying '*the principle of constancy of the velocity of light*'.
- The velocity of the train, which is expressed as v, equals half the speed of light, i.e. $0.5c$.
- The lightning strikes A and B simultaneously at 0 s.
- We denote the mirror placed at A as 'Mirror A', and that placed at B as 'Mirror B'.
- We denote the flash reflected by Mirror A as 'Flash A', and that reflected by Mirror B as 'Flash B'.

The following results are then obtained.

On the embankment:

- Flash A reaches M at 1 s.
- The interval of the light path of Flash A from Mirror A to M is $1l$.
- Flash B reaches M at 1 s.
- The interval of the light path of Flash B from Mirror B to M is $1l$.

On the train:

- Flash A reaches M' at 2 s.
- The interval of the light path of Flash A from Mirror A to M' is $2l$.
- Flash B reaches M' at 2/3 s.
- The interval of the light path of Flash B from Mirror B to M' is $2/3l$.

From the above, we can confirm that the observer placed at M (i.e., on the embankment) recognizes both flashes at the same time, in contrast to the observer placed at M' (i.e., on the train) who recognizes Flash B before Flash A. This result matches Einstein's explanation in Section 9 of [2] '*Observers who take the railway train as their reference-body must therefore come to the conclusion that the lightning flash B took place earlier than the lightning flash A.*'

Confirming the Pace of Time Progress

Now, let us describe the form 'time=distance/velocity' by using the numerical values obtained in the previous section. We note that 'interval of the light path of the flash' corresponds to 'distance' in 'time=distance / velocity'.

- For the interval of the light path of Flash A from Mirror A to M; $1\text{ s}=1l/c$.
- For the interval of the light path of Flash B from Mirror B to M; $1\text{ s}=1l/c$.
- For the interval of the light path of Flash A from Mirror A to M'; $2\text{ s}=2l/c$.
- For the interval of the light path of Flash B from Mirror B to M'; $2/3\text{ s}=2/3l/c$.

In '*the principle of the constancy of the velocity of light*', we can describe c as unity. Thus, the above expressions become:

- For the interval of the light path of Flash A from Mirror A to M; $1\text{ s}=1l$.
- For the interval of the light path of Flash B from Mirror B to M; $1\text{ s}=1l$.
- For the interval of the light path of Flash A from Mirror A to M'; $2\text{ s}=2\,l$.
- For the interval of the light path of Flash B from Mirror B to M'; $2/3\text{ s}=2/3l$.

As described above, the interval of the light path of the flash per second is $1l$ in every situation. This means that the ratio of the numerical value of the time interval of the motion of the flash to the interval of the light path of the flash is 1:1.

Overview

In [1], Einstein uses the universal equation 'time=distance/velocity' in order to establish the primary perspective of Einstein's time. In his study, 'velocity' in 'time=distance/velocity' corresponds to the velocity of a ray of light that obeys '*the principle of the constancy of the velocity of light*'. Using this principle, we can treat 'velocity' in 'time=distance/velocity' as unity. Thus, 'time=distance/velocity' can be expressed as 'time=distance' in his study. This fact leads us to recognize clearly that 'time' and 'distance' must be considered together all the way in [1]. In fact, every light ray in the universe is in motion by its nature. Hence, we cannot discuss the time interval for the propagation of light without considering the interval of the light path. Thus, any correct discussion always requires the expression '(numerical value of) time=(numerical value of) distance', when employing '*the principle of the constancy of the velocity of light*' in order to consider 'time' or synchronization of clocks by accounting for the round trip of the ray of light between the clocks. Then, we obtain the result that the interval of the light path of the ray per second is $1l$ in all situations, when considering numerical values in practical examples. Thus, the ratio of the numerical value of the time interval of the propagation of the light ray to the interval of the light path of the ray is 1:1. This means that the ticking pace of all the clocks is always identical.

Likewise, in the thought experiment described in [2], we can explain the fact that the observer placed at M' recognizes Flash B before Flash A by considering 'time=distance/velocity'. Then, the expression '(numerical value of) time=(numerical value of) distance' results in the interval of the light path of the flash per second being $1l$ on the embankment and also on the train, when numerical values in practical examples are considered. This fact implies that the pace of time progress is identical on the embankment and on the train. Previously, we assumed that pace of time progress can be confirmed by the ticking pace of a clock. Thus, we can say that the ticking pace of clocks is identical on the embankment and on the train.

Therefore, Einstein's time, namely '*Every reference body (co-ordinate system) has its own particular time*', should be rearranged as 'Every reference body (coordinate system) has its own particular measurement of the time interval for the propagation of light and, also it has its own particular measurement of the interval of the light path that must be used in order to calculate its time interval. The numerical value of the ratio of these two intervals is 1:1 always.'

Conclusion

From the claim in our study, we can assume that all clocks are synchronous in our universe unless the clocks are out of order. However, we know that in practical science experiments, the above assumption is disproved, such as in the famous Hafele–Keating experiment in 1971, which reported that atomic clocks loaded on an airplane indicate different values of time after twice around the world flights in the eastward and westward directions.

However, the conditions for this kind of experiment in the 21st century are much better than those in 1971. For example, the International Space Station (ISS) always goes round the Earth much faster than airplanes, and modern clocks are more precise than the clocks that were available in 1971. In addition, every experiment is required to be reproducible. Therefore, we expect that any experiment that uses cutting-edge technology, such as the 'Atomic Clock Ensemble in Space' (ACES), will surely provide clear evidence that disproves our claim. If a clock loaded on the ISS and another clock loaded on any stationary satellite indicate different values, we can retract our claim and the credibility of Einstein's time is completely established. If this kind of experiments by using the ISS or any other space ship have already been carried out, but the results are not published, we suggest that these results should be widely publicized immediately. If the science community withholds those actions, not only some people who do not believe in relativity, but also the general public may begin to doubt practitioners of modern science, which presently espouse Einstein's time. Therefore, we strongly suggest the above actions in order to maintain the credibility of science. We should not hold credibility of science to hostage in order to defend the concept of Einstein's time.

Finally, we note that, even if the claim in our study is correct, we should continue to study STR apart from Einstein's time, because, our claim, which is a results that the numerical value of ratio of 'time' to 'distance' (i.e., length of object in Lorentz factor) is 1:1, is the basis of the fact that 'ticking pace of the clocks are always identical' holds under the Lorenz factor as shown in the last half of section titled 'Considering "Time" and "Distance" in the Expression "Time=Distance"' in our study. In other words, even if Einstein's time is not correct, 'length' may shrink because of the Lorenz factor. Likewise, we can continue to study General Theory of Relativity (GTR) apart from Einstein's time. Hitherto, GTR may be describing that the ticking pace of clocks will be peculiar in curved space. However, the concept of curved space is not provided by Einstein's time, whereas GTR discusses both. Therefore, the issue of Einstein's time should be separated from the study of GTR, if the experiments by using the ISS or any other space ship prove that ticking pace of all clocks are identical.

Acknowledgments

I would like to thank Editage for English language editing, and Ms. Maxie Pickert for confirming the meanings of the German original text [2], Mr. Makoto Kawahara for confirming basic mathematics theory, and Ms. Noriko Teramoto for help with the English grammar.

References

1. Perrett W, Jeffery GB (1923) On the electrodynamics of moving bodies. The Principle of Relativity. Methuen and Company, UK.
2. Einstein A(1916) Relativity:The Special and General Theory. London: Methuen & Co, UK.

Fundamental Discovery of the Dialectical Unity Principle and its Consequences for Theoretical Physics

Kohut P*

Researcher, Czech republic

Abstract

The whole reality (Universe) represents the unity in its diversity. What is the basic mechanism that creates the unity of being? Many thinkers have expressed the Unity Principle by saying "everything is connected to everything else", but nobody has detected its essence. On the base of dialectical logic, the Unity Principle is discovered which illustrate the exact mechanism how the physical Universe may work at its macro and micro levels. Fundamental discovery of the dialectical Unity Principle is a base on which new fundamentals of theoretical physics are built in the field of cosmology and particle physics.

Keywords: Universe; Theoretical physics; Unity principle; Dialectical logic; Quantum dipole; Energy; force; Whole; Part; One; Many; Space; Time; Elementary particles

Dialectical Logic and the Unity Principle

Contemporary theoretical physics enters the deep crisis resulting from its positivistic and post-positivistic approach supposing reality to be mechanical and atomistic made of point-like particles or one-dimensional strings where the essence of matter, energy, space and time are undetectable mysteries. But the Universe (reality) is dialectical (relational) and so it is accessible by dialectical logic. Our aim is to show that the Universe is built of elementary bipolar relations of opposites (+/-) named quantum dipoles or quantum connections. Dialectical logic has achieved its apex in Hegel´s rational idealistic philosophy. His Absolute Idea represents a divine mind or the process of creative divine thinking. While Einstein was finding the mind of God in a form of the exact mechanism how the Universe works, G.W.F. Hegel already, at the beginning of 19-th century, disclosed almost completely the manifestation of divine Mind within his dialectical logic and its basic categories like unity of opposites, relations "being-nothing", "whole-part", "one-many", "repulsion-attraction" "continuity-discontinuity", "quantity-quantum-quality-measure", "finitude-infinity", "subject-object", etc. [1]. Nevertheless Hegelian revolution in philosophy and dialectical logic has been unfinished as Hegel could not come to the final simple solution – detection of the exact mechanism of the Unity Principle which discovery allowed us not only to finish dialectical logic but also build new theoretical physics (particle physics and cosmology) on the true base. Can we know the truth and the nature of our Universe? Yes, of course, we can. Hegel showed that there are no hidden secrets or realities inaccessible by our critical rational thinking. His philosophy was optimistic and his dialectical logic - very effective and promising instrument. Hegel disclosed brilliantly that the world is rational and dialectical and therefore accessible by dialectical logic. It is impossible not to come to the knowledge of the truth if we apply strong critical thinking and logical reasoning as well as contemporary level of knowledge following from successful quantum mechanics which shows a quantum character and mutual interconnectedness of reality. Although all indications exist already for a long time, theoretical physicists have not found the exact mechanism how mater, energy, space and time are quantized and structured at the basic quantum level. The basic question is: What is the basic elementary constituent of which the whole reality is made and how is it built of its basic constituents? What is the mechanism that creates the unity of being? If we want to know the truth we need to detect how the Universe as a whole looks like. The WHOLE is everything what exists. It is an ABSOLUTE - UNITY that manifests itself through almost infinite variety of its forms. Although the reality looks like disintegrated into many different and independent spheres, we feel intuitively that a great variety of existing forms should have a common basis.

If we look at the reality (existence) or the Universe as a whole, we can see that it is not a pure continuum, but it is structured. A pure unstructured continuum is nothing. So the whole Universe as space is structured and, at the same time, represents the unity in its internal structuration - diversity. As the Universe is structured, it must be built of its basic structural constituents. That is the reason why the Universe is quantized. But at the same time it represents the Unity. It means that its basic structural constituents must be interconnected. But connections are also structural constituents of reality (Universe). Does the Universe have many different basic building constituents or not? If we say yes, we must explain – why, what are these different constituents and what is the reason of their difference? If we say that only one basic elementary structural constituent is sufficient, we need only to find it and explain its essence. As connections are also structural constituents, they just represent what we are searching for. Connection is something that connects two aspects of reality, it means, it connects "something (one side)" to the "other (other side)" and, at the same time, it is created of both that sides. In dialectical logic they are named opposites and their mutual relation is the unity of opposites (Figure 1).

Everything what exists creates the whole reality in its unity. At the highest level of abstraction we know that something exists. But this something is nothing without its relation to the other. "Something" cannot relate to itself (self-relation, self-reflection) without its relation to the "other", otherwise it is nothing. The other (-) represents the limit of something (+), through which it determines itself as a difference. "Something" and its "other side" are not two independent entities but

***Corresponding author:** Kohut P, Researcher, Czech Republic
E-mail: pekohut@gmail.com

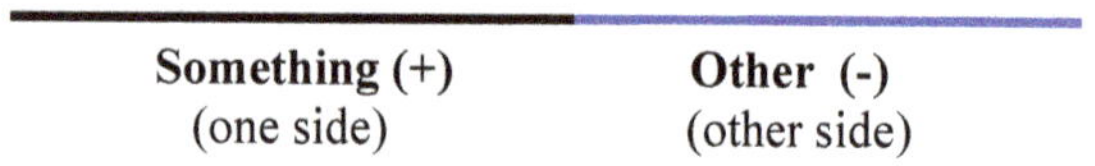

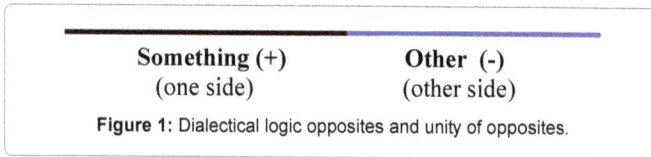

Figure 1: Dialectical logic opposites and unity of opposites.

only two sides (opposites) of the same "one". It is irrelevant what side is "something" or "other". Both they relate to each other in order to relate to themselves. The whole "one" is a self-relation (self-reflection) only because it is a mutual relation of its two opposite sides. Any of these two opposites reflects itself into itself through its other side as through its own limit (mirror). "Something" and "other" create a mutual positive and negative relationship which cannot be static, but only dynamic in the sense that "something" repels from itself its "other" side by repulsion (negation), but at the same time holds and attracts it to itself by attraction (negation of negation). Repulsion and attraction are two opposite forces through which both opposite sides of the same "one" are in a mutual dynamic relation manifesting by motion – vibration, oscillation. Motion is energy as a result of mutual attraction and repulsion of opposites. Energy is a measure of mutual attraction and repulsion of opposites. This dynamic bipolar relation (+/-) represents the elementary structural constituent of which the whole reality (Universe) is made. We can name it an elementary quantum dipole or elementary quantum connection. Known particles as well as space including vacuum are made of these quantum dipoles. The unity of the Universe means that all its aspects are made of the same constituents - quantum connections (dipoles). Elementary quantum dipole (connections) is an elementary quantum of space thereby the volume of space is given by the number of elementary quantum connections.

The Whole and the Part

The exact mechanism of the Unity Principle follows just directly from the analysis of dialectical relations "whole-part" and "one-many". Contemporary physics divides the whole reality into its parts mechanically. Mechanical separation of parts from the whole means the destruction of their mutual relations so that these parts can come to mutual interactions only through local touch (contact). Localism dominates in contemporary theoretical physics, where mutual interactions between "point-like" particles are presented as a result of mutual exchange of virtual point-like bosons moving with a limited speed of light. It is very strange that such a naïve mechanical interpretation of interactions between particles was incorporated into the Standard Model although non-locality results directly from quantum mechanics.

The dialectic relation between the whole and the part means that any part is separated from the whole only if it is separated from all parts of the whole. At the same time, its separation means its connection to the whole, so to all rest parts of the whole. As a result every Part is connected to all rest parts of the Whole. This must be valid also for elementary parts of which the whole reality is made. As connections are also parts of the whole, there is no difference between parts and connections at the elementary quantum level. Any elementary part is connected to all rest parts of the whole. Such is possible only if the elementary part is a bipolar relation of opposites (+/-) where every "+" is connected to all "-" of the Universe.

Elementary connections are elementary parts and elementary parts are elementary connections of two opposite sides (+/-). The Unity Principle means that everything is connected (in relation) to everything else in the sense that every positive pole "+" (i) of the Universe creates connections to all negative poles "-" (j), and so reciprocally. Bipolar connections +/- or (ij) represent the elementary quantum connections (dipoles) of which the whole reality (Universe) is made explained clearly in Figure 2.

Contemporary theories separate matter from space, supposing space to be only an empty or unstructured surrounding where material objects (entities) move. Space and time in Einstein´s relativity theories is a pure mathematical "space-time" continuum. Before, space was as an empty continuum in which all material bodies moved. In Einstein´s special relativity it was replaced by empty unstructured four-dimensional space-time continuum which was curved in general relativity thanks to presence of matter and energy. But this mathematical idealisation says nothing about the real quantum essence of space and time. Einstein´s space-time is not structured and quantized. It is a pure mathematical continuum. It is very strange that theorists having not found how space and time are quantized try to unite Einstein´s local theories with non-local quantum mechanics, although they mutually exclude each other. As a result string theories produce a huge number of absurd mysteries. Space is a basic attribute of every physical entity with its quantitative measure - volume. There are no entities without spatial volume. Point-like particles or one-dimensional strings are inappropriate at the quantum level even as mathematical idealisations because they deform the reality fatally. Space is not only a basic feature of everything, but at the same time it separates things from each other in the sense that it connects them together. Things can be mutually separated only if they are mutually interconnected. The internal structure of any thing is made of the same basic constituents as are connections through which things are interconnected. All things and their mutual connections are made of the same constituents - elementary quantum connections (dipoles). They are at the same time elementary quanta of space. Nothing exists in space as everything creates space. Objects do not move in space, they only move to each other thanks to their mutual quantum connections. Free space - vacuum - is made of their mutual connections. The Standard Model presents huge number of different point-like particles (fermions and bosons) placed in the vacuum, which essence is unknown. Vacuum is a mystery that can be arbitrarily used to solve all miracles of the Standard Model. For example, it gives enormous energy for very massive virtual gauge bosons in order to mediate a weak interaction in electroweak theory. In reality the vacuum is made of long and weak quantum connections comparing to the short and strong connections of which particles are made. So the vacuum cannot be a source of enormous energy needed for electroweak theory in particle physics in the sense of fluctuating vacuum producing undetectable virtual fluctuations.

Localism *versus* Non-Localism

There are only two basic interactions - non-local and local. Non-local interactions are manifested through mutual attraction and

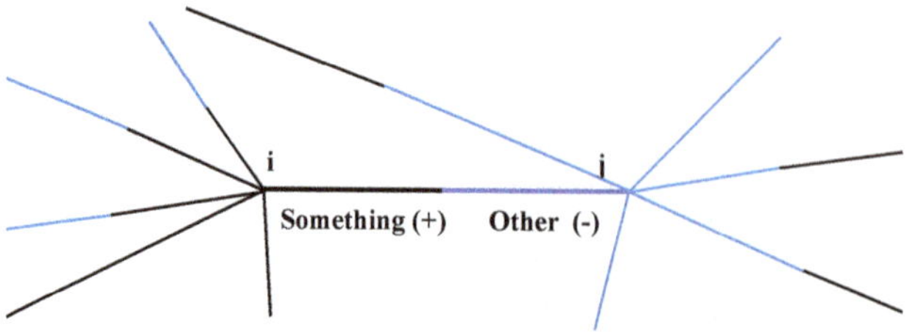

Figure 2: Bipolar connections representing the elementary quantum connections.

repulsion of opposite poles of quantum dipoles, while local interactions represent always repulsive forces acting through mutual local touch pressures between neighbouring elementary dipoles that push each other by their spaces. Attractive force is always non-local, while repulsive force can be non-local (e.g. in a photon) or local. Local force is always repulsive. Figure 3 shows two quantum dipoles acting locally by their mutual repulsive pressures. Elementary quantum connections (dipoles) represent elementary quanta of space, but differ by their lengths and energies. The left one is shorter, stronger having more energy and the right one is weaker, longer with less energy. As all flows coming from outside act to our basic senses (sight, hearing, touch, smell, taste) locally through touch interactions theoretical physics has a problem to accept "invisible" non-local connections although they result just directly from rational dialectical logic and quantum mechanics being confirmed experimentally and having a practical usage.

Photon as an Elementary Quantum of Existence

It is very strange that even a photon as an elementary quantum of light represents a mystery known as "wave-particle" dualism. Photon is a particle as well as a wave. How is it possible? What is the solution? Photon as an elementary quantum of free energy manifests clearly the bipolar essence of the whole being, but it has been a mystery so far. All we know that the motion of a classical harmonic spring oscillator creates a sinusoidal wave as a result of two forces with opposite orientation - attraction and repulsion. Sinusoidal wave is a consequence of both forces acting through harmonic oscillator. Photon creates sinusoidal wave during its flight. It means it must be a quantum oscillator which oscillations result from internal bipolarity of two opposite forces – attraction and repulsion. Photon is a quintessence of dialectical bipolar nature of reality. The greatest mistake of theoretical physics is the idea that elementary particles must be point-like entities without any internal structure and with zero volume. Even a photon as the simplest particle cannot be a point-like entity without internal structure. The photon is a simple quantum dipole consisting of two opposites (opposite poles) and consequently a holder of elementary quantum of space and energy. It is an elementary particle which, thanks to attraction and repulsion of its opposites, oscillates creating perpetually the sinusoidal wave during its flight which is manifested outside as an electromagnetic wave in relation to a measuring apparatus. Figure 4 explains the Photon γ (+/-) as elementary oscillating quantum dipole is the simplest particle. Photon as a quantum of radiation (light) is a free elementary quantum dipole +/- which, thanks to mutual attraction and repulsions of its opposite poles, performs a permanent oscillation (vibration, pulsation) manifesting outwards as an electromagnetic wave during a flight. This fact is a consistent and factual explanation of the "wave-particle" duality of the light as only a bipolar dynamic unity of opposites can result in oscillation (motion, energy) of a photon (Figure 5).

The photon is an elementary quantum oscillator. If we express its oscillation as rotation, its length is given by a diameter of rotating quantum dipole. Rotation projected to the perpendicular plane looks like oscillation. It is irrelevant if talking about rotation or oscillation (pulsation, vibration), as these motions are manifested outwards in the same way. Photon is an elementary quantum of energy. The essence of energy is also unknown for contemporary physics. Energy of a photon e_{ij} as a measure of its motion (frequency of vibrations ν_{ij}) can result only from mutual attraction and repulsion of its opposites.

Planck´s equation $e_{ij}=h\nu_{ij}$

Shows that energy of a photon is given by the speed of its vibrations (frequency). It is hardly believable that the essence of photon´s vibrations has not been detected so far. It is due to inappropriate idealisation of elementary particle as a point-like entity with its mysterious "particle-wave" dualism.

Photon performs two types of motion: horizontal and vertical. Horizontal motion represents its flight as a consequence of its dragging by cosmic expansion. Vertical motion is manifested by its oscillation (rotation) thanks to mutual attraction and repulsion of its opposite poles. Photon does not move "in" a free space-like vacuum, but thanks to its external quantum connections, it moves "towards" all other parts of the Universe. Simplicity of a photon allows its perfect oscillation (vibration) in a plane of its flight. As it is the simplest free quantum, it cannot resist its dragging by the expanding Universe, so it has no rest mass and its speed expresses the speed of cosmic expansion. Such is the nature of the speed of light as one of the basic physical constants unknown until now. The knowledge of the essence of Light is the way to understanding the essence of existence. There is no space and energy outside quantum connections as only they create the whole reality.

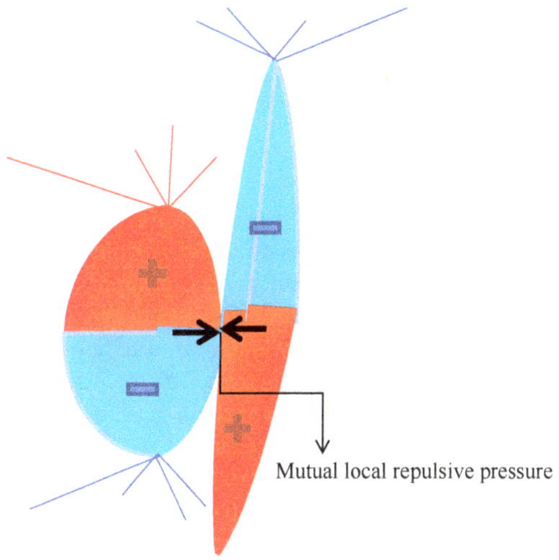

Figure 3: The basic interactions- non-local and local.

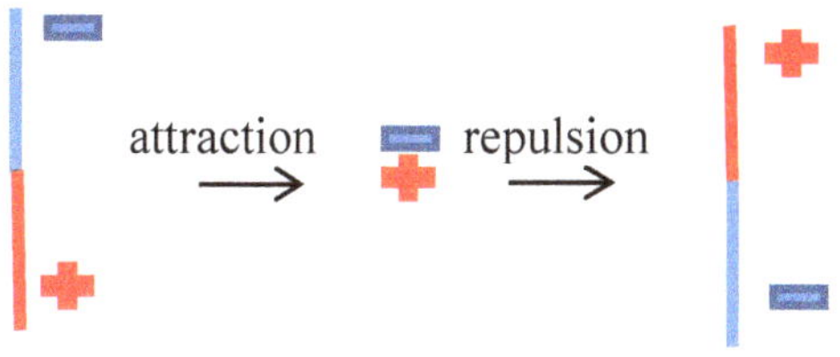

Figure 4: Photon γ (+/-) as elementary oscillating quantum dipole.

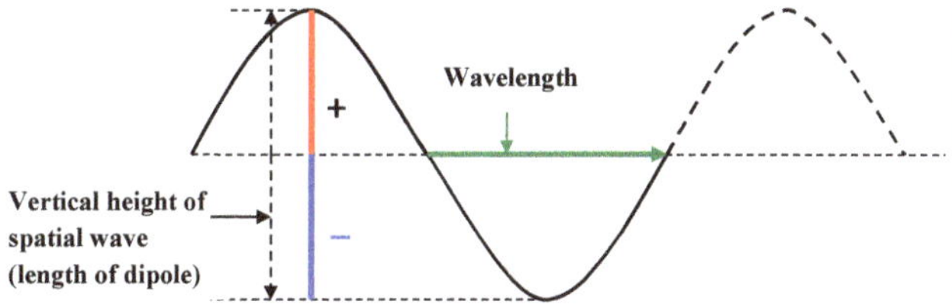

Figure 5: PHOTON free oscillating quantum dipole (+/-)

Quantum connections (dipoles) are not placed in space, but create it. They represent the elementary quanta of space with the elementary volume $v=3,87.10^{-45}$ m^3 as follows from my analysis explained by Kohut in [2]. Though elementary quantum dipoles are indistinguishable by their spatial volume characteristics, they differ from one another by energy e_{ij} and length d_{ij}, so that the following basic constant δ_t is valid [3] at the actual quantum state of the Universe:

$$\delta_t=e_{ij}d_{ij}=\alpha hc/\pi=4,61.10^{-28}\ kg.m^3s^{-2}$$

Where h: Planck´s constant, α: fine structure constant, c: speed of light. Very short quantum dipoles create the structure of basic particles (photons, electrons and protons), while long quantum dipoles create their mutual connections. Very long quantum dipoles connecting celestial bodies mutually create cosmic vacuum, so we name them vacuum quantum connections. The length of vibrating quantum dipoles like photons is given by the amplitude of vibration (oscillation). Photon´s oscillations can be presented as rotations of a quantum dipole with a circumferential velocity v:

$$v=2\pi r_{ij}/T_{ot}=\pi d_{ij}\nu_{ij}$$

T_{ot}: time of one rotation of a quantum dipole,

ν_{ij}: $1/T_{ot}$: frequency of quantum dipole oscillation,

r_{ij}: radius of dipole (half of its length),

d_{ij}: length of dipole.

$$\delta_t=e_{ij}d_{ij}=h\nu/\pi=\alpha hc/\pi=4,61.10^{-28}\ kg.m^3s^{-2}$$

The value $\delta_t=e_id_i$ is the same (constant) for every quantum dipole (connection) and represents the basic cosmic law from which other very important laws follow, e.g. Newton´s and Coulomb´s laws. It means the shorter the quantum dipole, the higher its energy. The longer it is, the lower its energy. Energy of very long quantum dipoles, connecting celestial bodies mutually and creating the cosmic vacuum, is very small, but their quantity is enormous. The vacuum is a holder of energy of quantum connections (dipoles) connecting physical objects mutually. Photon represents an elementary quantum dipole. As everything is made of elementary quantum dipoles (connections), we can say that everything is made of light (energy), which can exists in a form of free flying photons, or be bound in a form of basic particles (protons and electrons) as well as the vacuum. Not only the photon, but all particles oscillate, though their oscillations are more complicated Electron e^- (+/2-) created by two quantum dipoles shown in Figure 6 and Proton p^+ (3+/2-) made of six elementary quantum dipoles explained in Figure 7. All stable structures (particles) oscillate in one line (axis of oscillation) to the one common centre during attraction. All dipoles of a proton are very energetic (short and strong) so their forces of mutual attraction and repulsion are so strong that can compensate the mutual local repulsive pressures of spaces of quantum dipoles in such a way that the proton is the most stable composite structure. If structures are more complicated and composite, the mutual local pressures of dipole spaces destroy their compositions at the moment of their creation (so-called resonances). From the structure of a proton with three tops of positive poles is evident why the experiments in electron-proton scattering found that electrons scattered off three points inside the proton. It is not because of the quark structure but the bipolar essence of a proton.

One and Many

"One" is nothing without the other. "One" as a whole can create its relation to itself only if it divides itself into many ones. "One" creates its relation to itself through its relations to others. Through them it reflects itself into itself (self-reflection). "One" as a whole divides itself into many ones in such a way that they create the unity of the "One" in the sense that every "one" is connected to all other "ones". Through many ones the whole One is structured and quantized. Internal structuration means that the "One" repels from itself many ones by repulsion and, at the same time, holds them in a unity thanks to attraction. As the whole "One" represents a bipolar relation "something (+) – other (-)", its internal differentiation means that it gradually repeals from itself both opposites after one another. One as a whole comes from its unity to its diversity by internal structuration and at the same time it again and again reflects itself into its unity and so performs its self-reflection. "Many" as a negation of "One" is overcome by its return to its unity – negation of negation. Negation of negation is a self-reflection, meaning the One represents always the Unity which can exist only through its internal structuration, where everything is reflected in everything else, everything is connected to everything else and everything communicates with everything. This communication of everything with everything is an information process (software) based on physical processes at the quantum level of reality. "One" represents the self-creating and self-reflecting Unity of the highest complexity where everything communicates with everything else. Self-reflection of self-closed system of high complexity means the Life and Consciousness. Therefore the Universe as a whole is a self-closed, self-creating, self-knowing and self-aware system of the highest complexity and so represents the highest Consciousness manifesting itself as a "subject-object" relation. If we study the internal structuration and differentiation of the Universe without looking at its self-reflecting subjective unity, we see the whole reality as an expanding physical Universe, which differentiates itself in such a way that it expels gradually, step by step, new positive "+" and "-" negative poles "-" after one another. Continuing internal differentiation of the Universe, its plurality generation and structuration, means its cosmic expansion. The Universe is an expanding network of quantum dipoles (connections) transiting from its one quantum state to the following. At the beginning of expansion the Universe is only a simple quantum

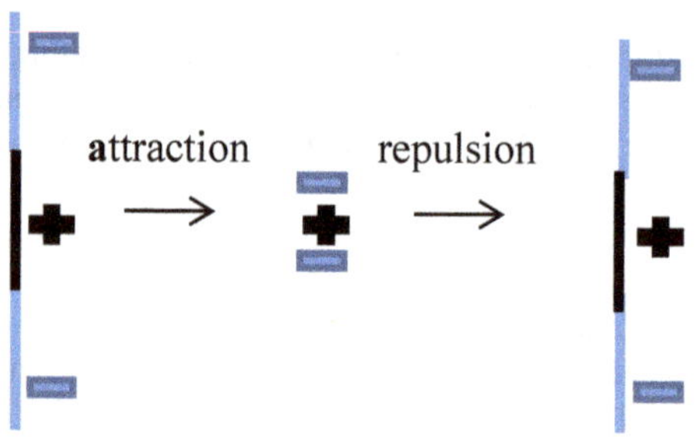

Figure 6: Electron e- (+/2-) created by two quantum dipoles.

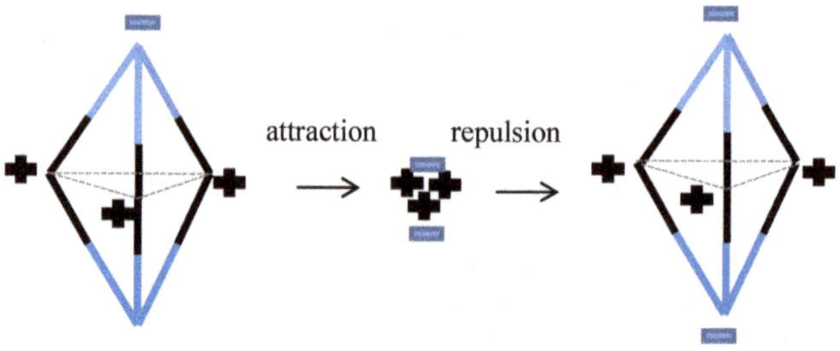

Figure 7: Proton p+ (3+/2-) made of six elementary quantum dipoles.

dipole (+/-), then he repels suppose firstly one positive pole (+) and next its negative one (-), so that after two elementary quantum jumps the Universe represents the structure (2+/2-). For simplification of our analysis we consider and number only quantum transitions between symmetric quantum states, when two new poles are expelled after one another. At the first quantum state the structure of the Universe is (+/-), at the second symmetric quantum states it is (2+/2-), at the third quantum state it is (3+/3-)…, at the k symmetric quantum state its structure is (k+/k-) and is created of $V_k=k^2$ elementary quantum dipoles (connections). The value $V_k=k^2$ represents the volume of space given by the number of elementary quantum dipoles (Table 1).

k: the number of positive respectively negative poles, as well as serial number of actual symmetric quantum state of the Universe representing cosmic time given by the number of elementary quantum jumps of the Universe from the beginning of its expansion. The Universe jumps from its one quantum state k to the following k+1 creating (expelling) new positive + and negative – poles with 2k+1 new quantum dipoles +/-. The internal structuration of the Universe resulting into its cosmic expansion can be easily described by the following basic quantum space-time equation:

$$V_k=k^2$$

This equation reflects the internal division and structuration of the Universe creating thus its own expanding space and flowing time. The Universe is quantized as its energy and space are localised in its elementary quantum connections and its time is given by its elementary quantum jumps. Elementary quantum jumps represent elementary changes of the Universe, its elementary quanta of motion (time) to which all other changes (motions, times) can be related. These elementary quantum jumps define cosmic time. Time is not a mystery but a manifestation of motion of the Universe. Time is a measure of motion. Every local motion can be compared to the universal cosmic motion. As explained by kohut in [4] contemporary one second corresponds to $(3/4)/(\pi c^5/2\kappa h\alpha)^{1/2}=8{,}144.10^{43}$ elementary quantum jumps of the Universe between two symmetrical quantum states, so we can allocate the time $\Delta t=(4/3)(2\kappa h\alpha/\pi c^5)^{1/2}=1{,}128.10^{-44}$ sec to one quantum jump ($\mathcal{K}$- gravitational constant). But it does not mean that the quantum jump has its duration. Time does not define the duration of elementary quantum jump, but just contrariwise, time is defined by the number of elementary quantum cosmic jumps. Every process (motion) and its duration can be compared to universal time. If some process takes one second, it means that it corresponds to $8{,}144.10^{43}$ elementary quantum jumps of the Universe. If the same process is dilated to two seconds (time dilation) because of high speed or strong gravity (big gravitational potential), it corresponds to $2x8{,}144.10^{43}$ elementary quantum jumps of the Universe. Cosmic time represents the universal base through which all local processes (motions, times) can be expressed. Both space and time are quantized and their quantitative characteristics can be numbered and expressed by integers.

Quantum state	j	1	2		k-1	k	k+1		n
i	Poles	-	-	....	-	-	-	....	-
1	+						+/-		
2	+						+/-		
							+/-		
k-1	+						+/-		
k	+						+/-		
k+1	+	+/-	+/-	+/-	+/-	+/-	+/-		
n	+								

Table 1: The table of increasing cosmic network of quantum dipoles during cosmic expansion.

If we allocate Δt sec to one quantum jump then the time of cosmic expansion is:

$$t=k.\ \Delta t$$

and the basic space-time equation of the Universe, where the volume V is expressed by m^3, obtains the following form:

$$V=z.t^2,\ \text{where}\ z=(d^2V/dt^2)/2$$

This is the basic equation of spatial dynamics of the Universe expressed by real dimensional units, where the spatial volume of the Universe is directly proportional to the square of time of cosmic expansion. In that form space and time are presented as continuous values, but we must remember that in reality they are quantized and can be truly expressed only by integers. Thus, if we want to study space and time from the viewpoint of cosmology, we can use them as continuous values, but such an approximation is inappropriate at the quantum level. Except of space-time characteristics the whole Universe is defined at the actual quantum state k also by the matrixes e_{ijk} or $d_{ijk,}$ where i – number of positive pole, j – number of negative pole and ij – quantum dipole with energy e_{ijk} and length d_{ijk} at the contemporary quantum state k of cosmic expansion. For the basic space-time equation of the Universe, derived from the mechanism of its internal structuration, the next relations are valid:

$$V=z.t^2,\ \text{where:}\ z=(d^2V/dt^2)/2$$

$$dV/dt=(d^2V/dt^2).t,$$

$$(dV/dt)^2=2\ V.d^2V/dt^2$$

The quantity d^2V/dt^2 is a fixed constant during the whole evolution of the Universe. All these equations express the space-time unity of the Universe. The speed of expansion of spatial volume dV/dt is directly proportional to the time of expansion. It accelerates unceasingly and this acceleration d^2V/dt^2 is constant. Three-dimensional space is self-closed therefore it can be imagined as an ideal three-dimensional surface of a four-dimensional sphere, for which the following formula is valid:

$V=2\pi^2r^3$ where r is a radius of spatial curvature.

From the relation for the circumference of the Universe $o=2\pi r$ and previous relations we obtain:

$$(do/dt)^2=-2o.d^2o/dt^2$$

The relations between spatial circumference o and time t are:

$$o=u.t^{2/3}$$

$$do/dt=(2/3)u.t^{-1/3}$$

$$d^2o/dt^2=-(2/9)u.t^{-4/3},$$

$$\text{where: } u=(2\pi d^2V/dt^2)^{1/3}$$

These equations show that the spatial circumference o increases in time but its speed do/dt decreases. The acceleration is negative. It means that the speed of cosmic expansion decelerates. The length of the longest quantum dipoles, representing the highest possible distances and connecting two opposite sides of the Universe, equals the half of circumference of the Universe o/2 and the speed of its increase, thanks to cosmic expansion, represents the highest possible speed of light c:

$c_{=}(do/dt)/2=o/3t$

$o/2=\pi r=(3/2)\,ct$

Speed of light represents the speed of cosmic expansion therefore it is the escaping speed for the whole Universe. As the speed of cosmic expansion decreases, so the speed of light decreases, too. But now theoretical physics accepts erroneously cosmic expansion to be accelerating and even Nobel Prize 2011 was awarded for this "discovery", although in reality acceleration of cosmic expansion is only a seeming phenomenon based on wrong dogma that the speed of light must be always the same in relation to the observer. This mistake has fatal consequences for contemporary cosmological theories as they postulate and search for mysterious dark energy as a source of accelerated cosmic expansion. This acceleration was deduced from observations showing that very distant supernovas look fainter and therefore, more distant than they should be by constant or decelerating cosmic expansion. But this interpretation is wrong and based on the misleading dogma that the light always moves towards us by a constant speed c.

The real situation is quite different, because the larger the distance from which the light travels, the slower is its speed towards us, as its actual speed c must be reduced by the speed v of extension of this distance thanks to cosmic expansion. If the light approaches us from the point of distance d, then this point moves away with the speed v thanks to cosmic expansion:

$v=H.d$, where:

H: Hubble´s constant,

d: actual distance of the light ray from us (observer),

then the light from the distance d approaches us by the speed $(c-v)=(c-Hd)$.

We need no dark energy to accelerate cosmic expansion as this acceleration is nonsense based on the wrong dogma. Time and trajectory, through which the light travels to us, are much greater than they would be by the constant light speed c approaching us. The larger the distance between us and the light, the slower is its speed towards us. So the cosmic objects (supernovae) seem to be much more distant and fainter than they are expected by a constant c.

Another reason why accelerating cosmic expansion is only an illusion is the deceleration of light speed during cosmic expansion. The speed of light expresses the speed of cosmic expansion, so the deceleration of cosmic expansion means at the same time the deceleration of the speed of light.

The "discovery" of accelerating cosmic expansion as a consequence of erroneous understanding of the speed of light leads to postulation of non-existent dark energy as a source of acceleration. Supporters of dark energy try to find its source in the vacuum. Of course, huge energy is contained in a vacuum consisting of an enormous number of elementary quantum dipoles, connecting mutually all visible material objects. The higher the number of material objects taken into the system, the more the number of mutual elementary quantum connections between them and the higher the whole energy of the system. The system with many objects has, thanks to their mutual vacuum connections, much more energy than is contained in visible matter. But it is not dark energy causing fictional acceleration of cosmic expansion. Even, dark energy together with dark matter is declared to carry about 96% of the whole energy (mass) of the Universe. Except of mysteries like virtual bosons, quarks, strings, hidden dimensions, multiverse, black holes, warm holes, imaginary time, false vacuum, etc., other great mysteries like dark matter and dark energy are included in "science". As we know celestial bodies rotate and their rotations also influence motions of other objects through non-local external quantum connections. Rotational motions of celestial bodies in cosmology result from oscillations (rotations) of elementary quantum dipoles. These rotational motions are sources of magnetic fields of rotating bodies.

The impact of rotational motions of torsion generators on other objects is studied deeply in theories of torsion fields of Russian physicists Akimov and Shipov and confirmed by many experiments including that by which the structure of molten metals is changed significantly by torsion (rotational) fields generated by electro-torsion generators. Certainly, their theories are strongly criticised. But, in reality, their torsion fields can be correctly interpreted only saying that they are mediated by direct non-local external quantum connections of rotating generators. Torsion fields are other significant evidence that non-locality and non-local instantaneous interactions represent a fundamental feature of reality removed from contemporary physical theories.

The Unity Principle and Occam´S Razor

We need to know the simple truth

It is impossible to have a simpler relation than attraction and repulsion of opposites (+/-) which the whole reality is made of, where every "+" is connected to all "-", and reciprocally. The Unity Principle is the clearest manifestation of Occam´s razor and results directly from dialectical logic of thinking reflecting the dialectics of existence. The highest complexity of the Universe is created of the highest simplicity of bipolar relations (+/-). Only direct non-local connections of everything to everything can allow the existence of life as a self-reflection (consciousness) of very complex self-closed structures - subjects. As everything is reflected in everything else, the whole Universe as a Unity represents a self-reflecting, self-mirroring, self-creating and self-aware system of the highest complexity – Consciousness. We do not need to create speculative theories like "Theory of Everything" or "Theory of Unified Field" but we need to understand deeply the exact and simple mechanism of the Unity Principle.

Basic Forces and Interactions

Energy and Force

There are only two basic forces – attraction and repulsion and two basic interactions – local and non-local. All known interactions: mechanical, electromagnetic, strong, week, nuclear and gravitational, are only their manifestations. Two basic forces – attraction and repulsion are always in a mutual dynamic equilibrium at all levels of hierarchy. At the level of elementary quantum dipole, attractive force of two opposites equals the repulsive force of quantum dipole, which can be manifested in two ways:

- repulsive force of opposites (non-local connection)
- local touch repulsive pressure of space of a quantum dipole on neighbour quantum dipoles.

In case of a photon (+/-), the dynamic equilibrium between two opposite forces is manifested as oscillation. In case of particles like proton, the high local repulsive force (pressure) between spaces of six elementary quantum dipoles, creating its structure (3+/2-), is compensated by strong attraction between opposites of quantum dipoles so that the whole structure of a proton is very stable.

The whole force of attraction and repulsion f_{ij} of a quantum dipole ij is:

$f_{ij}=f_{ija}+ f_{ijr}, f_{ija}=f_{ijr},$

where: f_{ija}: attractive force between opposites of a quantum dipole ij,

f_{ijr}: repulsive force of a quantum dipole ij.

Energy is a motion or potential for motion resulting from mutual attraction and repulsion of quantum dipoles. Forces of attraction f_{ija} and repulsion f_{ijr} acting through the entire length d_{ij} of a quantum dipole create, by multiplication with its length, the whole energy e_{ij} of a quantum dipole:

$e_{ij}=f_{ij}.d_{ij}$

$e_{ij}=(f_{ija}+f_{ijr})d_{ij.}$

If a quantum dipole changes its energetic level, it also changes its length. By losing a part of its energy it elongates, by its receiving it shortens. Quantum dipoles exchange mutually their energies as they are in a permanent mutual motion. The whole internal energy of a quantum dipole e_{ij} consists of its two parts: attractive e_{ija} and repulsive e_{ijr} which are always in a mutual equilibrium. While attractive part is manifested by attraction of opposite poles, the repulsive one by their repulsion or by the local pressure of a quantum dipole on the neighbours. In photons, the equilibrium between attractive and repulsive parts is manifested by oscillation. Quantum dipole, bound in a composite structures, cannot oscillate freely and so presses on neighbours, so its repulsive part of energy is manifested by its local pressure, which is in equilibrium with its attractive part between its opposite poles. In that case this attractive part of energy of a quantum dipole has a form of potential energy as it cannot cause the motion of quantum dipole because of local repulsive pressures of neighbour dipoles. As attraction is at equilibrium with repulsion, so the attractive part of energy of quantum dipole is equal to its repulsive one. The following relations are valid:

$e_{ija}=e_{ijr}$

$e_{ij=}e_{ija} + e_{ijr}=2e_{ija}=2e_{ijr}$

Any form of energy, e.g. kinetic or potential, is always energy of elementary quantum connections represented by the equilibrium of their two parts, attractive and repulsive, because attraction and repulsion are two sides of the coin, representing the dialectics of a quantum dipole as well as the whole Universe.

From the basic cosmic relation between energy and length of elementary quantum dipole

$\delta_t=e_{ij}d_{ij}=2e_{ija}d_{ij}$

we can derive the following relation:

$e_{ija=}\delta_t/2d_{ij}$

It is a classical Coulomb´s relation between potential energy of a dipole with elementary charges and its length:

$e_{ija=}(q^2/4\pi\varepsilon)/d_{ij}$, where: $\delta_t=q^2/2\pi\varepsilon$

q: Elementary electric charge,

ε: Dielectric capacitance

From the relation for the fine structure constant $\alpha=q^2/(2\varepsilon hc)$ and Coulomb´s relation we get:

$e_{ija}=\alpha hc/(2\pi d_{ij})$

where: - fine structure constant, h: Planck constant, c: speed of light

This Coulomb´s relation $e_{ija}=\alpha hc/(2\pi d_i)$ manifests a universal cosmic law:

$e_{ij}d_{ij}=\alpha hc/\pi$

which represents a dialectical relation between energy and length of elementary quantum dipoles. From this relation we obtain:

$f_{ij}=\delta_t /d_{ij}^2 = \alpha hc/(\pi d_{ij}^2)$

Attractive force f_{ija} of a quantum dipole which corresponds to its potential energy $e_{ija}=e_{ij}/2$ can be expressed as follows:

$f_{ija}=\alpha hc/(2\pi d_{ij}^2)$

It is a classical Coulomb´s law expressing the dependence of attractive force, acting between elementary electric charges, on their distance. It is at the same time the expression for the attractive force acting through the elementary quantum dipole with a length d_{ij}. This force is indirectly proportional to the square of its length.

Electrostatic Force

Particles or any physical objects with prevalence of positive poles are positively charged. Particles with prevalence of negative poles are negatively charged. Elementary charge is a minimal possible quantity of prevalence. Electron (+/2-) is the most well-known particle with a negative charge, proton (3+/2-) – with a positive one. Particles with a balance of positive and negative poles are neutral. Long quantum dipoles creating connections of material objects, are affected by attractive forces of their opposite poles. The sum of attractive forces of all quantum dipoles connecting two massive objects creates the whole attractive force between them. Let d is an average distance between two neutral objects. The first object contains k_1 positive and k_1 negative poles and the second one - k_2 positive and k_2 negative ones. The whole number of elementary quantum connections between two objects is $2k_1k_2$. So the whole attractive force f_a between both objects is a sum of attractive forces of all mutual quantum connections. If d is an average length of quantum dipoles, the next relation is valid:

$f_a=(\alpha hc/2\pi).2k_1.k_2 /d^2=(\alpha hc/\pi).k_1.k_2 /d^2$

This relation expresses the electrostatic attractive force between two electrically neutral objects and is directly proportional to the number of quantum dipoles connecting them. But, as we know, there is no attractive electrostatic force between electrically neutral objects. This force can be identified only if these objects are electrically charged and it is proportional to the multiplication of their charges. Indeed, this force affects all quantum dipoles connecting two material objects, but is fully compensated by repulsive spatial pressures of quantum dipoles coming out of these objects, so it looks like if there is no attractive force between them. If two objects are oppositely charged with charges q_1 and q_2, the attractive forces affecting their direct quantum connections are not fully compensated by pressures of outgoing external quantum dipoles, and so their uncompensated mutual attractive force is directly proportional to multiplication of their charges. If two objects have like charges, the missing mutual connections between them cause that the repulsive pressures of their external quantum dipoles prevail over the attractive forces of quantum dipoles connecting these objects, what is manifested as an electrostatic repulsive force directly proportional to multiplication of their like charges. Although Coulomb´s law is the same for expression of attractive and repulsive electrostatic forces, their reasons are different. The attractive electrostatic force is a consequence of non-local mutual attraction between opposite poles of quantum dipoles, while repulsive electrostatic force is caused by prevalence of local repulsive pressures of quantum dipoles over attractive forces as a consequence of deficiency of mutual non-local quantum connections. The indirect evidence for this statement is a mutual attraction between like charged particles, e.g. electrons, which can be manifested by certain conditions, e.g. by very low temperatures. Electrons are not point-like particles, but structures consisting of two quantum dipoles with one positive and two negative poles. By low temperature, when kinetic motions are very slow, electrons can create the bound compositions known as Cooper´s pairs. Their ability for mutual attraction allows the existence of superconductivity. Electrons in their basic (not excited) states represent structures with one positive and two negative poles (+/2-). The bound state of two electrons creating a Cooper´s pair is explained in Figure 8. Casimir´s phenomenon is another evidence for existence of attractive electrostatic force between neutral objects. This force acts between two neutral conducting plates. If approach them closely, the mutual attraction, known as Casimir´s attractive force, starts to act. This effect means that attractive forces between quantum dipoles, connecting both closely approached plates, are greater than repulsive spatial pressures of quantum connections coming out of them. There is no principal difference between electromagnetic force and others like strong and weak nuclear. They differ only by their intensity. In stable particles, the strong and weak forces are mediated by very short and energetic elementary quantum connections which can effectively compensate the great repulsive pressures of their spaces. Electromagnetic interactions can be converted into the strong ones only, if the barrier of huge repulsive pressures is overreached by a close approach, where long connections are dramatically shortened and attractive forces increased. Analogical is the opposite process, where strong interactions inside protons and antiprotons can be changed, after their annihilation, into elementary quantum dipoles - photons - carriers of electromagnetic energy.

If two particles are mutually approached to the certain distance and exceed the barrier of electrostatic forces, all mutual external quantum connections of both particles become internal and create a new particle. The mutual attraction increases to the level able to balance repulsive pressures of spaces of their quantum dipoles. If a stable equilibrium of these forces is achieved, the new microstructure (particle, atom) does not decay. But if this equilibrium is temporary installed by huge external energies, the repulsion of internal pressures of particle corrupts this equilibrium and particle decays immediately after its creation. Such a microstructure cannot keep its internal equilibrium of forces without great external energies and so it decays. The unstable short-living structures (resonances) supposedly occur thanks to great energies in particle accelerators-colliders.

Magnetic Force

Magnetic force is a consequence of mutually coordinated internal motions (oscillations) of quantum dipoles in atoms of magnetic materials (mostly metals) that can act to other materials with magnetic properties through their mutual external quantum connections. Magnetic are materials that can create mutually coordinated synchronized motions (oscillation) of quantum dipoles in their atoms (atomic dipoles) in the sense of their like orientations. Magnetic field of a magnet is created of its external quantum dipoles connecting the magnet with the whole Universe. Its external quantum connections reflect the internal coordinated motions of its inner dipoles in such a way that they can cause the mutual attraction between opposite magnetic poles, the repulsion between like poles and magnetisation of magnetic materials. Mutual attraction of opposite magnetic poles is a consequence of synchronized coordinated oscillations (rotations) of quantum dipoles inside magnets as shown in Figure 9. At the above picture we see two permanent magnets where the arrows show the same direction of synchronized oscillations (rotations) of atomic dipoles inside magnets. The external quantum connections coming out of both permanent magnets reflect these synchronized motions in the way that their motions become also synchronized (the same orientation) resulting in the decrease of their mutual local repulsive pressures so that the attractive force between opposite magnetic poles of both magnets prevails - magnets attract each other. From the above picture we see why the North Pole (N) is always at left side, while the South Pole (S) at right one independently of into how many parts is the magnet divided. Thus, we have disclosed why magnets have always two magnetic poles and why one pole cannot exist without the other as both magnetic poles result from the synchronic coordinated motions of their inner atomic dipoles. Mutual synchronized oscillations of atomic dipoles inside magnet are impossible without their mutual non-local quantum connections as just only through them atomic dipoles can synchronise their motions. Virtual photons as supposed mediators of magnetic interactions cannot explain this phenomenon in any case. This phenomenon is just a manifestation of quantum entanglement (non-local connections) through which the spins or magnetic moments of particles are coordinated. On the other hand Figure 10 illustrates, if like the magnetic poles are situated face-to-face, their internal atomic dipoles oscillate in mutually opposite directions what causes opposite orientation of motions of their external quantum connections coming out of both magnets resulting in the increase of their mutual local repulsive pressures which consequently prevail over their mutual non-local attractive forces so that like magnetic poles of permanent magnets repeal each other.

Magnetic force or field is mediated by non-local external quantum connections and so it is quantized in that sense. Coordinated synchronized oscillations of atomic quantum dipoles of magnet can influence, through mutual external quantum connections, internal motions of quantum dipoles in other magnetic materials in such a

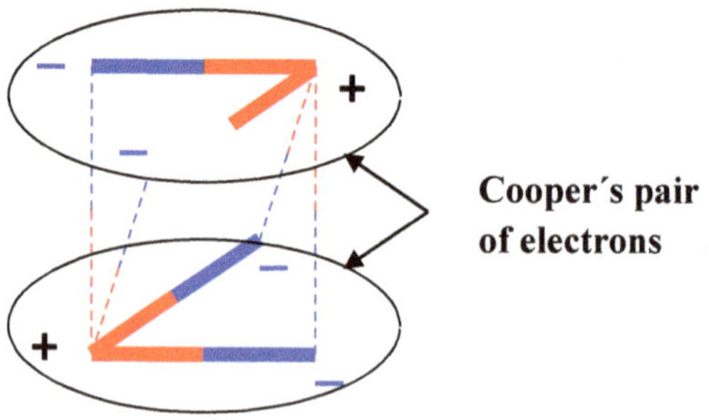

Figure 8: Issutration of Cooper´s pair.

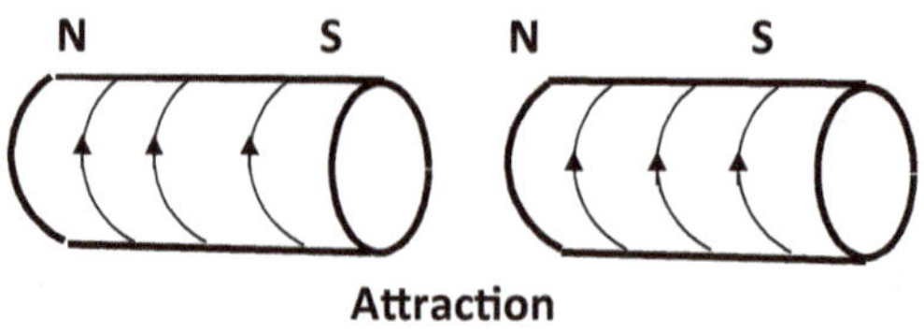

Figure 9: Permanent magnets showing attractions.

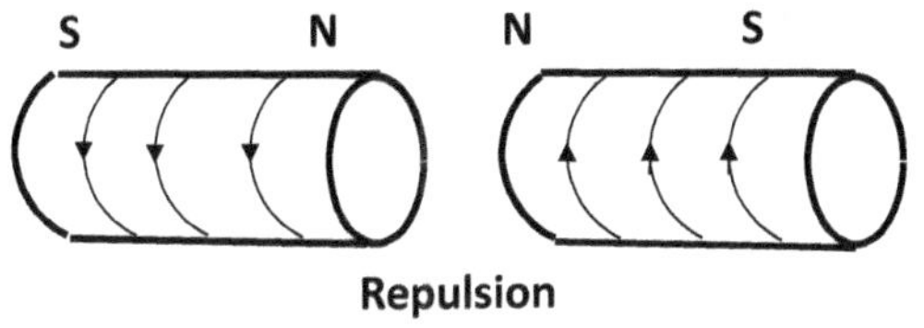

Figure 10: Permanent magnets showing repulsion.

way that these materials start to manifest their magnetic properties in the sense of coordinated oscillations of their internal atomic quantum dipoles. Magnetic as well as electrostatic forces are mediated instantaneously through non-local mutual quantum connections, but not through virtual photons moving with a limited speed.

Certainly, physicists do not know the essence of magnetic force or magnetic field as well as they do not know the essence of any force or field. They can describe their manifestations, but cannot interpret correctly the nature of these phenomena. All forces or fields are nothing more than mutual interactions between objects, e.g. particles, mediated by their mutual non-local quantum connections. Magnetic field (force) can be also produced by electric current as well as changing magnetic field can produce electric currents if applied to a conductor, but we are not going to analyse these electromagnetic phenomena now as they require a special individual approach. Theory of electromagnetism is well developed from the viewpoint of its phenomenology, but suffers from insufficient or wrong interpretation in the sense of ontology. It is declared that the photon is a quantum of electromagnetic field. Really, photon manifests its electrostatic properties because it is an elementary quantum dipole that unifies two opposite charges as well as magnetic properties through its internal motion-oscillation. Electromagnetic forces are mediated by elementary quantum dipoles, not in the sense of virtual photons moving with a limited speed of light, but of mutual non-local quantum connections. Electromagnetic interaction is a direct instantaneous non-local interaction.

Gravitational Force

Internal structuration of the Universe caused by its repulsive force is manifested by cosmic expansion. The certain part of the whole cosmic repulsive forces used for cosmic expansion is given by the relation derived [5]

$F_e=c^4/(16\kappa)=7{,}566.10^{42}$ N

Where: c: speed of light

$\mathcal{K}$: gravitational constant

Thus, we know the exact value of the force of cosmic expansion. As attraction and repulsion are two opposite forces in a mutual dynamic equilibrium, so the force of cosmic expansion has its own counterbalance in a cosmic gravitational force G, where:

$G=F_e=7{,}566.10^{42}$ N

Gravity is therefore a direct consequence and evidence of cosmic expansion. Attraction and repulsion are always in a mutual dynamic equilibrium at the level of every elementary quantum dipole as well as the whole Universe. Cosmic gravity affects all objects and all elementary quanta of space. It means that gravity, as a reaction to cosmic expansion, has a global as well as quantum character.

By derivation of Coulomb´s relation for the attractive force acting between two neutral massive objects `$f_a=(\alpha hc/2\pi)2k_1.k_2/d^2$ we have mentioned that this force is compensated by the repulsive force of pressures of quantum dipoles coming out of both objects. However, this compensation is valid only relatively, a certain part f_g of attractive force f_a is not compensated $f_g=\beta f_a$ and represents the attractive gravitational force f_g of bodies.

$f_g=\beta f_a=\beta(\alpha hc/2\pi)2k_1.k_2/d^2$

Uncompensated part of attractive forces by repulsive pressures of quantum dipoles is a consequence of deficiency of repulsive forces of the Universe caused by the fact that a certain part of these forces $F_e=7{,}566.10^{42}$ N is used for cosmic expansion. The total measure of this deficiency of repulsive forces and prevalence of attractive ones is manifested as gravity acting between bodies through their long mutual vacuum quantum connections. As gravitational force between celestial bodies is mediated by their mutual vacuum quantum connections, therefore it is a non-local instantaneous interaction in contrast with Einstein´s local theory, where gravity is a consequence of space-time curvature which local changes are propagated by gravitational waves with a limited speed of light. Newton´s theory of gravity is correct, because it is a relational theory, where gravity is a consequence of mutual instantaneous non-local interactions (relations) between physical objects, while Einstein´s theory of gravity is local non-relational theory. Newton´s theory needs only one small supplement: that the density of the vacuum, proportional the gravitational potential, causes the deceleration of processes in objects (time dilation), what is correctly accepted in Einstein´s theory. But Einstein´s gravity cannot explain naturally why rotations of galaxies are faster than they ought to be according to calculations of masses of the stars in them, so the existence of mysterious invisible dark matter is postulated. This phenomenon can be simply explained by Newton´s theory if we accept that galaxies, except of celestial bodies, contain also mutual non-local vacuum quantum dipoles connecting every object to all others in the galaxy, so that the galaxy is kept together despite its fast rotation. Of course, the mass of Galaxy is much bigger than the total mass of its celestial bodies, as a huge amount of energy (mass) is carried by mutual non-local vacuum quantum connections. Gravity is a global cosmic phenomenon as a direct consequence of cosmic expansion. Cosmic gravity acting between celestial bodies is a counterbalance compensating the repulsive force causing the cosmic expansion [6,7].

Strong Interaction

Before analysis of the strong interaction we will imagine the structures of all stable particles that oscillate in one main axis (line) with common centre of oscillation, where all tops of opposite poles come together during the phase of mutual attraction (contraction) shown in Figures 11 and 12. Muon and Tau have the same structure as an electron, only they are much more energetic and so shorter. They are unstable and change into electrons by transferring their energies into surroundings (Figures 13 and 14). If neutrino really exists, it represents a double-photon structure with specific internal motion. The neutrino is its own antiparticle, so neutrino and anti-neutrino represent the same particle. As the neutrino oscillates in one plane, it does not resist its dragging by cosmic expansion and therefore it has no rest mass and its speed is c. The same structure of quantum dipoles as a neutrino also other structures can have, e.g. double photon, mesons, neutral pions, but their internal motion is not so simple, so they do not represent the stable structures Illustrated in Figure 15. This structure of a double photon has two different centres of oscillation with different phases. A photon can associate with any particles without disturbing their internal structure and so bring them into excited states. It can also associate with itself without creating a new particle. Its spin j=1 means

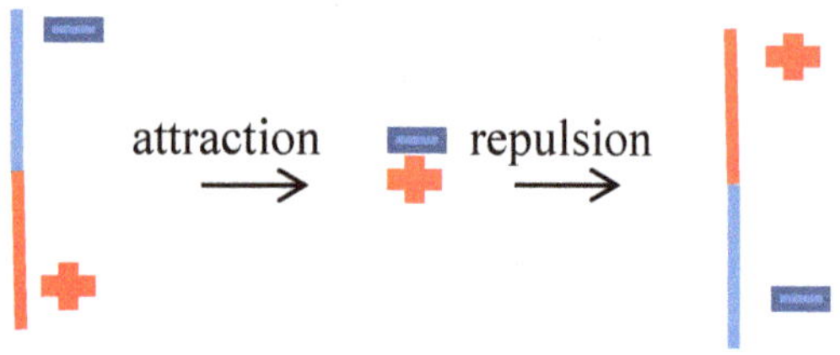

Figure 11: Photon γ (+/-) created by one oscillating quantum dipole.

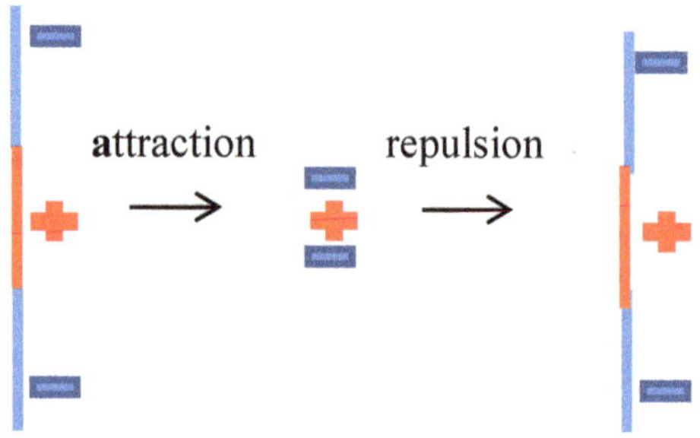

Figure 12: Electron e- (+/2-) created by two quantum dipoles.

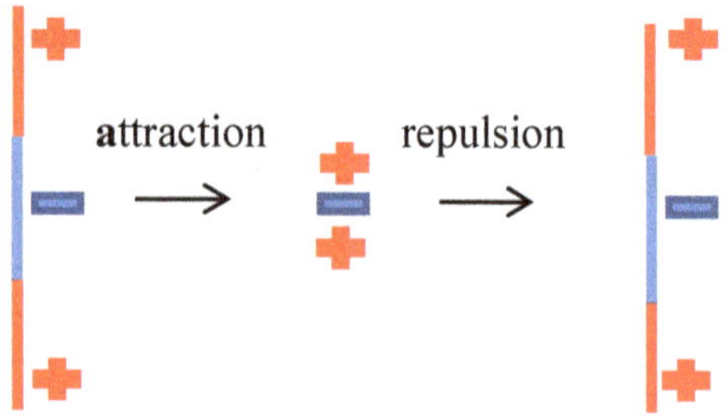

Figure 13: Positron e+ (2+/-) made of two quantum dipoles.

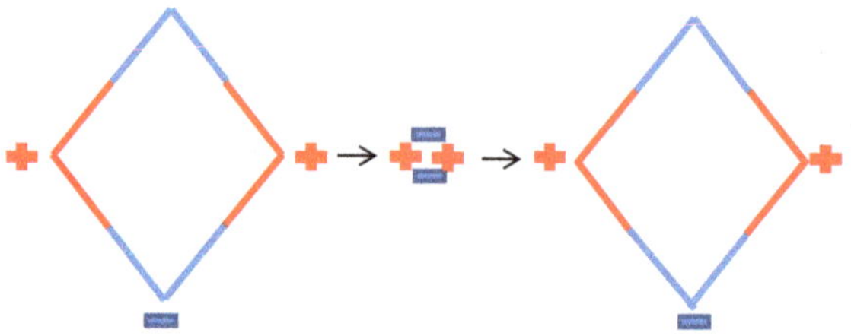

Figure 14: Neutrino ve (2+/2-) made of four quantum dipoles.

that the intermediate state known as a positronium, created after electron-positron collision, can decay either into two or three photons. A photon in relation to a magnetic field can deflect to the north or south magnetic poles or stay without any deflection. This means that the dipole is right-handed or left-handed, or performs both these motions simultaneously, meaning that it exists as a double-dipole, where one dipole is right-handed and the other left-handed with a neutral manifestation towards a magnetic field. The annihilation of electron (+/2-) and positron (2+/-) after their collision and consequent decay of intermediate positronium into two or three photons, are illustrated in Figure 16.

All stable structures (particles) oscillate in one line (axis of oscillation) to the one common centre (during attraction) Figure 17. All dipoles of a proton are very energetic (short and strong) so their forces of mutual attraction and repulsion are so strong that can compensate the mutual local repulsive pressures of spaces of quantum dipoles in such a way that the proton is the most stable composite structure. If structures are more complicated and composite, the mutual local pressures of dipole spaces destroy their compositions in the moment of their creation (so-called resonances). From the structure of a proton with three tops of positive poles is evident why the experiments in electron-proton scattering found that electrons scattered off three points inside the proton. It is not because of a quark structure but the bipolar essence of a proton. The proton can be destroyed only by its annihilation with an antiproton explained in Figure 18. Proton and antiproton represent the mutual mirror images so they attract each other very strongly creating the temporary high energetic composite structure of protonium (5+/5-), which, thanks to huge local repulsive pressures of dipole spaces, completely destroys the original structures of proton and antiproton with a definite release of 5 free photons γ at least. Of course, more photons are possible, because of excitation of initial particles before annihilation. In the structure of "protonium" (5+/5-) or (6+/6-), if excited by one photon, we can see some other substructures, which correspond to some mesons, so we can interpret the annihilation. As unstable neutral pions π^0, as well as eta mesons η, represent the bound states of two photons, both they decay into two photons 2γ:

$\pi^0 \rightarrow \gamma + \gamma$

$\eta \rightarrow \gamma + \gamma$

mega meson ω decays by the next way: $\omega \rightarrow \pi^0 + \gamma$

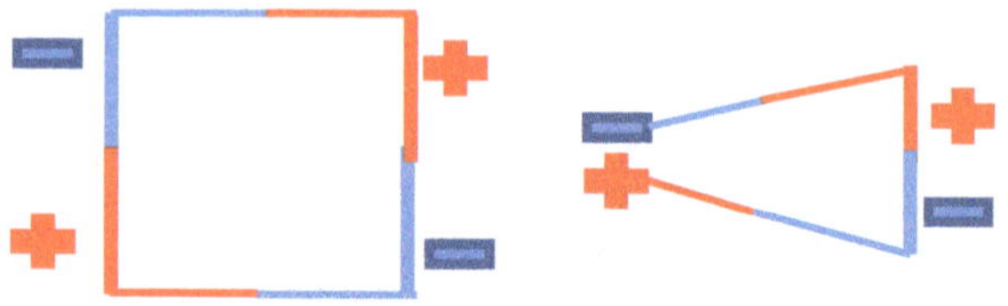

Figure 15: Representation of the stable structures.

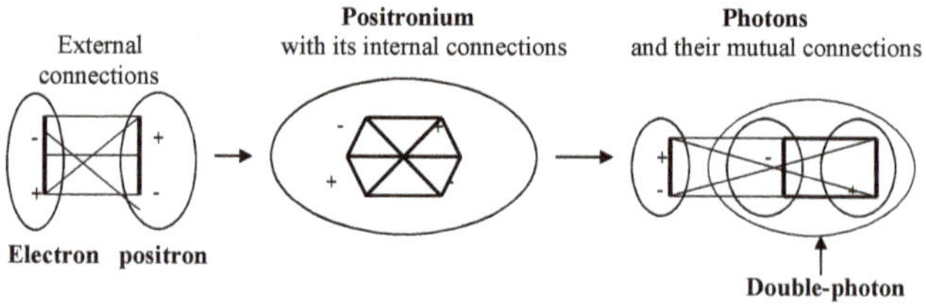

Figure 16: Collision and consequent decay of intermediate positronium into two or three photons.

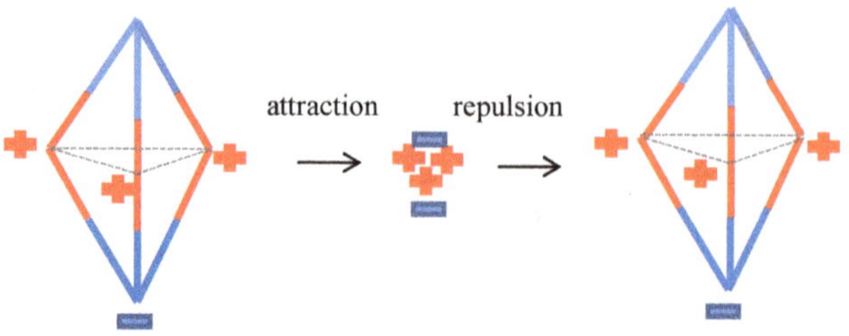

Figure 17: Proton p+ (3+/2-) made of six elementary quantum dipoles.

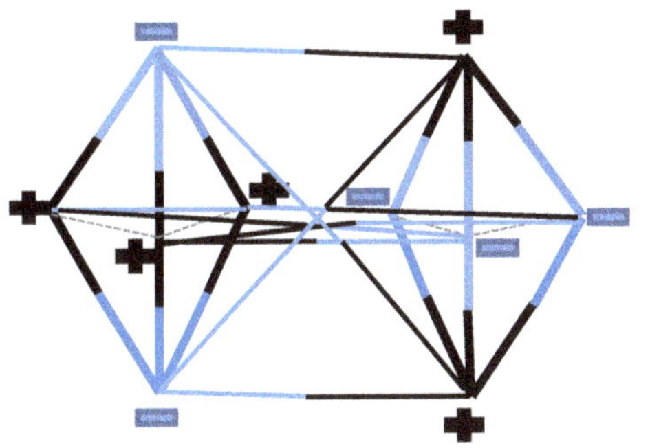

Figure 18: Proton – Antiproton Annihilation (p+p-) – protonium.

The annihilations by low energy collisions of proton and antiproton can be:

$p^+ + p^- \rightarrow \omega + \pi^0 \rightarrow \pi^0 + \gamma + \pi^0 \rightarrow \gamma + \gamma + \gamma + \gamma + \gamma$

$p^+ + p^- \rightarrow \pi^0 + \pi^0 + \pi^0 \rightarrow \gamma + \gamma + \gamma + \gamma + \gamma + \gamma$

$p^+ + p^- \rightarrow \pi^0 + \pi^0 + \eta \rightarrow \gamma + \gamma + \gamma + \gamma + \gamma + \gamma$

Contemporary theoretical physics supposes protons, neutrons and unstable baryons to consist of three quarks, while mesons of quark-antiquark pairs interacting by gluons. The quark model was invented to simplify the situation with a huge number of hadrons (baryons and mesons). Although it can help a little with classifications of these particles, it is mistaken by explanation of the real essence of micro-world. The problems of quark model are quite clear. Quarks cannot exist as individual entities, cannot be detected directly, they have unbelievable so-called "asymptotic freedom" and nobody can explain what is the reason for their different colours, flavours and other very strange properties. Let us take a look at how the quark model explains the decay of a neutral pion π^0. "The π^0 (neutral pion) is a quark – antiquark meson. The quark and antiquark can annihilate; from the annihilation come two photons. This just shows how the quark model complicates the very simple situation: We know that the pion decays into two photons. Why do we need the quark-antiquark annihilation in addition? Why do we not accept the pion as a bound state of two photons? Why photons, as elementary quanta of free energy, are not considered to be the basic constituents of all physical structures (particles and interactions)? Why do we not try to understand and detect the real nature of a photon but create so absurd constituents - quarks? Why do we complicate the situation so much if the truth is very simple? Now we know definitely that the neutral pion π^0 (2+/2-) represents a bound state of two photons and so its internal structure consists of four mutually interconnected quantum dipoles. We do not need any mystical undetectable quarks as we have real photons. Nothing is hidden and there are no mysteries in the physical Universe. Everything is clear and simple.

Weak nuclear interaction - Neutron beta decay

Inside a neutron we see the structure of a proton with very short end strong quantum dipoles which is clearly illustrated in Figure 19. One negative pole is connected to three positive opposites by much weaker and longer connections, so it can be released from this structure during beta decay Figure 20. We can see that the neutron (4+/4-) in its excited state with sixteen elementary quantum dipoles represents a bound state of a proton (3+/2) with six elementary quantum dipoles and an electron (+/2-) with two quantum dipoles. Eight quantum dipoles represent mutual quantum connections between the proton end electron structures. They are, at the same time, the constituents of the internal neutron structure. Neutron consists of a proton and an electron as well as their eight mutual quantum connections (dipoles) which are included into the neutron structure. If the proton and electron represent separate particles (e.g. in the structure of hydrogen atom), their mutual connections (being much longer and weaker) are external and represent their mutual vacuum or electromagnetic field. So the atomic vacuum is created by mutual connections between nucleons and electrons in the structure of atom. In 1920 Rutherford quite correctly supposed the existence of a neutral particle being a strong bound state of a proton and an electron, but this nice and clear idea was refused and the monstrous electroweak theory was postulated.

The neutron cannot be as stable as a proton as its structure and internal motion are more complicated and the neutron has more than one centre of oscillation. So the neutron (after its excitation by one photon) decays into a proton and an electron. Their mutual connections being before the constituents of neutron are now the external connections between a proton and an electron.

This decay is known as beta decay (β^- decay), because flying electrons represent beta (β^-) radiation and can be expressed as follows:

$n + \gamma \rightarrow p^+ + e^-$

"$n + \gamma$" represents the excited state of a neutron

Contemporary theoretical physics represents this decay, considering it to be a manifestation of the so-called weak interaction, by the following scheme:

$n \rightarrow p^+ + e^- + (\nu_e)?$

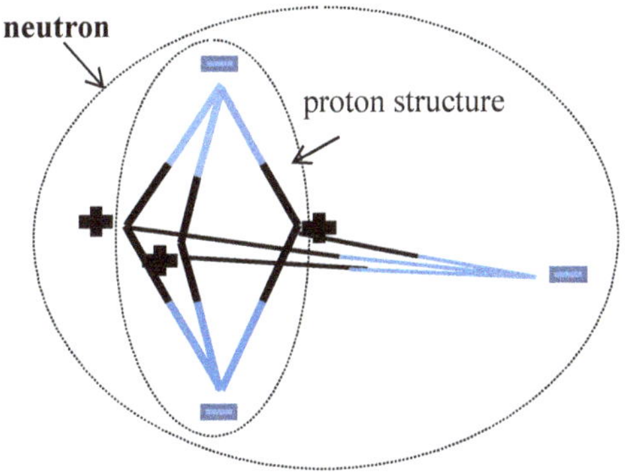

Figure 19: Neutron n (3+/3-) in its basic state (not excited) is created by nine quantum dipoles.

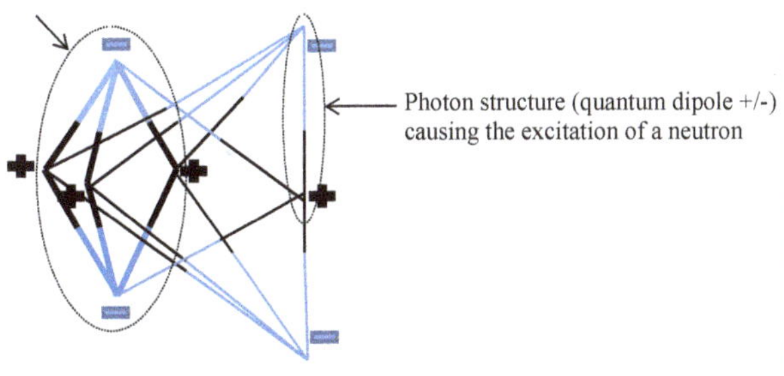

Figure 20: Neutron n (4+/4-) in its excited state created by sixteen quantum dipoles.

In addition to a proton and an electron the neutrino (antineutrino) ν_e is included. In our structural scheme the neutrino is missing. We do not deny the possible existence of a neutrino. The expression "(ν_e)?" only means that we cannot accept it to be a product of β^- decay in the presented form. It could be a product only if a neutron is bound in a heavy nuclei where nuclear forces and mutual repulsive pressures are enough strong to form a neutrino consisting of four strong, short and energetic quantum dipoles. Although a neutrino is not detectable during β^- decay its hypothetical existence was predicted as it seemed that some energy was missing and conservation of momentum, as well as angular momentum, was violated. Emitted electrons have a continuous kinetic energy spectrum, ranging from 0 to the maximal available energy of a few tens of MeV. A typical value is around 1 MeV. This continuous spectrum looks strange from the view-point of quantum theory. But continuous spectra of kinetic energy of electrons can be simply explained if accept that neutrons, before their decay, are excited by photons with any value of energy of continuous spectra, so the resulted electrons can also have kinetic energy of continuous spectra. We do not deny the possible presence of electron antineutrino (for us there is no difference between neutrino and antineutrino) in beta decay. We can only accept the excitation of a neutron, bound in a heavy nucleus, by three photons which, after catching a negative pole "-" from the neutron and changing it into a proton, consequently form one electron and one neutrino according to the following scheme:

$$n + 3\gamma \rightarrow p^+ + e^- + \nu_e$$

Our doubt about a neutrino as a product of beta decay without previous excitation of a neutron by photons follows also from the following consideration:

As emitted electrons have a continuous kinetic energy spectrum, if we want to receive the discontinuous energy spectrum, we must accept that energy carrying by a neutrino has also a continuous spectrum. But as the neutrino has no rest mass, we must accept the existence of neutrinos with internal energies of any value of continuous spectra, what means that their essence is analogical to that of photons, what can be possible as neutrinos represent bound states of two photons. Continuous spectra of photons exciting the structure of neutrons cause their decay by emitting electrons with energies of continuous spectra. The rule of the Standard Model that the lepton number must be conserved is wrong and artificial as we can see that the electron can be a substructure of an excited neutron. Only the charge number must be conserved as well as the number of nucleons (protons and neutrons), because proton is very stable and cannot be destroyed (except of annihilation). It can only be excited by an electron to the neutron, which can again decay into a proton and an electron.

It is supposed that the whole universe baths in a sea of neutrinos. In that case it looks much more likely that the decay of a neutron is caused by its previous excitation by a free neutrino, so the decay is as follows:

$$(n + \gamma) + \nu_e \rightarrow p^+ + e^- + \nu_e$$

Neutrinos before and after decay have different energy and momentum. The above mentioned scheme of β^- decay shows that neutrinos can easy interact with matter by a weak force. This looks much more likely than supposed very rare interaction of neutrinos with rest matter. In this case neutrinos behave like photons exiting the initial neutrons before they decay. So we suppose that β^- decay of a neutron can exist in two forms. If a neutron is excited only by one photon then the neutrino cannot be a product of decay. Only if a neutron is excited by three photons (or one photon and one neutrino) then the neutrino can occur as a product of beta decay. This could be the reason why the production of solar neutrinos is three times lower than predicted by the Standard Model. According to our understanding it looks very likely that only one of about three β^- decays produces a neutrino (in our understanding the double photon). So no neutrino oscillation is needed. According to the Standard Model three types of neutrino (electron, muon and tau) can exist with quite different energies (flavours) and they can mutually change into one another, so they oscillate. We do not deny that neutrinos can exist in different energetic states like photons can, but only an electron neutrino represents the stable state (like electron), other states are unstable and change into an electron neutrino. If we want to accept the Standard Model interpretation that the muon μ^- and tau τ^- decay into an electron, neutrino and antineutrino, it means they must consist of these structural constituents before decay. In that case neutrinos have the same property as photons to excite other particles. But much more real is that muon μ^- and tau τ^- are only more energetic versions of an electron e^-, they are unstable and convert into electrons after a very short time by transferring their internal energies into their external vacuum quantum connections. Of course, electron as well as muon and tau can be excited by photons.

Pions represent more complicated structures, so they decay. Pion π^0 (2+/2-) decays into two photons 2γ. Pion π^- (3+/4-) consequently can decay in one muon μ^- (+/2-) and neutrino ν (2+/2-). Muon μ^- consequently changes into an electron e^-. Pion π^+ (4+/3-) can decay into one muon μ^+ (2+/-) and a neutrino ν (2+/2-). Muon μ^+ then changes into a positron e^+ which annihilates with the nearest electron. Pions have structures analogical to those of excited protons (p^++γ) (antiprotons), but while protons are very stable, pions decay. The difference between positive pions and protons is in different mutual motions of their internal quantum dipoles and their different energy (mass). Positive pions are less energetic than protons (about seven times lesser) so their quantum dipoles are not enough strong to save the structure from its immediate decay. But the indirect evidence for the similarity between proton and positive pion structures is their similar momenta. The structures of a proton is (3+/2) while excited (4+/3-) is analogical to the structure of a positive pion π^+ (4+/3-). While proton is very stable, pion decays immediately into a muon μ^+ and a neutrino ν.

As neutrinos can be detected only indirectly, their role in beta decays is still opened and unclear. In any case, if we interpret all constituents of beta decay as structures of elementary quantum dipoles, the picture is becoming very clear and simple. But the so-called theory of electroweak interaction only complicates this situation very much.

Let us take a look at how the theory of electroweak interaction (TEWI) complicates the simple picture of neutron decay. As QED supposes virtual photon to be a mediator of electromagnetic interaction, so TEWI supposes that the weak interaction must also have a point-like mediator named W^- boson, which is very massive, but virtual at the same time. As it is almost 100 times as massive as the initial neutron - heavier than entire atoms of iron, it is supposed that W^- boson, for only a very short undetectable time, borrows high energy from the vacuum (this miracle is supposedly allowed by Heisenberg's uncertainty principle) and then, after making all needed miracles, returns it back to the vacuum. Another great miracle that W^- boson makes is the conversion of one down quark (charge of -1/3) of a neutron into the up quark (charge of +2/3) it means that a neutron consequently converts into the proton. This reversal of quarks is called "flavour change". After making this "important" conversion and returning borrowed energy to the vacuum, W^- boson subsequently decays. Feynman's diagram of β^- decay of a neutron according to the electroweak theory (Figure 21).

Although W^- boson is virtual during β^- decay and so undetectable, its real existence is also supposed. From the structure of electron (+/2-) and neutrino (2+/2-), the compound structure (3+/4-) of W^- with 12 elementary quantum dipoles can be created by high energy collisions as a short living structure (resonance). The same is valid also for compound structure of positron (2+/-) and neutrino (2+/2-), it means the structure (4+/3-) of W^+. But both these compound structures W^-, W^+ are not point-like bosons and appear only in very rare cases, in the high energy collisions. Electrons and unobserved electron neutrinos with enormous energy of about 40 GeV are supposed to be produced by decay of undetectable W^- bosons. It means that the neutrino with internal energy of only some MeV and zero rest mass must highly increase its internal energy to the value of 40 GeV. Using a monstrous 80 GeV boson to mediate low energy beta decay looks like killing the flies by atom bombs.

Insertion of virtual W^-, W^+ bosons into a simple picture of beta decay in order to create the electroweak theory is quite artificial and only complicates the simple situation. No virtual boson is needed, only real particles - neutron, proton, electron and maybe neutrino. No virtual processes are needed only real detectable interactions.

Theory of electroweak interaction tries to give together the electromagnetic interaction mediated by a virtual photon without rest mass with a weak interaction mediated by supposed very massive W^-, W^+ and Z bosons, so the so-called Higgs mechanism is required for breaking the electroweak symmetry and giving particles their rest mass. This hypothetical Higgs mechanism asks for the existence of very heavy Higgs boson which is declared to be found at LHC by energy of 125 GeV. If looking at the Higgs boson through one of its declared possible decay modes W^-, W^+, then it represents the basic structure (7+/7-), whicht same time represents the compound structure of electron, positron and two neutrinos. Only electron and positron are really detectable. Fictitious undetectable Higgs boson as a point-like particle is nonsense as well as mysterious Higgs field. Except for networks of elementary quantum connections (+/-) there are no other fields. Everything is made up of these connections. Photons as free elementary quantum dipoles (+/-) are the simplest particles having no rest mass. All other particles represent compound structures of two or more quantum dipoles with more complicated motions, so they local touch interactions with the vacuum (vacuum quantum connections) cause that they cannot be dragged by cosmic expansion and therefore they have rest mass as a measure of their resistance towards acceleration.

Another problem of Higgs boson is the conclusion that its "small" mass, although 126 times the mass of the proton, causes that the universe we live in is inherently unstable, it means that this mass is not enough to prevent the cosmic catastrophe. Unstable vacuum will result in cosmic cataclysm during transfer to the stable vacuum billions of years from now. If the Higgs mechanism breaks the symmetry between weak and electromagnetic interactions, what mechanism does break symmetries between strong and weak interactions and at what level of energy are all interactions united including gravity? If we are talking about symmetries the answer is very simple. The basic symmetry of all particles, interactions and fields means that all they are made of the same constituents-elementary quantum dipoles. We do not need huge accelerators and colliders in order to create such a big energy level, where all interactions ought to be unified, as the basic interaction is already known. Only real particles are detectable before and after high energy collisions, neither virtual bosons nor quarks. As all particles are made of elementary quantum dipoles, the picture of their mutual interactions is simple and clear without the necessity to include virtual undetectable realities there.

Nuclear Force

The nuclear force is an attractive one between two or more nucleons (neutrons and protons) binding them into atomic nuclei. Masses of light nuclei are less than the total mass of protons and neutrons which form them. According to contemporary quark model the nuclear force is a residual effect of much more powerful strong force (interaction) binding quarks by gluons. At the time before the quark model was created, the nuclear force was conceived to be transmitted by a neutral pion π^0. The most appropriate system for studying the nuclear force is a bound state of one proton and one neutron named deuteron being the nucleus of the deuterium atom named heavy hydrogen Figure 22. After synthesis of proton and neutron the photon is released taking out so-called binding energy Figure 23. In a bound state of nucleus it is not clear which of components is a neutron and which a proton as the negative pole is common for both nucleons. The compound state of one proton and one neutron in a deuteron (6+/5-) consists of 30 elementary quantum connections. If the photon is not released, the excited deuteron (7+/6-) consists of 42 elementary quantum dipoles. This structure represents factually the bound state of two protons and one electron Figure 24. In this structure we can see the substructures of neutron, proton, electron, but the deuteron is created not only of these structures but also of their mutual quantum connections being internal constituents of a deuteron. It is a clear manifestation of the holistic principle according to which the deuteron is not a simple sum of its structural components (protons and electron) but represents a higher quality defined also by their mutual quantum connections.

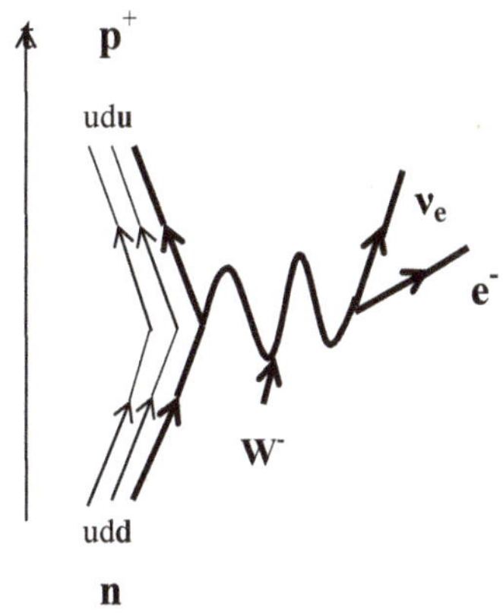

Figure 21: Feynman´s diagram of β- decay of a neutron according to the electroweak theory.

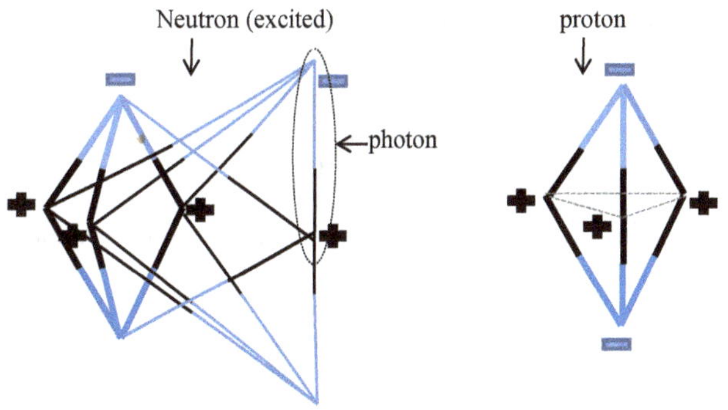

Figure 22: Synthesis of proton and neutron the photon is released taking out so-called binding energy.

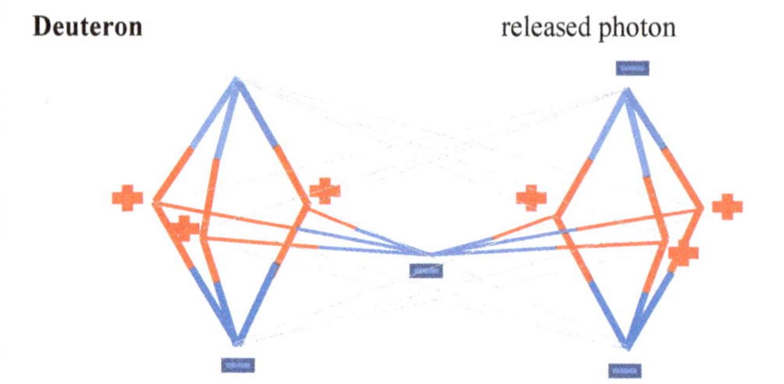

Figure 23: Synthesis of deuteron.

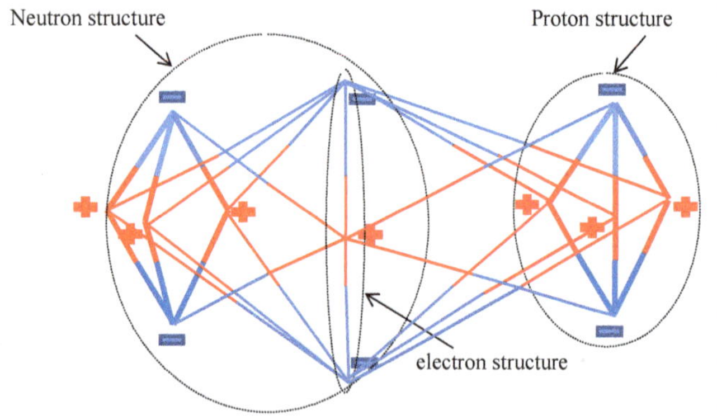

Figure 24: Structural representation of two protons and one electron.

The deuteron compositions (7+/6-) exist in heavier atoms with higher atomic numbers being sources of γ-rays during a radioactive decay. The evidence is the fact that the fusion of two nuclei with lower masses than iron generally releases energy, while the fusion of nuclei heavier than iron absorbs energy. So not only no photon is released but new free photons are absorbed in the structure of heavier nuclei. The opposite is true for the reverse process, nuclear fission. This means that fusion generally occurs for lighter elements only, and likewise, that fission normally occurs only for heavier elements. So, only the extreme astrophysical events can lead to short periods of fusion with heavier nuclei. This is the process that gives rise to nucleosythesis, the creation of heavy elements during events like supernovas. Synthesis of heavier nuclei is possible only by extreme energies which allow to compress nuclei very close, so that the mutual quantum connections become very short and strong able to overcome their mutual repulsive pressures. The claim that binding energy of nucleons in nucleus is given by energy needed to be released during their synthesis is limited only for lighter nuclei and so cannot be a dogma, because real binding energy of nucleons in nuclei is energy of their mutual quantum connections (dipoles). Creation of the required conditions for fusion on Earth is very difficult. Dipoles creating the internal structures of both nucleons (protons and neutrons) are very short, strong and energetic so they represent the strong forces, while quantum dipoles between both or more nucleons are weaker and represent the nuclear force connecting nucleons into a nucleus. Although the nuclear force is much weaker than the strong one, it is enough strong and short (the shorter - the stronger) to overcome the mutual local repulsive pressures between quantum dipoles. Now we see that the nuclear force is not a residual effect of a strong force binding quarks by gluons, but it is created, as well as a strong force, of elementary quantum dipoles, although much longer and weaker. The nucleus of a helium atom ${}_2He^4$, named α-particle, represents a bound state of 2 protons and 2 neutrons (12+/10-) consisting of 120 elementary quantum dipoles. The internal dipoles of nuclei are very short and strong (strong interaction) but their mutual connections are much weaker and can have different lengths and energies (nuclear interaction) Figure 25. Not all 120 mutual quantum dipoles (+/-) are imagined in the above picture, but we can see the difference between quantum dipoles creating the internal structure of 4 nucleons (strong interactions) and their mutual nuclear interactions.

The more nucleons are in nuclei, the heavier and less stable are the atoms as the number of mutual quantum connections dramatically increases with a consequent increase of their repulsive pressures. Atoms with a huge number of nucleons (protons and neutrons) in a nucleus are unstable and can decay. This so-called radioactive decay is a stochastic (random) process. The internal motion of quantum dipoles and they mutual pressures as well as impulses from outside can disrupt the equilibrium of attractive and repulsive forces and cause the atom spontaneously decays, where the huge amount of nuclear forces is released by emitting particles (α-particles, β-particles, γ-rays and others) which carry out high energies. The radioactive decay transforms the initial nucleus into another nucleus, or into a lower energy state. A chain of decays takes place until a stable nucleus is reached. An example of α-decay involves uranium (Figures 26 and 27):

$${}_{92}U^{238} \rightarrow {}_{90}Th^{234} + {}_2He^4$$

The process of transforming one element (e.g. uranium) into another (thorium) is known as transmutation.

The electron or positron represents the beta particle in beta decay. If an electron is involved, the number of neutrons in the nucleus decreases by one and the number of protons increases by one. An example of such a process is:

$${}_{90}Th^{234} \rightarrow {}_{91}Pa^{234} + e^-$$

In the nucleus with a big number of nucleons the local repulsive pressures of enormous number of mutual nuclear quantum dipoles

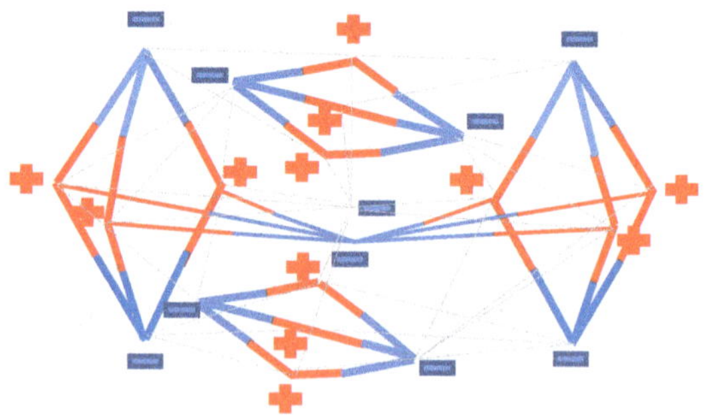

Figure 25: α-particle (nucleus of a helium atom ${}_2He^4$).

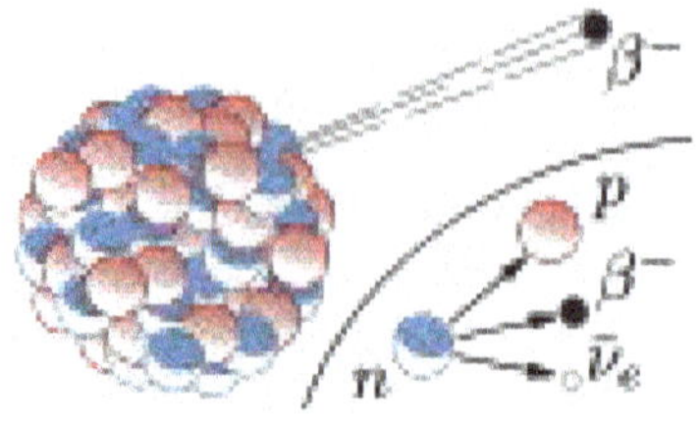

Figure 26: β-decay generally occurs in neutron rich nuclei.

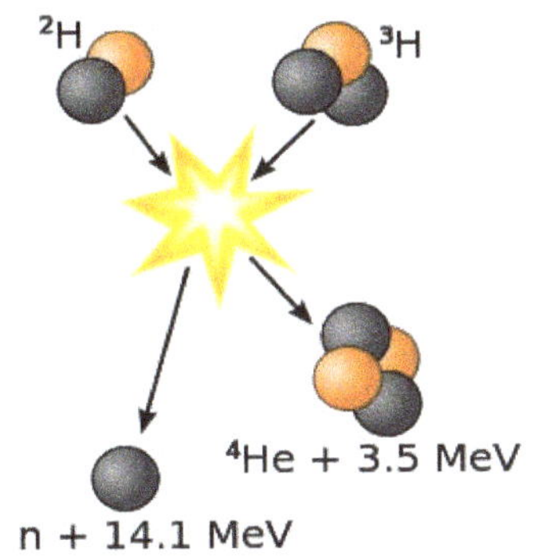

Figure 27: Nuclear fission of heavy nuclei.

between nucleons (protons and neutrons) is so high that the equilibrium between attractive nuclear forces of quantum dipoles and their repulsive pressures is very fragile and a small impulse is enough to cause the imbalance so that a radioactive decay can occur. This small impulse could be caused by excitation of the nucleus by a photon (or neutrino?), so that the number of mutual quantum connections in a whole structure of nucleus increases with a consequent increase of local repulsive pressures causing the radioactive decay. If the impulse is high, caused by interaction with energetic neutrons, the internal structure of radioactive nucleus of uranium increases the number and amount of repulsive pressures of quantum dipoles so dramatically that the nucleus is split in two nuclei with release of high energy particles like α, β, γ and neutrons, which can again cause the nuclear fission of other uranium nuclei and so generate the so-called chain reaction. On this principle the atom bombs are designed as well as nuclear reactors in nuclear power stations where the chain reaction is controlled. High energy can be released not only by nuclear fission of heavy nuclei, but also by synthesis (fusion) of light nuclei in thermonuclear reactions. At the picture taken from Wikipedia we can see the fusion of deuterium with tritium creating helium, freeing a neutron and releasing 17.59 MeV of energy. It takes considerable energy to force nuclei to fuse. Accelerated to high speeds (that is, heated to thermonuclear temperatures), they can overcome their local mutual repulsive pressures and get close enough for the attractive force to be sufficiently strong to achieve fusion. The fusion of lighter nuclei, which creates a heavier nucleus and often a free neutron or proton, generally releases more energy than it takes to force the nuclei together. Even when the final energy state is lower, there is a large barrier of mutual repulsive pressures that must be firstly overcome. It is called the Coulomb barrier. To achieve extreme conditions necessary for fusion, the initially cold fuel must be explosively compressed. Inertial confinement is used in the hydrogen bomb where the driver is x-rays created by a fission bomb. Long lasted research into developing controlled thermonuclear fusion is still unsuccessful.

All forces are nothing more than attraction and repulsion of quantum dipoles. Very short quantum dipoles create the strong attractive forces inside hadrons and leptons, nuclear forces are created by short and strong quantum dipoles between nucleons, electrostatic forces are formed by weaker and longer quantum dipoles, other forces between atoms and molecules are weaker than electrostatic ones, and the weakest are attractive forces of gravity between massive objects created by long mutual quantum dipoles representing a cosmic vacuum. Attraction and repulsion are always in a mutual equilibrium. Shortening and increasing of mutual quantum dipoles between nuclei during their fusions are at the same time accompanied by increasing of their mutual repulsive pressures, which overcoming is necessary for the successful fusion. The dynamic equilibrium of both opposite forces (attraction and repulsion) inside atoms and particles is manifested by internal motions (oscillations, vibrations, etc.)

Conclusion

From the viewpoint of dialectical logic contemporary theoretical physics failed. Now, when the Unity Principle is disclosed, new science can start its consecutive development on true fundamentals. The knowledge of the Unity Principle and its manifestations is a basis for new science including theoretical physics, biology, sociology, science of consciousness as well as philosophy. It represents a new scientific and philosophical paradigm and starting point of substantial great revolution in human knowledge giving true answers to all basic questions of our existence.

References

1. Hegel GFW (2010) The science of logic. Cambridge University press USA.
2. Kohut P (2011) God and the universe. Scientific god Journal 2: 056-067.
3. Kohut P (2014) Crisis of contemporary theoretical physics and the truth in the mirror of dialectical logic. Sntropy 1: 01-079.
4. Kohut P (2013) Unity principle: the truth in the mirror of dialectical logic. Scientific God Journal 5: 01-082.
5. Kohut P (2013) The unity principle. Syntropy 2: 220-242.
6. Kohut P (2012) The physical universe as a divine quantum information structure. Scientific god journal 3: 220-250.
7. Kohut P (2012) Social and spiritual development of mankind the way to freedom. Scientific god journal 3: 1015-1033.

To Find Different Gravity Laws... Proof Gravity Wave Equations are Real and Generate Gravitational Waves

Murad PA*

Morningstar Applied Physics, LLC Vienna, VA 22182, USA

Abstract

Various gravitational laws are examined. Binary pulsars imply unusual behavior as evidence for gravitational waves in contrast to Newtonian gravity. Gravity's intensity in the restricted three-body problem can significantly be reduced due to rotational centrifugal forces to explain shortcomings usually mentioned to invent dark matter. Although Libration Points are distinct, they represent a suitable testing ground. The Trojan asteroids at the Sun-Jupiter stable triangular points do not demonstrate that millions of asteroids reside at a singular point but with a very large scatter and do not asymptotically congeal due to attraction to create planetoids thereby suggesting Newtonian gravitation is not ample and has flaws. A phase-space trajectory methodology for an integral equation defines eigenvalues and a forcing function to test some different gravity laws. These findings indicate that a time-dependent law may not locally conserve energy and result in a celestial vibration, repulsive gravitation or produce gravitational waves. This provides insights for newer gravitational models possible suitable for using warp drive concepts. Furthermore, this effort warrants placing more space probes at the triangular Libration Points as cosmic anomalies.

Keywords: Pulsars; Gravity; Gravitation; Space; Time

Introduction

There are several different gravitational laws [1], which may exist for different applications and regions. The question is to find a suitable test to investigate if these models are correct or not. This may or may not require investigating anomalies. Moreover, if successfully tested, could a future model be created that possesses more granularity than an existing model with higher orbit accuracy?

Iorio [2] provides an excellent examination for defining specific anomalies, which are worth mentioning. There are currently accepted laws of gravitation applied to known bodies which may potentially pave the way for remarkable advances in fundamental physics. This is particularly important given where most of the Universe seems to be made of unknown substances dubbed Dark Matter and Dark Energy rather than pursuing a different model to validate a newer gravitational law other than Newtonian gravitation. Moreover, investigations can serendipitously enrich and find other solutions as well as other anomalies. The current status of some of these alleged gravitational anomalies in the Solar system are:

- Possible anomalous advances of planetary perihelia,
- Unexplained orbital residuals of a recently discovered moon of Uranus (Mab),
- The lingering unexplained secular increase of the eccentricity of the orbit of the Moon,
- The so-called Faint Young Sun Paradox,
- The secular decrease of the mass parameter of the Sun,
- The Flyby Anomaly,
- The Pioneer Anomaly, and
- The anomalous secular increase of the astronomical unit.

Obviously Newtonian gravitation is the most popular model and has worked exceedingly well while estimating orbits in the near-abroad for planets, moons, asteroids, and short-and long-range satellites. Its flaws, however, is if Einstein is correct and gravitational waves exist, Newtonian gravity is predicted by an elliptical partial differential equation and does not admit gravitational waves. This is problematic. Moreover, Einstein and other models would require time which satisfies a wave partial differential equation for the existence of gravitational waves. Current thinking of these separate models is they should all have to asymptotically satisfy the far-field boundary conditions based upon Newtonian gravitation. That is gravity should vanish at infinity. This immediately suggests the model should have some form of an inverse radial function with distance anchored at a celestial mass to satisfy this far boundary condition.

An existence of gravitational waves may not be as prevalent as one would assume. Gravitational waves are a means for allowing cosmic events to occur with some time delay as well as prevent the instantaneous responses whenever a supernova explodes. Einstein suggested there should be a time delay in such events. He attempted to generate a gravity model [3] with gravitational waves. In his initial model, the conclusion was gravity would be self-sustained and self-feeding. Hence gravity depends upon itself. Einstein felt this was not acceptable and changed his view to determining a geometric representation of the space-time continuum and gravity.

The other point worth mentioning is wave equations mathematically have real characteristics in their partial differential equation. If these characteristics with a hyperbolic partial differential equation, or a wave equation, confluence and join, we may have a situation analogous to fluid dynamics with respect to gravity. That is gravitational waves or gravitational shocks [4] may exist and can potentially use a propulsor to harness thrust for missions to the far-abroad. Thus this can become a crucial issue.

***Corresponding author:** Murad PA, CEO, Morningstar Applied Physics, LLC Vienna, VA 22182, USA, E-mail: pm@morningstarap.com

The question is why would we look at the Libration Points which have historically survived for a very long period of time? This is a reasonable request. Euler first made several contributions to include fluid dynamics. These involved using fluid particles in a stationary coordinate system in a control volume where you cannot differentiate about specific particles as well as look at fluid dynamic problems without the shear stress terms intrinsic to the Navier-Stokes equations. From an astrodynamic perspective, the Euler problem also involves the 'restricted' three-body problem. The rationale is when you compare celestial bodies and a space probe, the weight of the probe likened to the celestial bodies is insignificant in the order of the mass of the probe about 10-22 compared with the celestial body weights. Euler [5] simplified mathematically complex problems, which would defy solutions during the 18th century. This is understandable to rationally linearize nonlinear problems to gain some plausible insights.

This is very unusual in Figure 1. It shows the strength of the gravitational field of two celestial bodies. However, the important note is the effects of the coordinate system's rotation which has an overriding consequence when the combined effects are included. One of the arguments about dark matter is where galaxy spirals did not respond as expected. Either the gravitational law is not ample or there is a need for black or dark matter. As this figure shows, without understanding rotational effects, gravity can easily be altered. Moreover, Winterberg implies that the rotation of Neutron stars can alter its gravitational field with respect to its companion in binary pulsars. Thus we need to look at rotation between celestial bodies as well as body rotation effects from frame-dragging.

The Lagrangian or Libration Points are normally mentioned in a Cartesian Coordinate system rather than a polar coordinate system [*If a polar coordinate system is used for a two-body problem, the probe or second body sees a radial attraction due to gravity from the celestial body. There is no further force that would impact the angular momentum around the primary body and the problem is solved resulting in an elliptic, hyperbolic or parabolic trajectory. The problem is straightforward. For a three-body system, the probe body sees two attractions from the primaries' gravitation. However, the angular force does not vanish with a constant such as r2dθ/dt is constant as in the two-body problem. The mathematics is complex depending upon the geometric relation with the both primaries. This in polar coordinates creates an obvious problem*] to reduce difficult geometric circumstances. Motion between two large celestial bodies and the probe are confined to a two-dimensional plane. These are in a rotating coordinate system where a probe is not only influenced by the gravitational attractions for two bodies but also the centrifugal effect due to rotation about the coordinate system reference point. The origin reference point is the Barycenter of the two large primaries.

The other part of the Cartesian coordinate problem is with all of the solutions, the analysis comes down to 5 distinct points; three collinear points falling on a line joining the two primaries and two triangular points that are off to the left and to the right side of this line within a circle based upon a radius from the distance between the two primaries. One of the Libration Points falls on the same or close radius. This latter point about the collinear point is surprising and motivates to look at this problem purely with polar coordinates. The difference from this location would be dependent upon the mass ratios between the weights of the celestial primaries. Triangular points are considered stable points while the collinear points are considered as unstable. These points are selected based upon a point in the field where the probe's gravitational and centrifugal forces possess no force in the field. This is generated in a gravitational potential combined with rotational effects. The first derivative of this potential vanishes. What is not normally considered is the stability of the five points, which h can be determined by the Calculus of Variations. Here, the force field will be stable when the field is a relative minimum and unstable when it is a relative maximum. This requires dealing with second derivatives of the field as well as set the first derivatives to zero explained in Figures 1 and 2.

The question about different gravitational laws needs some insights. Newtonian gravitation is based upon the distance between two

$$U_{tot} = -\frac{Gm_1m_3}{r_1} - \frac{Gm_2m_3}{r_2} - \frac{1}{2}\frac{G(m_1+m_2)m_3}{l^3}r^2 \quad \Leftrightarrow \quad \frac{U_{tot}/m_3}{G(m_1+m_2)} = -\frac{(1-\mu)}{r_1} - \frac{\mu}{r_2} - \frac{1}{2}(x^2+y^2)$$

due to: gravity rotation gravity rotation

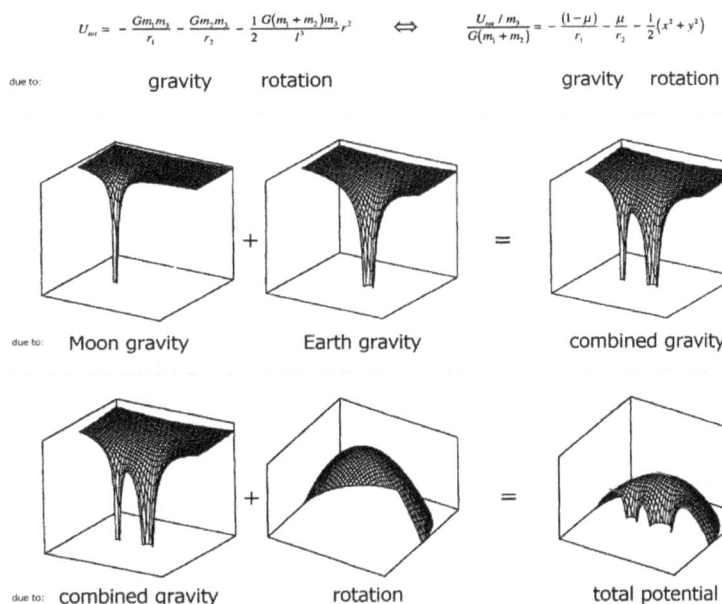

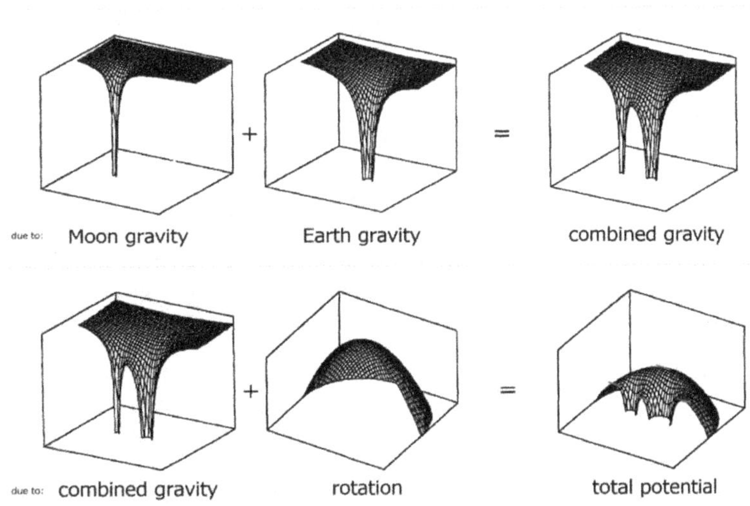

Figure 1: The three-body problem can be visualized as a combination of the potentials due to the gravity of the two primary bodies along with the centrifugal effect from their rotation. Note gravitation by itself does not respond due to these significant rotational effects.

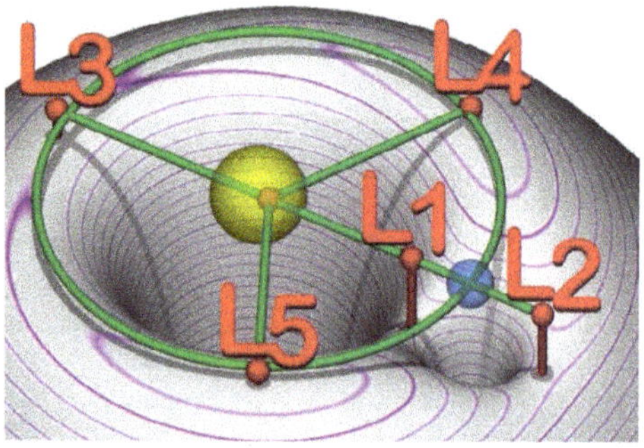

Figure 2: The three collinear Lagrange points (L_1, L_2, and L_3) were discovered by Leonhard Euler a few years before Lagrange discovered the remaining two triangular points [5]. The gravity wells or slight mounds at the Libration Points are not seen here.

attractive bodies. Jefimenko [6,7] implies the two bodies are attractive but gravity is also a function of velocity to create both an attractive force as well as generate angular momentum. Jefimenko not only defines a gravity law which asymptotically satisfies Newtonian gravity but supposedly includes the effects from Einstein's Theory of Relativity. To meet these requirements, he generates a co-gravity field to compensate for motion at speeds in the region of near and above the speed of light. Unfortunately, we are now entering a dangerous realm where we lack meaningful data or evidence to either clearly validate or deny any of these theories and models. Furthermore, according to Jefimenko, Newtonian is not a consequence of frame dragging, which includes tidal locking. This suggests that on a major planet, you will always see the same face of a large orbiting moon. If a wave equation is included in the model, gravity should also be a function of time. This requires some thought process to realize gravity is related in some manner with time. Moreover, the functional representation will involve the spatial coordinates as well as a function of time. How would this influence the mechanics in examining trajectories or relating energy that would normally accept as a constant but now would become a function of time? This suggests energy may either increase or decrease with time without any other factors which impact the dynamics. This will require some deeper consideration to really understand this point because it appears to violate conservation of momentum as well as energy. The question remains on how to define and validate a gravity law. This will have to be performed in the context of contemporary technology with some evidence. As an approach, the Libration Points will be examined from energy requirements followed by examining some trajectory performance for different gravitational laws. However, these Libration Points may provide evidence of the validity of one gravitational law over another.

Discussion

The question is why we should be concerned about the Libration Points. Basically, these locations using Newtonian gravitation imply that distinct but weak points exist. It is our intention to mention that triangular Libration points should also be considered as cosmic anomalies as the previously mentioned topics. Triangular Libration points were originally discovered based upon the Trojan asteroids considering the Sun-Jupiter system.

Data about the Trojan Asteroids

The term "Trojan" originally referred to the "Trojan asteroids" (Jupiter Trojans), which orbit close to the Lagrangian points of the Sun-Jupiter system. These were named after characters from the Trojan War of Greek mythology. By convention, the asteroids orbiting near the L_4 point of Jupiter are named after the characters from the Greek side of the war, whereas those orbiting near the L_5 of Jupiter are from the Trojan side. Later on, objects were found orbiting near the Lagrangian points of Neptune, Mars, Earth, and Uranus with respect to the Sun. Minor planets at the Lagrangian points of planets other than Jupiter may be called Lagrangian minor planets.

The current Trojan asteroids [8-16] are shown with a debris cloud between Mars and Jupiter's orbits: the L_4 swarm is believed to hold between 160,000 and 240,000 asteroids with diameters larger than 2 km and about 600,000 with diameters larger than 1 km. If the L_5 swarm contains a comparable number of objects, there is more than 1 million Jupiter Trojans 1 km in size or larger. For the objects brighter than absolute magnitude 9.0, the population is probably complete. These numbers are similar to comparable asteroids in the asteroid belt. The total mass of the Jupiter Trojans is estimated to be low at 0.0001 of the mass of Earth or one-fifth of the mass of the asteroid belt [*By contrast, the debris orbit between Mars and Jupiter has a weight, if all summed up, would be the size of the planet Mars. Thus these are relatively light and should be strongly influenced by large celestial bodies*]. The number of Jupiter Trojans observed in the L_4 swarm is slightly larger than observed in L_5. However, because the brightest Jupiter Trojans show little variation in numbers between the two populations, this disparity is probably due to observational bias. However, some models indicate the L_4 swarm may be slightly more stable than the L_5 swarm. Jupiter Trojans have orbits with radii between 5.05 and 5.35 AU from the Sun (the mean semi-major axis is 5.2 ± 0.15 AU), and are distributed throughout elongated, curved regions around the two Lagrangian points. Each swarm stretches for about 26° along the orbit of Jupiter, amounting to a total distance of about 2.5 AU. Jupiter Trojans do not maintain a fixed separation from Jupiter. They slowly librate around their respective equilibrium points, periodically moving closer to Jupiter or moving farther from it. Also note there is a collection of asteroids near L_3, which is an unstable Libration Point. The scatter at this location is as bad as the scatter from the L_4 and L_5 points which are supposedly stable. One wonders if the L_1 point has a similar scatter. Astronomers have found a Trojan-like asteroid, not far from the Earth [17], moving in the same orbit around the Sun. It sits in one of the 'Lagrange points', which are 60 degrees ahead of or behind the planets in their orbits. These are points of gravitational stability. Called 2010 TK7, the rock is about 80 million km from Earth and should come no closer than about 25 million km. It is suggested its orbit appears stable at least for the next 10,000 years. Carefully observe the magnitude of this exceptionally large motion. There is another point worth noting. These Trojan asteroids, as well as the orbital debris cloud between Mars and Jupiter, appear to be continually moving and not having confluence with each other to form planetoids. This movement may represent a transient change in the local gravitational field. These effects should be used to account for a gravitational model. Although the motion of the asteroids may be small, this should appear as a celestial vibration. Thus the movement from these asteroids should generate high-frequency gravitational waves (HFGWs). This point is also the same with the rings around Jupiter and Saturn. Here, there is asteroid debris moving in a specific trajectory. However, amongst the debris, there is an instantaneous motion that is not continuous and the bodies obviously collide within these belts. This difference in motion, as well as collisions, should be ample to also generate HFGWs. The issue for detection should be more severe because of the presence of a large celestial body.

This could create additional noise as well as alter the HFGWs. A cleaner detection would most likely occur with the Trojan asteroids previously mentioned. The prime ingredient is toward developing a reasonable test model to examine different gravitational laws. Obviously, this pictorial in Figure 3 implies the asteroids, at the stable triangular point with a huge scatter. Moreover, the motion is quite large about these points and again, there is generally no previously mentioned confluence of the asteroids to generate planetoids based purely upon mutual attraction at any of these Libration Points whether stable or unstable. The implication is that there is sufficient movement underway to prevent the mutual gravitational attraction through some yet to be determined repulsion or another phenomenon. Some comments suggest a different law would compensate for all of these differences explained in Figure 3.

Demonstrating the instability in the collinear points where a probe moves from one unstable collinear point to another for the Earth-Moon system is explained in Figure 4. If the triangular points were stable, one would supposedly assume the probe may capture the probe. This surprisingly does not occur. If anything, the motion is not near the triangular points but attracted to the gravitational pull of the Moon. This suggests the Libration Points are very, very weak and not as strong as, say the celestial bodies regardless of the rotating coordinate system.

Gravitational Models

Based upon the questions about the stability, wide scatter, and strength of the Libration Points, it would be reasonable to open the door to look at a different gravitational model from the conventional Newtonian gravity model.

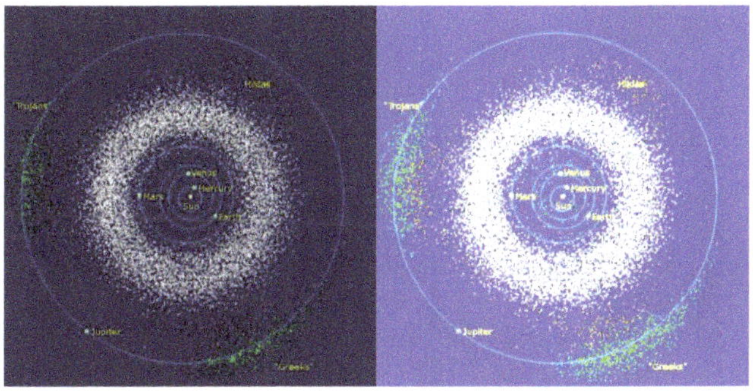

Figure 3: Estimates of the total number of Jupiter Trojans are based on deep surveys of limited areas of the sky. The false-color picture gives a better depiction of the Trojan asteroids near Jupiter's orbit and the asteroid belt.

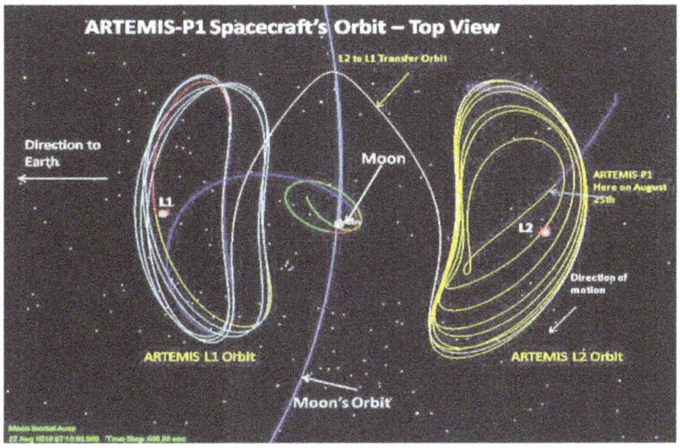

Figure 4: These results show the movement from the unstable Libration Points to each other. If stability and the influence were so strong, one would have assumed the craft would move to the triangular points. The issue is these locations are not very strong because of other to be determined effects.

There are many types of gravity laws explained in Figure 4. Basically, the *Newtonian gravity law* assumes there is an attraction that acts between two separate bodies. It is based on the masses of the two bodies and the separation distance between these bodies [18]. The gravitational field acting on one of the bodies is related to the mass of the bodies as well:

$$\bar{g} \approx -G\frac{m}{r^3}\bar{r} = G\frac{m}{r^2}\hat{r} \tag{1}$$

Of these, r is the distance as a vector between both bodies and the 'normal' vector r is in the radial unit direction, G is a constant value and m is the mass. In this model, the gravity asymptotically goes to zero at extremely large distances or disappears at infinity. Some of these laws are briefly described below. Winterberg's rule accounts for situations with pulsars and neutron stars. This implies the large rotational rate of these bodies somewhat change the strength of the gravitational attraction. Thus there is a consideration for angular momentum in the body itself as a source term. Winterberg does not prove this rationale and differences are observed between the companion star and the neutron star in binary pulsars. This difference in mass could be attributed as a similar effect as dark matter. Four-derivative theories [19,20] are a conformal gravity as an example from the theory of relativity. This means each term in the wave equation can contain up to 4 derivatives. There are pros and cons of 4-derivative theories. The pros are that the quantized version of the theory is more convergent and renormalisable. The cons are there may be issues with causality. A simpler example of a 4-derivative wave equation is the scalar 4-derivative wave equation:

$$\nabla^4 \varphi(r)=0 \tag{2}$$

This in a central field of force is explained through Table 1. The first two terms are the same as a normal wave Equation. Since this equation is a simpler approximation to conformal gravity, then m corresponds to a mass of the central source. The last two terms are unique to 4-derivative wave equations. It was suggested to assign small values to them to account for the galactic acceleration constant (also known as dark matter) and the dark energy constant. An equivalent solution to the Schwarzschild solution in General Relativity on a spherical source for conformal gravity has a metric with:

$$\phi(r) = g^{00} = (1-6bc)^{1/2} - \frac{2b}{r} + cr + \frac{d}{3}r^2 \tag{3}$$

The term $6bc$ is very small so it can be ignored. The problem is that now c is the total mass-energy of the source, b is the integral of density time's distance to the source squared. So this is a completely different potential to General Relativity and not just a small modification. The main issue with conformal gravity theories, as well as any theory with higher derivatives, is the typical presence of instabilities of the quantum version of the theory, although there might be a solution to this problem. Note these theories do not approach to Newtonian gravitation because of the r and r^2 terms. However, based on establishing different views about results during Pioneer 10 and 11, the existence of such terms may be explained. Iorio [2] performs an adequate definition where the attraction of the sun's gravity at a certain threshold distance becomes linear and a constant similar to these terms estimated at very small values. The *Jefimenko gravitational model* involves an attractive force between two bodies as well as creates angular momentum. The gravitational law is not only a function of distance between the two bodies but also includes the relativity velocity function between the two bodies. The Jefimenko model involves:

Gravity law	Assumptions	Gravitational rule
Newtonian gravitation	$\nabla \times \overline{g} = 0.$ and $\nabla.\overline{g} = -4\pi G\rho_g.$	$\overline{g} = -\nabla \phi and : \nabla^2\phi = 4\pi G\rho_g ; where : g \approx 1/r^2.$
Four-Derivative theories	$\phi(r) = 1 - 2m/r + ar + br^2,$ $\phi(r) = g^{oo} = (1-6bc)^{1/2} - \frac{2b}{r} + cr + \frac{d}{3}r^2.$	$\overline{g} = -\nabla\phi(r).$
Winterberg's rule	$\nabla.\overline{g} = -4\pi G\rho_g = 2\omega^2,$ Where $\rho_g = -\frac{\omega^2}{2\pi G}.$	$\overline{g} = -\nabla \phi and : \nabla^2\phi = -2\omega^2 ; where : g \approx 1/r^2.$
Jefimenko's gravity and co-gravity	$\nabla \times \overline{g} = -\frac{\partial \overline{k}}{\partial t} ; \nabla.\overline{g} = -4\pi G\rho_g ; \nabla.\overline{k} = 0.$ and: $\nabla \times \overline{k} = -\frac{4\pi G}{c^2}\overline{J}_g + \frac{1\partial \overline{g}}{c^2 \partial t}.$	$\frac{1\partial^2 \overline{g}}{c^2\partial t^2} - \nabla^2\overline{g} = 4\pi G[\nabla.\rho_g + \frac{1\partial \overline{J_g}}{c_g^2\partial t} - \frac{\nabla \times \overline{J_c}}{c}],$ $\frac{1\partial^2 \overline{k}}{c^2\partial t^2} - \nabla^2\overline{k} = 4\pi G\left[\frac{\nabla.\rho_g}{c^2} - \frac{1\partial \overline{J_c}}{c^3\partial t} - \frac{\nabla \times \overline{J_g}}{c^2}\right].$
Murad's modification of Jefimenko	$\nabla \times \overline{g} = -\frac{\partial \overline{k}}{\partial t} - \frac{4\pi G}{c_k}\overline{J}_k ; \nabla.\overline{g} = -4\pi G\rho_g ;$ $\nabla.\overline{k} = -\frac{4\pi G}{c_k}\rho_k and : \nabla \times \overline{k} = -\frac{4\pi G}{c^2}\overline{J}_g + \frac{1\partial \overline{g}}{c^2\partial t}.$	$\frac{1\partial^2 \overline{g}}{c^2\partial t^2} - \nabla^2\overline{g} = 4\pi G[\nabla.\rho_g + \frac{1\partial \overline{J_g}}{c^2\partial t} - \frac{\nabla \times \overline{J_c}}{c}],$ $\frac{1\partial^2 \overline{k}}{c_g^2\partial t^2} - \nabla^2\overline{k} = 4\pi G\left[\frac{\nabla.\rho_g}{c^2} + \frac{1\partial \overline{J_c}}{c^3\partial t} - \frac{\nabla \times \overline{J_g}}{c^2}\right].$
Murad's gravity law	$\nabla \times \overline{g} = \frac{i}{c}\frac{\partial \overline{g}}{\partial t} + \frac{4\pi\gamma G}{c}\overline{J}_g$ and $\nabla.\overline{g} = -4\pi\gamma G\rho,$ where $\gamma = \frac{1}{\sqrt{1-\frac{u^2}{c^2}}}.$	$\frac{1\partial^2 \overline{g}}{c^2\partial t^2} - \nabla^2\overline{g} = 4\pi\gamma G[\nabla.\rho_g + \frac{i\partial \overline{J_g}}{c^2\partial t} - \frac{\nabla \times \overline{J_g}}{c}]$

Table 1: Different Gravitational Laws which cover a spectrum of conditions of interest.

$$\overline{g} \approx -\frac{Gm}{r^3(1-\overline{r}.\overline{v}/rc)^3}\left[\left(\overline{r} - \frac{r\overline{v}}{c}\right)\left(1-\frac{v^2}{c^2}\right)\right] + \overline{r} \times \left[\left[\left(\overline{r} - \frac{r\overline{v}}{c}\right) \times \ldots\right]\right] \approx$$
$$\overline{g} \approx -G\frac{m}{r^3}\left[\left(1-\frac{v^2}{2c^2}\right)\overline{r}_0 - \frac{2rv^2}{3c^3}\overline{v}_0\right] \quad (4)$$

The last two laws in Table 1 are modifications to Jefimenko's laws from the author. The point here in the first system of partial differential equations is to obtain symmetry between the gravitational and co-gravitational fields. Whereas Jefimenko considers gravitational currents, the law considers that co-gravitational currents should also exist. The same holds for the co-gravity source term yet to be defined. A discipline of interest would be to examine the creation and experimental evidence of any of these currents and sources. The last law is an attempt to use this logic without co-gravitation. Mathematically, this involves using an imaginary term. One may hypothesize this as possibly representing an unknown dimension. Although it satisfies a wave equation and satisfying the Newtonian gravitational law, it would be considered as an improbable situation.

Test models

A reasonable test would involve specific regions that possess unusual trajectory performance or even result in singularities [21,22]. For example, the Libration Points are derived from a possible singularity region in the phase-space representation of the 'restricted' three-body problem. These points indicate that gravity attractions between the two rotating bodies compensate for each other as well as the motion of the third body. If a singularity, it would become infinite such as dividing by a distance where it goes to zero. Thus Libration or Lagrangian points would be interesting for testing different gravity models. We have to raise the question where there are five Libration Points using Newtonian gravitation. Countless analyses have looked at these points and others have assessed analytical means to generate enough stability to deal with these unstable points. The major issue, however, is that very few satellites or space crafts have really reached these locations. Moreover, valid models may have more of these points.

Analysis

This section will cover the conventional wisdom as well as variations which impact orbital trajectories or singular-like behavior. Again, the objective is to find plausible trends but not necessarily uncover specific results.

Preliminary results for orbits and trajectories

Let us consider a two-body problem using Newtonian gravity. With these assumptions, the radial and angular momentum equation are defined as:

$$\overline{F}_r = m\overline{a}_r = m\left(\frac{d^2r}{dt^2} - r\left(\frac{d\theta}{dt}\right)^2\right)\widehat{r}$$
$$\overline{F}_\theta = m\overline{a}_\theta = m\left(r\frac{d^2\theta}{dt^2} + 2r\frac{dr}{dt}\frac{d\theta}{dt}\right)\widehat{\theta} = \frac{m}{r}\frac{d}{dt}\left(r^2\frac{d\theta}{dt}\right)\widehat{\theta} \quad (5)$$

Subscripts for the forces in the LHS are not derivatives but actually the radial and azimuthal force directions respectively. Derivatives are functions of time. The radial force is based upon the gravitational attraction from the larger body. Moreover, the second equation assumes the azimuthal force vanishes for each of these bodies. The radial dimension is changed as the difference between the distances to the two objects. The problem is reduced to one free-variable with some definitions where μ is the total mass of both bodies (G (m1+m2)). This is considered as the gravitational attraction for this problem as follows:

$$\left(\frac{d^2r}{dt^2}-r\left(\frac{d\theta}{dt}\right)^2\right)=-\frac{\mu}{r^2}$$
$$m\left(r\frac{d^2\theta}{dt^2}+2r\frac{dr}{dt}\frac{d\theta}{dt}\right)=\frac{m}{r}\frac{d}{dt}\left(r^2\frac{d\theta}{dt}\right)=0\,h=r^2\left(\frac{d\theta}{dt}\right) \tag{6}$$

Clearly, the azimuthal gravitation disappears with the constant, h, that is the angular momentum per unit mass used to satisfy azimuthal acceleration. Thus, the second equation vanishes. A variable u is selected based upon an inverse function of the radius to simplify the problem:

$$\frac{d^2u}{d\theta^2}+u=\frac{\mu}{h^2} \tag{7}$$

The solution of this ordinary differential equation considering a geometric length l and eccentricity e is:

$$\frac{d^2u}{d\theta^2}+u=\frac{\mu}{h^2}+C\cos(\theta-\theta_0) \text{ or } r=\frac{l}{1+e\cos(\theta-\theta_0)} \tag{8}$$

Jefimenko's two-body problem

By applying the gravitational law, the effects of gravity will affect both the radial and angular momentum equations which become:

$$\left(\frac{d^2r}{dt^2}-r\left(\frac{d\theta}{dt}\right)^2\right)=-\frac{\mu}{r^2}\left[\left(1-\frac{v^2}{2c^2}\right)\overline{r_0}-\frac{2v^2r^2}{3c^3}\frac{dr}{dt}\right]$$
$$\left(r\frac{d^2\theta}{dt^2}+2r\frac{dr}{dt}\frac{d\theta}{dt}\right)=\frac{1}{r}\frac{d}{dt}\left(r^2\frac{d\theta}{dt}\right)=\frac{\mu}{r^2}\frac{2v^2r^2}{3c^3}\frac{d\theta}{dt}=\mu\frac{2v^2}{3c^3}\frac{d\theta}{dt} \tag{9}$$

Let velocity $\boldsymbol{v}$ of the spacecraft be assumed, albeit a bad assumption, as a constant. This is not as simple to alter the time derivative for angular changes. This would derive a relationship for the radial derivatives and a possible solution. First, find a solution by defining angular velocity:

$$For:\frac{1}{r}\frac{d}{dt}\left(r^2\frac{d\theta}{dt}\right)=\frac{2\mu v^2}{3c^3}\frac{d\theta}{dt}$$
$$then:\frac{d}{dt}ln\left(r^2\dot{\theta}\right)=\frac{2\mu r v^2}{3c^3},r^2\dot{\theta}\approx exp\left(\frac{2\mu r v^2}{3c^3}t\right) \tag{10}$$

Under restrictive assumptions, angular momentum is a function of the radius and time in a very small exponential term. Moreover, this exponential term may go to zero and this reverts to the Newtonian angular momentum case. However, if true, there is still an additional term in the radial momentum as an insignificant number. Furthermore, angular momentum can be altered as a function of time influencing trajectories. Note further the value of the exponential term may be close to zero for slow speeds analogous to the initial problem. It is interesting angular momentum is also a function of the rate of change of the angle. In a circular trajectory, this would be a constant term. However, for an elliptical orbit, this value changes based upon its location within the orbit. Unfortunately, there is no simple way to convert time into an angular function for this law. There is another option. Let us assume the angular momentum can be a function of the angle. This would include the terms in such expressions as in Jefimenko. For the simplest case, let:

$$Let:r^2\frac{d\theta}{dt}=h+\alpha(\theta-\theta_o),where:\alpha=\frac{2\mu v^2}{3c^3} \tag{11}$$

Note the value of $\boldsymbol{\alpha}$ is still an extremely small value because of the denominator. Using the inverse function for the distance $\boldsymbol{r}$ for $\boldsymbol{u}$, these equations become:

$$-h^2u^2\frac{d^2\theta}{dt^2}-\frac{\left(\alpha-\frac{2\mu r v^2}{3c^3}\right)}{(h+\alpha(\theta-\theta_o))}-h2u3=-\frac{\mu}{(h+\alpha(\theta-\theta_o))^2}\left(1-\frac{v^2}{2c^2}\right)u^2 \tag{12}$$

Simplifying this becomes:

$$\frac{d^2u}{d\theta^2}+u=+\frac{\mu}{h^2(h+\alpha(\theta-\theta_o))^2}\left(1-\frac{v^2}{2c^2}\right) \tag{13}$$

This is a nonlinear equation because of the denominator in the RHS. One possible solution is to use the Green's function with the initial value problem [15,16]. The solution becomes:

$$u(\theta)=\frac{\mu}{h^4}\left(1-\frac{v^2}{2c^2}\right)\int_0^\theta\frac{\sin(\theta-\xi)}{\left(1+\frac{\alpha}{h}\xi\right)^2}d\xi \tag{14}$$

Gravitational hyperbolic partial differential equation Model

Einstein's theory of relativity implies the gravitational model satisfy wave equations, say, such as:

$$\frac{1\partial^2\overline{g}}{c^2\partial t^2}-\nabla^2\overline{g}=4\pi G\nabla.\rho_g \tag{15}$$

If you look at the separation of variables, the homogeneous equation provides some variables to represent a function of time as well as the radial distance. The time term looks like:

$$T(t)=\alpha_1+\alpha_2t+\alpha_3\cosh(\lambda^2t)+\alpha_4\sinh(\lambda^2t)+\alpha_5\cos(\omega t)+\alpha_6\sin(\omega t),\ for\ g(t,r)=T(t).R(r) \tag{16}$$

These constants make non-dimensional values by initial or final conditions. The issue is how Newtonian gravitation can be correct without considering the time factors. Constants, either λ^2 or ω, depend upon real values. However, the hyperbolic sine and cosine terms are never observed since they most likely occur early during gravitational creation. This is not trivial. Moreover, the two functions asymptotically are large values as an exponential function of time and may approach a line that could cancel out the $\boldsymbol{t}$ term. This may be related to the coefficients α_3 if it is equal-α_4, where these two terms would vanish at large time values with:

$$T(t)=\alpha_1+\alpha_2t+\alpha_3\cosh(\lambda^2t)(1-\tanh((\lambda^2t))+\alpha_5\cos(\omega t)+\alpha_6\sin(\omega t) \tag{17}$$

The sinusoidal and cosine terms rise and fall with time while the coefficient of the *cosh* term will vanish after some considerable time. This implies the hyperbolic terms at zero time would have some value suggesting gravity initial exists, and then slowly decays. Here this initial value could be cancelled out by the α_1 term to compensate for this initial value unless gravity always existed before the Big Bang. However, these transient terms in this equation also imply several interesting features worth noting. These relations allow for the presence of sinusoidal behaviour. When we do this, we feel the magnitude of the constant factor α_1, is large with the coefficient for the sinusoidal term. If this second term is greater than the constant, it might lead to positive gravity or repulsion. The presence of these transient terms may explain why there is such a large variation in the Trojan Asteroids near the triangular Libration Points. The geometric functions indicate gravitational waves exist and may be of very small magnitudes. Values for α_1 may approach a negative unity while the α_2 constant may be very small. The two-body problem modified for a wave equation now would add a term similar to:

$$\left(\frac{d^2r}{dt^2} - r\left(\frac{d\theta}{dt}\right)^2\right) = -\frac{\mu}{r^2}T(t),$$
$$\left(r\frac{d^2\theta}{dt^2} + 2\frac{dr}{dt}\frac{d\theta}{dt}\right) = \frac{1}{r}\frac{d}{dt}\left(r^2\frac{d\theta}{dt}\right) = 0., h = r^2\left(\frac{d\theta}{dt}\right) \tag{18}$$

Here, there is no additional complexity for the angular momentum but only the transient radial momentum effects seen. This is an interesting point. For example, many planets and binary pulsars such as 1913+16 [19] have unusual rotation about the primary orbits. This may increase as a function of time. Here the latter example indicates the unusual trajectory motion is caused by the loss of energy in the neutron star generating gravitational waves. Motion is dominated by initial energy levels to define specific trajectories of the asteroids illustrated in Figure 5. The issue is how to relate time to spatial coordinates and especially angular changes. This is something which will not happen at the current time period. The only point is a comment made by Kozyrev who said *the sun is not a thermonuclear fusion device because it should possess a higher surface temperature.* When asked, what is a star? His response was: *a star is a machine that converts space-time continuum into energy!* Mass converting into energy, but how can we relate time to either angular momentum, energy or for that matter, mass? This is beyond our realm of technology but it deserves to raise the issue. Furthermore, transient terms suggest asteroids or other bodies near a Libration Point will be like a pot of boiling water always changing and altering the energy probably within the ZPE.

Preliminary results for Libration points

Examine the restricted three-body problem and look at finding the Libration Points using Newtonian Gravitation. For the three-dimensional Restricted Three-body problem, the kinematics are:

$$\ddot{x} - 2\omega\dot{y} - \omega^2 x = -V_x + F_x,$$
$$\ddot{y} + 2\omega\dot{x} - \omega^2 y = -V_y + F_{y,} \quad and \tag{19}$$
$$\ddot{z} = -V_z + F_z$$

The force F terms will be ignored; the dot term signifies time derivatives; where x, y, and z are elements of a Cartesian coordinate system, ω is the rotational rate, and the gravitational law requires:

$$V(x,y,z) = \frac{(1-\mu)}{r_1} - \frac{\mu}{r_2} \quad where: r_1^2 = (x-x^1)^2 + y^2 + z^2,$$
$$and \quad r_2^2 = (x-x^2)^2 + y^2 + z^2. \tag{20}$$

The value r is the distance between the major bodies based upon the subscript and the third body, μ is the normalized weight of the larger bodies. The energy expression is found by multiplying the first equation on x with $\dot{x}$ integrated in time. This is achieved with y multiplied by $\dot{y}$ and z by $\dot{z}$ integrated as a function of time. Libration points are based upon when the derivatives of V vanish. The resulting energy equation is:

$$E(x,y,z) = \frac{1}{2}(\dot{x}^2 + \dot{y}^2 + \dot{z}^2) - \frac{\omega^2}{2}(x^2 + y^2) - \frac{(1-\mu)}{r_1} - \frac{\mu}{r_2}$$
$$or: (x,y,z) = \frac{1}{2}(\dot{x}^2 + \dot{y}^2 + \dot{z}^2) - \frac{\omega^2}{2}(x^2 + y^2) + V(x,y,z). \tag{21}$$

Time does not appear explicitly. However, time may change related to variations with any of the other terms. For example, if gravity changes with time, then energy would also be a function of time. However, the question is where the energy comes from considering conservation. What does this mean? If unstable, a satellite placed at the collinear points would require some station-keeping propulsion to maintain its location whereas the stable points should require lower propulsion needs. The problem for Jefimenko's law as follows:

$$\ddot{x} - 2\omega\dot{y} + \left[\frac{2v^2}{3c^3}(\mu_1 r_1 + \mu_2 r_2)\right]\dot{x} = \omega^2 x\left(1 - \frac{v^2}{2c^2}\right) - \left[\frac{\mu_1(1-\mu_2)}{r_1^3} - \frac{\mu_2(x-\mu_1)}{r_2^3}\right]\left(1 - \frac{v^2}{2c^2}\right),$$
$$\ddot{y} - 2\omega\dot{x} + \left[\frac{2v^2}{3c^3}(\mu_1 r_1 + \mu_2 r_2)\right]\dot{y} = \omega^2 y\left(1 - \frac{v^2}{2c^2}\right) - \left[\frac{\mu_1 y}{r_1^3} - \frac{\mu_2 y}{r_2^3}\right]\left(1 - \frac{v^2}{2c^2}\right), \quad and \tag{22}$$
$$\ddot{z} = 0.$$

A correction term is *(1 - v²/c²)*. This is a very complicated situation with additional cross-coupling terms. Finding the energy term is difficult and finding the actual Libration Points. If the gravitational law involves a wave equation, a similar correction term is used as a function of time. The gravitational gradient looks like:

$$V(x,y,z) = -\left[\frac{(1-\mu)}{r_1} + \frac{\mu}{r_2}\right]T(t)$$
$$where: r_1^2 = (x-x_1)^2 + y^2 + z^2, \; and \; r_2^2 = (x-x_2)^2 + +y^2 + z^2. \tag{23}$$

Energy becomes:

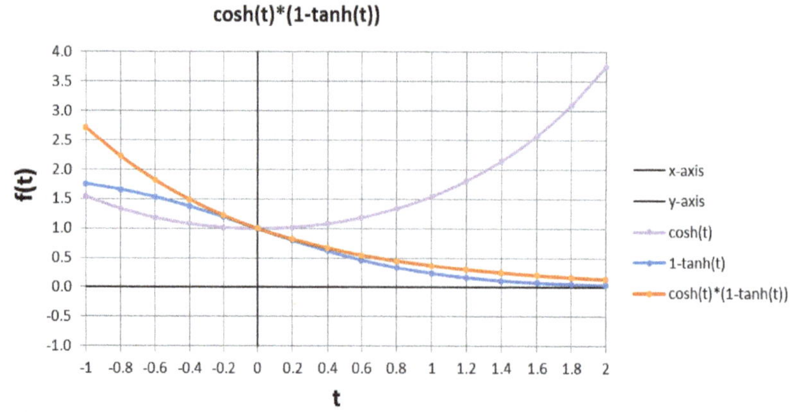

Figure 5: Specific trajectories of the asteroids.

$$E(x,y,z)=\frac{1}{2}(\dot{x}^2+\dot{y}^2+\dot{z}^2)-\frac{\omega^2}{2}(x^2+y^2)-\left[\frac{(1-\mu)}{r_1}+\frac{\mu}{r_2}\right]T(t)\ or$$
$$E(x,y,z)=\frac{1}{2}(\dot{x}^2+\dot{y}^2+\dot{z}^2)-\frac{\omega^2}{2}(x^2+y^2)+V(t,x,y,z). \quad (24)$$

In both of these cases, the energy is no longer a constant but includes a correction factor for the speed of light in the first state and a function of time in the second situation. This latter situation could explain why the collinear points are unstable requiring thrust for station keeping. Likewise, the speed of light factor suggests if a spacecraft increases, the influence of the primary bodies' gravity diminishes and has less of a navigation problem for influencing the spacecraft. Clearly, we have two different possible outcomes with contrasting views based upon the speed of the spacecraft. This may be ideal for a test function. For example, more knowledge needs to be found regarding the dynamics at these Libration Points to see if these additional terms are real or not. If not, then this gravitational law should be considered as invalid. Furthermore, the sense of energy varying as a function of time-based upon its influence needs to be confirmed. If not, one might jump to the conclusion equations are not valid and the gravity model for the Theory of Relativity may warrant some further investigations.

Libration points

The five Libration Points are found at the location where the effects from the primary bodies and the centrifugal motion are examined such that the attractions at the probe vanish where the first derivatives of the gravitational potential *V(x, y, z)* vanishes. Considering the time effect in *V (t, x, y, z)*, you essentially have the same five points but there is an addition based upon the time factor for the wave equation. This would be assumed as a sinusoidal type of change and should alter these locations. One could argue the reason for the unstable points is because they significantly move as a function of time. Similar comments are valid with the stable points.

For Jefimenko's law, which already satisfies a wave equation, there is an additional time term which includes an effect with velocity. Hence, the other transient terms are already buried in the other expressions in lieu of *V*. Thus, it warrants placing a probe at these locations and notes what occurs at these trajectories. There would be two separate issues. This involves first the motion and/or location of these points and then the impact of how the trajectories would be changed based purely on the gravitational law.

Phase space representation of the restricted three-body problem

The issue is how the probe's trajectory would be altered especially near the Libration Points. If chosen correctly the three-body problem can be formulated with a linear matrix as a viable solution. The approach is to look at some different gravitational laws based upon changing the eigenvalue solution or inclusion of a forcing function. This will provide some insights. A phase-space representation using eqn. (19), the equations of motion are:

$$\begin{aligned}&Let: u=\dot{x}, \dot{u}=\ddot{x};\ v=\dot{y},\ and\ \dot{v}=\ddot{y},\\&with: \dot{u}-2\omega v-\omega^2 x=-V_x\\&\dot{v}+2\omega u-\omega^2 y=-V_y\end{aligned} \quad (25)$$

These equations are rewritten as:

$$\frac{d}{dt}\bar{x}+\bar{\bar{A}}\circ\bar{x}=\bar{F},\ or\ \frac{d}{dt}\begin{bmatrix}x\\y\\u\\v\end{bmatrix}+\begin{bmatrix}0&0&-1&0\\0&0&0&-1\\-\omega^2&0&0&-2\omega\\0&-\omega^2&2\omega&0\end{bmatrix}\begin{bmatrix}x\\y\\u\\v\end{bmatrix}=-\begin{bmatrix}0\\0\\V_x\\V_y\end{bmatrix}. \quad (26)$$

Four eigen values for the *A* matrix are λ*i*= ± ω, ± ω; this includes repeated eigenvalues. The reason that the eigen value is important, as well as the forcing function, is that you are looking at a trajectory similar to:

$$\ddot{x}+2\varsigma\omega_n\dot{x}+\omega_n^2 x=F(t) \quad (27)$$

Thus, by noting the differences in the frequency and the damping factor, different responses should be seen from those with different factors. However, in this problem, there is no damping function. Thus we need to establish the eigenvalues and the forcing functions, which would result in different trajectories. The solution to this problem is:

$$\bar{x}(t)=\bar{x}_0 e^{\bar{\bar{A}}t}-\int_0^t e^{\bar{\bar{A}}(t-\xi)}\left(\bar{F}(\xi)-\bar{F}(0)\right)d\xi. \quad (28)$$

Initial conditions are included with the forcing function at zero time, which may not be zero. The exponential matrix is based on the eigenvalues such as:

$$e^{\bar{\bar{A}}t}=\beta_1\bar{\bar{I}}+\beta_2\bar{\bar{A}}+\beta_3\bar{\bar{A}}^2+\beta_4\bar{\bar{A}}^3. \quad (29)$$

One may replace the matrix with the eigenvalues. The identity matrix is the first term and there are four equations to be found. The values are:

$$\begin{bmatrix}e^{\lambda_1 t}\\e^{\lambda_2 t}\\e^{\lambda_3 t}\\e^{\lambda_4 t}\end{bmatrix}=\begin{bmatrix}1&+\lambda_1&+\lambda_1^2&+\lambda_1^3\\1&+\lambda_2&+\lambda_2^2&+\lambda_2^3\\1&+\lambda_3&+\lambda_3^2&+\lambda_3^3\\1&+\lambda_4&+\lambda_4^2&+\lambda_4^3\end{bmatrix}\begin{bmatrix}\beta_1\\\beta_2\\\beta_3\\\beta_4\end{bmatrix} \quad (30)$$

The ***βi*** terms are found by using the inverse matrix. Clearly, this logic demonstrates the impact of the eigenvalues **λ*i*** as well as the forcing function ***F*** upon the solution of the probe's eventual trajectory. With repeated eigenvalues, this yields:

$$\begin{bmatrix}e^{\lambda_1 t}\\e^{\lambda_2 t}\\te^{\lambda_1 t}\\te^{\lambda_2 t}\end{bmatrix}=\begin{bmatrix}1&+\lambda_1&+\lambda_1^2&+\lambda_1^3\\1&+\lambda_2&+\lambda_2^2&+\lambda_2^3\\1&+\lambda_1&+\lambda_1^2&+\lambda_1^3\\1&+\lambda_2&+\lambda_2^2&+\lambda_2^3\end{bmatrix}\begin{bmatrix}\beta_1\\\beta_2\\\beta_3\\\beta_4\end{bmatrix} \quad (31)$$

Note the effects of the repeated eigenvalues that can be expected with the value of the inverse matrix. This solution is a time of function in lieu of an angle orientation as would normally be considered for the two-body problem. Also, the exponential matrix provides a sinusoidal-like function of time multiplied by time, which should be very interesting.

Wave gravitational equation: The impact of involving a wave equation for gravity affects the forcing function as a time function. Here the equation used will look like:

$$\bar{x}(t)=\bar{x}_0 e^{\bar{\bar{A}}t}-\int_0^t e^{\bar{\bar{A}}(t-\xi)}\left(\bar{F}(\xi)T(\xi)-\bar{F}(0)T(0)\right)d\xi. \quad (32)$$

This solution does *not* impact the eigenvalues or the basic nature of the types of trajectories involved. However,

time will alter trajectories and implies the Libration Point locations will be different.

Jefimenko gravitational law: Extra terms using velocity with the correction speed of light factor is intrinsic to this law. The equations change in:

$$\frac{d}{dt}\begin{bmatrix}x\\y\\u\\v\end{bmatrix}+\begin{bmatrix}0 & 0 & -1 & 0\\0 & 0 & 0 & -1\\-\left(\omega^2+\frac{\mu}{r^3}(1-\frac{v^2}{2c^2})\right) & 0 & -\frac{2v^2}{3r^2c^3}(\mu_1r_1+\mu_2r_2) & -2\omega\\0 & -\left(\omega^2+\frac{\mu}{r^3}(1-\frac{v^2}{2c^2})\right) & +2\omega & -\frac{2v^2}{3r^2c^3}(\mu_1r_1+\mu_2r_2)\end{bmatrix}\begin{bmatrix}x\\y\\u\\v\end{bmatrix}=0 \quad (33)$$

The distance r might be confusing also recall r_1 and r_2 are functions of x and y. In this, r may be the distance of the probe measured from the barycentre of the two primaries. Despite this, there are many problems with this equation. The solution is not the same due to nonlinearity in the matrix with the distances from the probe to the two primaries as a function of both x and y. Hence the nonlinear effect exists. Moreover, this does not provide the solution previously mentioned. Despite this vector equation, we can make some judgments regarding a solution. The eigenvalues change from the matrix which indicates the equation will be different due to damping factors and frequency responses of the matrix. However, the coefficient for these additional terms is a very small number. If at high speed of the order of the speed of light, some terms would diminish or disappear. This trajectory as well as in the wave equation will provide notable differences compared to the Newtonian results.

If some of these variables are used as forcing functions, the matrix can be simplified as in:

$$\frac{d}{dt}\begin{bmatrix}x\\y\\u\\v\end{bmatrix}+\begin{bmatrix}0 & 0 & -1 & 0\\0 & 0 & 0 & -1\\-\omega^2 & 0 & -\frac{2v^2}{3r^2c^3}(\mu_1r_1+\mu_2r_2) & -2\omega\\0 & -\omega^2 & +2\omega & -\frac{2v^2}{3r^2c^3}(\mu_1r_1+\mu_2r_2)\end{bmatrix}\begin{bmatrix}x\\y\\u\\v\end{bmatrix}=\begin{bmatrix}0\\0\\+\frac{\mu}{r^3}(1-\frac{v^2}{2c^2})x\\+\frac{\mu}{r^3}(1-\frac{v^2}{2c^2})y\end{bmatrix} \quad (34)$$

Note this simplifies the eigenvalues of the previous matrix; however, there is a nonlinear forcing function in the matrix whereas it vanished in the previous matrix. Restrictive assumptions are made regarding the time factor for both matrices. Unfortunately, in this formulation, the final results cannot be established because the matrices contain functions of a radius and thus, time but all we can do is estimate trends in the eigenvalues.

Results

These results show some unusual trends regarding Newtonian gravitation. This is obvious because of maturity, where newer gravity relationships need to be determined. Some of this difference is the energy equations and results in either more kinematic components or time. There is a need to continue and exploit these relationships to more final conclusions. Successful results will be if Lagrangian points can be predicted, and how much of a variation exists with the Newtonian Libration Point locations. The most puzzling terms are the inclusion of time functions for wave equations and the use of a gravity model for relativity. Eqns. (15)-(18) are crucial because it demonstrates, as Jefimenko suggests, gravity includes distance as a relationship for velocity especially when using the initial-value problem with one of the boundary conditions. If there is a comparison between Jefimenko and relativity, more so than what Jefimenko claims, this use of a partial differential wave equation would be a valuable contribution.

Other functions of time should also look at eqn. (30). This may provide a different insight with gravity to integrate into space propulsion traveling with the far-reach. These are some important considerations for these situations. Although we are defining trajectories for these different gravitational laws, place the probe at the Libration Points and examine what occurs. Considerable evidence suggests there would be some variations in these locations as well as their movement as a function of time. For the wave equation, you have the same Libration Points but these points will shift as a function of time; the time factor is most likely sinusoidal with a large scatter. Thus it implies where gravity may grow and fall as a function of time. Energy might not be conserved but energy is also a function of time. Where do this energy and the momentum occur to provide conservation? In other words, an asteroid reaches the zone near the triangular point with a given initial value. The asteroid moves in locations that possess less than or equal to the energy of the asteroid. Thus the asteroid with its given energy can go into zones with lesser energy. Although the total energy may be conserved, values may change from one asteroid to another. Regarding Jefimenko's gravity, the Libration Points may be at different locations from Newtonian gravitation.

The effect of time should also be imbedded in these results in which Jefimenko's gravity and co-gravity equations are also wave equations supporting gravity waves. Here, there should be additional terms to include *(t)*. Thus, these trajectories should also move about these points. Similarly there is a relationship for these equations if a relationship can be established to account for the angular rotation rate with time.

Conclusion

This investigation stressed if different gravitational models exist, there should be some evidence to test and validate these separate models. Some models require increasing energy as a function of time. For the Libration Points, the existence of time due to using a Relativistic model, may explain the unstable collinear points. Some solutions for simple problems create mathematical issues which do not create simple or current trivial trajectory solutions. If anything, the use of Newtonian gravitation leads toward viable and simple mathematical representations; however, if we are to treat with dark matter and the unknowns in the far-abroad to discover unusual phenomenon, will we be capable of investigating other possible models that may be required? Thus, the need for placing a probe near these points is warranted to obtain data treating these points as anomalies.

References

1. Murad PA (2014) To Find Different Gravity Laws. STAIF II 2 Albuquerque, New Mexico.
2. Iorio L (2014) Gravitational Anomalies in the Solar System. Cornell University Library.
3. Lorentz HA, Weyl H, Minkowski H (1952) Einstein: The Principle of Relativity Dover Publications, USA.
4. Murad PA (2014) Gravitational Shocks, Shock Waves, As an Artifact for Exotic Space Propulsion. STAIF II, Albuquerque, New Mexico.
5. https://en.wikipedia.org/wiki/Theory_of_relativity.
6. Jefimenko OD (2006) Gravitation and Co-gravitation. Electret Scientific Company Star City, USA.
7. Jefimenko OD (1997) Electromagnetic Retardation and Theory of Relativity. Electret Scientific Company, Star City, USA.
8. Yoshida F, Nakamura T (2005) Size distribution of faint L_4 Trojan asteroids. The Astronomical journal 130: 2900-2911.
9. Nicholson, Seth B (1961) The Trojan asteroids. Astronomical Society of the Pacific Leaflets 8: 239-246.
10. http://www.minorplanetcenter.net/iau/TheIndex.html
11. Jewitt DC, Sheppard S, Porco C (2004) Jupiter's Outer Satellites and Trojan's. Cambridge University Press, India.
12. Sheppard SS, Trujillo CA (2006) A thick cloud of Neptune Trojans and their colors. Science 313: 511-514.
13. NASA's WISE Mission Finds First Trojan Asteroid Sharing Earth's Orbit 27 July 2011.

14. Connors M, Paul W, Christian V (2011) Earth's Trojan asteroid. Nature 475: 481-483.

15. Marzari F, Scholl H, Murray C, Lagerkvist C (2002) Origin and Evolution of Trojan Asteroids. Asteroids III.

16. Jewitt DC, Trujillo CA, Luu JX (2000) Population and size distribution of small Jovian Trojan asteroids. The Astronomical Journal 120: 1140-1147.

17. Murad PA (2013) Gravitational Shocks, Shock Waves, and Exotic Space Propulsion. 2013 SAP Journal and STAIF II.

18. Murad PA (2013) A Tutorial to Solve the 'Free' Two-Body Binary Pulsar Celestial Mechanics Problem. STAIF II.

19. Mannheim PD (2005) Alternatives to Dark Matter and Dark Energy. Prog Part Nucl Phys 56: 340-345.

20. Mannheim PD (2006) Solution to the ghost problem in fourth order derivative theories. Found Phys 37: 532.

21. Battin RH (1987) An Introduction to the Mathematics and Methods of Astrodynamics. AIAA Education Series, USA.

22. Baker RML (2017) Astrodynamics - Applications and Advanced Topics. Academic Press, New York.

18

Unitary Representations of the Translational Group Acting as Local Diffeomorphisms of Space-Time

Moffat J*, Oniga T and Wang CHT

Department of Physics, University of Aberdeen, King's College, Aberdeen AB24 3UE, UK

Abstract

We continue to develop further a new mathematical approach to the quantisation of general field theories such as general relativity and modified gravity. Treating quantum fields as fibre bundles, we discuss operators acting on each fibre that generate a 'Fibre Algebra'. The algebras of two types of operators are considered in detail, namely observables as generic physical variables and more specialised quantum operators suitable for describing particles, symmetries and transformations. We then introduce quantum states of these operators and examine their properties. By establishing a link between the commutativity and group cohomology of the translational group as a local gauge group, we show that this leads to unitary representations. of the local gauge group of diffeomorphisms under very general topological conditions; as well the construction of generalised symmetric quantum states invariant under this group action. Discussion of these results in the context of loop quantum gravity and other current theories highlights constraints on the local nature of space-time.

Keywords: Algebra; Lie algebra; Quantisation; Fibre bundle

Introduction

The development of perturbation methods in quantum field theory by Dyson, Feynman and others, was an essentially *ad hoc* answer to the problems generated by the process of taking renormalisation limits for point particles. The further development of the renormalisation group based on ideas of self-similarity, has led us to develop new approaches to the quantisation of paths in spacetime [1,2].

One way of eliminating the infinities of renormalisation is through assuming a form of supersymmetry (SUSY) in which additional 'spinor charge observables' are added to the Lie algebra of the Poincaré group of spacetime symmetries. In an irreducible representation of the resulting 'graded algebra', the centre of the algebra is reduced to multiples of the identity operator. It turns out that all particles in this representation have equal mass, as in the Wigner theory of irreducible representations of the Poincaré group and its Lie algebra [3]. However, the spins of the particles in this 'super'-representation are not fixed at a common value. In other words, the representation corresponds to a multiplet of particles, all with the same mass, but with different spins (equal numbers of fermions and bosons called partners and superpartners). At a given perturbation loop level, this generates, in a naïve application of the theory, equal mass contributions to the calculation but with loop contributions of opposite sign which cancel, giving a finite result. In practice, this symmetry is broken, resulting in different masses for the partners and superpartners, giving only approximate cancellation [4]. This relates to non-commutative spacetime and gravity through the superfield formalism [5-7].

Theories successfully describing the electroweak and strong forces take a similar approach. They use a group action acting on the Lagrangian which is symmetric. However, in contrast to global SUSY, this gauge group action is assumed to be local. If we apply this approach to SUSY, in the linear representation, the translation operator is represented by $P_\mu=\partial_\mu=\partial/\partial x^\mu$ corresponding to local translations in space-time between arbitrary coordinate frames i.e., a theory of 'supergravity' (SUGRA). We expected that this gravitational theory should involve a spin-2 graviton and its fermionic spin 3/2 gravitino companion particle. However, initial survey results at CERN using the TLAS detector [8], confirmed by recent postings, indicate no significant deviation from the Standard Model at 13 TeV, implying that the mass symmetry is badly broken; the mass-energies of the superpartners (if they exist) are possibly much larger than initially thought and limit their ability to control loop divergence. This leaves open the possibility of other potentially finite but more algebraic approaches including noncommutative geometry [9,10], and modified theories of gravity such as scalar-tensor gravitation [11-13].

The principle of local gauge symmetries inherent in interacting quantum fields is indeed a fundamental geometric origin of nonlinearities that are difficult to treat with a perturbation theory. Since the starting point of such a theory normally requires a background configuration with gauge fixing that could not only limit but also potentially affect the consistency of quantum dynamics. In contrast, gauge principle offers a natural underlying conceptual simplicity especially formulated in terms of modern fibre bundles, at the expense of introducing symmetry-redundant degrees of freedom.

The canonical quantisation of interacting systems without artificially restricting gauge symmetries have been pioneered by Dirac [14,15] in terms of an extended Hamiltonian formalism with 'first class' constraints that distinguish dynamical degrees of freedom while preserving and generating gauge transformations. When the extended Hamiltonian analysis is applied to general relativity, the Hamiltonian of gravity turns out to be completely constrained in terms of the Hamiltonian constraint and the momentum constraints [16]. The latter are also known as the diffeomorphism constraints since at least classically they generate spatial diffeomorphisms. There are two possible ways to 'process' these constraints. The first procedure uses them to generate the quantum dynamics of the spatial geometry using Dirac

***Corresponding author:** Moffat J, Professor, Department of Physics, University of Aberdeen, King's College, UK, E-mail: james.moffat@abdn.ac.uk

quantisation [17,18]. The second procedure is based on the reduced phase space method [19-21], where all constraints would be eliminated at the classical level in order to obtain a nonvanishing effective Hamiltonian that generates the evolution of certain unconstrained geometric variables. As discussed above, in this process, gauge is fixed classically and so it is not clear whether any resulting quantisation would preserve the full gauge properties of general relativity.

The conceptual and technical benefits of Dirac's constraint quantisation have recently been demonstrated through applications in gauge invariant interactions between generic matter systems and their weakly fluctuating gravitational environments leading to quantum decoherence of matter and spontaneous emissions of gravitons [22-27]. There have also been ongoing efforts to achieve full Dirac quantisation of general relativity in a background independent and non-perturbative approach using loop quantum gravity [28-31,17].

Our aim, following Dirac, is to initiate and develop a finite mathematical framework based on a new algebraic approach to quantum field theory and quantum gravity derived from foundational work by Moffat [32-36] and later developments including [1,2]. We specifically address the question of diffeomorphism invariant quantum states as the diffeomorphism constraints in the current formulation of loop quantum gravity could not be implemented through unitary representations. While this work does not invoke a particular approach to quantum gravity, such as loop quantisation, we discuss possible future prospects of incorporating our present development in the context of current quantum gravity research.

Algebras of Quantum Observables

A commutative operator algebra of observables is equivalent to the set of continuous functions on a compact Hausdorff space. This equivalence arises through the Gelfand transform of a quantum operator A, an observable in a set of commuting observables. This maps $A \rightarrow \hat{A}$ with $\hat{A}(\rho) \rightarrow \rho(A)$ and ρ a continuous complex valued homomorphism.

Since the Gelfand transform $A \rightarrow \hat{A}$ is an algebra isomorphism, the spectrum (the set of measurement eigenvalues) of the operator A, denoted (A), equals $\sigma(\hat{A})$. More generally we take as general context for this the modern algebraic theory of relativistic quantum fields [1,2,37,38] in which we denote the set of all â€˜local observablesâ€™ $O(D)$ as representing physical operations performable within the space-time constraints D. In particular, if $O(D)$ is weakly closed (i.e., a von Neumann algebra) and A is an observable in $O(D)$ then we assume that A is both bounded and self-adjoint, thus A generates an abelian subalgebra of $O(D)$ containing its spectral projections corresponding to the measurement process.

If $g \to \alpha_g$ is the corresponding mapping from the Poincare group P to the group of all automorphisms of $O(D)$ then $\alpha_g(O(D))=O(gD)$. We also assume that $D_1 \subseteq D_2$ implies $O(D_1) \subseteq O(D_2)$. The union of all the local algebras generates an algebra of all observables. The closure in the ultraweak operator topology generates the 'quasi-local' von Neumann algebra of all observables.

The idea behind [9,39] consists of finding non-commutative generalisations of the structure of the Einstein-Hilbert action which can be applied in order to gain greater insight into the Standard Model action. The approach taken introduces the theory of fibre bundles. A *fibre bundle* is a pair (E,π) where E is a topological space and π is a projection onto a subset M called the base space. For a point x in M, the set $F=\{y \in E \cap \pi^{1}(x)\}$ is the fibre at the point x, and it is assumed that the bundle is locally a product space about each point x. E is a vector bundle if each fibre has the structure of a vector space. In this case we can go further and define the fibre at x as consisting of all bases of the vector space attached to the point x; and define the group G of changes of basis, acting on the fibre space F. This is an example of a 'principal bundle'. A section or lifting s through the fibre bundle E with projection function π mapping E to the base space M is a continuous right inverse for π.

We thus assume that each local algebra $O(D)$ should correspond to the algebra of sections of a principal fibre bundle with base space a finite and bounded subset of spacetime, such as the interior of a bounded double-cone [38]. The algebraic operations of addition and multiplication are assumed defined fibrewise for the algebra of sections. Additionally, the group of unitaries $U(x)$ acting on the fibre Hilbert space is our gauge group. For example if the fibre $F(x)$ has a two-component basis, as a Hilbert space, of Clifford variables; then the spinor structure of left and right handed Weyl 2-spinors, and a gauge group including SU(2), form the basis of our current understanding of electroweak unification.

At the algebra level, this representation consists of the section algebra $\mathbf{A}(x)$ of 2×2 matrix operators acting on the fibre spinor space $F(x)=\pi^{-1}(x)$ with fibrewise multiplication and addition. It includes all such matrices and is thus a von Neumann algebra which is both norm separable (equivalently, finite dimensional) and with a trivial centre. We assume that this applies more generally and define such a norm separable matrix algebra with trivial centre to be a *quantum operator algebra*.

Let us now introduce a *quantum state of* $\mathbf{A}$ to be a norm continuous linear functional f acting on $\mathbf{A}$ which is real and positive on positive operators and satisfies $f(I)=1$. To distinguish these from the more restricted vector states of the form $\omega_x(A)=$ áx, Axñ, in Dirac notation, we may refer to f as a *generalised* quantum state. The state space is a convex subset of the Banach dual space $\mathbf{A}^*$ of $\mathbf{A}$ and thus inherits the weak* topology from $\mathbf{A}^*$. The unit ball of $\mathbf{A}^*$ is weak* compact and the state space is a closed and thus compact, convex subset of the unit ball. Application of Krein and Milman [40] shows that it is the weak* closed convex hull of its extreme points. These extreme points are called pure quantum states. A normal state is a state which is also continuous for the ultraweak operator topology. The ultraweak topology on the algebra $\mathbf{A}$ can in fact be derived from the h of $\mathbf{A}$ as the dual space of the set of normal states of $\mathbf{A}$, so that, formally, $(A,f)=(f,A)$ for each element A of $\mathbf{A}$ and each normal state f; i.e., A is a non-commutative Gelfand transform [41-44].

The Translational Group Action on the Fibre Algebra

The fibre algebra $\mathbf{A}(x)=\{A(x); A(x) \textit{ is an operator acting in the Hilbert space } F(x)\}$ is assumed to be a quantum operator algebra, and we select as our local gauge group the translational group T acting as a group of automorphisms of $\mathbf{A}(x)$. The translational group T is a subgroup of the Poincaré group P, a locally compact, connected Lie group, and is generated by the 4 basic linear translations in space-time plus 6 rotations and boosts. These define a smooth manifold, and each element g of P is of the Lie group form:

$$g = T(\varphi_1, \cdots, \varphi_n)$$

where T is the corresponding point on the manifold surface, and with n=10 parameters in our case. The origin of the manifold T (0,0,¼,0) corresponds to e, the group identity, denoted *id*.

Denote $\phi=(\phi_1,,\phi_n)$ and assume ϕ is a first order infinitesimal. This means that ϕ^2 is negligible, and that $T(\phi)$ is close to the origin e of the group. If U is a unitary representation of P on a Hibert space H then to first order:

$$U(g)=U(T(\varphi))=I+i\pi(\varphi^{\mu}t_{\mu})=I+i\varphi^{\mu}\pi(t_{\mu})$$

with π a mapping from the Lie algebra of generators t_{μ} of P into the set of linear operators acting on H. $U(g)$ a unitary operator implies that $\pi(t_{\mu})$ is a Hermitian linear operator , an observable.

Additionally, let ϕ become a second order infinitesimal so that ϕ^2 is not negligible but ϕ^3 is negligible. Since U is a group representation we have $U(g_1)U(g_2)=U(g_1g_2)$. From this expression, it follows that:

$$\pi[t_{\alpha},t_{\beta}]=if_{\alpha\beta}^{\gamma}\pi(t_{\gamma})$$

with f the antisymmetric structure constants of the group [45-48]. This then gives rise to a representation π of the corresponding Lie algebra generators. Consider now the translational group T acting locally as unitaries $U(g)$. From ref. [38] and our previous discussion, we know that if g is an element of the translational group T then g has the form $g=\exp(-iP(g)\cdot a)$ where a is the corresponding group parameter, $P(g)$ is the Lie algebra generator of g, and $P(g)\cdot a=P_{\mu}(g)a^{\mu}$. Hence under the homomorphism π, $U(g)=\exp(-i\pi(P(g)\cdot a))$.

Group Extensions and Mackey Theory

A group G is called an extension of a group C by a group B if C is a normal subgroup of G and the quotient group G/C is isomorphic to B. We assume in what follows that the group C is abelian. These assumptions imply [49] that the sequence:

$$0\to C\to G\to B\to 0$$

is a short exact sequence with the mapping from C to G being the identity map, and from G to B being (up to isomorphism) the quotient mapping from G to G/C. The converse is also true in the sense that any short exact sequence of groups can be identified with the extension of a group.

Given an abelian group K (say), and an arbitrary group Q, we assume that we have mapping γ and η such that : $Q\to\text{aut}(K)$ is a group homomorphism from Q to the group of all automorphisms of K and η: $Q\times Q\to K$ is a system of factors for Q and K. Together these two mappings can be used to define a multiplication on the set $K\times Q$ giving it a group structure:

$$(\xi_1,y_1)(\xi_2,y_2)=(\xi_1\gamma(y_1)(\xi_2)\eta(y_1,y_2),y_1y_2)$$

$\forall\xi_1,\xi_2\in K, y_1,y_2\in Q$. Defining $\theta:\xi\to(\xi,e)$ and $\theta:\xi\to(\xi,e)$ with e the identity of Q then $K\times Q$ with this group structure is an extension of K by Q, normally denoted as $K\eta Q$. It follows that the sequence:

$$0\to K\overset{\theta}{\to}K\times Q\overset{\delta}{\to}Q\to 0$$

is a short exact sequence. If Q and K are groups with borel measure structures, and η is borel mapping then we say that $K\eta Q$ is a borel system of factors for Q and K. Given these conditions we have the following result [50].

There exists in the group extension $K\eta Q$ a unique locally compact topological structure relative to which $K\eta Q$ is a topological group such that the identity map from $K\times Q$ (as a product borel space) into $K\eta Q$ is a borel measurable mapping. The map θ is a bicontinuous isomorphism from K onto a closed normal subgroup of $K\eta Q$. The isomorphism from $K\eta Q/\theta(K)$ onto Q defined by the mapping δ is also bicontinuous. Moreover, $K\eta Q$ is a separable group.

We will refer to the above as Mackey's theorem. An early version of this result was conjectured by Moffat [34] in relation to representations of the Poincare group.

If α:$g\to g$ is a group representation of the Poincaré group P as automorphisms of the quantum operator algebra **A** and there is for each $g\in P$ a unitary U_g implementing α_g then $U_g U_h$ and U_{gh} both implement $\alpha_{gh}=\alpha_g\alpha_h$. There is thus a phase factor $\lambda(g,h)$, a complex number of modulus 1, with $U_g U_h=\lambda(g,h)\, U_{gh}$. By considering in a similar way the group product $ghj=(gh)j=g(hj)$ these phase functions $\lambda(g,h)$ satisfy the requirement to be 2-cocycles of P.

No that if we set $V_g=v(g)U_g$ with $|v(g)|=1$ then for $g\in P$, V_g also implements α_g. We define a lifting as the choice of an appropriate $v(g)$ so that $g\to v(g)U_g$ is a group representation i.e., the 2-cocyles are trivial. In cohomology terms the expression $v(g)v(h)/v(gh)$ defines a coboundary. A local lifting is a lifting which is restricted to a neighborhood of the identity of the group.

The Poincaré group P is locally Euclidean and the underlying topology is locally compact. The Poincaré group was the first locally compact non-abelian group in which the later Mackey methodology was initially developed by Wigner in his theory of unitary irreducible representations of P. [3,44]. The group thus has an associated Haar measure m. The translation subgroup T and each of the rotation groups $R(x^{\mu})$ around a single spacetime axis x^{μ} are abelian path-connected subgroups of P.

A representation of group T as automorphisms of a quantum operator algebra $\mathbf{A}(x)$ is a group homomorphism α:$g\to\alpha_g$ from T into aut $(\mathbf{A}(x))$, the group of all automorphisms of $\mathbf{A}(x)$. This representation is *norm measurable* (resp. *norm continuous*) if the real valued mapping $g\to\|\alpha_g-i\|$ is Haar-measurable (rep. continuous) as a function defined on T. From the Appendix, a weakly measurable representation of this form is norm measurable.

We have the following result:

Let $g\to\alpha_g$ be a group representation of the translational group T as automorphisms of the quantum operator algebra $\mathbf{A}(x)$ which is norm measurable. Then the mapping $g\to\|\alpha_g-i\|$ is norm continuous i.e., the representation is norm continuous.

Proof. From the Appendix we can infer that the set W defined below is measurable, since the representation is norm measurable:

$$W=\left\{g\in T,\|\alpha_g-i\|\leq\frac{\varepsilon}{2}\right\}.$$

Each * automorphism α_g is an isometric linear map, thus if $g\in W$ then from the illustrative calculations in the Appendix, with $\alpha(g)=\alpha_g$ we have

$$\|\alpha_{g^{-1}}-i\|\leq\|\alpha_g-i\|.$$

Hence, if $g\in W$ then $g^{-1}\in W$, and if g, $h\in W$ then, exploiting again the isometry of each automorphism, from the Appendix we have:

$$\|\alpha_{gh}-i\|\leq\|\alpha_h-i\|+\|\alpha_{g^{-1}}-i\|\leq\varepsilon.$$

Thus we have

$$W^2\subset\{g\in T;\|\alpha_g-i\|\leq\varepsilon\}.$$

Now W contains the identity id in the Lie group T and if $g\in W$, g not equal to id, then $g=\exp(\theta k)$ where K is an element of the Lie

Algebra. Thus W contains intervals of the form $g(\theta)=\{\exp(\theta k),0<\theta\leq \theta_1\}$ and, it follows, must have strictly positive measure. Since T is locally compact, there is a compact set $C\subset W$. Then the set CC^{-1} contains a neighbourhood N of *id.* [45]

It follows that $N\subset CC^{-1}\subset W^2\subset\{g\in T;\|\alpha_g - i\|\leq \varepsilon\}$ and the mapping $g\rightarrow\|\alpha_g - i\|$ is continuous at the origin. By translation it is continuous everywhere on T.

Although not assumed in the above result, typically, these automorphisms $\alpha_g;g\in T$ will be inner, and they will be generated by a unitary (gauge) representation $g\rightarrow W_g$ in the fibre algebra A(x). In this case, since A(x) is norm separable, it follows that the subset $S=\{W_g;g\in T\}$ is also norm separable, and we can select a countable dense subset $\{W_{gn};n\in Z\}$ of S. Defining UW as the neighbourhood

$$UW=\left\{W_g;g\in T,\|W_g - I\|\leq\frac{\varepsilon}{2}\right\}$$

of the unitary gauge group in A(x) complementary to the neighbourhood W of the automorphism group, we can choose, for each group element g in T, a unitary W_{gn} with $\|W_g - W_{gn}\|\leq\varepsilon/2$. This implies that

$$\|W_{gn^{-1}g} - I\|=\|W_g - W_{gn}\|\leq\frac{\varepsilon}{2}$$

and thus $(gn)^{-1}\, g\in UW$ and hence $T=\bigcup_n(gn)(UW)$. From this, it follows that the neighbourhood UW also has positive measure.

A *unitary representation* the translational group T is defined to be a group homomorphism $U: g\rightarrow U_g$ from T into the group of unitary operators acting on the fibre Hilbert space $F(x)$. Let α be a group representation $g\rightarrow\alpha_g$ of T as automorphisms of the quantum operator algebra $A(x)$ acting on $F(x)$. Then a unitary representation $U: g\rightarrow U_g$ of T implements α if $\alpha_g(A)=U_g^*AU_g$ for all g in T and all A in $A(x)$.

We have the following key result:

Let α: $g\rightarrow\alpha_g$ be a representation of the translation group T as automorphisms of the quantum operator algebra $A(x)$, which is norm measurable. Then α is implemented by a norm continuous unitary representation $U: g\rightarrow U_g$ from T to the group of unitary operators acting on the Hilbert space $F(x)$.

Proof. If α: $g\rightarrow\alpha_g$ is a group representation of the translation group T as automorphisms of the quantum operator algebra $A(x)$, which is norm measurable, then we already know that α: $g\rightarrow\alpha_g$ is norm continuous. The translation subgroup T is thus is an abelian path-connected, hence connected, group with a norm continuous representation as automorphisms of the von Neumann algebra $A(x)$. Thus by the main theorem in Moffat the result follows [35].

This is an extremely powerful result due to its generality. To briefly discuss:

- The original result [34] is valid in fact for any norm continuous automorphic representation of a connected abelian topological group acting on a general von Neumann algebra. There is aways a norm continuous unitary implementation in this case.
- The specific application to a weakly measurable (hence strongly continuous representation of the translational group T, comes about because of the result that any such automorhic representation is actually norm continuous. Topology goes hand in hand with algebra.
- This generality allows its application to (for example) the weak closure of any GNS representation derived from a generalised quantum state.
- The key step in the original proof was the novel application of ultrafilter theory to generate a lifting from the carrier space of the centre to the carrier space of a set of comuting unitaries. this exploits the duality between an observable and its Gelfand transform.

From the our discussion so far, we can also deduce the following;

Suppose there is a homomorphism π from the Lie algebra of T into $A(x)$ and a corresponding representation τ of the translational group T as automorphisms of the quantum operator algebra $A(x)$. Then the representation τ: $g\rightarrow\tau_g$ can be implemented by a strongly continuous unitary representation.

Proof. A strongly continuous automorphic representation of the translational group T is weakly measurable. Thus there is a norm continuous unitary representation implementing τ. A norm continuous unitary representation is certainly strongly continuous, and the result follows.

Examination of the proof developed by one of us (Moffat) in ref. [35] on which this and the previous result depend, indicate in fact a two stage process at play.

Step 1. Exploit the spectral properties of a set of implementing unitaries to show that these unitaries commute, generating a commutative algebra.

Step 2. Use the Gelfand transform to select out a 2-cocycle set giving a unitary representation.

Discussion of Group Extensions and the Local Flatness of Spacetime

In our second result above, W_g has the form $W_g=\exp(-i\pi(P(g).a))$. It follows that $\{W_g;\forall\, g\in T\}$ is a commuting set if and only if the translational group T is abelian (i.e., spacetime is locally flat).

Let $N=Ker(\tau)=\{g\in T;\ \tau_g=id$ the identity$\}$. Replacing T by T/N we may assume that $\tau_g=\tau_h$ implies $g=h$.

Let O be the unit circle. Since $A(x)$ is a quantum operator algebra, the unitary group of the centre of $A(x)$ is isomorphic to O. It is then easy to see that $\Gamma=\{\lambda W_g;\ \lambda$ is in the unit circle, and $g\in T\}$ is an abelian subgroup of the unitary group of $A(x)$.

Define the mapping $\gamma\,(\lambda)=\lambda I$ which maps the unit circle O into the subgroup Γ, and the mapping $\eta(\,\lambda W_g)=g$ which maps the group Γ to the translational group T. Then the sequence:

$$0\rightarrow O\overset{\gamma}{\rightarrow}\Gamma\overset{\eta}{\rightarrow}T\rightarrow 0$$

is short exact. . Thus Γ is an extension of O by T. We can in fact identify the group Γ with [45]

$$\eta'(\lambda,g)=\eta(\lambda W_g)=g$$

Since the Hilbert space $F(x)$ on which the unitaries act is separable, by assumption: it is sufficient to demonstrate that the mapping $g\rightarrow\langle\xi, U_g\,\xi\rangle$ is a borel measurable mapping for all $g\in T$, $\xi\in F(x)$. From the discussion above we can identify the group Γ with the borel system of factors $O\eta'T$.

Let the mapping J denote the identification $O\times T\leftrightarrow\Gamma\leftrightarrow O\eta'T$ then J is a borel mapping by Mackey's theorem, from our earlier discussion of group cohomology. Thus the mapping $W: g\rightarrow W_g$ is a borel measurable mapping from the translational group T into the group Γ. Hence the mapping $g\rightarrow\langle\,\xi,\, W_g\xi\,\rangle$ is a measurable mapping for any ξ in the Hilbert space $F(x)$. In addition, the character $\sigma: \Gamma \to C$ is a continuous mapping, thus $\sigma \circ W : g \to \sigma(W_g)$ is borel. Combining these together it follows that the mapping $g \to \langle \xi, U_g \xi \rangle = \sigma(W_g)^* \langle \xi, W_g \xi \rangle$ is borel measurable.

Invariant States of the Algebra of Observables

In this section we change the focus to recall the overall structure of the von Neumann algebra R of all observables, consisting of the weak closure of the sets of all finite unions of local algebras $O(D)$. The structure of R is much more general than that of a quantum operator algebra. For example, R may well have a non-trivial centre. We further investigate the structure of R by examining its invariant states. Each section algebra consists of operators $\{A \in O(D); A|_{F(x)} = A(x)$ *for all* $x \in M$ which are quantum fields, and each such section algebra is a subalgebra of $\mathbf{R}$.

Let f be a generalised quantum state of the algebra $\mathbf{R}$, and $\alpha: g \to \alpha_g$ be a group representation of the Poincaré group P as automorphisms of $\mathbf{R}$. The state f is said to be Poincaré invariant if $f(\alpha_g(A)) = f(A)$ for all $g \in P$ and A in R.

We call a *T-symmetry state*, a quantum state of **R** which is invariant under the translational group T. In ref. [2] we characterised the condition for such states to exist. As discussed earlier, there is a vector of Lie algebra generators of these translations a^μ denoted by the abstract vector P_μ. Acting on **R**, a translation a^μ maps the local algebra $O(D)$ into the local algebra $O(D+a^\mu)$. It is worth noting here that the vacuum state of a quantum system (the state of lowest energy) on the Minkowski flat background is in general translation invariant with constant a^μ [38]. Since translational group T is the local gauge group for diffeomorphisms, in this sense we identify T-symmetry states as diffeomorphism invariant states. Such symmetry states turn out to be not uncommon, as we now prove.

This takes us to the final result:

Let $\tau : g \to \tau_g$ be a group representation of translational group T as automorphisms of the algebra of observables **R** which is weakly measurable. Then there is always at least one quantum state of the algebra which is a T-symmetry state.

Proof. Let f be a state of **R** and define the 'induced' transformation $v_g(f) = f \circ \tau_g$. This means that $v_g(f)$ is also a state and $v_g(f)(A) = f(\tau_g(A))$.

Since the subgroup T is abelian, and the mapping $g \to v(g)f$ is a group homomorphism, the set $\{v(g);\, g \in T\}$ is a continuous group of commuting transformations of the dual space $\mathbf{R}^*$. If f is a quantum state of the algebra then define E to be the weak * closed convex hull of the set $\{v(g)f; g \in T\}$. Then E is a weak * compact convex set and each $v(g)$: $E \to E$. By the Markov-Kakutani fixed point theorem it follows that E has an invariant element. Thus each quantum state generates a symmetry state which is in the closed convex hull of the set of all translations of f [46].

Concluding Remarks

Motivated by modern diverse developments of quantum field theory, quantum gravity, and modified gravity, including supersymmetry, supergravity, noncommutative geometry, scalar-tensor gravity, and loop quantum gravity, we have initiated a new algebraic approach to background independent quantisation of gravity and matter as a potential unified approach to quantising gauge theories. Here we specifically address the existence and construction of diffeomorphism invariant quantum states. For this purpose, an analysis of the group cohomology of the translational group as the local gauge for diffeomorphisms is carried out that proves to provide useful tools for the presented analysis.

Crucially, by establishing a link between the commutativity and group cohomology of the translational group, we show that a strongly continuous representation of the translational group can be implemented by a norm, hence strongly continuous unitary representation. This result leads to the construction of quantum states invariant under the action of the translational group, as the local gauge group of diffeomorphisms, with unitary representations. Applied to the GNS representation of an invariant state, many of which exist, as we have shown, this result leads to the construction of quantum states invariant under the action of the translational group, as the local gauge group of diffeomorphisms, with strongly continuous unitary representations.

It will be of interest to apply our framework to specific quantum gravity programmes involving further gauge properties, such as spin symmetry in standard loop quantum gravity and additional conformal symmetry, which may involve additional scalar fields, in conformal loop quantum gravity [17,47-49,31]. The details of such extended investigations and resulting invariant states relevant for the non-perturbative quantisation of general relativity, modified gravity, and gauge theories [28,30] will be reported elsewhere [50-56].

Acknoledgment

We are grateful for financial support to the Carnegie Trust for the Universities of Scotland (T.O.) and the Cruickshank Trust and EPSRC GG-Top Project (C.W.).

References

1. Moffat J, Oniga T, Wang CHT (2017) A New Approach to the Quantisation of Paths in Space-Time. J Phys Math 8: 232. doi: 10.4172/2090-0902.1000232.
2. Moffat J, Oniga T, Wang CHT (2017) Ergodic Theory and the Structure of Non-commutative Space-Time. J Phys Math 8: 229. doi: 10.4172/2090-0902.1000229.
3. Wigner E (1939) Unitary Representations of the Inhomogeneous Lorentz group. Annals of Mathematics 33: 1074-1090.
4. Balin D, Love A (1996) Supersymmetric Gauge Field Theory and String Theory. Taylor and Francis, London.
5. Salam A, Strathdee J (1978) Supersymmetry and Superfields. Fortschr Phys 26: 57-142.
6. Ferrara S (2016) Linear Versus Non-linear Supersymmetry in General. J High Energy Phys 4: 65.
7. Nakayama K, TakahashiF, Yanagida T (2016) Viable Chaotic I nation as a Source of Neutrino Masses and Leptogenesis. Phys Lett B 757: 32.
8. Potter C (2013) Supersymmetry Searches with ATLAS, Overview and Latest Posted Results LHC Seminar CERN.
9. Connes A (1994) Noncommutative Geometry. Academic Press.
10. Seiberg N, Witten E (1999) String Theory and Noncommutative Geometry. J High Energy Phys 4: 101-120.
11. Wang CHT (2011) Parametric instability induced scalar gravitational waves from a model pulsating neutron star. Phys Lett B 705: 148-151.
12. Wang CHT (2013) Dynamical trapping and relaxation of scalar gravitational fields. Phys Lett B 726: 791-794.
13. Wang CHT (2016) A consistent scalar-tensor cosmology for inflation, dark energy and the Hubble Parameter. Phys Lett A 380: 3761-3765.
14. Dirac PAM (1958) The Principles of Quantum Mechanics. Oxford Univ Press Oxford.
15. Dirac PAM (1964) Lectures on Quantum Mechanics. Yeshiva University, New York.
16. Wang CHT (2005) Conformal geometrodynamics: True degrees of freedom in a truly canonical structure. Phys Rev D 71: 1-10.
17. Wang CHT (2005) Unambiguous spin-gauge formulation of canonical general relativity with conformorphism invariance. Phys Rev D 72: 087501.

18. Marsden JE, Ratiu TS, Scheurle J (2000) Reduction theory and the Lagrange-Routh equations. J Math Phys.

19. Marsden JE, Weinstein A (2001) Quantization of singular symplectic quotients. Springer.

20. Sniatycki J, Weinstein A (1983) Lett Math Phys 7: 155.

21. Pons JM, Shepley LC (1995) Evolutionary laws, initial conditions and gauge _xing in constrained systems. Classical Quantum Gravity 12: 1771.

22. Oniga T, Wang CHT (2016) Quantum gravitational decoherence of light and matter. Phys Rev D 93: 044027.

23. Oniga T, Wang CHT (2016) Spacetime foam induced collective bundling of intense fields. Phys Rev D 94: 061501.

24. Oniga T, Wang CHT (2017) Quantum dynamics of bound states under spacetime uctuations. J Phys Conf Ser. 845: 012020.

25. Oniga T, Wang CHT (2017) Quantum coherence, radiance, and resistance of gravitational systems.

26. Quinones DA, Oniga T, Varcoe BTH, Wang CHT Quantum principle of sensing gravitational waves: From the zero-point fuctuations to the cosmological stochastic background of spacetime.

27. Oniga T, Mansfield E, Wang CHT, Cosmic Quantum Optical Probing of Quantum Gravity Through a Gravitational Lens.

28. Ashtekar A, Rovelli C, Smolin L (1991) Gravitons and loops. Phys Rev 44: 1740.

29. Ashtekar A (2005) Gravity and the quantum. New J Phys 7: 198.

30. Thiemann T(2007) Modern Canonical Quantum General Relativity. Cambridge Univ Press Cambridge.

31. Campiglia M, Gambini R, Pullin J (2017) Conformal loop quantum gravity coupled to the standard Model. Classical Quantum Gravity 34: LT01.

32. Moffat J (1973) Continuity of Automorphic Representations. Math Proc Camb Phil Soc 74: 461.

33. Moffat J (1974) On Groups of Automorphisms of the Tensor Product of von Neumann Algebras, Math. 34: 226.

34. Moffat J (1974) Groups of Automorphisms of Operator Algebras PhD Thesis. University of Newcastle upon Tyne.

35. Moffat J (1975) Connected Topological Groups Acting on von Neumann Algebras. J London Math Soc 2: 411.

36. Moffat J (1977) On Groups of Automorphisms of von Neumann Algebras, Math. Proc. Camb. Phil. Soc 81: 237.

37. Haag R (2012) Local Quantum Physics: Fields, Particles, Algebras. Springer.

38. Araki H (2009) Mathematical Theory of Quantum Fields. Oxford Univ Press, Oxford.

39. Chamseddine A, Connes A, Marcolli M Gravity and the Standard Model with Neutrino Mixing. Corenl University Library.

40. Krein M, Milman D (1940) On extreme points of regular convex sets. Studia Mathematica.

41. Kirrilov A (2008) An Introduction to Lie Groups and Lie Algebras. Cambridge Univ Press, Cambridge.

42. Kurosh A (1956) The Theory of Groups. Chelsea Publishing Company, New York.

43. Mackey G (1957) Ensembles Boreliens et Extensions des Groupes. J Math Pure et Appl 36: 171.

44. Mackey G (1957) Borel Structure in Groups and their Duals. Trans Amer Math Soc 85: 134.

45. Hewitt E, Ross K (1963) Abstract Harmonic Analysis. Springer.

46. Reed M, Simon B (1980) Functional Analysis. Academic Press, London.

47. Wang CHT (2006) New phase of quantum gravity. Phil Trans R Soc A 364: 3375.

48. Wang CHT (2006) Towards conformal loop quantum gravity. J Phys Conf Ser 33: 285.

49. Veraguth OJ, Wang CHT (2017) Immirzi parameter without Immirzi ambiguity: Conformal loop quantization of scalar-tensor gravity. Cornel University Library.

50. Fischer AE, Marsden JE (1979) Isolated Gravitating Systems in General Relativity. J Ehlers North-Holland, Amsterdam.

51. Isenberg J, Marsden JE (1982) A slice theorem for the space of solutions of Einstein's equations. Phys. 89: 179.

52. Isenberg J (1987) Parametrization of the Space of Solutions of Einstein's Equations. Phys Rev Lett 59: 2389.

53. Gotay MJ (1986) Constraints, Reduction and Quantization. J Math Phys 27: 2051.

54. Douglas RG (1972) Banach Algebra Techniques in Operator Theory. Academic Press, New York

55. Loomis LH (2011) Introduction to abstract harmonic analysis. Courier Corporation.

56. Wang CHT (2005) Nonlinear quantum gravity on the constant mean curvature foliation. Classical Quantum Gravity 22: 33-35.

19

Gravitational Shocks, Shock Waves, and Exotic Space Propulsion

Murad PA*

Morningstar Applied Physics, LLC Vienna, VA 22182, USA

Abstract

Several different gravitational laws can be derived that fall within a spectrum that covers an extreme from elliptical partial differential equations for Newtonian gravitation to hyperbolic or wave equations demonstrated by other laws from Jefimenko to Einstein's relativity. If each of these equations is valid for specific conditions at a considerable distance from space, then there is an interesting counterpoint with similar physical mathematical behavior that may be analogous between gravity and say, fluid dynamics. Here, Newtonian gravitation appears mathematically similar to subsonic flow while the other laws mathematically are comparable to supersonic flow. This evaluation advocates identifying experiments that may observe creating an inhomogeneous gravitational field that mathematically result in producing gravitational shocks or waves embedded in regions with merging different distinct strength gravitational fields. If such shocks are feasible, exploiting these gravitational shocks in a propulsion system would create thrust to possibly shadow or repel gravitation. Variations in energy to generate mass may create these distinct and separate gravitational fields for these gravitational shocks. Such an investigation is warranted for mankind to exploit developing this embryonic technology that potentially may develop an exotic space propulsor capable of moving faster than light (FTL).

Keywords: Gravitational waves; Gravitational laws; Elliptical/hyperbolic equations; Fluid dynamics

Nomenclature

B=magnetic field

c=the speed of light

E=electric field

e=electric charge

F=thrust or force

G=gravitational constant

g=gravity field

J=current

m=mass

t=time

W=weight

V=velocity or volume

u, v, w=velocity in the x, y, z directions

x, y, z=coordinate directions

Symbols

φ, Φ=potential

ρ=density

ω=rotation rate

Subscripts

e=electric

m=magnetic

Introduction

This paper will discuss some different principles that offer insights as well as discover possible views that may exist but are yet to be established. One may view this as fantasy but clearly we do not fully understand gravitational models especially at considerable distances beyond our current grasp because of too many uncertainties. For example, consider Dark Matter that may be altered by imploring a different gravitational law other than Newtonian gravitation. If these approaches are relevant, they may possibly allow for the development of a warp drive engine. The desire to traverse across the cosmos is unfortunately only a dream rather than a concrete reality.

Current realities using conventional propulsion technologies to cover such vast distances are basically impractical. What are some of the requirements to develop a warp drive? Clearly conventional rocketry where mass is judiciously expelled using either chemical or nuclear energy to generate long-term high-speed thrust is insufficient to meet the need to go to the far- beyond. There may not be an ample amount of mass to reach these high enough speeds to reach near faster than light, to economically reach a long distance necessary to exploit some cargo or information, and then return. Thus mankind would be constrained on this beautiful blue marble rock forever to only explore in the near-field solar system without outwardly discovering ventures that would create untold intellectual growth beyond our own imagination.

Are there realistic propulsion technologies to potentially achieve these incredible distances within a feasible time period? The Alcubierre drive, the Krasnikov tunnel [1,2] and other means offer some interesting and tantalizing possibilities. This also includes some work by Recami [3] and his associates investigating a simulation of using quantum tunneling by experiments with classical evanescent waves, which they claim were predicted to be Superluminal on the basis of

***Corresponding author:** Murad PA, CEO, Morningstar Applied Physics, LLC Vienna, VA 22182, USA, E-mail: pm@morningstarap.com

extended relativity. Recami [4] defines a waveguide with a segment of a 'photonic barrier' in an undersized waveguide (evanescence region). These computer simulations are based on Maxwell equations only. In other words, they claim verifying the actual possibility of superluminal group velocities, without violating the so-called (naïve) Einstein causality. Moreover, the phenomenon of a one-dimensional non-resonant use of tunneling is analyzed through two or more successive (opaque) potential barriers, separated by intermediate free regions exploiting solutions to the Schroedinger equation [5].

The problem, despite all of these possibilities, is that most of these approaches are purely theoretical constraints with only some limited or none-existent experiments. To examine physical phenomena, these events may still *not* offer a realistic approach to move from one point to another amongst the heavens. Clearly limiting technologies let alone defining suitable experiments could definitely be the problem. Obviously mankind has not performed enough research to look at specific experimental problems for some understanding to meet these possibilities.

Two specific issues are also of concern. That is to reach beyond space either moving faster than the speed of light or to travel back and forward into time. Is this a technology transition as simple as changing a switch or does this require developing a separate apparatus with disjointed separate devices to perform these unique functions? Gertsenshtein attempted to answer these questions to look at induced singularities in the metric equations for Einstein's field equations [6-12].

The author mentioned in a technical paper about the possibility of living in a world that includes an additional dimension. We normally think of new dimensions in our mind only as a linear extension as an additional spatial extension of a coordinate system [13]. Here, we coincide and converge in a multi-verse with both linear time and exponential time simultaneously; linear time for the present while the exponential time is used for quickly reaching out into either the past or the future.

Then, there is time reversal and what do we do about it? We could believe that to prevent a chaotic universe, there must be a cause and an effect where the growth of entropy provides the 'arrow of time' to move forward. How can you use this realistically in a meaningful space-time metric with acceptable technology?

The actual physical mechanizations for these theoretical hypotheses to create these effects, however, are lacking. How can you induce a singularity in the space-time continuum? If we are talking about gravitational effects, is the problem simple enough to use several different weight scales? No. This is a trivial response and the problem is far more complex. The best we could do is to discover some 'new' possible physical phenomenon.

The issue clearly lies within the framework that we need to look at these problems but also to explore a different set of scientific instruments with specific sensitivities to support such investigations. For example, one may ask fifty years ago why a satellite would have any value incorporating an infrared or ultra violet sensor when you can easily examine the earth using only visual cameras. Why would you need or obtain more information than you already receive from visual sensors?

The differences obviously reveal details of different worlds that are totally unexpected that include information from a host of coupled sensors. This provides a better understanding of the Earth. In other words, one type of sensor may have shortcomings with a particular database whereas another sensor operating within a different spectrum may complement an initial sensor to further exploit data or cancel the initial sensor's vulnerabilities.

Such new types of sensors [14,15] will be unusual regarding finding gravitational waves per Baker. They will need to identify specific characteristics that clearly as of yet, are totally unknown. A new stage of investigations to just identify these instruments is required as a separate effort.

What is needed for instrumentation to go faster than the speed of light as well as go either forward or backward in time? What is a device that can calibrate such a 'solar' calendar to identify the past? The only way would be to locate the orientation of the Earth triangulated with a navigational relationship to the sun and adjacent reference stars from say, using pulsar timing, to note the location for a specific timeline in the past, present and future.

How do you use the technology, instrumentation or implementation in such a spacecraft? These are all very sophisticated questions that need examination for the future. The problem for moving faster than light is different in that you could measure different energy levels. However these energy levels would not give you unique values unless you have specific knowledge that the target was either slower or faster than the speed of light. This result concerning energy is nonlinear and self-defeating demonstrated in Figure 1 from Meholic [16]. This shows symmetry revealing several different energy levels that are the same value at different velocities. Another interesting idea would be to develop a space-time curvature sensor but this too has limitations. Obviously the point is to realistically look at gravitation and space travel with a body of instrumentation that currently exceeds our technological depth.

Clearly the problem is that before we can talk about near light speed, instrumentation technology must be developed. This is addressed only with the concept of examining gravitational problems, which is the focus of this effort.

Discussion

We need some basic ground-rules similar to ideas with Recami

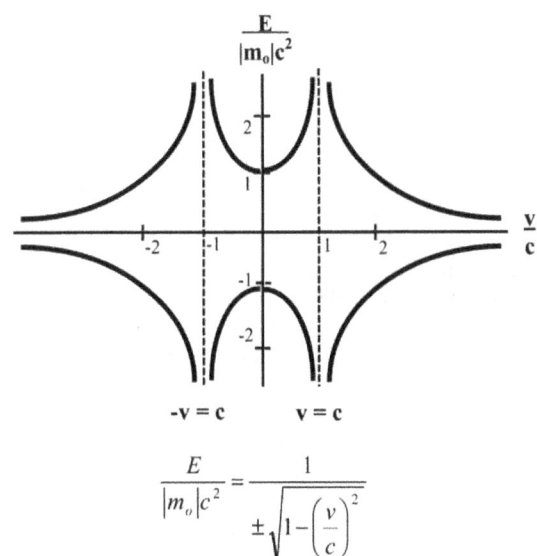

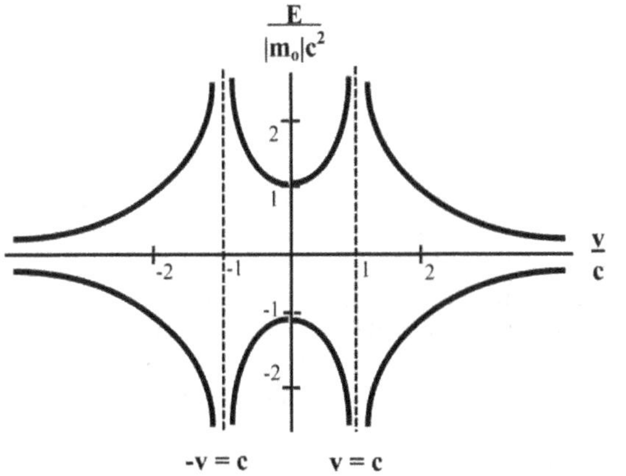

$$\frac{E}{|m_o|c^2} = \frac{1}{\pm\sqrt{1-\left(\frac{v}{c}\right)^2}}$$

Figure 1: Normally the axis is used only above energy value but in reality, a possibility is also negative.

to treat with this problem regarding assumptions, and premises [17]. There are basic natural postulates with some caveats such as:

- The laws of electromagnetism and mechanics are valid not only for a particular observer but for the whole class of the "inertial" observers.
- Space and time are homogeneous and space is considered as isotropic. The exception is that the propulsion system can induce anisotropic or non-homogeneous effects to produce thrust.
- The total energy of an ordinary particle increases when its speed **v** increases except as shown by Figure 1. Energy tends to infinity when **v** tends to **c**. Thus, infinite forces occur for a particle or spacecraft to reach the speed **c**. This generates the popular opinion that speed **c** can be neither achieved nor overcome. However, as speed **c** photons exist which are born, live, and die always at the speed of light (without any need for accelerating from rest to light speed), so we will assume that objects can exist that are endowed with speeds **v** possibly larger than **c**.
- Let us add that, still starting from the above Postulates, the theory of relativity can be generalized to accommodate also Superluminal objects; such a non-restricted version of SR is sometimes called "extended relativity". (Murad [18] actually performs an analysis that extends the SR viewpoint where a particle can go faster than light without terms that are imaginary).

One real approach other than chemical energy for space travel is to raise questions about either generating a small amount of thrust such as an ion engine or using a Woodward Machian drive for an infinite amount of time. The other option is to look into generating fields that interact inherent to space. If we agree with Dirac's approach where particles are instantaneously created and annihilated continually, we would need to capture the fields of these particles when they achieve electric, magnetic, and gravitic fields that are also first created and then, somehow undergo the decaying portion of the particle's cycle of life and death. This demands understanding the frequency switching level of the ZPF, which may be a far smaller wavelength than we can possibly measure and then determine a switching technology to electronically capture the time rise or decay. This does not appear promising.

Let us explore gravity as the major concern. To do this, we need some understanding of inertia. If we use the Puthoff approach, a particle's inertia is an electromagnetic force that acts against the zero-point field [19]. The only problem is at constant speed, the force creating inertia would still exist and this violates Relativity as well as Newtonian law ignoring other forces acting on the particle. This argument would be acceptable if inertia disappeared when the particle moves at constant velocity without any other forces. Let us assume that the electromagnetic attraction acts omni-directional at constant speed with the ZPF where the attractions generally cancel in all directions similar to Figure 2. However, if acceleration or deceleration occurs, the inertia vector operates in a definable manner directly opposite to motion. This might be an acceptable option. Other explanations about inertia exist but are not as comprehensible in this author's views as Puthoff's. If we look at fields, we could talk about a Poynting motivator for a candidate Propulsor [20] where force is

$$\bar{F} = e(\bar{E} + \bar{v} \times \bar{B}) + \frac{1}{4\pi}\bar{E} \times \bar{B} = e\bar{E} + \bar{J}_e \times \bar{B} + \frac{1}{4\pi}\bar{E} \times \bar{B} \qquad (1)$$

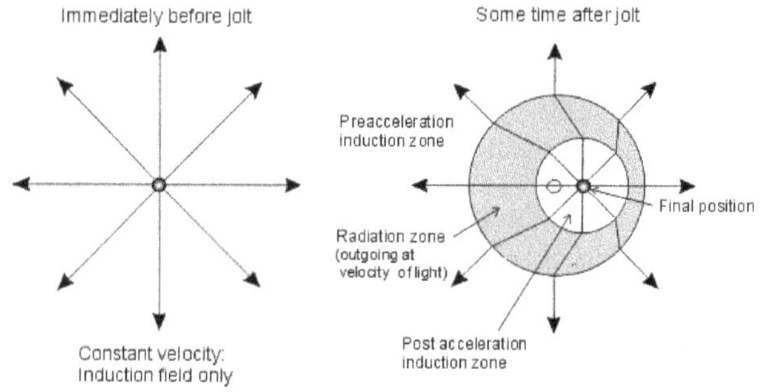

Figure 2: A directional vector form can use inertia to define what occurs during changes in acceleration.

This would ignore the first and second terms in the RHS of the equation, which also depends upon the charge and current that is dependent upon the presence of electrons or ions that produce the charge. Significant amounts of energy for the last radiation term would be required. Such a motivator, for example, depends strictly upon the Poynting contribution [20] in the last term of the RHS. This radiation term is not usually included; however, it will create a force. This would require considerable amounts of electrical and magnetic fields with sufficient orientation to generate thrust. The major point is that the device will most likely require a significant amount of electrical energy so let us presume that a spacecraft possesses an on-board hybrid nuclear reactor involved that uses a neutron generator or an aneutronic Fusor to drive a fissile reactor bed. This system uses the best conceptual efficiency to generate nuclear energy [21]. Coolants from the basic reactor would generate electricity to create these terms as well as induce the Poynting field.

Gravitational issues

A controversy exists where current understandings of some events in space do not satisfy gravitation on a galactic scale. Thus, there may be a need for either a new gravitation law or a different kind of matter. Dark matter that is essentially invisible meets some of this need and has supposedly no electric or magnetic charges. The sense is that dark matter compensates for resolving cosmological problems. An alternative view suggests that the standard Newtonian gravitational law should be altered and reconsidered to adjust for these events without resorting to using dark matter magic. Still another alternative suggests that rotation of a celestial body may alter the gravitational field for highly rotating neutron stars resident with pulsars [22]. The issues with Dark Matter involve:

- First, Einstein's field equations fail to explain the dark matter and dark energy, and the equations are inconsistent with the accelerating expansion of the galaxies [23-25].
- Second, we can prove that there is no solution for Einstein's field equations for the spherically symmetric case with cosmic microwave background (CMB).
- Third, from Einstein's equations, it is clear that R=4πG c4 T, where T=gijTij is the momentum energy density tensor. Discontinuities of the tensor T give rise to the same discontinuities of the theory of dark matter and dark energy curvature as well as the discontinuities of space-time. This is certainly an inconsistency that needs to be solved.
- Fourth, it has been observed that the universe is highly non-homogeneous as indicated by e.g., the "Great Walls", filaments and

voids. However, the Einstein field equations do not appear to offer a good explanation of this inhomogeneity.

If additional mass that creates gravity could provide an explanation that generates the required forces to occur, there is still yet another possibility. What if matter exists that applies repulsive or anti-gravitic effects in the opposite direction? By symmetry, this is also a reasonable hypothesis for dark matter. Again, the solution is most likely a realistic change in the gravity model, which we may eventually discover when a spacecraft goes further past the Oort cloud.

If dark matter exists, one question is how would this be altered to create propulsion at high speeds? If there are no charges, dark matter could *not* possibly be excited by either electric or magnetic fields expelled at high speeds necessary to provide thrust [25,26]. It is not intuitively obvious that dark matter propulsion has any feasibility at all so we can possibly ignore this capability unless dark matter can be chemically reconstituted. The other part of the problem is if a spacecraft moves at high speed, how would the craft structurally survive in collision with dark matter? This could be a very serious structural problem that further prevents mankind going into space at superluminal speeds. Let us assume that understanding galactic problems have solutions based upon the gravitational laws in lieu of dark matter.

Analysis

If we look at gravitational relations, Newtonian gravitation is the simplest representation. Here, gravity is an inverse distance law which when involved as a force becomes a gradient that is an inverse square law dependent between two attractive bodies, say a spacecraft and a celestial body. If the mass or gravity qualifies with inertia that is altered, then this force between these two bodies would result in an adjustment. If mass could become a negative quantity, then we would experience repulsion.

The attractive force becomes:

$$\overline{F} = G\frac{m_1 m_2}{r_{1-2}^2} \tag{2}$$

Where **G** is a gravitational constant, the masses are for the separate bodies, and ***r*** is the difference between the two bodies and the larger mass body acts immoveable with respect to the smaller spacecraft mass.

Questions regarding gravity models

Gravity obeys certain laws and maybe we are not taking full advantage of these capabilities. We need to walk away and look at these problems with a different lens to see this challenge. Our understanding about gravity is only due to the local environment in the region of our planet and the near-term solar system. Clearly the Dark Matter hypothesis already stresses these capabilities and suggests the need for new gravitational laws. Moreover, if a study in relativity, gravity warrants a change from Newtonian gravity if gravitational waves exist. These different gravitational laws are shown in Table 1 which displays elliptical and wave equations.

In these gravitational laws, these may be considered as speculative but are offered to consider the possibility for creating gravitational waves as well as the far-field. One point should be considered for these wave equations. If other than taking part in a supernova, the transient term of these wave equations has a coefficient of $1/c^2$, which is exceptionally small. In other words, the transient term is likely insignificant regarding the temporal term with transient derivatives.

Gravity law	Assumptions	Gravitational rule
Newtonian gravitation	$\nabla \times \overline{g} = 0.$ and $\nabla.\overline{g} = -4\pi G\rho_g.$	$\overline{g} = -\nabla\phi and: \nabla^2\phi = 4\pi G\rho_g; where: g \approx 1/r^2.$
Four-Derivative theories	$\phi(r) = 1 - 2m/r + ar + br^2,$ $\phi(r) = g^{oo} = (1-6bc)^{1/2} - \frac{2b}{r} + cr + \frac{d}{3}r^2.$	$\overline{g} = -\nabla\phi(r).$
Winterberg's rule	$\nabla.\overline{g} = -4\pi G\rho_g = 2\omega^2,$ Where $\rho_g = -\frac{\omega^2}{2\pi G}.$	$\overline{g} = -\nabla\phi \ and: \nabla^2\phi = -2\omega^2; \ where: g \approx 1/r^2.$
Jefimenko's gravity and co-gravity	$\nabla \times \overline{g} = -\frac{\partial \overline{k}}{\partial t}; \nabla.\overline{g} = -4\pi G\rho_g; \nabla.\overline{k} = 0.$ and: $\nabla \times \overline{k} = -\frac{4\pi G}{c^2}\overline{J_g} + \frac{1\partial \overline{g}}{c^2\partial t}.$	$\frac{1\partial^2 \overline{g}}{c^2\partial t^2} - \nabla^2\overline{g} = 4\pi G[\nabla.\rho_g + \frac{1\partial \overline{J_g}}{c_g^2\partial t} - \frac{\nabla \times \overline{J_c}}{c}],$ $\frac{1\partial^2 \overline{k}}{c^2\partial t^2} - \nabla^2\overline{k} = 4\pi G\left[\frac{\nabla.\rho_g}{c^2} - \frac{1\partial \overline{J_c}}{c^3\partial t} - \frac{\nabla \times \overline{J_g}}{c^2}\right].$
Murad's modification of Jefimenko	$\nabla \times \overline{g} = -\frac{\partial \overline{k}}{\partial t} - \frac{4\pi G}{c_k}\overline{J_k}; \nabla.\overline{g} = -4\pi G\rho_g;$ $\nabla.\overline{k} = -\frac{4\pi G}{c_k}\rho_k \ and: \nabla \times \overline{k} = -\frac{4\pi G}{c^2}\overline{J_g} + \frac{1\partial \overline{g}}{c^2\partial t}.$	$\frac{1\partial^2 \overline{g}}{c^2\partial t^2} - \nabla^2\overline{g} = 4\pi G[\nabla.\rho_g + \frac{1\partial \overline{J_g}}{c^2\partial t} - \frac{\nabla \times \overline{J_c}}{c}],$ $\frac{1\partial^2 \overline{k}}{c_g^2\partial t^2} - \nabla^2\overline{k} = 4\pi G\left[\frac{\nabla.\rho_g}{c^2} + \frac{1\partial \overline{J_c}}{c^3\partial t} - \frac{\nabla \times \overline{J_g}}{c^2}\right].$
Murad's gravity law	$\nabla \times \overline{g} = \frac{i}{c}\frac{\partial \overline{g}}{\partial t} + \frac{4\pi\gamma G}{c}\overline{J_g}$ *and* $\nabla.\overline{g} = -4\pi\gamma G\rho,$ *where* $\gamma = \frac{1}{\sqrt{1-\frac{u^2}{c^2}}}.$	$\frac{1\partial^2 \overline{g}}{c^2\partial t^2} - \nabla^2\overline{g} = 4\pi\gamma G[\nabla.\rho_g + \frac{i\partial \overline{J_g}}{c^2\partial t} - \frac{\nabla \times \overline{J_g}}{c}]$

Table 1: Different Gravitational Laws that cover a spectrum of conditions of interest.

For this reason, Newtonian gravity is fortuitous despite that it maybe fatally flawed if gravitational waves exist.

Let us first briefly review some potential gravity models as well as Newtonian gravity. Gravity plays a crucial role and is a concern to the propulsion specialist. The simplest is Newtonian gravitation [24,25], that is adequate to predict satellite motion and traverse to other celestial bodies in our solar system. The major issue is to carefully map the gravitational potential on a celestial body to consider effects due to concentrated mountain ore, mountain ranges or the presence of oceanic liquids.

Here, the Newtonian potential is φ and ρ is density with the terms previously mentioned that account for local gravitational effects throughout the body. Since the curl is zero, the gravity vector could be represented by a potential function. Boundary conditions for gravity vanish at infinity and asymptotically go to zero due to the inverse radius solution term. Although suitable for predicting motion where planetary speed is far lower than light speed, it does not mathematically support gravitational wave phenomena because time does not explicitly appear. If gravitational waves exist, they would have to move at infinite speed using this Newtonian model.

In his initial paper on Relativity, Einstein developed a model for gravity that modifies Newtonian theory. He claimed his equation supported the existence of gravitational waves. Einstein was concerned about the premise that under the Newtonian paradigm, we should instantaneously feel effects created by stars at infinity such as super novas, which he felt was wrong. He implied that some time lag should exist. His formulation also felt that a gravitational field was self-sustaining. Einstein went in an entirely different direction

Einstein went in an entirely different direction to develop a model where curvature of space-time geometry takes into account gravitational effects. Misner provides an example of this theory in deriving a photon world line that results in predicting the bending of light by gravitational forces [25]. Petkov suggests, in deference to Einstein's space-time theory, it is the anisotropy of space- time that causes inertia and gravity that is against our initial assumptions [26]. Haisch et al. by contrast suggest gravitational wave propagation is not rigorously consistent with space-time curvature [19].

An object with negative mass would repel ordinary matter, and could be used to produce an anti-gravitic effect. Alternatively, depending upon the mechanism assumed to underlie the gravitational force, it may seem reasonable to postulate a material that shields against gravity or otherwise interfere with a gravitational force.

A Magnetar, for example in Table 1, is a neutron star with an extremely strong magnetic field generated by the convection of hot nuclear matter produced as a consequence of nuclear reactions. Murad looks at a laboratory analogue of a geodynamo or Magnetar that involves a rapidly rotating liquid metal [22]. Neutron stars typically have the mass of 1.40 times the sun; however, the rotation rates are considerably higher from 10 to 600 cycles per second. The source term ρ can be negative indicating a repulsive mass density.

If a gyroscope is placed at 45° on a table and let go, the gyroscope falls on the table. However, if the rotor is spinning, it is capable of remaining aligned at this initial angular orientation. As the rotor speed decays, the gyroscope starts to precess rotating in a circumferential direction. When the rotation drops below a certain limit, the gyroscope falls to the table top. The rotation may induce a repulsive gravitational source that levitates the gyroscope according to this equation in contrast to using couples. Another way is that a gravitational field would repulse negative mass. Such a source can be considered as negative matter. If Winterberg is correct, then the inertial mass of the neutron star has to be greater than the companion star to compensate for the loss of gravitation due to spin that compensates for rotation.

The neutron star source term in a binary pulsar and weight is greater by:

$$\rho_p = \rho_c + \frac{\omega_p^2}{2\pi G}, \; W_p \approx \frac{V_p}{V_c}\left[1 + \frac{\omega_p^2}{2\pi G \rho_c}\right] W_c \text{ and}$$

$$\omega_p = \pm\sqrt{\left[2\pi G \rho_c \left(\frac{V_c}{V_p} - 1\right)\right]} \tag{3}$$

The subscript 'p' is for the neutron star while the 'c' is the value for the companion star; the V value is the volume for each star. The neutron star may be located on an elliptical trajectory with the companion star based upon their mutual attraction with each other. The theoretical value for the equilibrium rotation rate for both bodies assumed to have equal weight creating a circular orbit will be ω_p. The gyroscope analogy may be correct; however, for a Magnetar or a neutron star in a binary pulsar may have other attributes that need to be considered which amplifies gravitation such as the spinning magnetic field, which may have a considerable contribution to gravitic changes.

The discovery of apparent gravitational energy loss by the Hulse-Taylor pulsar, PSR 1913+16, provides indirect evidence of the existence of gravitational Jefimenko in Table 1, introduces gravitation and a cogravitation field used to predict force defined by the equation: waves. Thus other laws than Newtonian gravity show exist and support the view of gravitational waves [27-31].

Physicists normally use Einstein's field equations to discuss gravity that is mainly a main diagonal element on a tensor. A vector, usually used by engineers, could represent the main diagonal of such a tensor, which is what we will confer here. If rotational effects were to include gravity, these could appear as off-diagonal terms in this gravity tensor. This is important and makes the problem considerably more complex in Einstein's field equations.

Jefimenko, introduces gravitation and a cogravitation field used to predict force defined by the equation (Table 1) [29-31]

$$\bar{F} = m[\bar{g} + \bar{u} \times \bar{K}] \tag{4}$$

Gravity and a cogravity fields are defined similar as a Lorentz force relationship. This is a crucial analytical finding based upon Heaviside's 1893 paper where equations governing gravitation are considered somewhat similar to Maxwell's equations. Jefimenko uses cogravity field K to account for relativistic effects acting upon a rest mass and introduces time into the equations. He includes gravitational currents and sources. Mass is a gravitational source.

This equation resembles the Lorentz force acting upon an electromagnetic particle. For relativistic effects, he carries the terms one step further and defines gravity as:

$$\begin{aligned} \bar{g} &= -\frac{G\ m}{r^3(1 - \bar{r}.\bar{v}/rc)^3}\left[\left(\bar{r} - \frac{r\bar{v}}{c}\right)\left(1 - \frac{v^2}{c^2}\right) + \bar{r} \times \left[\left(\bar{r} - \frac{r\bar{v}}{c}\right)\right] \times \ldots\right] \approx \\ \bar{g} &\approx -G\frac{m}{r^3}\left[\left(1 - \frac{v^2}{c^2}\right)\bar{r}_o - \frac{2rv^2}{3c^3}\bar{v}_o\right] \end{aligned} \tag{5}$$

The leading coefficient on the RHS adjusts for Newtonian gravity while the first term in the parenthesis is a light speed correction as a function of distance and before leaving this subject, these laws also satisfy a relationship between the gravitational source term and the gravitational current. This is: velocity. Any gravitational law should asymptotically component depends upon speed and is essentially a torque. Such a torque in a field involving planets may cause planetary rotation in the direction of the orbital velocity of an adjacent planet. According to Jefimenko, this explains why the same side of the Moon always faces the Earth, which was a major motivation factor to look into developing a new gravitational model. In fact, Lavrentiev found that this was also true for all of the large planets in the solar system [32]. In the words of Mark Twain:

"Everyone is a moon, and has the dark side which he never shows to anyone."

Here, the gravitational attraction is proportional to both the radial distance and the speed of the body with respect to a planet producing the field. This model by Murad in differs upon Jefimenko's original equations due to additional source and current terms [33-40]. Differences in these equations suggest the partial differential equations are also hyperbolic or wave equations (Table 1).

The issue of self-feeding gravity can be taken one step further and may be related to the high rotational rate within the spiral arms of galaxies. If rotation induces a torsion field, a galaxy could create such fields as it spins or rotates about its axis. This torsion field could induce a cogravitation field as well as create electromagnetic radiation that, by an inverse Gertsenshtein effect and later with Forward, may induce a localized gravitational field. Once this field is established, the induced angular momentum effect may, by Jefimenko's gravitation laws, further increase the galaxy's rotation rate that increases the strength of the initial torsion field [41,42].

To extend Newtonian gravitation, let us use a simpler rationale for a new gravitational model. Here the relativistic factor operates upon both currents and source terms [43]. Because the curl does not vanish, gravity cannot be represented by a potential as in Newtonian gravitation. Thus, gravitation is not only a function of gravitational currents but is also a function of itself similar to what Einstein originally proposed.

This is surprisingly the desired wave equation. For no source or current terms and steady-state conditions, this equation asymptotically approaches Newtonian gravity. The gravitational current time derivative may be troublesome from a physical perspective due to the imaginary multiplier. If gravitational currents exist, the current acts as a complex variable having the imaginary term. Gravity currents could be a function of time as well as spatial dimensions for this model.

Before leaving this subject, these laws also satisfy a relationship between the gravitational source term and the gravitational current. This is

$$\frac{\partial \rho_g}{\partial t} + \nabla . \bar{J}_g = 0 \tag{6}$$

Note that the first term appears but Newtonian gravitation never discusses any issues about gravitational currents.

If gravitational currents exist, then gravitational waves would exist. The current is quite interesting. It implies that time rate changes in the mass due to changes in the core or with the presence of nuclear explosions would generate changes in the current. Likewise if there were no time changes, the gradient would be equal to zero which also implies that a gravitational current could still exist and change it self upon environmental conditions.

A view of gravity and fluid flow

Thus, gravity which may obey either an elliptical or hyperbolic wave partial differential equation may be mathematically analogous to the partial differential equations obeyed by a subsonic and supersonic fluid [44-48]. The governing subsonic flow equations and Newtonian gravitation are both an elliptical partial differential equation. This implies that all solutions depend upon a closed boundary in the domain of interest. Under this circumstance, the body that is exposed to a series of several celestial or smaller bodies will, at low relative speeds be linearly additive for summing up the gravitational effects.

Gravitational waves must satisfy a wave equation that includes time. What does this mean and can phenomenon be examined that demonstrate this capability? Basically wave equations do not depend upon the entire boundary as previously mentioned for Newtonian gravity. With wave equations, this is still a viable boundary condition that gravity vanishes at distances. The boundary influences the domain only based upon a specific region that depends upon a signal zone. Again, if the boundary for both is the same, no differentiation can be found if an elliptical or hyperbolic event occurs. This is not normally considered in a space mechanic problem.

Another part of the problem is that for a propulsion device, how can we look at gravity? For example, gravitational waves have not yet been observed or are defined to date. These waves need to be defined if they are part of a propulsor.

Fontana raises an interesting propulsor using gravitational waves. Here wave generators are used at several places azimuthally aimed about the nose section of the spacecraft [49]. These generators are aimed at a single focus to a point in front of the spaceship. These gravitational waves intersect at the same point and could, theoretically create a singularity that would alter the gravitational field in this region. This singularity would tend to draw the spaceship by *pull* instead of *push*! Again this raises questions about how to generate a singularity.

Generating gravitational fields

Let us define portions of a propulsor that might be suitable into the development of a warp drive. If an energy difference can create a gravitational field, it implies that energy is converted as some form of mass. The presence of mass will induce gravity. Thus we are looking at this with a different perspective that exists and is confined within a geometric volume of the propulsor. The important part is to define the magnitudes of the energy levels required to convert different and separate gravitational fields. This is not trivial. Once within the propulsor, how does the propulsor or subsequent gravitational effects interact between its internal capabilities with its external environment to produce thrust?

A technical paper was reviewed written by an individual that lives in a prison in Illinois [50]. He interestingly raises the question of what happens with interactions between two different inhomogeneous gravitational fields. Normally we only consider a homogeneous gravitational field that exists with these laws previously mentioned. Here, the view is what occurs if there are, say holes in the field and what are any propulsion implications. When these fields merge, are they gradual or additive as one would intuitively expect if the fields that are both weak for a satellite that depends linearly between the Earth and the Moon, or what if one field is intensely stronger? This is a weak-

weak interaction in the former. The last possibility is that these fields would create unique regions with distinct boundaries if the fields were very strong similar to the gravitational field of several separate celestial bodies but with much more strength. Moreover, if these fields are distinct, what occurs if one travels along this boundary between the two different fields? Can you use this knowledge to violate moving either forward or back into time? Without any solid proof or experimental data, it is difficult to accept this premise at face value although the notion is indeed intriguing.

If Newtonian gravity is mathematically analogous to subsonic flow, and these other gravity laws obey wave equations, we have covered both edges of the spectrum. Thus it is possible that gravity can operate in a similar mathematical fashion to sustaining regions of mixed flow with subsonic flow and supersonic flow for a gas. This suggests that a body may travel at certain speeds near the region of speed propagation; say below light speed, which is comparable to the speed of sound. The interesting situation is that the speed of propagation can be altered; is it feasible that the propagation speed for gravity is not a constant but may also be altered by the propulsion specialist? Consequently gravity could also obey a similar rule of mixed flow but no such gravity law for this counterpart currently exists.

A fluid may reach transonic flow similar to regions that contain high subsonic flow where shocks are formed and this generates regions actually reaching supersonic flow. The specification for a generic Tricomi-like partial differential equation is provided as follows:

$$\frac{\partial^2 \Phi}{\partial t^2} + x\frac{\partial^2 \Phi}{\partial x^2} = f(x,t) \text{ or like } \frac{\partial^2 \Phi}{\partial t} + \left(1 - \frac{v^2}{c^2}\right)\frac{\partial^2 \Phi}{\partial x^2} = f(x,t) \qquad (7)$$

The second equation is a candidate for a new type of gravitational law. Here these equations are elliptical whenever x or the multiplier that uses velocity is positive. It becomes a parabolic differential equation when x or the multiplier (v=c) is zero and the equation becomes a wave equation when x and the multiplier is negative. This could be a crucial ingredient to using such a warp propulsor. Can we demonstrate such equations to represent gravity?

If the Pioneer satellites were altered when the electricity circuitry was changed from directing internal electricity using sensors to a heat sink, the change in motion would exhibit behaviour that indicates gravity might have changed [39]. Some views are that the heat sink radiated energy away from the environment and this created propulsion. Each Pioneer moving at different directions relative to the sun, reached a specific distance when the Newtonian gravity suddenly changed. This anomaly was a peculiar or strange deviation from the common rule. The problem is that three other long-range satellites demonstrated similar behaviour where gravitational changes were observed that did not have the same satellite architecture as the Pioneer satellites.

One may argue that this could be a function due to reaching a reference distance. Another satellite for several years stayed within the early planet regions and did not show such an anomaly. Thus, time is not a function of the phenomena. What we are presuming is that there may be a region where a satellite reaches a certain velocity or juncture distance from the sun that is like the ***x*** in the above Tricomi equation that suddenly possesses a different gravitational law.

An aerodynamic analogue

There is a rationale that may show how to deal with these different gravity laws. Each may be correct at specific locations from the sun. For example, near the Earth and within the near-term solar system, Newtonian gravity is correct; however, at considerable far-term situations, a wave equation may be satisfied in Figure 3.

There is an aerodynamic analogue that may be useful. In fluid gas flow for a rocket nozzle, the pressure increases within the combustion chamber. As the flow leaves the diverging section of the nozzle, pressure decreases while the gas particle velocity increases reaching sonic speed or transonic flow (Tricomi equation) at the throat of a converging/diverging portion of the nozzle. This point achieves a maximum flow rate. Under these conditions using no shocks that occur within the nozzle, the flow pressure decreases in the diverging portion of the nozzle further converting pressure into kinetic motion to reach supersonic speeds.

Liepman and Roshko show an ideal Prandtl- Meyer expansion in the Hodograph plane. Here, Figure 4 indicates that subsonic flow occurs from the origin to where the Mach number 1 appears and the expansion flow for a wave equation goes to infinity [52]. An Epicycloid diagram from Sears shows results using a Method of Characteristics for a similar nozzle type of event [53]. A result on the LHS is at lower speeds below sonic conditions where higher and supersonic speeds are evident in the RHS. The rationale is that the Pioneer satellites may have demonstrated a difference in gravity analogous to passing the propagation speed with the gravity wave equation formulation at further distances. This also implies the previous comment that different regions may exist near other regions with variations in the strength of gravitation.

Creating gravitational shocks and propulsion

This is the essence of the problem. Before we make a warp drive, specific experiments are needed to understand the physics of the pieces that would work within a propulsive architecture. Many have spent considerable time to nullify inertia and these look promising. However, what comes to pass after you control inertia? Does the system allow travel in a geodesic that reaches faster than the speed of light?

The issue is raised in creating a gravitational shock. Part of the problem is to discover means for either creating gravitational waves or detecting such waves. These efforts have found that the sensitivity for such instruments are far lower than currently measured and requires an investigation just to define suitable instrumentation. For example, LISA requires several satellites with sufficient separation. These are

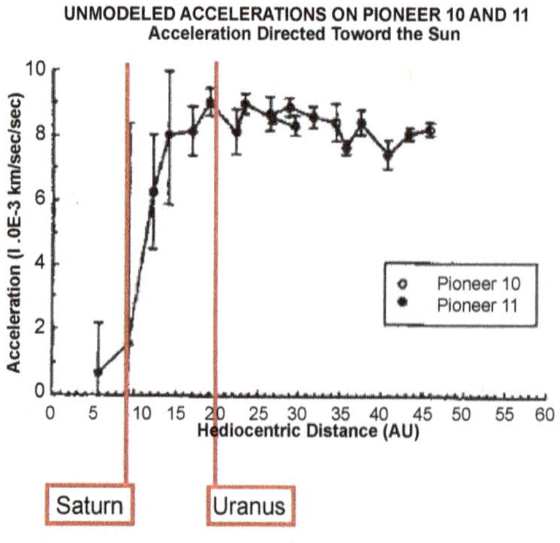

Figure 3: This could be the point where a Tricomi-like gravity equation characteristic changes for a different type of gravity law [51].

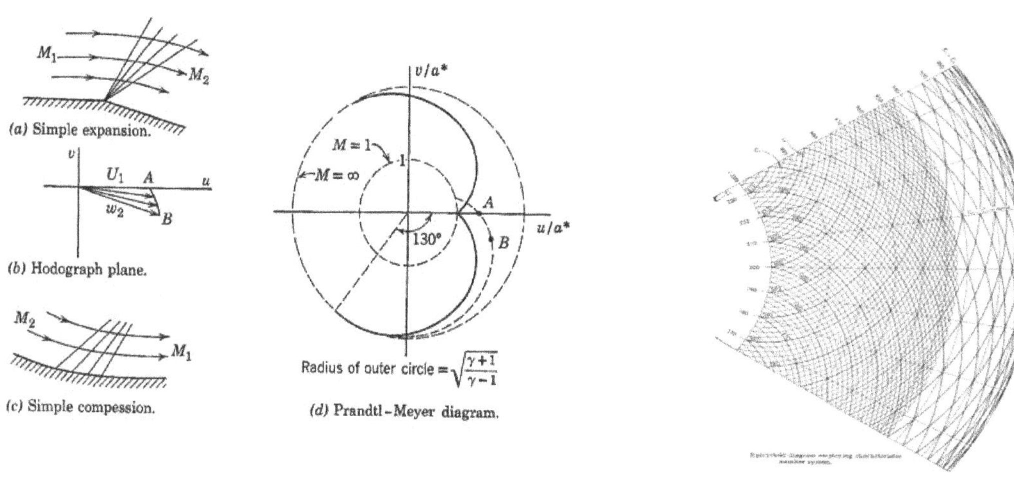

Figure 4: The effects of flow moving from subsonic to supersonic speeds or characteristics that could explain similar events with gravity.

oriented with each other using lasers. If the satellites are perturbed because of a gravitational wave, the assumption is that the laser misalignment would be sufficient to detect such events. This obviously assumes that gravitational waves move at the speed of light.

If you accept some experiments performed by Kosyrev and commentary made by Podkletnov, as well as examining black holes with no accretion disk that generates jets, it is feasible that gravity or gravitational waves might greatly exceed the speed of light [54,55]. If true, then LISA results would be fatally flawed when all of these satellites might simultaneously jump together if perturbed by a gravitational wave and therefore may show no meaningful results.

With these thoughts to create an experiment, how can we resolve this problem? The only way we currently have would be to use mass. In other words if large masses interact with each other at sufficient speed, they should generate a gravitational wave disturbance. The notion is to convert energy into mass similar to deBroglie matter waves. Such disturbances could resemble a nuclear explosion that would generate a radial expansion for creating a gravitational shock.

The notion of a shock requires some explanation. If there are wave equations within the gravitational law, they will create characteristics based upon the mathematical physics of the problem. The problem is to form characteristics to confluence or join and this would create what we are referring to as a shock. Such merged characteristics will create some response such as a radiation pressure and this may induce propulsion.

The issue of using a radial effect with these merging characteristics would allow experimentation but not really provide sufficient capabilities to develop a propulsor. The ideal model would be to have a cylinder within a propulsive duct used to create a gravitational disturbance and then allow it to move down a cylinder in a specific direction to hopefully provide directional thrust. This pumping could generate a sudden increase in a large energy input. The question is we are not expelling mass but somehow mass is consumed or altered in this process to generate energy.

The other issue for the propulsion system is if shocks are created, should they be pulses or continuous? Obviously starting on such a system, initial effects will be to examine pulses. When the technology reaches maturity, the ability to generate a continuous gravitational shock should be easier to produce.

The desired system would employ generating a normal or an oblique gravitational shock. These gravitational shocks would appear similar to signals and cones that are normally addressed in these problems. Other shocks such as a normal shock should exist as well. The effects could generate several alternatives. The body that generates the shock may appear outside of the body that will generate thrust and accelerate on a specific trajectory or geodesic. The flight path could be tailored by altering the shock's appearance especially if an intersection would occur near a celestial body. Another alternative is that the shock could be repulsive from the gravitational field of the surface on a celestial body that might create a gravitational reflection. If so, this could generate thrust away from the body into another direction. Altering the geometry of the shock where its symmetry is perturbed, the adjustments would allow control thrust changes in the trajectory.

Such issues will also assume that embedded regions can be placed near the body's shock as well as immediately in front of or behind such a shock. To do this with technology is obviously beyond our contemporary knowledge but this should be feasible. The other possibility is internal to the propulsor. What value would be achieved if a steady-state gravitational shock exists? This is an interesting problem if your propulsor generates a large stationary gravitational field in an oblique shock within a small region. This question is important as well as how to determine creating the steady-state gravitational shock.

The author apologizes for not having solutions to these questions but part of the problem is to first understand what the issues are considering new thoughts about phenomenon or feasibility. Can such a device provide propulsion based upon the proximity of other celestial bodies and what is the sensitivity that would be observed by these effects? For example, this may work in the vicinity of a celestial body but may have significant problems if there is no field to pull or push against. One such region might be in intergalactic space where motion is at considerable distances from such fields and that the fields within intergalactic space may be so weak to be less than insignificant.

Conclusion

The problem warrants investigation to generate a new type of physical phenomena such as creating a gravitational shock. Some of the previous wave equations depending upon gravitational laws, indicate that such phenomenon should mathematically exist. Can we perform experiments for investigating gravitational shocks that currently is beyond the realm of embryonic technologies as well as provide instrumentation? Moreover, we need to develop instrumentation to create a gravitational shock outside of a spacecraft or internal within a propulsor. This type of propulsion can become a critical element to develop a warp drive to travel faster than light flight. This is necessary if mankind is to achieve some semblance of both our own destiny and understand the role we play within the cosmos.

Acknowledgment

The author appreciates some notions and positive comments by Dr. Robert Baker who was also a co- author on several other papers concerning the intricacies of gravitation and multiverse. The author also appreciates Garett Volk who helped with the text.

References

1. Murad PA (2005) Warp-Drives: The Dreams and Realities Part I: A Problem Statement and Insights. AIP Conference Proceedings, USA.
2. Murad PA (2005) Warp-Drives: The Dreams and Realities Part II: Potential Solutions. AIP Conference Proceedings, USA.
3. Recami E (2008) A homage to E. C. G. Sudarshan: Superluminal objects and waves (An updated overview of the relevant experiments). Cornell University Librabry.
4. Recami E, Barbero APL, Hernandez-Figueroa HE (2000) Propagation Speed of Evanescent Modes. Physical Review E.
5. Recami E (2004) Superluminal tunneling through successive barriers: Does QM predict infinite group velocities. Journal of Modern Optics 51: 913-923.
6. Gertsenshtein ME (1962) Wave Resonance of Light and Gravitational Waves. Soviet Physics JETP14:84-85.
7. Gertsenshtein ME (1966) The Possibility of an Oscillatory Nature of Gravitational Collapse. Soviet Physics JETP 51:129-134.
8. Gertsenshtein ME, Ingel K, Fil'chenkov ML(1978) Problem of the Physical Singularity in the General Theory of Relativity. Soviet Physics Journal 21: 841-845.
9. Gertsenshtein ME, Melkmova E Yu. About the Singularity Criterion in General Relativity. Soviet Physics Journal 30: 174-178.
10. Gertsenshtein ME (1998) Time Machine and The General Theory of Relativity. Russian Physics Journal 41: 104-106.
11. Gertsenshtein ME (1998) Elementary Particle Physics and Field Theory: New Problems in General Relativity. Russian Physics Journal.
12. Gertsenshtein ME, Braginskii VB (1967) Concerning the Effective Generation and Observation of Gravitational Waves. ZhETF Pis'ma. 9:348-350.
13. Murad PA (2011) Consequences of Unusual Behavior in Einstein's Field Equations for Advanced Propulsion Schemes. Space, Propulsion & Energy Sciences International Form.
14. Baker RML (2007) Chinese High-Frequency Gravitational Wave Research Program. Space Technology Applications International Forum (STAIF-2007), Albuquerque, New Mexico, USA.
15. Baker RML(2009)The Peoples Republic of China High Frequency Gravitational Wave Research Program. Proceedings of the Space, Propulsion and Energy Sciences International Forum.
16. Meholic GV (2004) A Novel View of Spacetime Permitting Faster-Than-Light Travel. AIP Conference Proceedings.
17. Recami E (2008) A homage to ECG Sudarshan:Superluminal objects and waves (An updated overview of the relevant experiments. Cornell Library University.
18. Murad PA (2002) Faster Than Light Travel Versus Einstein. Galilean Electrodynamics.
19. Haisch B, Rueda, A, Puthoff HE (1998) Advances in the Proposed Electromagnetic Zero-Point Field Theory of Inertia. Cornell Library University.
20. Murad PA Brandenburg JE (2012) The Murad-Brandenburg Equation- A Wave Partial Differential Conservation Expression for the Poynting Vector/Field. Aerospace Research Central.
21. Brandenburg, JE (2011) The Hybrid Fusion-Fission Reactor - The Energy Crisis is Solved. Journal of Space Exploration.
22. Murad PA (2012) Understanding Pulsars to Create a Future Space Propulsor. Aerospace Research Central.
23. Ma T, Wang S (2012) Gravitational field Equations and Theory of Dark Matter and Dark Energy. Cornel Library University.
24. Lorentz HA, Weyl H, Minkowski H (1952) Einstein: The Principle of Relativity, Dover Publications.
25. Misner CW, Thorne KS, Wheeler JS (1973) Gravitation. WH Freeman and Company, New York.
26. Petkov V (1999) Propulsion Through Electromagnetic Self-Sustained Acceleration. Aerospace Research Central.
27. Taylor JH (1994) Binary Pulsars and Relativistic Gravity. Rev Modern Physics 66: 711-719.
28. Baker J, Campanelli M, Lousto CO, Takahashi R (2002) Modeling gravitational radiation from coalescing binary black holes. Phys Rev D65:124012.
29. Jefimenko OD (2006) Gravitation and Cogravitation. ElectretScientific Company Star City, ISBN 0-917406-15-X.
30. Jefimenko OD (1997) Electromagnetic Retardation and Theory of Relativity. Electret Scientific Company, StarCity, West Virginia.
31. Jefimenko OD (1992) Causality Electromagnetic Induction and Gravitation. Electret Scientific Company, Star City, West Virginia.
32. Lavrentiev MM, Dyatlov VL, Fadeev SI, Kostova NE, Murad PA (2000) Rotation Effects of Bodies In Celestial Mechanics, Proceedings of the Science 2000 Congress, St. Petersburg, Russia.
33. Murad PA (2005) Warp-Drives: The Dreams and Realities, Part I: A Problem Statement and Insights, STAIF-2005, Albuquerque, New Mexico.
34. Murad PA (2005) Warp-Drives: The Dreams and Realities, Part II: Potential Solutions", STAIF-2005, Albuquerque, New Mexico.
35. Murad PA, Baker jr, RML (2003) Cosmology and the Door to Other Dimensions and Universes", Paper AIAA 2003-4882 presented at the 39th AIAA/ASME/SAE/SAEE Joint Propulsion Conference, Huntsville, Ala.
36. Murad PA, Baker Jr, RML (2003) Gravity with a Spin: Angular Momentum in a Gravitational-Wave Field. Presented at the First International High Frequency Gravity Wave Conference, Mitre Corporation.
37. Murad PA (2003) Its All Gravity Presented at STAIF 2002. AIP Conference Proceedings 654: 932-939.
38. Murad PA (2010) An Anzatz about Gravity, Cosmology, and the Pioneer Anomaly. February 2010 SPESIF Meeting at John Hopkins University.
39. Murad PA (2007) Exploring Gravity and Gravitational Wave Dynamics. Part I: Gravitational Anomalies. STAIF-2007, Albuquerque, New Mexico.
40. Murad PA (2007) Exploring Gravity and Gravitational Wave Dynamics, Part II: Gravity Models. STAIF-2007, Albuquerque, New Mexico.
41. Gertsenshtein ME (1962) Wave Resonance of Light and Gravitational Waves. Soviet Physics JETP 14: 84-85.
42. Forward RL (1963) Guidelines to Antigravity. American Journal of Physics 31: 166-170.
43. Murad PA. Revisiting Gravitational Anomalies and a Potential Solution.
44. Murad PA (1995) A Framework for Developing Navier- Stokes Closed-Form Solutions. AIAA Paper No. 95-0479, 33rd Aerospace Sciences Meeting and Exhibit, Reno, NV.
45. Murad PA (1999) An Extended Navier-Stokes Algorithm and The Challenges of Relativistic Fluid Dynamics. AIAA Paper No. 99-0562, 37th Aerospace Sciences Meeting, Reno, NV.
46. Murad, PA (2006) Closed-Form Navier-Stokes Solutions to Fluid Dynamic Equations Related to Magnetohydrodynamics. 42nd AIAA/ASME/SAE/ASEE Joint Propulsion Conference and Exhibit, Sacramento,California.

47. Murad PA (2006) Closed-Form Solutions to the Transient/Steady-State Navier-Stokes Fluid Dynamic Equations. STAIF, Albuquerque, New Mexico.

48. Murad PA (2008) Closed-Form Navier-Stokes Solutions to the Transient/ Steady-State Fluid Dynamic Equations Related to Magnetohydrodynamics. AIAA/ASME/ASEE Joint Propulsion Conference.

49. Giorgio Fontana, Robert ML, Baker Jr, Murad PA (2007) Hyperspace for Space Travel. STAIF-2007, Albuquerque, New Mexico.

50. Private conversations with Dennis Bushnell, Chief Scientist, NASA Langley.

51. Frank Dodd, Smith Jr (2009) Our Conformal Keplerian Solar System.

52. Liepmann HW, Roshko A (1957) Elements of Gasdynamics. Galcit Aeronautical Series, John Wiley &Sons.

53. WR Sears (1954) General Theory of High Speed Aerodynamics. High Speed Aerodynamics and Jet Propulsion, Princeton, New Jersey.

54. Kosyrev NA (1971) On the Possibility of Experimental Investigation of the Properties of Time. Time in Science and Philosophy 111-132.

55. Podkletnov E, Modanese, G (2006) Study of Light Interaction with Gravity Impulses and Measurements of the Speed of Gravity Impulses. Gravity-Superconductors Interactions: Theory and Experiments.

Treatments of Probability Potential Function for Nuclear Integral Equation

Matoog RT*

Department of Mathematics, Faculty Applied Sciences, Umm Al-Qura University, Kingdom of Saudi

Abstract

Here, the nuclear fractional integro differential equation is considered. Then, the numerical solution of the linear fractional integro differential operator $L\Phi = \lambda_m \Phi_m$ is discussed. In addition, the eigenvalues (EVs) and the corresponding eigenfunctions (EFs) are obtained for the nuclear integral equation (NIE). Finally, the relation between the fractional coefficient and the potential function is obtained.

Keywords: Fractional integro differential equation; Nuclear integral equation; Eigenvalues; Eigen functions

Introduction

The theory of fractional calculus (FC) is a mathematical field, which unify and generalize classical calculus for non-integer order of derivation thus dealing with derivatives and integrals of arbitrary and complex orders. One can state that, the whole theory of fractional derivatives (FD) and integrals were established in the second half of the 19th century. The generalization of the concept of derivative and integral to a non-integer order has been subjected to several approaches and some various alternative definitions of fractional derivatives appeared in ref. [1]. Surveys of the history of the theory of fractional derivatives can be found in ref. [2].

For three centuries the theory of FC developed namely as a pure theoretical field of mathematics useful only for mathematicians. However, in the last few decades FC has been a fruitful field of research in science and engineering. In fact, many scientific areas are currently paying attention to the FC concepts and we can refer its adoption in viscoelasticity and damping, diffusion and wave propagation, electromagnetism, chaos and fractals, heat transfer, biology, electronics, signal processing, robotics, system identification, traffic systems, genetic algorithms, percolation, modeling and identification, telecommunications, chemistry, irreversibility, physics, control systems as well as economy, and finance [3,4]. It has been shown that new fractional-order models are more adequate than previously used integer order models.

Many authors pointed out that FC provided an excellent instrument for the description of properties of various real materials in particular the description of memory and hereditary properties of various materials and processes. This is the main advantage of FD in comparison with classical integer-order models, in which such effects are in fact neglected. The advantages of FC has been produced a successful revolution to modify many existing models of physical processes, e.g., the description of rheological properties of rocks, mechanical modeling of engineering materials such as polymers over extended ranges of time and frequency [5,6]. Fractional order models often work well, particularly in heat transfer and electrochemistry, for example, the half-order fractional integral is the natural integral operator connecting the applied gradients (thermal or material) with the diffusion of ions of heat [7].

The theory of application of the IEs is an important subject within applied mathematics. The IEs are used as mathematical models for many varied physical situations. In addition, The IEs occur as reformulations of other mathematical problems.

The area of the IEs is quite old, going back almost 300 years, but most of the theory of IEs dates from the twentieth century. An excellent presentation of the history of Fredholm IEs can be found in Bernkopf [8], who traces the historical development of both functional analysis and IEs and shows how they are related.

There are many well-written texts on the theory and application of solving IEs analytically. Among such, we note Muskhelishvili [9], who developed the theory of singular IE, Green [10], Hochstadt [11], Knawel [12,13] Kress [14], Michelin [15], Smirnov [16] and Tricomi [17].

The state of the art before 1960 for the numerical solution of IEs is well described in the book of Kantorovich and Krylov [18]. From 1960 to the present day, many new numerical methods have been developed for the solution of many types of IEs, such as the Toeplitz matrix method, the product Nystrom method, the Galerkin method, Runge-Kutta method and Block-by-block method (Abdou et al. [19], Baker and Miller [20], Dzhuraev [21] and Delves and Walsh [22]). There are many numbers of texts on the numerical solutions of the different types of IEs; we note especially Delves and Mohamed [23] Atkinson [24,25], and Golberg [26].

In other way, the theory of EVs and EFs is playing now an important role in solving the IE, especially when the IE in the homogeneous case, or when the kernel takes a singular form. The linear combination of the EVs and EFs is called the spectral relationship. Many different methods are used and derived to establish these spectral relationships. For this aim, the reader can obtain more information for the spectral relationships with different applications can be found in the work of Popov [27], Mkhitarian and Abdou [28,29], Abdou [30-36] Abdou and Ezz-Eldin [37], Abdou and Salama [38], Jiany [39], Jeanine and Barber [40] and Pang [41]. Consider L is a linear fractional integro-differential operator

***Corresponding author:** Matoog RT, Professor, Department of Mathematics, Faculty Applied Sciences, Umm Al-Qura University, Kingdom of Saudi
E-mail: rmatoog_777@ yahoo.com

$$L\Phi(x;t)=\lambda\Phi(x;t)$$

$$i\frac{\partial^{\alpha}}{\partial t^{\alpha}}\left(\frac{\partial\Phi(x;t)}{\partial x}\right)+P(x)\Phi(x;t)-\int_a^b k(|x-y|)\Phi(y;t)dy=\lambda\Phi(x;t) \quad (1)$$

$$\left(\frac{\partial\Phi(x;t)}{\partial x}=UV(x);\Phi(a;t)=\Phi(b;t)=A(t);0<\alpha<1;i=\sqrt{-1}\right.$$

The function P(x) is a continuous function, while $k\ (|x\text{-}y)|$ is a discontinuous function. U is a constant and is a parameter

The equations of the form $L\varphi=\lambda\varphi$ arise in many mathematical physics problems. It is often true that, the special solutions called EFs or characteristic functions. These EFs must not identically zero and satisfy one or more conditions that are supplementary related to the problem being solved. The EFs exist only for special values of the parameter λ; these values of λ are called EVs or characteristic values.

Consider Φ_m is a solution of the equation $L\Phi_m=\lambda_m\Phi_m$, λ_m is an eigenvalue, which is not identically zero and which satisfies the supplementary conditions, then Φ_m is called an EF belonging to the EVλ_m. Here, in this paper, the asymptotic behavior of the EVs and EFs for the linear operator will be discussed. In addition, some results will be considered. In section 2 the basic concepts of the linear integral operator is considered. Moreover, the Riemann - Liouville, and the Caputo derivatives of order fractional integral are considered. In section 3, using the separation variable method and the linear differential method, we obtain the general solution of the nuclear integro differential equation. Some important results for EVs and EFs are considered.

Basic Concepts

Definition (1): The integral operator we were $K\phi=\int_a^b k(|x-y|)\phi(y)dy$ where $k(|x,y|<M$ or satisfies $\int_a^b\int_a^b k(|x-y|)\ |^2dxdy=c^2$ in $L_2[a,b]$, is bounded and continuous. Moreover, the integral operator is called compact.

Lemma 1: The kernel $k\ (x-y),a\le x,y\le b$, can be decomposed in an infinite number of ways, into the sum of the suitable degenerate kernel $k_0\ (x\ ,y)$ and another continuous kernel $k_1\ (x\ ,y)$ whose norm $||k_1||$ can be made small as we wish i.e.

$$k(|x-y|)=\sum_{t=1}^{n}a_t(x)b_i(y)+k_1(x,y)$$

Where $a_i(x)$, $b_i(y)$ and $k_1(x,y)$ are uniformly continuous and hence bounded in the interval $[a,b]$.

Theorem1: Let H be $L_2(a,b)$ and K a degenerate integral operator

$$K\Phi=\sum_{t=1}^{n}a_t(x)\int_a^b b_t(y)\Phi(y)dy$$

K is a compact operator, if $a_i(x)$ and $b_i(x)$ belong to $L_2(a,b)$ for all i.

Theorem 2: Let (K_n) be a sequence of compact operators on a Hilbert space H, such that for some operator K we have $\lim_{n\to\infty}\|K-K_n\|$. Then, K is also compact i.e., $K=\int_0^1 k(|x-y|)\phi(y)dy$ is a compact operator on $L_2[0,1]$.

Definition (2): The Riemann -- Liouville fractional integral of order α>0, of the function $f(\ 0,\infty\ )\to g(\infty)$ is given by

$$I_{0^+}^{\alpha}=\frac{1}{\Gamma(\alpha)}\int_0^t(t-s)^{\alpha-1}f(s)ds$$

Provided that the R.H.S is point wise definition (0,)

Definition (3): The Caputo derivatives of order α>0 of a continuous function f:(0,∞)→ g(∞) is given by $D_{0^+}^{\alpha}f(t)=\frac{1}{\Gamma(n-\alpha)}\int_0^t\frac{f^{(n)}ds}{(t-s)^{\alpha-n+1}},\ n<\alpha\le n+1$

Theorem 3: Let K be a compact operator on the Hilbert space H, and let $\{\psi_n\}$ be a linearly independent sequence of EVs corresponding to some nonzero eigenvalue μ, that is $K\psi_n=\mu\psi_n$ for all n. Then, $\{\psi_n\}$ contains a finite number of elements.

Method of Solution

Consider the general solution of (1) in the form:

$$\Phi(x,t)=A(t)\phi(x) \quad (2)$$

Then, using (2) in (1), with aid of the boundary conditions, we have

$$A(t)=\sum_{n=0}^{\infty}\frac{Ut^{n\alpha}}{\Gamma(n\alpha+1)},\ 0\le t\le T<1,\ 0<\alpha<1 \quad (3)$$

$$\phi(x)=e^{i\int_a^x(\lambda-P(u))du}-ie^{i\int_a^x(\lambda-P(u))du}\left\{\int_a^x e^{i\int_a^{\zeta}(P(u)-\lambda)du}.\int_a^b k(|\zeta-u|)\phi(|u)dud\zeta\right. \quad (4)$$

Hence, the general solution of (1) is given as:

$$\Phi(x,t)=\left(\sum_{n=0}^{\infty}\frac{Ut^{n\alpha}}{\Gamma(n\alpha+1)}\right)\times\left[e^{i\int_a^x(\lambda-P(u))du}-ie^{i\int_a^x(\lambda-P(u))du}\left\{\int_a^x e^{i\int_a^{\zeta}(P(u)-\lambda)du}.\int_a^b k(|\zeta-u|)\phi(|u)dud\zeta\right\}\right] \quad (5)$$

The formula (5) represents a unique solution of the integro differential equation. To prove the boundedness and the orthogonally of the integro differential operator, we must know some properties of the fractional integro differential equations and the famous properties of the IE. Therefore, we consider eqn. (1) where $p(x)$ is a real continuous function in (a,b), the kernel $k\ (x-y)$ is continuous in the same interval. Moreover we have i.e., $k(|x-y\ |)=\overline{k(|x-y\ |)}$ and λ is real.

Definition 3: The kernel $(t-\tau)^{\alpha-1},\forall t,\tau\in[0,T],T<1,0<\alpha<1$ satisfies for every continuous function $h(\tau)$ and all $0\le\tau_1\le\tau_2\le t$ the integrals $\int_{\tau_1}^{\tau_2}(t-s)^{\alpha-1}h(\tau)d\tau$, $\int_0^t(t-s)^{\alpha-1}h(\tau)d\tau$ are continuous functions of t, i.e., $\left|\int_0^t(t-s)^{\alpha-1}h(\tau)d\tau\right|\le M;M>0$ is bounded.

Lemma 2: For the integro differential operator L of (1) and for every λ, the EFs of L, is bounded under the conditions $|p(x)|=m_1\int_a^b\int_a^b k(|x-y|)\ |^2dxdy=m_2<1$;(m_1, m_2 are constants),

Proof: Taking the norm of both sides of (2)-(4) then using the two conditions of $p(x)$ and $k(x,y)$ with the famous relation $\left|e^{i\int_a^x(\lambda-P(u))dt}\right|=1$, we have

Hence, we have $|\ \phi\ \|\le\frac{1}{1-m_2}$ and after using definition (3), we have $||A||\le M$

$$|\ \phi\ \|\le\frac{M}{1-m_2} \quad (6)$$

The formula (6) proves the boundedness of the function $\Phi(x,\lambda;t)$ for all values of λ and $x\in[a,b],t\in[0,T\ ],T<1$.

Lemma 3: The EFs of the operator (1) corresponding to distinct EVs are orthogonal.

Proof: Assume that Φ_1 $(x; t)$ and Φ_2 $(x; t)$ are two real EFs corresponding to the two different eigenvalues λ_1 and λ_2 respectively i.e., we have

$$L\Phi_i(x;t) = \lambda_i\Phi_i(x;t) = \lambda_i\phi_i(x)A(t)$$
$$A(0) = U; \quad \Phi_i(a;0) = \Phi_{ii}(b;0) = U, i = 1,2 \tag{7}$$

Integrating (1) with respect to t, we get

$$-i\left(\frac{\partial\Phi(x;t)}{\partial x} - \frac{\partial\Phi(x;0)}{\partial x}\right) + \frac{1}{\Gamma(\alpha)}\int_0^t (t-\tau)^{\alpha-1}(P(x)-\lambda)\Phi(x;\tau)d\tau$$
$$+\frac{1}{\Gamma(\alpha)}\int_0^t (t-\tau)^{\alpha-1}\int_a^b k(|x-y|)\Phi(y;\tau)dxd\tau \tag{8}$$

Assuming that Φ_1 $(x; t)$ and $\overline{\Phi}_2(x;t)$ satisfy eqn. (8). Then, multiplying the result by $\overline{\Phi_2(x,t)}$ and by Φ_1 $(x; t)$. Then subtracting the results and integrating with respect to x from a to b, then using the symmetric of the kernel $k(|x-y|) = \overline{k(|x-y|)}$ Finally, we obtain

$$-i\int_a^b\left\{\overline{\Phi_2(x;t)}\left[\frac{d\Phi_1(x;t)}{dx} - \frac{d\Phi_1(x;0)}{dx}\right] - \Phi_2(x;t)\left[\overline{\frac{d\Phi_2(x;t)}{dx}} - \overline{\frac{d\Phi_2(x;0)}{dx}}\right]\right\}dx$$
$$= \frac{(\lambda_1-\lambda_2)}{\Gamma(\alpha)}\int_0^t\int_a^b (t-\tau)^{\alpha-1}\Phi_1(x;t)\overline{\Phi_2(x;t)dx\,d\tau} \tag{9}$$

After integrating and using the boundary conditions, the term in the left hand side will vanish. Hence, we get

$$\int_0^t\int_a^b (t-\tau)^{\alpha-1}\Phi_1(x;\tau)\overline{\Phi_2(x;t)dx\,d\tau} = 0 \tag{10}$$

The formula 10) represents the condition of the orthogonally of the EFs in the space $[L_2\,[a,b]\,XC\,(0,T\,),T<1$. Moreover, the function of time $(t-\tau\,)$,0 <α<1 is called the weight function.

Lemma 4: The EVs of the integro-differential operator (1) are real

Proof: Write the integro-differential operator L in the form of two operators, one is differentiable and the second is integrable i.e.

$$L = L_0 + K\,, L_0\Phi = \left(\frac{1}{i}\frac{\partial^\alpha}{\partial t^\alpha}\frac{d}{dx} + p(x)\right)\Phi,\quad K\Phi = \int_a^b k(|x-y|)\Phi(y;t)dy$$

Since L_0 is self adjoint operator with the condition $\Phi(a;t)=\Phi(b;t)$, see Abdou [45]. In addition, the kernel of the second integral operator is symmetric then K is self adjoint. In many mathematical physics investigation and computation of a function $f(\lambda)$ the neighborhood of a finite point λ_0 or in the neighborhood of the point at infinity is connected with considerable difficulties, the expression of the neighborhood of the point at infinity means for λ→∞ or λ→∞. These difficulties may often be overcome by means of an asymptotic formula that substitutes a simpler function for the given function $f(\lambda)$. This simpler function is chosen in such a manner that it can be investigated and computed in an easier way than the original function $f(\lambda)$. Moreover, it approximates to an arbitrary degree of accuracy when λ tends to λ_0 or approaches infinity.

Now, let we have the interval $\ell=(c,d)$ and consider the functions $f(\lambda)$, $g(\lambda)$, $h(\lambda)$, $m(\lambda)$,... , defined on the interval (c,d). So we can define the following:

(1) If $\lim\limits_{\lambda\to\lambda_0}\dfrac{f(\lambda)}{g(\lambda)} = 0$ $\lambda\in(c,d)$, we say that $f(\lambda)$ is of higher orders of smallness than $g(\lambda)$ on the interval (c,d) for $\lambda\to\lambda_0$ and write

$$f(\lambda) = O(g(\lambda)) \text{ for } \lambda\to\lambda_0 \text{ on (a,b)} \tag{12}$$

(2) if there is a constant β, $0<\beta<\infty$, such that for all $\lambda\in(c,d)$ belonging to a sufficiently small neighborhood of λ_0, the inequality $\left|\dfrac{f(\lambda)}{g(\lambda)}\right| < \beta$ holds, we say that, $f(\lambda)$ is of the order of $g(\lambda)$on the interval (c,d) and write

$$f(\lambda) = O(g(\lambda)) \text{ for } \lambda\to\lambda_0 \text{ on } (c,d) \tag{13}$$

(3) If the inequality $\left|\dfrac{f(\lambda)}{g(\lambda)}\right| < \gamma, 0<\gamma<\infty$ is full-filled for all $\lambda\in(c,d)$, we say that $f(\lambda)$ is of the order of $g(\lambda)$

(4) We say that the two functions $f(\lambda)$ and $g(\lambda)$ are equivalent for $\lambda\to\lambda_0$ on the interval (c,d) and write

$$f(\lambda) = (g(\lambda)) \text{ for } \lambda\to\lambda_0 \text{ on } l \tag{14}$$

The relation (14) can be adapted in the form

$$f(\lambda) = g(\lambda)[1+o(1)] \tag{15}$$

The formula (15) is called an asymptotic representation of $f(\lambda)$ in the neighborhood of the point λ_0 on the interval (c,d). The reader must know that the asymptotically equivalent functions on unbounded interval ℓ in the neighborhood of the point at infinity, i.e., for λ→∞, is defined similarly.

Theorem 4: Let $g(x)$ be an integrable function in the interval $[y,\beta]$ and μ be a parameter, then

$$\int_\alpha^\beta e^{\pm i\mu x}g(x)dx \to 0 \quad \text{as} \to\infty$$

Theorem 5: For the boundary value problem of (1), the EVs and EFs are asymptotically equivalent on the interval $[a, b]$ to the EVs and EFs of the boundary value problem

$$i\frac{\partial^\alpha}{\partial t^\alpha}\left(\frac{\partial\Phi(x;t)}{\partial x}\right) + P(x)\Phi(x;t) = \lambda\Phi(x;t)\;\; \Phi(a;t) = \Phi(b;t) = A(t) \tag{16}$$

Proof: Write the solution of (5), after using (2) and the condition $\Phi(a;t)=1$ in eqn. (5) we have

$$A(t) = [1-i](\sum_{n=0}^{\infty}\frac{Ut^{n\alpha}}{\Gamma(n\alpha+1)})e^{i\int_a^x(\lambda-P(u))du}\left\{\int_a^x e^{i\int_a^\zeta(P(v)-\lambda)dv}.\int_a^b k(|\zeta-z|)\phi(|u)dzd\zeta\right\} \tag{17}$$

Using the notations

$$\int_a^x P(u)du = B(x) \quad \int_a^b k(\zeta-z)\phi(z)dz = F(\zeta) \tag{18}$$

then, using the second condition, $\Phi(b;t)=A(t)$, in eqn. (5) we get

$$A(t) = [1-i]e^{i\lambda(b-a)-iB(b)}(\sum_{n=0}^{\infty}\frac{Ut^{n\alpha}}{\Gamma(n\alpha+1)})\left\{\int_a^b e^{i\lambda(\zeta-a)+iB(\zeta)}F(\zeta)d\zeta\right\}) \tag{19}$$

The second term in the right hand side (19) consists of the functional $e^{i\{\lambda(b-a)-B(b)\}}$ which is bounded in the interval (a,b), also the function $e^{iB(\zeta)}$ $F(\zeta),\zeta\in(a,b)$ is an integrable function in (a,b). Then, by theorem (5), as λ→∞, the second term of (17) tends to zero. Thus for large value of λ, the formula (17) becomes

$$1 = e^{i\int_a^b(\lambda-P(t))dt+}\quad O(1) \tag{20}$$

Therefore, for large value of λ the roots of (20) becomes

$$\lambda_m = \omega + \frac{2\pi m}{b-a},\ \omega = \frac{1}{b-a}\int_a^b P(u)du \quad m=0, \pm 1, \pm 2, \tag{21}$$

This can be adapted in the form

$$\lambda_m = \frac{2\pi m}{b-a} + \omega + O(1) \tag{22}$$

The corresponding EFs are

$$\Phi_m(x;t) = \left(\sum_{n=0}^{\infty}\left[\frac{2\pi m}{b-a} + \omega\right]\frac{Ut^{n\alpha}}{\Gamma(n\alpha+1)}\right)\left[e^{i\int_a^x\left[(\omega - P(u) + \frac{2\pi m}{b-a})\right]du}\right] \quad m=0, \pm 1, \pm 2, \tag{23}$$

The two formulas (22) and (23) represent, respectively the EVs and EFs of (1), and the general solution is given by:

$$\Phi_m(x;t) = \left(\sum_{n=0}^{\infty}\left[\frac{2\pi m}{b-a} + \omega\right]\frac{Ut^{n\alpha}}{\Gamma(n\alpha+1)}\right)\left[e^{i\int_a^x\left[(\omega - P(u) + \frac{2\pi m}{b-a})\right]du}\right] \quad \omega = \frac{1}{b-a}\int_a^b P(u)du \tag{24}$$

Conclusion

Figure 1 represents the 3D-plot for the EFsΦ with different values α and m. For fixed values of x=0.5, the increasing of fractional parameter $\alpha \in (0,1)$ implies to increase the profile of the EFs. Figure 2 represents the 3D-plot for the EFs Φ with different values α and x. For fixed values of m, the increasing of fractional parameter $\alpha \in (0,1)$ implies to increase the profile of the EFs, the peak of the vibration waves at large values of α be greater than others. Figures 3 and 4 represent the 3D-plot for the EFs Φ with different values α and x at (m=10000, m=0) respectively. For fixed values of, by comparing the result in Figure 2 with the result in Figure 3, the effect of fractional parameter $\alpha \in (0,1)$ has the same effect

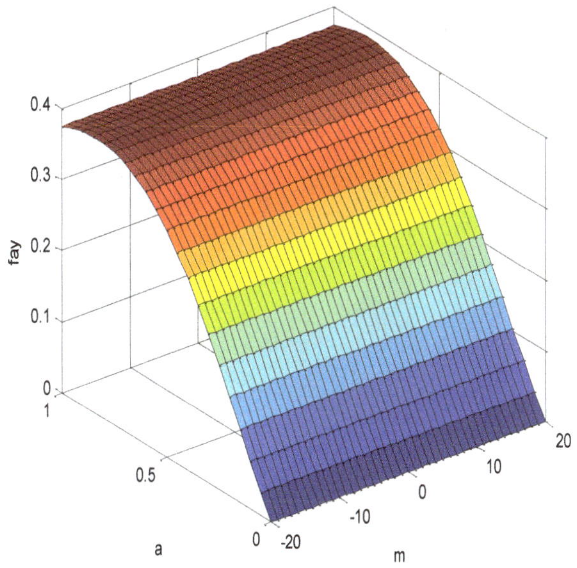

Figure 1: 3D plot between (Φ, α, m) at x=0.5.

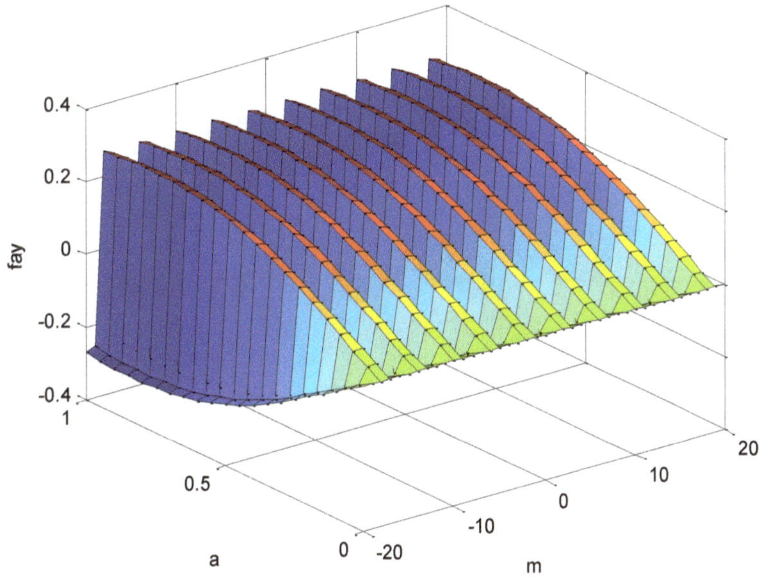

Figure 2: 3D plot between (Φ, α, x) at m=0.0.

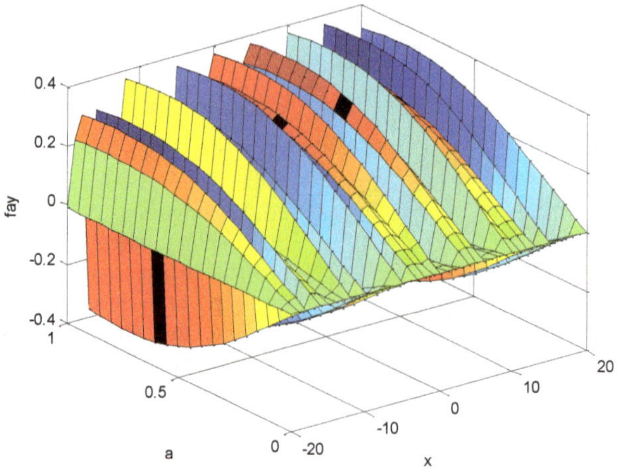

Figure 3: 3D plot between (Φ, α, x) at m=10000.

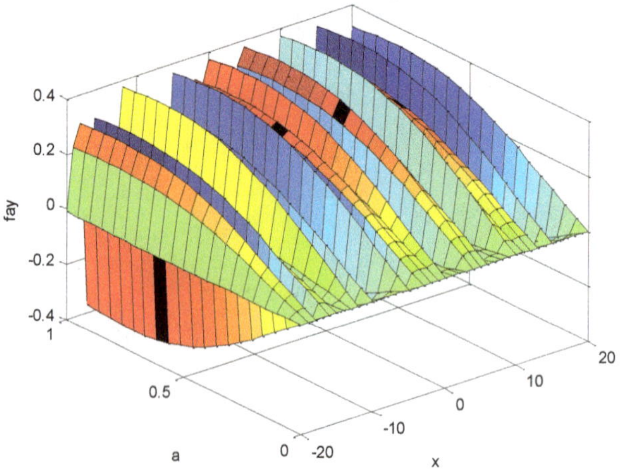

Figure 4: 3D plot between (Φ, α, x) at m=0.0.

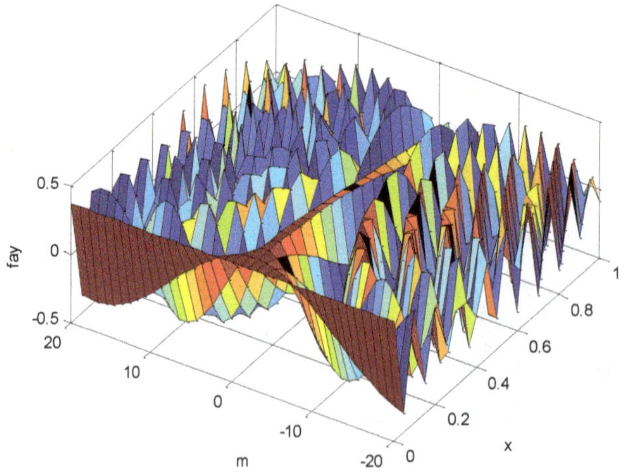

Figure 5: 3D plot between (Φ, m, x) at α=1.0.

on the EFs and the peak of the vibration wave. Figures 5-8 represent the 3D plot between (Φ, m, x) at different values α=1.0, 0.9,0.5,0.1 of on respectively. The effect of the fractional order parameter is very much prominent.

Future Work

In the future work the solution of nuclear integral equation in the homogeneous case, will be discussed and proved.

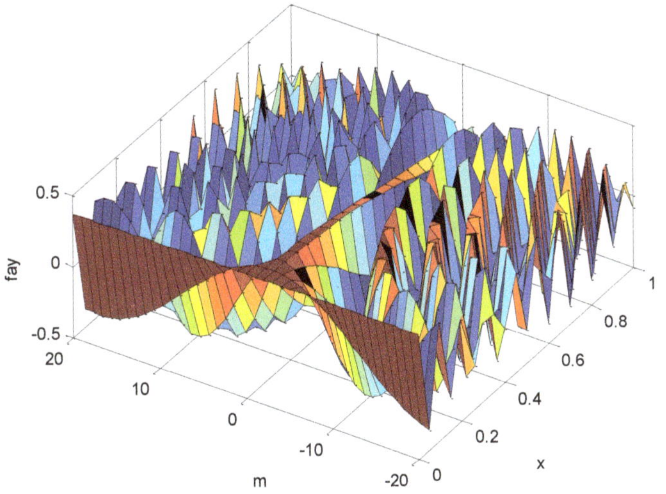

Figure 6: 3D plot between (Φ, m, x) at α=0.9.

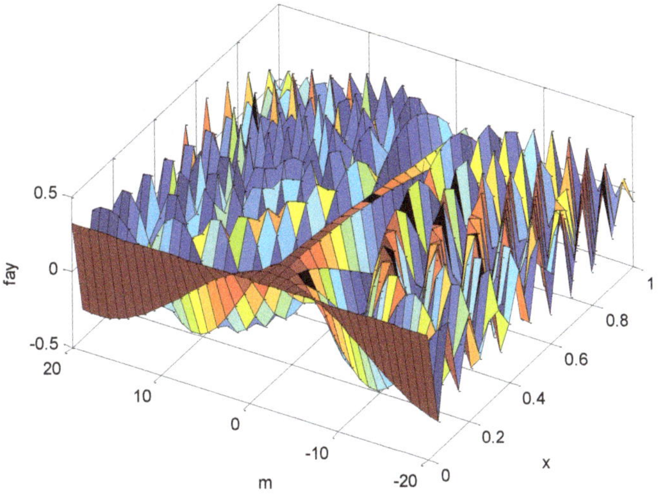

Figure 7: 3D plot between (Φ, m, x) at α=0.5.

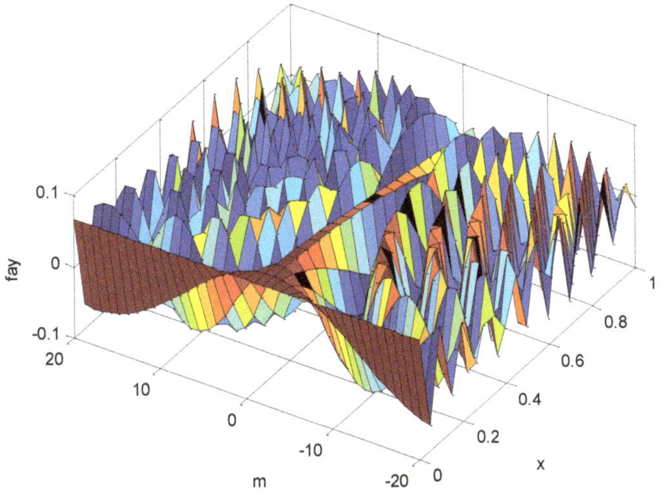

Figure 8: 3D plot between (Φ, m, x) at α=0.1.

References

1. Podlubny I (1999) Fractional differential equations, an introduction to fractional derivatives, fractional differential equations, to methods of their solution and some of their applications. ACADEMIC Press INC.
2. Miller KS, Ross B (1993) An Introduction to the Fractional Calculus and Fractional Differential Equations: Wiley Interscience.
3. Hilfer R (2000) Applications of Fractional Calculus in Physics: World Scientific Publishing Company, Singapore.
4. Machado JT, Silva MF, Barbosa RS, Jesus IS, Reis CM, et al. (2009) Some applications of fractional calculus in engineering. Mathematical Problems in Engineering.
5. Caputo M, Mainardi F (1971) Linear models of dissipation in anelastic solids. La Rivista del Nuovo Cimento 1: 161-198.
6. Mainardi F (2010) Fractional Calculus and Waves in Linear Viscoelasticity: An Introduction to Mathematical Models. Imperial College Press, UK.
7. Gorenflo R, Mainardi F (2007) Time-fractional derivatives in relaxation processes: A tutorial survey. Fract Calculus Appl Anal 10: 283-299.
8. Bernkopf M (1966) The development of function space to with particular reference to their origins in integral equation theory. Arch Hist Exact Sci 3: 1-96.
9. Muskhelishvili NI (1953) Singlar Integral Equations: boundary problems of function theory and their application to mathematical physics. Springer, Berlin.
10. Green CD (1969) Integral Equations Methods. Thomas Nelson publishers, Britain.
11. Hochstadt H (1973) Integral Equations. John Wiley, New York.
12. Kanwal RP (1991) Linear Integral Equations. Academic Press, New York,
13. Kanwal RP (1996) Linear Integral Equations Theory and Technique. Springer Science & Business Media, Boston.
14. Kress R (1989) Linear Integral Equations. Springer-Verlag, Berlin.
15. Mikhlin SG (1964) Integral Equations: Their applications to certain problems in mechanics, mathematical Physics and Technology. Pergamon Press, Oxford, UK.
16. Simirnov VI (1964) Integral Equations and Partial Differential Equations. A Course of Higher Mathematics. 4: 572-574.
17. Tricomi FG (1985) Integral Equations. Dover Publishers, New York.
18. Kantorovich L, Krylov V (1964) Approximate Method of Higher Analysis. Interscience Publishers, Netherlands.
19. Abdou MA, Mohamed SA, Darwish MA (1998) A numerical method for solving the Fredholm integral equation of the second kind. Korean J Comp Appl Math 5: 251-258.
20. Baker C, Miller G (1982) Treatment of Integral Equations by Numerical Methods. Academic Press New York.
21. Dzhuraev A (1992) Method of Singular Integral Equations. London, New York.
22. Delves LM, Walsh J (1974) Numerical Solution of Integral Equations. Clarendon Press, Oxford.
23. Delves LM, Mohamed JL (1985) Computational Methods for Integral Equations. Cambridge University Press.
24. Atkinson KE (1976) A Survey of Numerical Methods for the Solution of Fredholm Integral Equations of the Second Kind. Society for Industrial and Applied Mathematics.
25. Atkinson KE (1997) The Numerical Solution of Integral Equations of the Second Kind. Cambridge Univ Press.
26. Golberg MA (1990) Numerical Solution of Integral Equations. Plenum Press, New York.
27. Popov GY (1982) Contact Problem for a Linearly Deformable Base. Kiev Odessa.
28. Mkhitarian SM, Abdou MA (1990) On various method for the solution of the Carleman integral equation. Dakl Acad NaukArm SSR 89: 125-129.
29. Mkhitarian SM, Abdou MA (1990) On different method of solution of the integral equation for the planer contact problem of elasticity. Dakl Acad NaukArm SSR 89: 59-74.
30. Abdou MA (1996) Fredholm integral equation of the second kind with potential kernel. Comput Appl Math 72: 161-167.
31. Abdou MA (2000) Fredholm integral equation with potential kernel and its structure resolvent. Appl Math Comput 107:169-180.
32. Abdou MA (2003) Fredholm-Volterra integral equation and generalized potential kernel. J Appl Math Comput 131: 81-94.
33. Abdou MA (2001) Spectral relationships for the integral operators in contact problem of impressing stamp. J Appl Math Comput 118: 95-111.
34. Abdou MA (2001) Integral equation with Macdonald kernel and its application to a system of contact problem. J Appl Math Comput 118: 83-94.
35. Abdou MA (2002) Spectral relationships for the integral equation with Macdonald kernel and contact problem. J Appl Math Comput 118(2002)93-103.
36. Abdou MA (2003) Fredholm-Volterra integral equation of the first kind and contact problem. J Appl Math Comput 173: 231-243.
37. Abdou MA, Ezzeldin NY (1994) Krein's method with certain singular kernelfor solving the integral equation of the first kind. Per Math Hung 28: 143-149.
38. Abdou MA, Salama FA (2004) Volterra-Fredholm integral equation of the first kind and spectral relationships. J Appl Math Comput 153: 141-153.
39. Jiang TF (2008) Calculation of atomic hydrogen and its photoelectron spectrain momentum space. J comput phy commun 178: 571-577.
40. Jiayin LI, Barber JR (2008) Solution of transient thermo elastic contact problems by the fast expansion method, J Wear 265: 402-410.
41. Pang M (2008) Stability and approximations of eigenvalues and eigen functions for the Neummann Laplacian part 2 J Math Anal Appl 345: 485-499.

A New Approach to the Quantisation of Paths in Space-Time

Moffat J*, Oniga T, and Wang CHT

Department of Physics, University of Aberdeen, UK

Abstract

A discrete path in space-time can be considered as a series of applications of the translation subgroup of the Poincare group. If there is a local mapping from this translation group into a neighbourhood of the identity of a quantum Weyl algebra fibre bundle, then the whole classical path can be lifted into the fibre bundle to form a unique quantum field as a section through the fibres. Under the further assumptions of scale relativity, we also show that a discrete closed loop in space time, corresponding to two classical paths sharing the same end points, is renormalisable and the finite limit has anomalous dimension equal to the fractal dimension. We end by introducing the possibility of a 'push forward' connection on the bundle *O*(*D*) of Ehresmann type.

Keywords: Quantum Weyl algebra; Fibre bundle; Quantum Paths; Entropy

Introduction

Einstein's field equations for gravity are an 'effective' theory at the classical level, relating mass-energy flux to the changes in local space-time. At the underlying quantum level, we assume space-time is non-commutative due to the existence of additional non-commutative algebraic structure at each point x of space-time, forming a quantum operator 'fibre algebra' A(x). This structure then corresponds to the single fibre of a fibre bundle. A gauge group acts on each fibre algebra locally, while a 'section' through this bundle is then a quantum field of the form

$\{A(x);x\in M\}$ with M the underlying space-time manifold. In addition, we assume a local algebra $O(D)$ corresponding to the algebra of sections of such a principal fibre bundle with base space a finite and bounded subset of space-time, D. The algebraic operations of addition and multiplication are assumed defined fibrewise for this algebra of sections. The region D corresponds to the space-time constraints within which only that subset of $\{A(x);x\in M\}$ with $x\in D$ is of local physical interest or is capable of measurement. Alternatively, it can represent the set of sections $\{A(x);x\in M\}$ with compact support in D.

We relax the requirement of that each fibre algebra is norm separable and hence finite dimensional [1]. Here we assume that each fibre algebra has a faithful representation as a von Neumann algebra with trivial centre (a 'factor') acting on a separable Hilbert space via the Gelfand, Naimark, Segal (GNS) construction. This further implies that each fibre algebra is countably decomposable; every set of orthogonal projections is countable; and thus the fibre algebra has a faithful normal quantum state.

We assume, for now, that background space-time is globally flat. The union of all the local algebras generates algebra of all observables defined on the subset of space-time $\bigcup_\alpha D_\alpha$. The closure in the ultraweak operator topology of this set of local algebras generates the 'quasi-local' von Neumann algebra R of all observables. Choosing the ultraweak topology on R ensures that it contains an identity element. A key benefit of reformulating quantum field theory in this way as a'local algebra' formalism is the ability to consider coherently the many inequivalent irreducible representations, each corresponding to the Gelfand-Naimark-Segal (GNS) construction. These essentially represent different 'projections' of the same underlying algebraic structure.

Quantum Paths in Space-Time

Wald places representation of the Weyl form of the CCR at the centre of his approach to quantum field theory, generating the 'fundamental observables' and corresponding states [2]. We can add to that approach by interpreting it from the fibre bundle perspective. Thus we start by considering classical phase space. Given a dynamical system, entropy is defined through considering the phase space of the system. The emergent behaviour of this classical system gives rise to regions of phase space, each corresponding to similar macro-level behaviour. The number and variation in size of these regions reflects the overall complexity of the system. This identification is known as 'coarse graining'. The entropy of such a coarse grained region is essentially a count of all of the different micro-configurations constituting that region. A system starting in a low entropy state will tend to wander into larger coarse grained volumes; hence thermodynamic entropy tends to increase over time if the system is isolated, giving rise to the second law of thermodynamics. The structure of classical phase space is such that each set of initial conditions (x_μ, P_μ) generates a unique solution $S(x, P_\mu)$. For a Hamiltonian system it is possible to reformulate classical mechanics as a symplectic vector space of solutions, or equivalent initial conditions of location and momentum, equipped with a bilinear form Ω which ultimately derives from Hamilton's equations of motion;

With $\xi=(x, p)$ a 6-dimensional vector we have $\frac{\partial\xi}{\partial t}=\Omega\frac{\partial H}{\partial\xi}$ where $\Omega=\begin{bmatrix}0_{3\times3} & I_{3\times3}\\ -I_{3\times3} & 0_{3\times3}\end{bmatrix}$.

This is of the form of a symplectic vector space $V\oplus V^*$ with V a real finite vector space and with dual V^*. The skew-symmetric rank 2 tensor Ω then takes the general form;

$$\Omega(x\oplus\eta,\ x'\oplus\eta') = \eta'\cdot x - \eta\cdot x'.$$

In our case V is the configuration space, V^* the (dual) momentum space and $V\oplus V^*$ the phase space, a product vector bundle over V with

***Corresponding author:** Moffat J, Professor, Department of Physics, University of Aberdeen, King's College, UK; E-mail: james.moffat@abdn.ac.uk

fibre V^*. By choosing particular values such as (1,0,0,0,0,0) we can pull out particular elements; $\Omega(x \oplus \eta\ ,\ 0 \oplus \eta') = \eta' \bullet x = \eta'_1$.

From this point of view the Dirac canonical quantisation of elements of phase space such as η'_1 is equivalent to the canonical quantization; $\Omega \to \hat{\Omega}$ as a (not necessarily bounded) linear operator, and this form of canonical quantisation extends smoothly to countably infinite phase space. The approach is particularly transparent in flat space-time [2]. Given the canonical quantization; $\Omega \to \hat{\Omega}$ we can form the Weyl unitaries $\hat{W} = \exp i\hat{\Omega}$. Then closure of linear combinations of these unitaries and their adjoints in the normed operator topology is then a C*-algebra called the Weyl algebra.

In classical mechanics, given a particular dynamical relationship, we can select out the subset of phase space consisting of initial values. Each initial value vector *(x(0), p(0))* generates a unique solution *S={(x(t), p(t)); t>0}* propagating through phase space as a function of time *t*. In fibre bundle terms the solution space is also a product bundle with bundle projection *π(S)=S(0)=(x(0), p(0))* onto phase space. We can then extend the definition of Ω to the space of solutions *S* as Ω(S(1), S(2))=Ω(π(S(1), π(S(2)). This allows us to define an inner product on *S* as $\langle S(1), S(2)\rangle = -i\Omega\left(S(1)^*, S(2)\right)$ where S(1)* is the complex conjugate solution. It turns out (Wald, 1994) that this defines an inner product on *S* relative to which a one particle Hilbert space can be defined. For a quantum system of bosonic harmonic oscillators we can then assemble a symmetric tensor product Fock space in the usual way, using creation and annihilation operators.

An example of the Weyl form in a two dimensional locally flat space-time is now given for a local algebra *O(D)*, having a representation as observables acting on the Hilbert space $L^2(x,t)$ with Lebesgue measure.

For a small increment of space-time (δx,δt) we consider the Poincare Translation subgroup element $T(\delta x, \delta t):(x,t) \to (x+\delta x, t+\delta t)$ and define; $U_{T(\delta x,\delta t)} f(x,t) = f(x-\delta x, t-\delta t)$ for $f(x,t) \in L^2(x,t)$

Then *U* is a local group homomorphism of the translation group *T* as observables acting on $L^2(x,t)$. Define also;

$$
\begin{aligned}
&V_{\delta t} = U_{T(0,\delta t)}; V_{\delta x} = U_{T(\delta x,0)} \\
&\Rightarrow V_{\delta t} V_{\delta t'} f(x,t) = U_{T(0,\delta t)} U_{T(0,\delta t')} f(x,t) = f(x, t-\delta t' - \delta t) \\
&= U_{T(0,\delta t'+\delta t)} f(x,t) = V_{\delta t+\delta t'} f(x,t) \Rightarrow\ V_{\delta t} V_{\delta t'} = V_{\delta t+\delta t'}
\end{aligned}
$$

A similar result applies for $V_{\delta x}$,by symmetry.

Now introduce a deformation of the form; $T(\delta x, 0) \to Z_{\delta x} f(x,t) = \exp(it\delta x) f(x-\delta x, t)$. The mappings V and Z are unitary representations on $L^2(x,t)$ and so also is their product $V \times W : (\delta x, \delta t) \to T(\delta x, \delta t) \to V_{\delta t} Z_{\delta x}$

Then we have;

$$
\begin{aligned}
V_{\delta t} Z_{\delta x} f(x,t) &= V_{\delta t} \exp(it\delta x) f(x-\delta x, t) \\
&= \exp(it\delta x) f(x-\delta x, t-\delta t) \\
Z_{\delta x} V_{\delta t} f(x,t) &= Z_{\delta x} f(x, t-\delta t) \\
&= \exp(i(t-\delta t)\delta x) f(x-\delta x, t-\delta t) \\
&= \exp(-i\delta t\delta x) V_{\delta t} Z_{\delta x} f(x,t) \\
Z_{\delta x} V_{\delta t} &= \exp(-i\delta t\delta x) V_{\delta t} Z_{\delta x}
\end{aligned}
$$

Then $T(\delta x,\delta t) \to (Z_{\delta x}, V_{\delta t})$ is a local Weyl representation of the CCR on $L^2(x,t)$. By the Stone-von Neumann theorem, the resulting C*-algebra and its weak closure as a von Neumann algebra must be unitarily isomorphic to Wald's equivalent 'algebraic approach' to quantum field construction and his Weyl Algebra, since we can assume all relevant Hilbert spaces are separable in application to observed real systems [3].

In summary, this example indicates that a continuous local group homomorphism from a neighbourhood of the identity of *T* to a neighbourhood of the identity of the set of observables in *O(D)* exists as a Weyl representation of the CCR. It can be easily extended to 4 dimensions by replacing *x* by the 3-vector $\mathbf{x}=(x_1, x_2, x_3)$.

Definition: A discrete path in space-time is one which consists of linked causally directed intervals in space-time, each of the same Euclidean 4-dimensional length. More formally, we define the path as a series of *n* linked increments **a**(*j*) each of the same Euclidean interval length |**a**(*j*)| but varying direction such that the path begins at **x(0)** and ends at **x(1)**, with *T*(a(*j*)):x→x+a(*j*) elements of the translation subgroup *T*. The total path is then generated by the product $\prod_{j=1}^{j=n} T(\mathbf{a}(j))\mathbf{x}(\mathbf{0})$ with the final end point $\mathbf{x}(\mathbf{1}) = \mathbf{x}(\mathbf{0}) + \sum_{j=1}^{j=n} \mathbf{a}(j)$. For a fixed initial point **x(0)** we can identify this path with the finite group product $\prod_{j=1}^{j=n} T(\mathbf{a}(j)) \in T$.

Theorem 1: Let there be given a continuous local group homomorphism from a neighbourhood *V* of the identity of *T* to the neighbourhood *W* of the identity of the set of observables in *O(D)* as a Weyl representation of the CCR. Then a discrete classical path *CP* in space-time can be lifted to a quantum section *QP* through *O(D)*.

Proof: We construct the section iteratively, following a method suggested by Pontryagin [4]. Let CP be a discrete classical path in space-time. From the definition, the path is a series of *n* linked increments a(*j*) each of the same Euclidean interval length |a(*j*)| but varying direction such that the path begins at x(0) and ends at x(1), with *T*(a(*j*)):x → x+a(*j*) elements of the translation subgroup *T*. We can thus, as noted earlier, characterise CP as the product mapping $T^n(a(j))$ in the translation group *T* for a given x(0). We have, by assumption, a continuous local group homomorphism $\phi:T(a\ ) \to U_{T(a)}$ from *V*, a neighbourhood of the identity of T to *W*, a neighbourhood of the identity of *(O(D))*. Clearly, by redefining the number of links in CP if necessary, we can assume that |**a**(*j*)| is sufficiently small so that *T*(a(*j*)) ∈ *V* for all *j* and by choosing appropriate units we can assume that CP consist of *n* links each of length *1/n*.

We also assume that for *n* large; $|\mathbf{x}(t_2) - \mathbf{x}(t_1)| < \frac{1}{n} \Rightarrow T(\mathbf{x}(t_1))^{-1} T(\mathbf{x}(t_2) = T(\mathbf{a}(t_1)) \in V.$

Let *m* be a positive integer strictly less than *n*, and suppose that the path *QP* has been defined such that its initial value is *QP(0)*=A(x(0)). We proceed by induction. Assume that *QP* has been defined for all values of x(*t*) with $0 \le t \le \frac{m}{n}$, and satisfies, for all such *t* with $0 \le t \le \frac{m}{n}$;

(a). Fixed endpoint; *QP(0)*=A(x(0));

(b). Local lifting to the Weyl algebra near the identity; If **x**(s), **x**(t) in *CP* satisfy $|\mathbf{x}(s) - \mathbf{x}(t)| \le \frac{1}{n}$ then $T(\mathbf{x}(s))^{-1} T(\mathbf{x}(t)) \in V$ and $\varphi\left(T(\mathbf{x}(s))^{-1} T(\mathbf{x}(t)\right) = QP(\mathbf{x}(s))^{-1} QP(\mathbf{x}(t)) \in W$

We now extend the path *QP*(**x**(t)) with $0 \le t \le \frac{m}{n}$, stepping forward one additional link on CP so that $t = \frac{m}{n} + \frac{1}{n}$, by the following construction;

$$
QP\left(\mathbf{x}\left(\frac{m+1}{n}\right)\right) = QP\left(\mathbf{x}\left(\frac{m}{n}\right)\right) \varphi\left(T\mathbf{x}\left(\frac{m}{n}\right)^{-1} T\mathbf{x}\left(\frac{m+1}{n}\right)\right) \tag{1}
$$

From eqn. (1) the extension of the path QP still satisfies (a): QP(0)=A(x(0)) since φ acting on the identity of the group local translations T is the identity operator in O(D). We need to show that the extension under induction still satisfies (b).

Let h be a real number with $|h| \le \frac{1}{n}$. If h is positive then by induction h satisfies the extension shown in equation (1). Thus we have;

$$QP\left(\mathbf{x}\left(\frac{m}{n}+h\right)\right)=QP\left(\mathbf{x}\left(\frac{m}{n}\right)\right)\varphi\left(T\mathbf{x}\left(\frac{m}{n}\right)^{-1}T\mathbf{x}\left(\frac{m}{n}+h\right)\right)$$

If on the other hand h is negative then setting $\mathbf{x}(s)=\mathbf{x}\left(\frac{m}{n}+h\right)$ and $\mathbf{x}(t)=\mathbf{x}\left(\frac{m}{n}\right)$; since s, t are both equal to or less than $\frac{m}{n}$ by the inductive hypothesis they therefore satisfy;

$$\varphi\left(\left(T\mathbf{x}(s)\right)^{-1}T\mathbf{x}(t)\right)=\left(QP\mathbf{x}(s)\right)^{-1}QP\mathbf{x}(t) \text{ with } t=\frac{m}{n}+h \text{ and } s=\frac{m}{n}$$

$$\Rightarrow \varphi\left(\left(T\mathbf{x}\left(\frac{m}{n}\right)\right)^{-1}T\mathbf{x}\left(\frac{m}{n}+h\right)\right)=\left(QP\mathbf{x}\left(\frac{m}{n}\right)\right)^{-1}QP\mathbf{x}\left(\frac{m}{n}+h\right)$$

$$\Rightarrow QP\mathbf{x}\left(\frac{m}{n}+h\right)=QP\mathbf{x}\left(\frac{m}{n}\right)\varphi\left(T\mathbf{x}\left(\frac{m}{n}\right)^{-1}T\mathbf{x}\left(\frac{m}{n}+h\right)\right)$$

Thus equation (1) holds for both positive and negative values of h.

It follows that;

$$\left(QP\mathbf{x}\left(\frac{m+1}{n}\right)\right)^{-1}QP\mathbf{x}\left(\frac{m}{n}+h\right)=\left(QP\mathbf{x}\left(\frac{m}{n}\right)\varphi\left(T\mathbf{x}\left(\frac{m}{n}\right)^{-1}T\mathbf{x}\left(\frac{m+1}{n}\right)\right)\right)^{-1}\left(QP\mathbf{x}\left(\frac{m}{n}\right)\varphi\left(T\mathbf{x}\left(\frac{m}{n}\right)^{-1}T\mathbf{x}\left(\frac{m}{n}+h\right)\right)\right)$$
$$=\varphi\left(T\mathbf{x}\left(\frac{m+1}{n}\right)^{-1}T\mathbf{x}\left(\frac{m}{n}+h\right)\right)$$

Thus the path extension satisfies both requirements (a) and (b) completing the inductive step $\frac{m}{n}\to\frac{m+1}{n}$, provided we satisfy the local topological constraints, namely;

We have that if n is sufficiently large then there is a neighbourhood U such that;

$$|x(t_2)-x(t_1)|<\frac{1}{n}\Rightarrow \text{by continuity } T(x(t_1))^{-1}T(x(t_2)=T(a(t_1))\in U^{-1}U\subset V$$

and $\varphi\left(T(x(t_1))^{-1}T(x(t_2)\right)=\varphi(T(a(t_1))\in W$

Setting, for small h>0;

$$t_1=\frac{m+1}{n};t_2=\frac{m}{n}+h\Rightarrow|\mathbf{x}(t_1)-\mathbf{x}(t_2)| \text{ small}$$

$$\Rightarrow T\mathbf{x}\left(\frac{m+1}{n}\right)^{-1}T\mathbf{x}\left(\frac{m}{n}+h\right)=T\left(\mathbf{x}(t_1)\right)^{-1}T\left(\mathbf{x}(t_2)\right)=T\left(\mathbf{x}(t_2)\right)T\left(\mathbf{x}(t_1)\right)^{-1}=T\left(\mathbf{x}(t_2)-\mathbf{x}(t_1)\right)\in V$$

$$\varphi\left(T\left(\mathbf{x}\left(\frac{m+1}{n}\right)\right)^{-1}T\mathbf{x}\left(\frac{m}{n}+h\right)\right)\in W \text{ as required.}$$

Since ϕ(identity of T)=identity of $O(D)$ the induction hypothesis is true for $\frac{m}{n}=0$; $h=\frac{1}{n}$. By induction the path QP(t) can be extended in O(D) for all discrete steps m less than or equal to n.

Theorem 2: The constructed quantum path QP is unique.

Proof: The initial point of QP is unique by condition (a). If $QP(\mathbf{x}(t))$ is unique for all

$t\le t_0$ then let $t_0<t\le t_0+\varepsilon$. Then;

$$T\mathbf{x}(t_0)^{-1}T\mathbf{x}(t)\in V\Rightarrow\varphi\left(T\mathbf{x}(t_0)^{-1}T\mathbf{x}(t)\right)=QP\mathbf{x}(t_0)^{-1}QP\mathbf{x}(t)\Rightarrow QP\mathbf{x}(t)=QP\mathbf{x}(t_0)\varphi\left(T\mathbf{x}(t_0)^{-1}T\mathbf{x}(t)\right)$$

Thus the path QP is uniquely determined for all points $t<t_0+\varepsilon$. The result follows by induction.

Theorem 3: There is a projection π from the fibre bundle $O(D)$ mapping the quantum path back to the translation subgroup T

Proof: From condition (a) this is clear for the initial point. We again use induction to prove the general case. If the projection π maps the observable QP $\mathbf{x}(t)$ back to $T\mathbf{x}(t)$ for $t\le t_0$ then let $t_0<t\le t_0+\varepsilon$. Then as before we have;

$$QP\mathbf{x}(t)=QP\mathbf{x}(t_0)\varphi\left(T\left(\mathbf{x}(t_0)\right)^{-1}T\mathbf{x}(t)\right)$$

$$\text{thus } \pi QP\mathbf{x}(t)=\pi QP\mathbf{x}(t_0)\pi\circ\varphi\left(T\left(\mathbf{x}(t_0)\right)^{-1}T\mathbf{x}(t)\right)=T\mathbf{x}(t_0)\left(T\left(\mathbf{x}(t_0)\right)^{-1}T\mathbf{x}(t)\right)=T\mathbf{x}(t)$$

The result follows by induction.

Renormalisation of Discrete Fractal Paths in Space-Time

The principle of relativity is captured within the assumptions of the Riemannian geometry of 4-manifolds, where formulae equating a tensor expression to zero remain invariant under covariant and contravariant coordinate transformations. It is a natural extension of these ideas to additionally postulate that the scales of measurement inscribed on the clocks or measuring rods used by an observer should also not be absolute. Mathematically this can be captured by the additional requirement that the tensor formulae should be invariant under transformations of scale [5]. From this perspective a relativistic quantum system is a *scale free system* as first defined by James [6].

The derivation of a particular quantum relationship has to be inferred, in a rather *ad hoc* way, from the context and can be captured in the abstract by a function Φ linking system inputs and outputs.

Definition: A system is scale free if observers using different scales observe the same functional relationship Φ [6].

Definition: A system input variable is dimensionally independent if it cannot be described dimensionally by a combination of other inputs; otherwise it is described as dimensionally dependent [7].

Assume we have a scale free system with output value a, functional relationship Φ; k dimensionally independent input variables a_1, a_2,........, a_k and 2 dimensionally dependent input variables b_1,b_2. Given the mathematical relationship linking inputs to output; $a=\Phi(a_1,\ldots a_k,b_1,b_2)$ it is possible to vary the arguments $a_1,\ldots a_k$ using arbitrary positive numbers so that:

$$a_1'=A_1a_1,\ldots,a_k'=A_ka_k$$

By definition, the dimensions of a,b_1,b_2 may be represented as power monomials in the dimensions $a_1,\ldots a_k$ for example:

$$[b_1]=[a_1]^{p_1}\ldots[a_k]^{r_1}$$
$$[b_2]=[a_1]^{p_2}\ldots[a_k]^{r_2}$$
$$[a]=[a_1]^{p}\ldots[a_k]^{r}$$

We therefore obtain the transformations:

$$b_1'=A_1^{p_1}\ldots A_k^{r_1}b_1$$
$$b_2'=A_1^{p_2}\ldots A_k^{r_2}b_2$$
$$a'=A_1^{p}\ldots A_k^{r}a$$

The above transformations form a group of continuous gauge

transformations with $A_1 \ldots A_k$ as the parameters. For a scale free system, our physical relationship can then be represented as a relationship between gauge transformation group invariants:

$$\Pi = \Phi(\Pi_1, \Pi_2)$$

These invariants are given by:

$$\Pi_1 = \frac{b_1}{a_1^{p_1} \ldots a_k^{r_1}}$$

$$\Pi_2 = \frac{b_2}{a_1^{p_2} \ldots a_k^{r_2}}$$

$$\Pi = \frac{a}{a_1^{p} \ldots a_k^{r}}$$

The invariants Π_1 and Π_2 are similarity parameters and the functional relationship Φ has the equivalent form;

$$\Phi(a_1,\ldots a_k, b_1, b_2) = a_1^p \ldots a_k^r \Phi'\left(\frac{b_1}{a_1^{p_1} \ldots a_k^{r_1}}, \frac{b_2}{a_1^{p_2} \ldots a_k^{r_2}}\right)$$

Three possibilities are available for this system under renormalisation of one of the similarity parameters; [7]

a) Φ tends to a non-zero finite limit as $\Pi_2 \to 0$ This means that Φ can be replaced by its limiting expression, with complete separation of variables and the functional relationship is a product of powers whose values can be determined by dimensional analysis.

b) Φ has power law asymptotics of the form $\Phi = \Pi_2^{\alpha_1} \Phi'\left(\frac{\Pi_1}{\Pi_2^{\alpha_2}}\right)$, as $\Pi_2 \to 0$. The power law form of the limiting expression still leads to separation of variables, but with characteristic exponents equal to the 'anomalous' fractional dimensions of a form of renormalisation [8,9].

c). neither a). nor b) holds; Φ has no finite limit different from zero and no power-law asymptotics.

In summary, for scale relativity, as a scale, free system, application of a renormalisation group is mathematically equivalent to the intermediate asymptotics approach. We can exploit this equivalence to prove the following result.

Theorem 4: Under the assumptions of scale relativity, a discrete closed loop in space-time; corresponding to two discrete non-oriented paths sharing the same end points, is renormalisable and has a finite limit as the step size of the curve tends to zero.

Proof: Assume that we have a fractal closed loop *L* in space-time with Euclidean diameter *d*. We approximate *L* by a discrete closed path $L(\eta)$ where η is the Euclidean length of each segment of $L(\eta)$. Standard dimensional analysis shows that $N(\eta)$, the number of segments in the path $L(\eta)$, is a function of the form $f\left(\frac{d}{\eta}\right)$. We will establish the nature of this function and its renormalisation limit, following a suggestion of Barenblatt and Isaakovich [7].

The fractal, self-similar nature of the discrete path implies that if we consider a finer segmentation of segment length ξ, then $N(\xi) \propto N(\eta) N(\xi \mid \eta) = \dfrac{f\left(\frac{d}{\eta}\right) f\left(\frac{\eta}{\xi}\right)}{f(1)^2}$ where $N(\xi|)$ is the number of segments of length ξ in a segment of length η, and $N(d) = N(\eta \mid \eta) = f(1)$.

It follows that $\dfrac{f\left(\frac{d}{\xi}\right)}{f(1)} = \dfrac{f\left(\frac{d}{\eta}\right) f\left(\frac{\eta}{\xi}\right)}{f(1)^2}$ thus $f\left(\frac{d}{\xi}\right) = \dfrac{f\left(\frac{d}{\eta}\right) f\left(\frac{\eta}{\xi}\right)}{f(1)}$

This implies that *f*, for the limiting case, must be of the form $f\left(\frac{x}{y}\right) = C\left(\frac{x}{y}\right)^D$ with *C* and *D* constants; *C=f(1)*. Thus we have;

$$f\left(\frac{d}{\eta}\right) = f(1)\left(\frac{d}{\eta}\right)^D \Rightarrow N(\eta) = \left(\frac{d}{\eta}\right)^D \text{ and } L(\eta) = \eta\left(\frac{d}{\eta}\right)^D$$

Locally along the limiting smooth form this implies $L(\xi \mid \eta) = \xi\left(\frac{\eta}{\xi}\right)^D = \eta^D \xi^{1-D}$.

The renormalisation limit is thus finite and we identify *D* with the path fractal dimension. We end by introducing the possibility of a 'push forward' connection on the bundle *O(D)* of Ehresmann type.

Definition: If π is the projection map from $O(D) \to D$, let **x** be an element of *D* and *p* an element of the fibre $\pi^{-1}(\mathbf{x})$, so that $\pi(p)=\mathbf{x}$. The 'push forward' of π, denoted π_* is a connection we define as follows. Let $t \to A(x(t))$ be a section passing through the point p in the fibre $\pi^{-1}(x(t_0))$, so that $p=A(x(t_0))$ and $\pi(p)=x(t_0)$, a point on the curve $t \to x(t)$ defined on the base space *D* and passing through the point $x(t_0)$ with velocity $v_{\mathbf{x}} = \frac{\partial \mathbf{x}}{\partial t}|_{t_0}$. If $v_p = \frac{\partial (Ax(t))}{\partial t}|_{t_0}$, with v_p the velocity of the curve $t \to A(x(t))$ at *p*, then $\pi_*(v_p)=v_x$

We now define the tangent space to *O(D)* as the linear space *TO(D)* generated by the set $\left\{\frac{\partial A(x(t))}{\partial x_\mu}; t \to A(x(t)) \text{ is a path in } O(D) \text{ and } x_\mu = t, x, y, z\right\}$

Similarly, we define the tangent space to *D*, denoted *TD*, as the linear space generated by the set;

$$\left\{\frac{\partial(\mathbf{x}(t))}{\partial x_\mu}; t \to \mathbf{x}(t) \text{is a path in } D \text{ and } x_\mu = t, x, y, z\right\}$$

With these definitions, we see that the push forward connection π_* maps *TO(D)* to *TD*.

Conclusion

Through the paper we explained the series of applications of the translation subgroup of the Poincare group. In the end, we introduced the possibility of a 'push forward' connection on the bundle *O(D)* of Ehresmann type.

Acknowledgments

CW is grateful to the Cruickshank Trust, Scotland for financial support.

References

1. James M, Oniga T, Wang CHT (2016) Group cohomology of the Poincare group and invariant quantum states, Cornel University Library.
2. Wald RM (1994) Quantum field theory in curved spacetime and black hole thermodynamics. University of Chicago Press.
3. Rosenberg J (2004) A selective history of the Stone-von Neumann theorem. Contemporary Mathematics 365: 331-333.

4. Pontryagin LS (1996) Topological Groups. Gordon and Breach, New York
5. Nottale L (2011) Scale Relativity and Fractal Space-Time: A New Approach to Unifying Relativity and Quantum Mechanics. Imperial College Press, London.
6. James M (2006) Mathematical modelling of information age conflict. Advances in Decision Sciences.
7. Barenblatt, Isaakovich G (1996) Scaling, self-similarity, and intermediate asymptotics: dimensional analysis and intermediate asymptotics, Cambridge University Press.
8. Gell-Mann M, Low FE (1954) Quantum electrodynamics at small distances. Physical Review 95:1300.
9. Goldenfeld Nigel, Martin O, Oono Y (1989) Intermediate asymptotics and renormalization group Theory. Journal of Scientific Computing 4: 355-372.

An Implicit and Untested Premise of the Special Theory of Relativity

Asokan S.P.*

Independent Researcher, New No 9, Old No 2, T.S.V Koil Street, Mylapore, Chennai, Tamil Nadu, India

Abstract

The derivation of Lorentz Transformation Equations in the Special Theory of Relativity, besides being explicitly based on the two postulates of that theory is also critically based on an implicit premise that the detection of a light signal/particle at a point in space at an instant of time is an event that is not exclusive to any inertial reference frame but it is capable of being measured by the observers in other inertial reference frames as well. This paper explains how this untested paradoxical premise vitiates the entire theory and suggests its replacement with a contrary postulate that the detection of light in an inertial reference frame is an event that is exclusive to that frame. This paper shows that if that contrary postulate is accepted, then the absoluteness of space and time declared by Newton can coexist with the absoluteness of the speed of light in vacuum declared by Einstein without any conflict between them.

Keywords: Lorentz transformation; Non-synchronization of moving clocks

Introduction

Of the happening of any observable event, it can be said that there is a near universal agreement about the fact that it has happened somewhere in the space at some point of time, even though there may be disagreements on where and when it happened. Indirectly, this indicates universal agreement on the existence of absolute space and absolute time notwithstanding the difficult, if not impossible, task of fixing an absolute inertial frame of reference from which the absolute measurements of space and time distances could be made. All observers in one inertial frame of reference, say S, agreeing on a common origin event, say O, would define that event with values of x-space coordinate and a time coordinate say (x,t). But the observers in another inertial frame of reference, say S' moving with a uniform velocity v relative to S in the positive X-direction, would define that event (*if it is observed by them also*) with different values of x-space coordinate and a time coordinate say (x', t') even though they had agreed on the same origin event O. This difference results from a universal "*ignorance*" that the observers in every frame of reference assume that their frame is at rest while only all other inertial frames are moving. Fortunately for us, the first postulate of the Special Theory of Relativity (STR) ensures that this ignorance does not hinder our scientific pursuits in any way. According to the STR, the aforesaid two sets of coordinates defining one and the same event (in the absolute space and time spectrum) by the observers in the frames S and S' are connected by the following two equations, known as Lorentz Transformation Equations [1];

$x' = a(x - vt)$

$t' = a(t - vx/c^2)$

Where $a = 1/(1 - v^2/c^2)^{1/2}$

It will be shown in the following sections that the derivation of the above equations is explicitly based on the following two postulates of the Special Theory of Relativity [2]:

(i) The laws of nature have the same mathematical form in all inertial reference frames;

(ii) The speed of light in vacuum is the same for all inertial reference frames.

An important fact is that Einstein, while deriving Lorentz Transformation Equation from the two postulates of the STR, had critically relied on a premise that the detection of a light signal/particle at a point in space at an instant of time is one and the same event that is observed by both observers–each moving with a uniform linear velocity relative to one another-, and they differ only in the values of the spatial location and the time of occurrence of that event measured by each of them. In other words, the detection of a light signal/particle at a point in space at an instant of time is an event that is not exclusive to any inertial reference frame but it is capable of being measured by the observers in other inertial reference frames as well. Presumably, Einstein would have intuitively taken it as an obvious truth that needed no express statement. As a matter of fact there would be necessity for a transformation equation only when one and the same event is observed by two observers and such an equation is obviously meaningless when the nature of the event is of such a manner that it is observable only by one observer and not by the other observer. The following sections of this article examine the validity of the said premise.

A Straight Forward Derivation of Lorentz Transformation Equations

Starting with Galilean Transformation Equation, which is in agreement with the first postulate of the STR, and making necessary alterations to make it agree with the Second Postulate also, Lorentz Transformation Equations can be easily derived by any novice without involving any advanced mathematics.

Suppose for an event of detection of a light signal at a particular point in space and at a particular instant of time by the Light Detector L(D), which is stationary in the frame S is assigned the coordinates (x,t) by a stationary observer S in that frame. It is obvious x=ct where c is the speed of light in vacuum explained in Figure 1.

Suppose the observer S' using a light detector L'(*D'*) that is stationary relative to him, detects the light signal at a distance **x'** from him at the very same instant. We may conclude from Law of Constancy of the Speed of Light that *x'=ct*.

***Corresponding author:** Asokan SP, Independent Researcher, TSV Koil Street, Mylapore, Chennai, Tamil Nadu, India, E-mail: spasokan@gmail.com

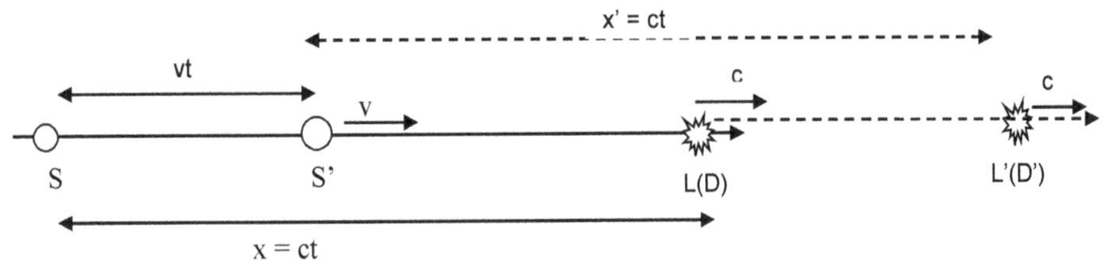

Figure 1: Illustration of instant of time by the Light Detector.

The expectation of the author of this article is that the Light Detector L(D) stationary in the frame S and the Light Detector L'(D') stationary in the frame S' will receive the light at the same instant of time (t) but at different spatial locations separated by the distance vt. But, according to the STR, both Light Detectors L(D) and L'(D') will receive the light at the same instant of time (t) at the same place where Light Detector L(D) is located. While deriving *Lorentz Transformation Equations* from the two postulates of STR, Einstein proceeded on the basis that that particular light signal was capable of detection at the very same point in space and at the very same particular instant of time by the observers in both S and S'. In other words the one and the same event of detection of light at a at a particular point in space and at a particular instant of time can be observed by the observers in S as well by observers in S' even though the values of space and time coordinates (x',t') they would assign to that event would be different. This author prefers to give this assumption a name "*Premise three*" for future references in this article. Though this premise has not been expressly stated anywhere in the Special Theory of Relativity, it forms the core of the STR.

According to the *Premise three*, which claims that the same event can be detected by the stationary observers stationed in the frame S' also and the Galilean Transformation

$x'=x-vt$

Since x=ct for the event under consideration, the above equation becomes

$x'=ct-vt$

The Second Postulate requires that when the same event is expressed by the coordinates (x',t') in the frame S', then they should satisfy the equation x'=ct' in obedience to the law of constancy of the velocity of light in all inertial reference frames, which is the Second Postulate. This means

$t'=x'/c$

$=(ct-vt)/c$

$=t-(vt/c)$

$=t-vx/c^2$ [because t=x/c]

Now the following two equations

$x'=x-vt$ and

$t'=t-vx/c^2$

can serve as Transformation Equations satisfying the second postulate as well as the Premise three. If one does not miss Physics for Mathematics, one will note that the above expression for t' gives only the part of the total time t that was taken by the light signal to travel the distance beyond the origin point of the frame S' (that is the distance x' in S') to reach the point at a distance x in S. But in the STR Physics was sacrificed for the sake deriving a mathematical equation that correctly transforms events of detection of light signals from one inertial frame to the other and it was taken that *t'* gives the time between the origin event and the measured event as measured by the observers in S' whereas *t* gives the time between the very same events as measured by the observers in S.

The consequence of taking t' as the counterpart of t in the frame S, instead of taking it as a part of t, is that, according to an observer in frame S, the clocks attached to S' will be showing different time at an instant of time depending on their relative distances from the origin point whereas all the clocks attached to S will be showing an identical time; the equation $t'=t-vx/c^2$ gives the time shown by the clock attached to S' that is at a distance x from the origin point of S as measured in S at an instant of time which is shown as t by all clocks attached to S. We may call this consequence "*Non-synchronization of moving Clocks*"

Introducing "*Premise three*" with its inevitable consequence of *Non-synchronization of moving Clocks*, Einstein satisfied the second postulate of the STR. But that has cost him the first postulate. Einstein crossed this hurdle in an ingenious way. Adding any constant factor, say **a**, on the RHS of both equations would not disturb the only requirement x'/t'=x/t=c that is needed for upholding the second postulate of the STR. So we can modify the equations as

$x'=a(x-vt)$ and

$t'=a(t-vx/c^2)$

Let us now choose a suitable value for *a* so that the equations satisfy the first postulate of the STR also.

From the above equations it can be derived

$x=[1/a(1-v^2/c^2)]\ (x'+vt')$

$t=[1/a(1-v^2/c^2)]\ (t'+vx'/c^2)$

So to satisfy the first postulate

$a=1/a(1-v^2/c^2)$

$a^2=[1/(1-v^2/c^2)]$

Therefore

$a=\pm 1/(1-v^2/c^2)^{1/2}$

The negative value is ignored and it is taken

$a=1/(1-v^2/c^2)^{1/2}$

The final forms of the equations are

$x'=a(x-vt)$ and

$t'=a(t-vx/c^2)$

Which are equivalent to

$x=a(x'+vt')$ and

$t=a(t'+vx'/c^2)$

Where $a=1/(1-v^2/c^2)^{1/2}$

Here again, if one does not miss Physics for Mathematics, one will note that this addition of factor **a** to RHS of the equations is responsible for the physical consequences of

(i) Length Contraction ($x'=x/a$ when $t'=0$); and

(ii) Time Dilation ($t'=t/a$ when $x'=0$)

[As already stated, "*Premise three*" is responsible for the physical consequence of *Non-synchronization of moving Clocks.]*

The above derivation of *Lorentz Transformation Equations* may appear to be crude, unsophisticated and even artificial. But careful dissections of derivations of those equations given by various authors including the one given by Einstein himself in his original 1905 German-language paper published as zur Elctrodynamik bewegter Korper, in Annalen der Physik 17:891, 1905) [3] would show that those derivations followed, in essence, the same logic that we have used in the above derivation even though the derivations of those authors may have been clothed in sophisticated formats reflecting the scholarliness and mathematical geniuses of those authors. This straightforward derivation of *Lorentz Transformation Equations* has been preferred in this article because it shows not only the respective parts played by the two postulates of the STR in that derivation but also how those equations critically depend on *Premise three.*

It may also be seen that out of infinite events that have been taking place at infinite points of space at infinite instants of time, only the events of the light signal being at a particular point in space at a particular instant of time during its transmission alone were considered. It has been generalised that the same equations would govern even the other infinite events that have nothing to do with the transmission of light. Obviously, the validity of this generalization depends on experimental verification of the predicted consequences of the STR. It may also be noted that the *cause-effect relations* between events, which constitute the bedrock of all natural laws, has been left to the mercy of the impossibility of communication at a speed greater than that of light.

Physical Consequences of Lorentz Transformation

For a better understanding of the physical consequences of *Lorentz Transformation Equations,* they can be rewritten in the following format;

$x'=(x/a)-vt'$ (Or) $x=(x'/a)+vt$

$t'=(t/a)-(vx'/c^2)$ (Or) $t=(t'/a)+(vx/c^2)$

The first term in RHS of the First Line Equations indicates *Length Contraction.*

The first term in RHS of the Second Line Equations indicates *Time Dilation.*

The Second term in RHS of the Second Line Equations indicates *Non-synchronization of moving Clocks*

The Second term in RHS of the First Line Equations indicates *Relative Uniform Motion between the frames.*

Premise Three is Untested

Though there have been claims of experimental verification of *Time Dilation,* so far no one has claimed to have verified *Non-synchronization of moving Clocks,* which was a direct consequence of Premise three. Even in the famous *Mu-Meson Experiment* [4], only the *Time Dilation* of moving mu-mesons/clocks was claimed to have been verified and it was not verified whether at a given time instant in earth frame, the moving mu-mesons/clocks showed different times i.e., different stages of decay. Premise three can be said to have been experimentally verified only if a single event of detection of a light signal at a particular point in space and at a particular instant of time is observed by both of the two observers belonging to different inertial frames of reference.

It does not seem to be possible for any observer measure the location of a photon at different instants during its movements over a distance and draw a trajectory of its transition. It is only possible for any observer to know the location of a photon as and when it impinges on a light detector. The spatial location of that photon at the instant when it impinges on the light detector is the spatial location of the light detector at that instant; and the time of such impinging is the time as per the clock fitted with that light detector. Obviously a light detector can be stationary only in one inertial frame of reference and its clock is synchronised with the stationary clocks of that frame. Let us assume that the light detector L stationary in the frame S and another light detector L' stationary in frame S' happen to be at one spatial point at the instant of time when a ray of light reaches that point. Now according to *Premise three,* the detection of light by the light detector L and the t detection of light by the light detector L' must be simultaneous events as viewed an any frame of reference since both the spatial distance and time difference between the two events were zero. Only if that simultaneity is experimentally proved, *Premise three* can be said to have been proved.

It may also be seen that since the time **t'** measured by the moving observers corresponds to the time t measured by the moving observers is the counterpart of **t,** the Clocks of the moving observers, which were found to have been synchronized before the commencement of the relative motion between the two systems, became non-synchronized even at the very commencement of the relative motion between the two systems. [When $t=0$, $t'=-vx'/c^2$ for all values of x']. It is our experience that any physical change takes place either gradually or in small quantum jumps; and hence the alleged instant change in the times shown by clocks in the range of 0 to ∞ (x' ranges from 0 to ∞) appears to be unrealistic and improbable. Suppose the small time taken to change the velocity of the moving observer from zero to a non-zero value is also taken into consideration. The alleged change in consequence of time changing from zero to values between +∞ to -∞ in a very small time required to accelerate the moving observer from the velocity zero to v, appears to be in conflict with the prediction of the STR that no cause can have its effect at a place whose distance from the cause is more than the distance that light may travel during the time interval between the said cause and the said effect.

Premise Three is Paradoxical

We have seen that *Non-synchronization of moving Clocks* is a direct consequence of Premise three. Let us imagine two infinitely long rulers, say S and S' lying parallel to one another. Let us imagine that an observer with a clock is sitting on each mark on both rulers. Let us synchronize those clocks so that all clocks held by the observers in both rulers show the same time at any instant of time. Let us now impart a constant velocity v to one of the two rulers. Now those two rulers with their respective clocks would constitute two inertial frame of reference say S and S'.

Lorentz Transformation Equations derived on the basis of Premise three predicts that the observers in the frame S at a given instant of time i.e., when all their clocks show the same time the clocks in the other frame S' would show different times ranging from -∞ to +∞, the time difference between two clocks separated by a distance 'x' would be equal to '-vx'/c^2'.

Similarly, the observers in the frame S' at a given instant of time i.e., when all their clocks show the same time the clocks in the other frame S would show different times ranging from -∞ to +∞, the time difference between two clocks separated by a distance 'x'' would be equal to '+vx/c^2'.

Since clocks include all kinds of clocks including biological clocks, the human observers sitting on the marks of the rulers are also clocks. This conceptualisation gives rise to a *Multiple Twin Paradox*. Let us place one-month old twins–one on a ruler-mark in S and the other on the coinciding ruler-mark in S' when both rulers are at rest in relative to one another. Similarly, let us place a twin on each ruler-mark of S and his twin brother on the coinciding ruler-mark in S'. Let us assume that all twins so placed on the rulers are of equal age, say one month. Now, as soon as a uniform relative velocity imparted between the two rulers the following paradoxical situation would arise, as a direct consequence of Premise three.

While all babies in S will observe that they are still 1 month old, they will observe that 'babies' in the frame S' have attained different ages ranging from -∞ to +∞ depending on the spatial distance between one another; the age difference between two 'babies' separated by a distance 'x' would be equal to '-vx'/c^2'. (Thus the observers in S will be equal to "demy-gods' seeing the entire past, present and future of one stream of events as an infinite time spectrum).

Similarly, the babies in the frame S' at a given instant of time i.e., when all of them are of the same age the 'babies' in the other frame S would be at different ages ranging from -∞ to +∞ depending on the spatial distance between one another; the age difference between two 'babies' separated by a distance 'x'' would be equal to '+vx/c^2'.

Paradoxically the babies in each frame would claim that all babies in their frame continue to be of the same age while those in the other frames have suddenly acquired different ages ranging from -∞ to +∞ years depending on the relative spatial differences between them.

It may be seen that the above observed phenomenon of one-month old babies in a moving frame acquiring ages ranging from -∞ to +∞ is not a gradual process it happens almost instantaneously at the very moment when a relative velocity becomes operative between the frames as a result of acceleration given to one of the frames for a very short time . After the relative motion is settled with a uniform velocity, there will be no more sudden jump in ages and each baby will age at the same uniform rate though the uniform rate of their clocks will be slower than that of the other frame by a factor equal to 1/a.

Suppose one baby, say B stationary in the frame S acquires acceleration and starts moving with a constant velocity v. Now B no longer belongs to the frame S and has become a new member of the frame S'. What will be its age after this change of frame? If we apply Lorentz Transformation Equations, the age of B will make an instantaneous jump from t to a(t –vx/c^2).

The usual Twin Paradox presented in books on STR is only a case of one particular pair–one of the pair say A on a ruler-mark in S and its counterpart say A' on the coinciding ruler-mark in S' when both rulers are at in relation to one another. After the imparting of uniform relative velocity v between the rulers, at any instant of time, they will be separated by a distance of vt for the observers in S and vt' for the observers in S'. The observers in S including A would observe at any point of time, say t that A' is younger than them by (t–t/a) units. But the observers in S' including A' would observe at any point of time, say t' that A is younger than them by (t'–t'/a) units. Who is really younger **A** or **A'**? This is a paradox. (Incidentally, many authors of books on STR extend the story still further and imagine that A' takes a u-turn and travels back with the same speed to meet A. To deal with this extended story let us imagine a third ruler say S^{-v} which is moving with a uniform velocity –v relative to S. Now A' has to jump to the ruler S^{-v} to return back to meet A. If A makes that jump when the time in S is t, than the ruler-mark in S to which he jumps would read 2avt and the clock on that ruler-mark would be showing time as (2at-t/a), according to Lorentz Transformation Equations. This means that instantaneously the age of A' will increase from (t/a) to (2at-t/a). This is at least unrealistic, if not a paradox. A' will take further time (t/a) to reach A. So when A' returns back to A, his age will be 2at and the age of A will be 2t. At the time of their reunion A' will be older than A. But in many books on STR it has been claimed that A' will be younger than A at the time of their reunion).

A Third Postulate Suggested in Lieu of Premise Three

It is suggested that in the place of the untested and paradoxical Premise three, the following statement may be adopted as the third postulate of the STR.

"The detection of light by an inertial reference frame is an event that is exclusive to that frame."

The above postulate will make it clear that there is no question measuring the same event by another frame and hence there is no necessity to derive transformation equation for that event. Only the absence of this postulate led Einstein to start on a premise that one reference frame can measure the instant of the detection of light signal in another reference frame.

A corollary of the above postulate may be derived to be the following.

"The speed of light relative to any inertial reference frame cannot be measured by any observer that is not stationary in that frame."

The above corollary is important since it directly disproves Lorentz Transformation.

With the inclusion of the above third postulate in the STR, we can retain the Galilean Transformation

$x'=x-vt;\ y'=y;\ z'=z;\ t'=t$

and Galilean Velocity Transformation formula

$u'_x=u_x-v$

with an addition of an exception clause that the above formula will be $u'_x=u_x$ in a special case where u_x=c, the speed of light in vacuum.

A Hypothesis Suggested to Conceptualise the Third Postulate

A hypothesis that may help one to conceptualise the aforesaid third postulate may be that each inertial frame has its own space with all such spaces of inertial frames overlapping over one another. When a

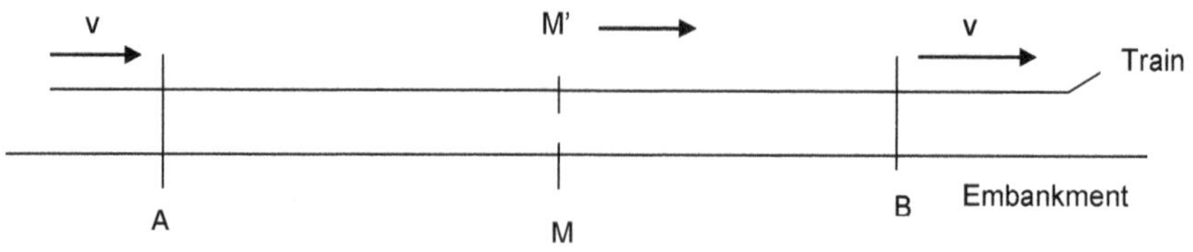

Figure 2: Two strokes of lightening A and B.

light source emits light, the light spreads in all these spaces and the speed of light in each space is c relative to the inertial frame attached to that space; and any observer/object stationary in an inertial frame can detect light that spreads in his/its inertial frame and he/it will be insensitive to the light spreading in the spaces attached to other inertial frames.

An Experiment Suggested to Prove/Disprove the Premise Three

The following is an extract of a thought experiment stated by Einstein in Chapter 9 of his book "Relativity–The Special and the General Theory" to "prove" that the events which are simultaneous with reference to one inertial frame are not simultaneous with respect to another inertial frame explained in Figure 2 [5].

"Are two events (e.g., the two strokes of lightening A and B) which are simultaneous with reference to the railway embankment also simultaneous relative to the train? We shall show directly that the answer must be in the negative.

When we say that the lightening strokes A and B are simultaneous with respect to the embankment, we mean that the rays of light emitted at the places A and B, where the lightening occurs, meet each other at the mid-point M of the length A B of the embankment. But the events A and B also correspond to positions A and B of the train. Let M' be the mid-point of the distance A and B on the travelling train. Just when the flashes (as judged from the embankment) of lightening occur, the point M' naturally coincides with the point M but it moves towards the right in the diagram with the velocity v of the train. If an observer sitting in the position M' in the train did not possess this velocity, then he would remain permanently at M and the light rays emitted by the flashes of lightening A and B would reach him simultaneously, i.e., they would meet just where he is situated. Now in reality (considered with reference to the railway embankment) he is hastening towards the beam of light coming from B, whilst he is riding on ahead of the beam of light coming from A. *Hence the observer will see the beam of light emitted from B earlier than he will see that emitted from A.* Observers who take the railway train as their reference-body must therefore come to the conclusion that the lightening that the lightening flash B took place earlier than the lightening flash A. We thus arrive at the important result:

Events which are simultaneous with reference to embankment are not simultaneous with respect to the train, and vice versa (relativity of simultaneity). *Every reference-body (co-ordinate system) has its own particular time*; unless we are told the reference-body to which the statement of time relates, there is no meaning in a statement of the time of an event'.

In the opinion of this author that if the above experiment is actually conducted it will disprove the conclusion of Einstein that Simultaneity of events is relative. In other words it will disprove Premise three. The experiment will reveal that contrary to Einstein's prediction the rays of light emitted at the places A and B will meet each other at the mid-point M' (Let us call it Event M') in the train also besides meeting at the mid-point M of the embankment. (Let us call it Event M). Both events M and M' will happen simultaneously since light travels with the same speed c in both frames. But they will happen at different places.

To verify the above conclusion one has to place two light detectors fitted with accurate clocks at the mid-point M–one to receive light ray from A and another to receive light ray from B and let those clocks record the exact time of receipt of light at the light detector attached to it. Similarly one has to place two light detectors fitted with accurate clocks at the mid-point M'–one to receive light ray from A and another to receive light ray from B and let those clocks record the exact time of receipt of light at the light detector attached to it. It will be seen that the rays of light emitted at the places A and B will meet each other at the mid-point M' in the train also besides meeting at the mid-point M of the embankment. Thus one can experimentally disprove Premise three. The underlying logic is that the ray of light observed in frame are different from that observed in the other frame though both rays move with the same speed c in their respective frames. Contrary to Einstein's statement "*Every reference-body has its own particular time*", the observable truth is that "*Every reference-body has its own transmission of light.*"

Conclusion

The absoluteness of space and time declared by Newton can coexist with the absoluteness of the speed of light in vacuum declared by Einstein without any conflict between them provided that we reconcile ourselves to the observed fact that a light detector can detect only light signals that impinge on that detector with a velocity c relative to that detector, where c is the universal constant denoting the speed of light in vacuum measurable by any observer.

References

1. Rajam JB (2002) Atomic physics: S. Chand & Company Ltd, India.
2. Fredrick JK, Edward GW, Malcom JS (1993) Physics- classical and modern Physics. Mcgraw-hill inc , USA.
3. Einstein A (1905) On the electrodynamics of moving bodies. Viva books private ltd, India.
4. Ajay G (2009) Special Theory of Relativity. Viva books private ltd, India.
5. Einstein A (2008) Relativity – the special and the general theory. Pigeon books, India.

Distance and Mass Correction on Logarithmic Metrics for High Z's, Due to the Non-Cero Angular Momentum in an Isotropic Rotational Universe Hypothesis

Pons-Rullán B*

The Stenbock House, Rahukohtu 3,15161 Tallinn, Estonia

Abstract

Dark can't be on the same sentence of a Hypothesis. Let's surmise here an alternative model in which the time turn around the space with non-zero angular momentum. Then, there should be a center and we may observe anisotropies, and indeed although they are not explicit in space, but are observed over time, and we identify a beginning and an arrow-of-time. If so, to preserve the metrics invariance through scale, estimated distances to deep galaxies may be corrected by replacing the FLRW metrics by another logarithmic based, according to now a days observations in supernovae Ia range up to 0.2<Z<2, predicting how much further over brightness adjustment might be. Universe may seems to accelerate its expansion because the assumptions, not because the Ghost Energy and as corollaries either G, c, h, α or Ho, seems to be from our point of view as observers, logarithmic variable in look-back-time. That means that together with the isotropy in the metric expansion of space-time, it requires a x5,75 measurement of the mass in CMB, as observed, the near "keplerian" velocity distribution in very ancient galaxies, and a non-dependent on the radius distribution of the velocities of stars and galaxies in our locality, with enough time and appropriate normalization. As corollaries it may explain the weakness of gravity relative to the other forces; the Horizon Problem; simultaneity; baryonic asymmetry; the homogeneity and granularity of the CMB; the extra-large size of the galactic black holes; the very fast configuration and the narrow range of dimension of the mass of galaxies and stars; the absence of small G, K, M, L, T, orange and red stars from Population III; the age of quasars; the highest density than expected of stars and heavy metals in the very deep space; the lowest density than expected of brown and black dwarfs; the absence of slow neutron stars; the decreasing change of the critical mass for black holes and its progressive deactivation; that the Universe is younger and older than estimated - according to the observer - and alternative descriptions to Hyperinflation, Big Bang - it was a long boring and non-forceful process.

Keywords: 6D twistor space-time; Rotation of the universe; Logarithmic decelerating expansion; Hypersphere space-spindle time-spins; Over-redshift and over-rotational and virial speed; Alternative hypotheses to the big bang; Universe rotates

Introduction

We will develop the following from the reinterpretation of the Accelerated Expansion of the Universe, based on the measurements of supernovae Ia, with the apriorism of a constant and isotropic metric; from the alternative assumption of Decelerated Expansion - so that matter and energy will add 0 - and compare it with several non-linear metrics that may make the measured values ?? compatible. With this alternative approach the result is a logarithmic scale, which can also afford temporary covariance - and though the energy conservation -, but requires anisotropy that is not observed in space unless an additional space-time dimension is conceived.

Energy

If we assume a metric that does not need dark energy and preserve the classic first principle of thermodynamics, the logarithmic metric with isotropic space, implies a multiple rotational time coordinates. Neither GR nor QM are compatible but they both are deterministic, and this is enough to ensure that at some scale, they are false. If cdt changes by dt/ln(t), then GR evolves in confluence with the non-linear and irreversibility paradigm.

Modern stochastic-chaotic-dissipative paradigm contemplates linearity as an exception, and the GR or QM have not been adapted to upgrade them to norm: neither to non-equilibrium, nor to irreversibility, nor to non-ergodicity. By Lyapounov, space and time do not have pair evolution; while by Poincaré, the recurrence may be in conservative systems. Birkhoff summarized it: only in linear relations, space and time comply with ergodic property. The evolution of complex systems in space is exponential in time and what for Relativity is a minus sign, for non-linearity is an exponential relationship. Adding a second temporal coordinate, incorporates non-linearity into the GR and transforms the light cones into hyperboloids of revolution, and following this path, through singularities, chaos, stochastic and irreversibility. To suppose that a divine observer would understand the temporal dimensions of different nature as spatial dimensions by their properties of symmetry, is a hypothesis as valid as the opposite, which has the disadvantage of not being able to play the dice.

Adding spatial dimensionality doesn't change the symmetry group or the flatness, but adding a single temporal dimension changes the properties of curvature - and though of gravity - in the very beginning and the future as changes the path of the proper time (even the constancy of Λ). Space and time are both coordinates, and while they have different sign on flat, GR doesn't know anything about the irreversibility of the arrow: GR needs to play dice. They are more

***Corresponding author:** Pons-Rullán B, Independent Researcher, The Stenbock House, Rahukohtu 3,15161 Tallinn, Estonia
E-mail:bart@bartolo.com.es

different indeed: we ride into space but on top of time. Space is the landscape and time the horse. The hypothesis brings a way to solve the Cauchy problem in a 4D deterministic space-time [1]. Odds are apparently against: [2].

Campbell's Theorem states that any analytical solution of the Einstein field equations in N dimensions can be locally embedded in a Ricci-flat manifold in (N+1) dimensions. A circumference of two-spatial and one-time dimensions (2+1), spins around a point. A sphere turn around a one-dimensional axis. A 5-dimensional hypersphere rotates over a symmetric three-dimensional space, presented as a dual space manifold or in a space-time of 6 dimensions. If the hypersphere expands and spins, dimensions might grow in logarithm rate to preserve the invariance to scale and to describe the different nature of the axis and hypersphere, we'll call the non-isotropic dimensions different because their symmetry properties: time.

The number of arbitrary constants sizes the distance of a model to the fundamental reality. If the universe has a conservative non-zero momentum, a model considering this Λ as constant, would not be fundamental. If time is another coordinate like space, expands as any other spatial dimension and spins, there might be more non-isotropic dimensions than a single one and nor the speed of light and the gravitational constant, and hardly no constant, are arbitrary, but only their relations in units changing in each moment of time. G and c remain constant in every observation, for every observer, in every time, in all space -"soft cosmological principle"-, but an observer measuring them in other epoch will use his clock, not the clock he would use at that time. The isotropy will be in space and the anisotropy in time, but maybe some lateralization clues, beyond the arrow of time, would be identified [3-8].

GR equations do not show any preference for any particular shape, scale or metrics. A universe in which everything, including elementary particles, is twice as large (or small) as in our universe and in which the duration of a second is twice (or half) as long, would be completely equivalent to our universe. If the expansion is linear or logarithmic, the metrics would be different, but laws of physics would be the same [9]. If the Universe twists around a 3D-space axis (with the remaining inertia? - conserves the momentum -, after the particle generation?), the three-dimensional space axis expands isotropically with an also expanding revolution time-surface related both to the spin, where the radius represents one dimension of time t_r (Figure 1). For an outsider observer any distance and any time grows on a logarithmic scale and dimensions do not have the same real or imaginary nature than a human insider observer. Maybe a new perspective about Bergson and Einstein 1922 time conception could fit here (time as itself or proper, as a path on a dimensional time dimensionality) [10].

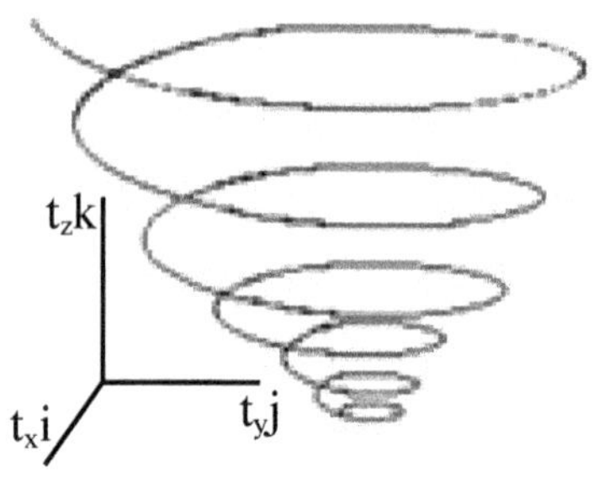

Figure 1: Path on a dimensional time dimensionality.

The authors ventured a universe with non-zero momentum, and from time to time, some body tries to rescue the hypothesis [11-13]. If we could measure the tangential speed of the temporal spiral -t_a'- from our inertial reference, it would be as a stone at the end of a "temporary sling" growing 3D space in proportional volume to the cube of the radial time (t_r) increase [14-21]. Expansion would be related to the down speed of radial time, t_r', apparently against our experience which feels time as constant and measures an accelerating growth. An outsider observer may "see" it that way, but the human sizing would be determined by its limitation in the embeding dimensional perception, that makes the clock seems to have constant speed.

For Abbot's Flatland citizens embebed in our reality, if their time could be represented as our third space dimension, we would see all their time as a dimension and we should maybe seem to be for them Chiral Gods. They would describe us as our own projections on their space. A drawing of a right hand can be switched for the left hand – reflection - only with in a lower group of transformations – rotation - if we access to extra dimensional spaces. So, one extradimension repairs one symmetry, and their asymmetric time would be for us their destiny seen as a hole. Any broken symmetry can be restored from the point of view of an "growing block" observer living in more dimensions and with no esoteric purposes. A Chiral God would improve invariance from a complete Hilbert space-time with one or more dimensions, and tensor-geometrical description in a manifold is only a transport of the references from the ortoND to the (N-1)D hypersurface with N-1 free degrees. So asymmetries that configures our reality can be restored to the Nothing, considering more dimensions, which may not be folded up in our reality, but projected. Recovering J.L. Borges poetry - "God moves the player and the player plays the chess/, which God behind which God began the game?" -, we'll qualify this hypothetic outsiders as chiral and ergodic divinities and ourselves as mortal enantiomers.

GR is preserved with a soft principle of constant c: while c remains as a constant relation, not necessarily as an absolute value. Divine observer would measure the variability of speed of causality and mortal observer would size the expansion, both exegeses because their own references, but they would not agree about which one is constant. For a divine observer all 6 would be orthogonal dimensions, but from the human point of view inside the manifold shape of the space-time, the perspective is limited in representation 3 perpendicular contravariants referring to a fourth frame – time - taken as constant, distinguished from the others because it only happens in a non-commutative sense, but it's a summary of 3 time covectors (time is an obvious anisotropy of space-time).

For an outsider observer, in this simulation, classical mechanics applies to what we call surface of temporal dimensions deemed for descritpion purposes as spatial euclidean dimensions, some of them symmetric and some anti-symmetric, (as a simplification we model a cylinder rotating and expanding). As we measure identical expansion in all dimensions of the axis - distance/time -, we do not need more than one space dimension to emulate mathematically the divine observer perspective in his conceptual reference to our space. A single space-dual time diagram model of a single real dimension - isotropic spindle - and two imaginary dimensions - temporal surface - all transformed into real, and all referred to a time frame.

$$ds^2=-cdtr^2-dta^2+dx^2+dy^2+dz^2?-¢d\tau^2+dx^2 \quad (1)$$

In other words, to simplify the simulacrum 2 symmetric-spatial dimensions will be replaced by 2 anti-symmetric-temporal dimensions, with no need of the complexity of far beyond twister algebras (3,3) or

(4,4) if we consider also scale and ?, as dimensions [22]. This closed and anisotropic model (isotropic for a resident in 1D spindle-space), describes a universe that emulates ours as a resident observer would assume if he lives in 6D.

In this space time diagram simulation, time perceived by the mortal observer as constant, is the spiral path of time: the vector addition of temporary increases

$$dt_r+dt_a=(t_r+t_r')(\sin\omega+\cos\omega). \tag{2}$$

To the mortal enantiomeric observer resident in that single spatial dimension, t' -the rate of progression of time- would be constant, but not for the outsider. In the beginning, the difference was important, but it happens that hundreds of billions of years later, sinω long ago is negligible in relation to cosω, and for both observers $t'= t_r'$.

Different masses, like different t_a', generates different never-crossing spiral paths, and no mass will have as its path just the radius itself. As the interpretation for rotation is spatial, Hawking 1969: "These models could well be a reasonable description of the universe that we observe, however observational data are compatible only with a very low rate of rotation" [15]. But is not if the time spins, not the space.

The incorporation of Hyperinflation and Dark Energy to the models to fit the observations results a Hubble "roller coaster" function, Hubble Flux, Ho(t), with bizarre loops, accelerations, decelerations, to avoid resignation of the Conservation of Energy Law - inflatons? -. If the Universe as a whole in its rotation preserves the angular momentum (Λ, Cosmological Constant), it would be a very fundamental configuration parameter of reality. Taking the classic descriptive rotational deterministic model, the areal velocity may be kept:

$$\pi\omega tr^2=\Lambda, \tag{3}$$

and the tangential speed of time would be

$$ta'=\omega tr=?(G?^2/8tr^3). \tag{4}$$

To preserve momentum, the angular time speed of growth decreases faster than the growth of t_r; and the ratio between the derivatives of both time dimensions is not constant. Therefore, at least a divine observer, in any way won't conceptualize this rotating speed, - perceived time -, as constant. In the Beginning it would be very obvious for any observer, but now even Him/Her should be very accurate in assessing the decline, and we can compromise that $t'=t_r'$ in current universal scale, but not at the first moments of time, near the Big Bang. From a divine outsider observer measuring the expansion in a n-dimensional euclidean geometry and angular time would decrease faster than the radial expansion. As the ice dancer extends his arms in Figure 2, from the audience perspective, and rotates more slowly than he "expands" the length of his arms. Assuming Λ & m as constants, the angular velocity

$$\omega\alpha\sqrt{(G/tr^3)} \tag{5}$$

and the total amount of gravity would be time-dependent and may be constant only in the space dimensions:

$$G=(8\Lambda^2/\pi^2m^2)/t_r \tag{6}$$

From an unimaginable divine existence in 6 or 8 dimensions (quaternions +Λ&h) [23], all of them are perpendicular, but some are symmetrical -axis- and other – rotating - are not, depending if they are spinning or not. From another limited perspective, if we conceptualize this model as a single dimension space which the finger imaginatively draws in the air - curvature $\Omega k=1/t_r$-, it'll expands, but more slowly. The growth of the one-dimensional simulated space, with a single degree of freedom: up or down-, would be the maximum for any movement of an impossible inhabitant living in this "needle on the vinyl record" (analogy to explain that asymmetrical time dimensions twist rigidly in a path depending of the mass) because in case of being overtaken the ant could jump forward to the expansion itself. In a dynamic system paths can't cross out the attraction of "fixed point", and that limitation holds the causality maximum speed: c.

The ant in the dancer finger (resident in the axis+spiral time), which also gets weight while the dancer extends his arm (perceiving the time and the strength from a one-dimensional space), can't go further than the finger and can't spin faster, but has its legs inevitably attached to the end of the nail and only investing energy he could break against the centrifugal acceleration and walk to the center, drawing a different spiral path.

$$tr''=2\omega^2 tr, \tag{7}$$

which is obviously the same as saying that

$$tr''=ta'^2/tr \text{ or } tr''=2\Lambda^2/\pi^2tr^3. \tag{8}$$

For a divine observer either space and time expands alike in its three coordinates, allowing mathematically replace and integrate in reference to radial time, and for the divine observer the speed of expansion will be $\alpha 1/tr^2$. In our role of divinity for the ant at the end of the finger of a dancer, Expansion has proportional growth to the radial space-time

$$2 \times \pi \times t'\ (2\pi t_r') \tag{9}$$

locally, but for an observer resident in the one-dimensional space. For us a unit of time is always the same unit of time, i.e., the dancer turns spinning down for the public sitting in the stands of Olympus. The ant in his nail rides on t_r, and imposes its reference measuring constant Ho, to what divine observer consider only as relations between spatial dimensions units relative to units of temporal dimensions, c. This makes mortal observer perspective "see" movement with the assumption that time is linear (as watching the landscape moving from inside a train) $\alpha 1/t_r$, and clears out from the equation

$$\pi Hotr=4c\Lambda^2. \tag{10}$$

The relation between time of the observers keeps a transformation and c must decrease $\alpha 1/t_r$, from the point of view of an observer that has a different clock. But, from the outsider observer point of view, if c is constant, Ho α1/tr!

In this model any particle with mass is time-dimensional: since its creation has a past and though is not a point but a temporary spiral path.

Figure 2: Calculation of length and the angular velocity of the dancer (Gα1/tr!).

So analogous a photon would be a perpendicular dimensional particle in space, not in time: simultaneous in each sincronized manifold, with no past and no future. Two spaces like points in the hypersurface of the present are each outside of the light-cone of the other, so each one has access only to the past of the other: to preserve causality, a simultaneous photon broadcast may not be seen by the other. If we "see" a photon is because our cone intersects with another cone in a hole trazoid manifold and cross to the time Zone explained in Figure 3. Not near of a hudge mass, cones grows parallel and as time goes on, the points in space-time tends to be not linear but hyperbolically related in causality between them. So, the causality relation draws a null hypersurface between simultaneous coordinates

A photon dilutes linearly with Expansion, and the speed of causality -c- would be the expansion from our point of view (the speed of the landscape from the perspective of the traveler). The speed of light for a divine observer, would be only the expansion ratio between space and time units, that is, adjusting the system measures a matter of scale (1sg always equivalent to 4,775 109 m to the outsider observer). Why there should be a "speed of light" if it's only a convention of units of measurement? It's just a way to express our perception of expansion from the perspective of an observer measuring from inside the Universe with a clock that measures a non-constant time G, T, P, c, α α1/tr!

Time is a length derivative for a photon and its emission would be simultaneous to observation, but both paths, photon and mass-observer, have not the same invariance axis. The point of emission of a photon follows a temporal spiral path increasing both temporal and angular time, but the space where the photon expands follows only temporal time. In the next instant other spinning paths - stories- will intersect with expanding paths – movements -, but would never meet again. The divine observer "sees" a spatial dimension that expands homogeneously and only conceives the speed of time expansion. For the mortal enantiomeric observer who does not "see" that space and time are expanding together with him inside, the speed of Causality is the increase of radial distance in time, rather than the expansion rate.

Acceleration could be as well represented as a temporary vector. If we measure the space in terms of our clock, it would be as if the ant moves to a temporary lower pitch spiral, where finds a past field and share the reality with other ants which remained at rest to us some time ago (being reality the full path since the beginning of time to now, and therefore radially remembered). Our apparent c, inversely proportional to radial time, mathematically re-integrating, results a model with logarithmic scale factor:

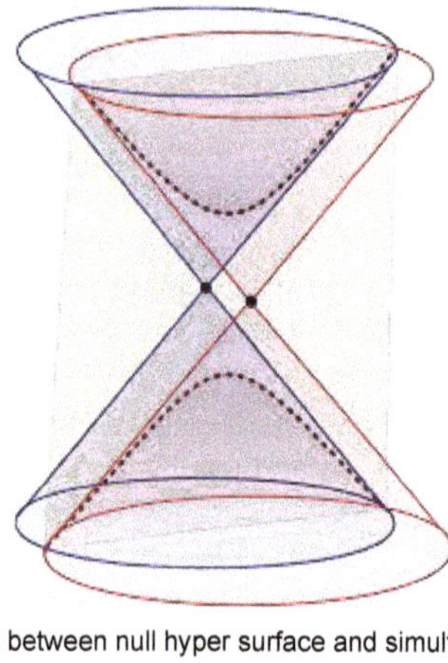

Figure 3: Relation between null hyper surface and simultaneous coordinates.

$$a(t)=\ln(t_r). \tag{11}$$

That is, as it was said by Teller [24], from the equations of a rotating system, the space-time expands at logarithmic scale.

If we change the metric considering the time as a variable according to a(t), so that the speed of causality is constant from the perspective of a divine observer, our measure of c can't be constant, and we would measure the distance to an object in light years variables regarding this scale factor (each megaparsec -cosmological correction -, but also every second would be smaller in look-back-time, retaining the ratio, while older). We measure distances calibrated on the brightness according to the parameters of Cepheid in six galaxies, applying to larger scales what it works in our galactic environment, assuming the metric in which the time -and thus the speed of light with "mortal" references - remains constant. If the Universe preserves a non-zero angular momentum, to be spatial-isotropic but temporal-anisotropic, while angular time remains negligible to radial time, a logarithmic metric would be finest than a FLRW metric.

In the absence of a reference step that affirms or denies, beyond the local super cluster of galaxies, a distance expressed in current light-years admit the vagary of increase with the average speed of light ¢, corrected by the scaling factor, as we would have considered the cosmological expansion, but not temporary expansion. Expressed in relation to our present time $t_{ro}=1$, taking it as a unit:

$$D_c=D_L\ln(1/t_r)/(1-t_r)=D_L¢. \tag{12}$$

D_c, proper distance corrected.

D_L, proper distance estimated under the assumption of constant time.

¢, average of cover the logarithmic scale

Standard candles supernovae Ia distances estimation, may include other variables like metallicity - more luminosity with less heavy particles, that may be more common the more further they are and needs a better knowledge of the processes. According metrics normally in use the Z=1, is given in $t_{ro}/2$, but according to a metric in which the time is also affected by the scale factor, radial time would be transported to an axis as

$$\ln(t_r)=1/(Z+1), \tag{13}$$

and would also apply the above D_c correction, plus a correction of the time scale transporting an axis with a linear scale -cdt- by

$$t_r=e^{1/Z+1}/e=e^{-Z/Z+1}. \tag{14}$$

According to this metrics the speed of light at every moment of the past seems inversely proportional to the time, because it's referenced to constant time, but still remains constant for a divine observer, as the relation between t_r and D_c: 2π, is constant.

If the assumptions were correct and model fits with the choice of what is relevant, regardless of the values of Λ, m, c and H_0, we should be measuring a redshift corresponding to the difference between D_c in each metric with its cosmological correction of its proper distance dependent on the above parameters. The question is: regarding what model applies the correction? Linear, de Sitter, ΛCDM, Benchmark?

It happens that when calculating the distance correction

$$\int_z^0 cdt/a(t) \text{ for } c=1/t_r \text{ and } a(t)=e^{1/1+z} \tag{15}$$

is identical result in the model of linear expansion, (m=0, Λ=0, with c constant and $a(t)=t/t_0$) and the logarithmic simulation; i.e., in both cases proportional to ln(1+Z). So if we compare them both, we can save mathematical rhyme and verse and ignore at purposes of apportioning the cosmological correction. By the way, we can save relativistic corrections because the comparison of measurement criteria has nothing to do with distance themselves in absolute terms, or the use recessional velocities discussed in Table 1

To reach those conclusions, we have not needed any parameters or any universal constant! Below Z=0.2, in the few billions of light-years closer, but beyond the reach of the methods of direct estimation of distances, the difference is less than 1%, so the metric models FLWR would gain an application limit, being optimistic about the ability of future predictable astronomical techniques, Z<0.2.

"%overZ" means the extra-redshift, i.e., we'll overestimate in 12% the distance at Z=1 vs. a linear no-matter model. Any other Z may have to be correct with a precise ratio.

Up to here, we have assumed that Λ is a fundamental constant and precisely that invariance fixes a limit in the application of the logarithmic scale, and may not happen when ω begin unless negligible compared to tr. Then we could speculate a reformulation of a more complex metric in the same time dimensional basis, wherein the reference time considers ω. Mikowski space-time will increase dimensions and split dt in dtr and dta, then c would not be constant for all observers, even for divine, and other constants are treated as what they are: variables. The implications of non-conservative magnitude would be deep: Noether asymmetry. Some Λ(t) variable anisotropic models has been proposed as maths jokes, meaning also α, c o G. In any case, the logarithmic scale has not much observational sense above Z=11.1, unless maybe could apply until Z> 1,089, although to guess and follow a dark path until another better model [25-27].

Following the example of Z=1, what we believe according to FLWR metric it happened in 0,5tr (without cosmological correction), it was less time ago according to the way we measure it now: 0,61tr (also without correction of Dc). When we measure the brightness of supernovae Ia 0,5tr, they seem to be further than we expect because in fact they are further away than where they might be. It can't balance because the metric used as implicit assumption applies a bias. In 0,5tr, we assume from the equations above the distance is 1.27 times estimated when it is 1.39 times, up 12%. In both cases it was considered c constant from our perspective: elongating the wavelength, but not the time in which it happens; and having no direct references we can't know but in relative terms, how distant supernovae are in each Z, or its youth, but about what we assume on them.

Up to Z=2 (max. 1,914) this simulation is consistent with measurements from Perlmutter, Riess, Schmidt, which has been used to predict an accelerated expansion, as the metric has been taken for granted (Figure 4). The current mainstream explanation of the paradox of the bizarre behavior of the expansion brings us to a bigger and darker problem than the one that is trying to solve: the "fifth-

Z	0.5	1	2	3	5	10
t lineal	0.67	0.5	0.33	0.25	0.17	0.09
DL	1.21	1.39	1.65	1.85	2.13	2.64
t log	0.71	0.61	0.51	0.47	0.43	0.4
Dc	1.18	1.27	1.37	1.42	1.48	1.53
% over	3	12	28	43	71	112

Table 1: Over Redshift of recessional velocities.

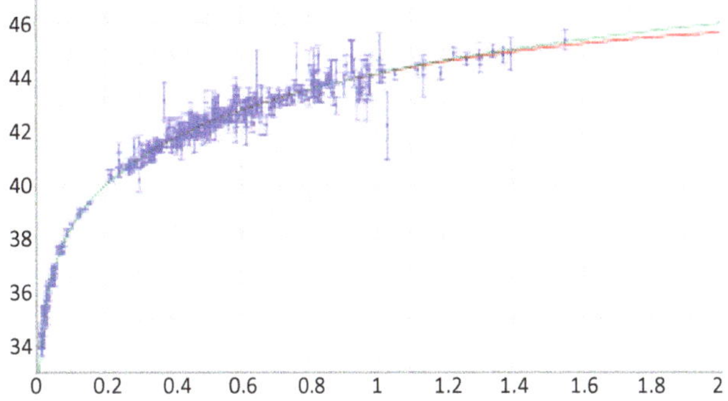

Figure 4: Illustration of Amanullah's 2010 Compilation.

column" Constant Λ, or the Quintessence scalar field Λ(t). Perhaps the convergence of the results is a fluke, perhaps a Confirmation Bias, but can also be considered an argument for upgrading the Conjecture of an spinning universe to Hypotheses - since has been deduced from an isotropic Universe with non-0 Angular Momentum -, and maybe in the future, with better tools, we'll refine and measure older Z's and confirm or reject this model.

It means not only that the expansion is decelerating, but also that Transparency corresponding to Z=1089 would be in a younger age of the universe expressed in current seconds although a larger distance, which means nothing because the units of space and time are conditioned by our observation criteria. From the perspective of an outsider observer with metric in which the time jointly expands as the space, look back time tends to infinity

$$(Z+1)\,(\ln(Z+1)-1) \tag{16}$$

and time of the first phase of the expansion was immeasurably large (although sized in very tiny units). It has probably taken longer to reach from the Beginning to Decoupling, - in this diminishing space-time metric referring to our time units - that from that Era to us (in a clock running from there to now). Then h is also variable as the scale factor:

$$\alpha \ln(t_r). \tag{17}$$

From our metric we interpret as explosive a process that a divine observer or an observer who lived in those first seconds and years of life of the Universe, would understand as a very peaceful and progressive development, and would not have any sense to set out a Problem of Isotropy or Hyperinflation, because there is no Problem to fix. Calling the process Big Bang is just a bias in the criteria of the observer: it seems explosive to us because we compress in our clock that huge time as we show an entire dramatic series concentrated in a split of a second we would not understand the history and all would seem dizzying to us. According to our patterns the Universe would be 37% younger and 58% larger, taking as reference the age that we date today: 13,720 billion years (the radius with the Guth Hypothesis would be x3 [28]). Nothing burst and we are not only slowing, but almost stopping!

So, while the very long decoupling era -+/- 115.000 our-years-, an observer from then had time to extend the poly-tropic star model to the whole Universe itself. He could model an adiabatic expansion of a ionized and degenerated plasma with high specific heat in thousand degrees magnitude, as it might be just beyond CMB. We would expect to identify some kind of pattern similar to the surface of the Sun: convective bubbles in the transition phase from convection to conduction in transparency, and explain the slight inhomogeneity and its rotational nature. We know the size of central galactic singularities can't be explained by the same process of massive stars collapse, because the Eddington limit, unless then that limit was then 3 to 5 times larger as it's obvious from this hypothesis perspective. Then the galaxies birth from the inhomogeneity of a convection surface would be easy and natural, with stars not hundreds but millions times more massive. The same reason would explain the why the IMF modeling doesn't apply to the early youth and we are not able to find any orange, red or brown Population III small stars with clean spectrum.

Exotic Matter

The overlap between predictions of the Bariogenesis and calculations by the resonant analysis of the cosmic background radiation, estimations by virial theorem, by rotational speeds in super clusters and galaxies, by X-rays pressure gradients in hydrostatic equilibrium of hot gas, by espectrography of rotational gas in superclusters, by computer simulation of the rate of formation of structures (Sachs&Wolfe), soft&strong gravitational lensing, by Siuyaez and Zeldovich Effect in the CMB, Boomerang project provides consistency to the hypothesis of dark matter, but can a hypothesis be called theory containing the word "dark"? Maybe we may think otherwise: change the paradigm, as we bury the geocentrism, heliocentrism, the estate of rest, the linear motion, determinism, and also bury anthropocentric, mathematical totalitarian, fine adjustment, conservation information principles. The best argument for darkness is do not find a better explanation.

Dark Matter includes cold and relativistic baryon components (intergalactic gas, gaseous filaments, neutrinos, planets, planetoids and belts, brown dwarfs, black dwarfs, neutron stars, "strange" stars, black holes, halo, hydrogen, water,) which existence is observable and extensible by the Cosmological Principle, where we do not observe them (OGLE & MACHO surveys). Although is difficult to "weigh" and if so, with a significant margin of error, even in the best estimation, baryonic matter is not enough to explain the measurements obtained. Mainstream solves the "dark matter problem" with a more complex hypothesis that the question: implicit assumption that if the measurement is correct, there must be exotic matter that interacts only gravitationally. Another alternative option would be that the implicit metric FLRW of constant time, distorts measurements, and will be the approach here

We do not know how much mass contains the black holes of Population III, we did not even know of the existence of dwarf galaxies like Segue1, with apparently 3400 times more invisible matter than visible. What part of dark matter is unknown matter for the resolution of our devices? Extrapolating the models to the early Universe, there should be more massive and far-flung protoplanetary belts than we assume in our solar system, more black dwarfs, many more brown dwarfs (IMF), much many more slow neutron stars, and who knows if much many, many, many more black holes (because the massive gap between the center of the galaxies and binary systems and because the unnova silent decay modeling).

No margin for exotic matter remain in Nucleogenesys. The question is not whether or not dark matter exists, but what's the nature of the exotic matter, whether it is cold or relativistic, and which is its distribution. On one side, baryonic matter may spread out (neutrinos, gas,), but in any case also with higher density as close it is to the center (black holes, brown and black dwarfs, orphans exoplanets, neutral hydrogen). Dark matter hypotheses leads to bizarre distribution of exotic matter: "profile NPW" [29] or other non-singular isotherm distribution. La Silla Observatory, has mapped the orbits of more than 400 stars in a volume four times greater up to now in more than 13,000 light years from the Sun. "The amount of mass fits well with what we see -stars, dust and gas-. This leaves a narrow room for extra matter -dark matter- that we expected find. Our calculations show that should have been clearly seen in our measurements, but simply, it wasn't there!", says the team leader Bidin et al. [30].

There are clear evidences of galactic halos. Observing Bullet clusters collision occurred 150 million years ago explained by Clowe [31], dark matter would be associated with normal matter and not with gas, and measuring the deflection by gravitational lensing, halo is left behind after rubbing, which limits to the inner visible galaxy the presence of dark matter. Other measurements such as MACS J0025.4-1222 reconfirm the effect [32], although estimates its gravitational collapse limit their mass, and it's not enough to explain gravity effects. If massive halos would be really associated with faster rotational discs, therefore with brighter galaxies, we may expect a correlation between the speed of a satellite galaxy to the main galaxy and its disk rotation speed. The dwarf galaxies Fornax and Sculptor, describes a uniform distribution of dark matter. Nor gaseous filaments justify the peculiar distribution needed to explain dark matter (Eckert in, AEE, XMM-Newton telescope) [33]. With our sizing hardware most part of Dark Matter seems to be exotic and bizarre but more than this: strange. In every new approach we find new hidden mass: the very old Segue1 does not have 5 or 6 times more dark than visible matter, but x3.400 [34,35]. It seems that the most of the stars are older out of the main sequence, the more dark matter referred in older galaxies.

We start from the prejudice of seeing galaxies as gravitationally consistent systems, not aware about intergalactic expansion. The problem of rotational or virial speeds ceases to exist as such, accepting stars and galaxies are in fact escaping from their gravitational systems at lower speeds than recessional as shown in Figure 4, following a hyperbolic, logarithmic, golden or other runaway spiral path pattern, still remaining tens, hundreds, or thousands of galactic orbits, before dispersion is evident in some tail as a comet.

According to constant-time metrics, since the formation of the very first galaxies the Sun has completed about 50 orbits to the Milky Way -70 or 80 in logarithmic metric-, so maybe it's not that the rotational speeds require dark matter for justify why stars remain in galaxies, but are simply escaping in elliptical orbits. Something holds but not kidnap them, and there is not enough time to scatter through intergalactic space, which expands faster together than dispersed locally. This would implies that the recessional velocity measures apparent expansion speed, which is unlike what intergalactic space-time relative to intragalactic expands, resulting in a younger universe than estimated, which may be consistent with the previous argument that estimates 1/3 less age explained in Figure 5.

Mainstream means that galaxies are gravitationally bounded

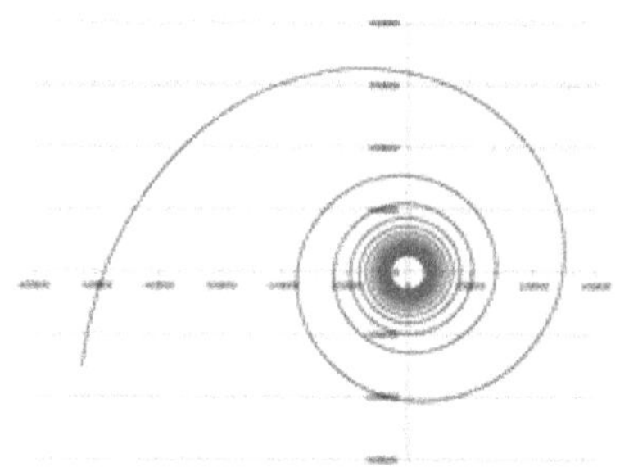

Figure 5: Azcorra proposed model.

systems: do not change their volume with expansion. Is a cluster a gravitationally bounded system, and so is a super cluster even wider filamentary structures and walls, with no graduality? The theoretical effect of expansion in the Sun-Earth system is 44 orders of magnitude smaller than the internal system gravitational forces. Applied to the solar orbit around the galactic center, the effect is 11 orders of magnitude less than the acceleration due to own gravitational effects of ligation. Even at giant galaxy cluster scales, the effect of the expansion is 7 orders of magnitude smaller than the acceleration due to internal self-gravitation of the cluster itself but this is today! If G depends on the inverse of time, the effect of modification of the gravitational ligation should be noted in high Z's, and also in the fine structure.

The "problem of the big numbers" -strange coincidences in the power of 10 in the macro and micro constants- led the terse Dirac to speculate on the variability of G as the inverse of the time [36] and proposed the LNH, describing its incompatibility with FLRW. He was criticized by Zwichy [37] and also supported by Chandrasekhar [38]. Sciama [39], and later Brans and Dicke with a scalar-tensor VSL theory [40], developed the conjecture including the Mach Principle. Gamow first derided, and later call the hypothesis smart [12], and proposed to verify Sommerfeld's constant [17]. The alternative chosen by the Mainstream has been the "teleological principle" as usual: rescuing the Boltzmann rectification (which maybe delayed physics some decades), updated with Guth [28] and reformulating hypothesis following the Anthropic Principle. We want to be special.

Why gravity would be related to inverse to the square of the distance –conceptually, a surface- and not to distance -proportional- or volume -cubed-? Newton did not know the value of G, but GxM, and according to this hypothesis, G outstanding decreases linearly, but may only be measurable from Z> 0.2. Inevitable but consciously influenced by the narrative confirmation bias, considering the metric logarithmic time, gravitational density has been diluted

$(\alpha 1/tr^3\ \alpha(Z+1)^3)$ "pari passu" to the photon density (αT^3). (18)

Considering these factors, is not in super symmetric particles, right-handed antineutrinos, axions or WIMP´s, where we find the explanation about exotic matter. In look-back-time, the apparent gravity would be

$\alpha G/tr^3,\ G(Z+1)^{1/3} = (Z+1)^{1/4}$ (19)

higher than its projection to our time, that means 5,746 times more from the detachment to today, coinciding with the proportion of exotic matter calculated with the resonant CDM model.

If so, galaxies should be larger and increasingly less dense in relation $>(Z+1)^{1/3}$, because intragalactic expansion is maybe less but anyway expanding like intergalactic space-time itself, and because in fact they are in runaway spirals. On a sample in Z~4, Ferguson notes that galaxies were smaller, irregular and more massive than our days. In the same way, comparing large galaxies in SDSS, between Z=0.2 and Z=1, those with more than 1.5 kpc radius multiplied by 500 [41]. Sizing galaxies with dispersion speeds -σ-, Osiris have found 6 times higher densities than current average (slightly less than expectable $(Z+1)^3$), in four elliptical galaxies in Z~1, half of size ("a bit larger" than expected Z+1) and only an increase of 1.8σ (slightly less than Z+1 expectable, if the space intragalactic expands similar to intergalactic) [42]. GN-Z11 is 25 times smaller with a 1% of the stellar mass of our Milky Way.

Brightness is more or less on the fourth power of maximum speed (Tully-Fisher) or stellar velocity dispersion (Faber- Jackson): " grosso modo " its squared mass. With the I band specter, we can estimate distance (we have already analyzed its bias) and weight. The mass calculated by the virial theorem depends on the radius of the cluster, the square of the velocity dispersion and inversely to G, but we have seen that both G and radius are linearly dependent to radial time. As their dependence is canceled one against the other, it would be proportional to $\Delta\sigma^2$, which for the above example Z=1, means a mass of 3.24 times over-estimated. In any case virial weighing of galaxies should include a further decrease $(Z+1)^{1/3}$ of baryon/exotic ratio, which reaches to 4:1.

Distribution of stellar rotational speed in the Milky Way, is observed as expected p^2/r^3 by the keplerian laws up to 10,000 light-years - on the same order of magnitude as its thickness shown in Figure 6, while mass distributed further than 20% of its radius is "running away", followed for an increase according to the expected for a spiral orbit escape, but it still remains 2/3 radius with approximately constant rotational speeds, when according to the classical model should be decreasing if we did not consider the Expansion! In a deep galaxy far away, rotational speed $-v_t-$ won't be proportional to 1/?r but

$v_t\,\alpha 1/?(rc/¢)$, (20)

which makes them dependent on the distance of the galaxy -further- and the reference radius for measuring -smaller-. Four articles in 2017 based on the European telescope in Chile -ESO- point to fit in more "Keplerian" velocity curves in distant galaxies, [43], just as it is expected to be following the present hypothesis, and it is waiting to expand data with the VLT and JWST telescopes.

A unit of length taken by an observer with straightedge billions of years ago, was shorter if we measure according to the straightedge of our clock, but it was "the same" for the rules of physics then, if we transport then our space-time metric. Although most of the mass of the galaxy is in spiral orbit not enough to be retained by gravity, while the expansion exceeds "the runaway spiral", the distance of the peripheral mass grows but less, than the grew of the length-unit taken as a standard at every moment of the life of the Universe.

If so, the rotational speeds should be kept by expanding space-time as they were billions years ago, being the distribution according to Kepler's third law a brake, but may not be the model. We assume centripetal force $F_g = GMm/r^2$ must be balanced with the centrifugal "force"

$F_c = 2mr\dot{\omega}^2 = 2mv_t^2/r$, (21)

and then we define the Kepler's rotational speeds expected. What if the measure did not apply constant time metrics? r could not be canceled on both sides of the equation at different times in history, r_c ? r_g. "Dark matter had less influence on the early universe. Observations of distant galaxies carried out with the VLT suggest that they were dominated by

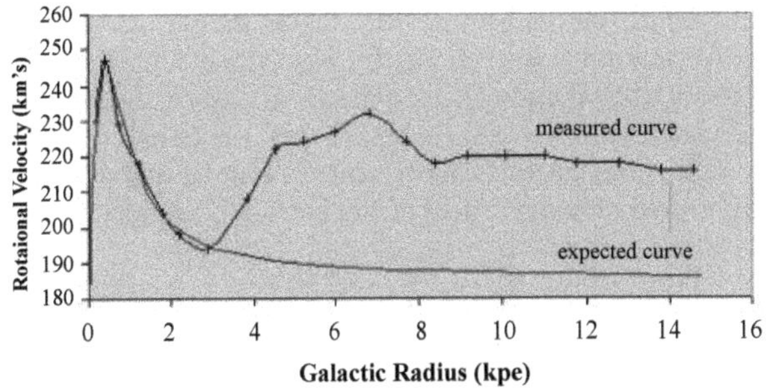

Figure 6: Graphical Illustration of VLT and JWST.

ordinary matter". (ESO 1709). Rotational speeds distribution would be a fossil record of the maximum speeds, corrected by logarithmic metric

$$r_c/r_g = c_r/c = Z+1. \quad (22)$$

which provides graduality and fits with the near "keplerian" velocities distributions in very far away galaxies -0,6<Z<2,6-. [43] Extrapolating in look back time, exotic matter should become more and more exotic. Does baryonic matter decay in exotic matter or is it created?

We could say the same for the virial speed in super clusters. The outline of a galaxy is fuzzy and to adjust r_c to $\ln(t_r)$ on the same axis and system of units, we need a common reference - normalization-: a galactic radius R such as holds horizontal line

$$Vt\alpha\sqrt{(1+z)/R} \quad (23)$$

If intragalactic behaves metrically the same as intergalactic space-time, gravity can't be a brake dependent of the distance between mass, which is only possible if G decreases with distance (in fact decreases with time all over the space the same), closing the argumentative circle. As an example, in 0,61tr, when the space-time units were half of our time, the velocity distribution curve of the Milky Way was "keplerian" to 6 kpc (twice than in our time). In other words: an outer star in our galaxy has the same v_t than when it was at 13,000 million light years away, and covers the same amount of space units per unit of time, but both have grown proportionately and in fact it escapes following a spiral path.

If G decreases as so, the Chandrasekhar limit may increase in look-back-time arrow, and with enough time single white dwarfs would decay beyond the electronic exclusion pressure even without feeding, and maybe before their evolution to "diamond" black dwarf. In high Z's, that adds another correction to the distance estimated by supernovae, because then the critical mass was higher (i.e., Z=0,5, x 1,67; Z=1, x2,1; or Z=2, x2,75); and so was the luminosity, but not its spectrum. In any case, fit the correction because the variability of G, c or h, is beyond the scope of this paper.

The constant $8\pi G/c^4$, -which relates the second derivative of the metric tensor $G^{\mu\upsilon}$, to the relativistic energy-momentum tensor, $T^{\mu\upsilon}$-, would be dependent on the volume of the Universe: as the Universe Expands, less mass is needed for the same effect in curvature. Galaxies configuration around a black hole were built in a size according to the change in the order of magnitude of the Eddington limit: central galaxy mass versus stellar black holes mass may have grown per millions due to the proportionality of the critical mass to the volume, -αtr^3-, and its multiple stellar origin intercollision and migration, is not a reasonable hypothesis.

History left us its footsteps on distribution: peaks during hunger activity periods of its central black hole (the more outward the more frequent should be, we must appreciate even a tendency to increase v_t with the radius); the more pronounced valleys, the more closer near the center (decelerating expansion $\alpha 1/tr^2$, gravity may force the curve to the classical $\alpha 1/?r$, escaping the rest of stars as the tale of a comet); and for the same reason above, these "fossil peaks" will be lower (it will not split on the radius, but as the movement of a whip), being consistent to those observed without the need of non-baryonic Dark Matter. So, G is comover, not proper.

Parallel grups in Moscow, Otago, Wuppertal and Washington sizes slight approach in G's, between -0,1% y el +0,7%. G'(tr) may decrease if Expansion is decreasing. There has not been evidences with frequencies of atomic clocks or decreases in the rotation speeds of the bodies (although and beyond its "relativity" gravitational interpretation) or orbital decay (LLR, with mirrors on the Moon); but also alternative scenarios (if the brightness depends on G7 and radius of 1/G, temperatures may boil the ocean few hundred years ago, [24]). Jordan 1971 even proposed Relativity modifications [44], but had not impact either. Belinchón collects different methodologies of rotational transformations, in a Bianchi IX categories frame [27].

Kaluza and Klein related for any extra-dimension the color of light with G:

$$\alpha \sim G/Я^2, \quad (24)$$

So, being the Sommerfeld constant $\alpha 1/\varepsilon hc$, as $h\alpha 1/t$ and $c\alpha 1/t$, we may either size very small direct variations (depending if h variation is or is not "pari passu" with c), or try to interpret indirect observations. Since then observational astrophysics has tried to verify the hypothesis of the constancy of G through the constancy of fine structure constant, α, -DEEP2 confirms the color of oxygen up to Z=0,7-0,9 [45]-. Color modification depends on the electron's charge, e, on the permittivity ε, on the Planck constant, h, and on the speed of causality, c, -$\alpha\ e/\varepsilon hc$-; if ii would not be so, either Cosmological Principle is wrong at large scale, or charges, or h or c, changes with time and expansion. In this hypotheses, h changes with the scale factor ln(tr)=1/(Z+1), and c with the inverse of radial time, so

$$\alpha\ tr/\ln(t_r):1-(e^{z/z+1}/(Z+1)) \quad (25)$$

The effects of G' in the gravitational dynamics, in stellar behavior, in the light of stars, in the rotation of binary pulsars, or the Hubble constant itself, may be less than 1% variability on color spectrum in the last 1.5 billion years, Z~0,35; and in 0,5<Z<1, so we may be able to observe 7%<|G'/G|<18%, which means less than 10^{-11} or 10^{-12} per year [2]. Analyzing the Oklo natural reaction, Shlyakhter [46] supports a change even less than $10^{?17}$ per year, over the last two billion years. Lamoreaux 1994 found a 4,5 parts in 10^8, decreasing of α, [47]. Webb et al. [48] from 1997 to 2011 edit several papers analyzing the changes of the spectrum of light from distant quasars, trough metal clouds, with a so slight effect that it would only be clear to Z's high -Δ1/100.000- , in a consistent range with in this Conjecture and with differences depending on the direction, which may indicate laterality.

Einstein 1911: "The principle of the constancy of the speed of light can be kept only when one restricts oneself to space-time regions of constant gravitational potential". All this sounds familiar to the STM or MOND [18] and its relativistic expansion TeVeS, but much more fits with the MOG [49] or STVG [50] models: assuming G/c as a constant; not by its possibilism, but because both depend analogously on 1/tr,

$$1+Z=c_r/c. \quad (26)$$

Interaction between mass is not given by its proper distance, which depends on c and a(t), and therefore excluding expansion. Gravitational force at large scale changes to comoving criteria. Relationships between variables that depend alike proportional to time would be constant, T/G, P/T, c/G, c/T, Λ/?(?) because their values are only adjustment of units. If G is inversely proportional to the radial time, rewinding the film looking back in time, gravity was in the same order of magnitude as the other interactions almost a trillionth of a trillionth of what, in terms of our measurement prejudices, was the first second.

This alternative interpretation to the exotic Dark Matter leads us to reconsider the acting distances of "Newtonian" forces as "comovers" and the geometrical measuring depending of c, as "proper" length. From our point of view the gravitational constant weakens with the

Expansion, but from the "comoving" point of view, it doesn't. It doesn't matter how much energy will increase the hadron colliders: they will simulate a very partial view about what happened on these temperature ranges in terms of GTU theories, because particles behavior would not be only a matter of temperature, but also depends on the space-time density forces range and the angular speed.

If Dark Energy were the answer, it would be split to the different questions: Initial Hyperinflation, Acceleration of constant or variable Expansion with respect to volume or to time, in its vacuum or scalar versions. $\Lambda'(t)$. How the universe conserves energy? Is the amount of dark energy related to the amount of dark matter? If Dark Matter grows with time, does Dark Energý grows in look back time? Granularity of the CMB draw tipical boiling emerging patterns of auto-similar structures of 1° and frecuency around 200, compatible with flat curvature and null critical density (as a condition for the energy-matter balance to be null).

From Statistical Mechanics, Dark Matter comes from the incompatibility with the Virial Theorem. From Classical Mechanics, from it comes from the incompatibility with Kepler's Laws. From General Relativity, it is the focus of gravitational lens, but with tremendous variance. From Electromagnetism, it is an interpretation of the harmonic analysis of inhomogeneities in the CMB. From Quantum Mechanics, it is an alternative explanation of the missed mass that for CP asymmetry and arbitrary mass values ?? balance. From the Numerical Calculus, it is the output of the tautological adjustment of the galactic genesis. From Special Relativity, the Dark Energy (Cosmological), comes from the incompatibility with the limit of the rate of causality, c. From Thermodynamics, different versions deals with the compatibility of Increasing Expansion with Energy Conservation Law.

If the Dark Matter is the answer, it would not be neither the same to the different questions. To explain the rotational speed in galaxies we need a profile distribution inverse to the barionic matter, but in a cluster it seems to be associated to baryonic matter. There are different proportions according to the method of weighing used and the object analyzed (they measure three orders of magnitude between x3,5 as a minimum measurement by gravitational lens and x3.400 of Seguel) [34]. The oldest galaxies have velocity distributions "keplerians" like, and means that exotic itself would be of baryonic origin, so we double the task: find exotic matter, why and how migrates to the halo, and how it transforms the one in the other; in another roller coaster "deus ex machina". If so, the CMB would not contain that much of darkness.

It is not the same Dark Matter to justify masses of galactic holes, stellar mass distributions or B-modes. The distribution, proportion and decay are not compatible to justify masses of galactic holes, distributions of stellar masses, B-Mode and simulations of galactic genesis. We need two or three types of dark energy and even some more dark matter forms, we need non gravitational migration dynamics to the halo and decays from barionic to exotic, which are also incompatible with each one arbitrary evolution without an identifiable pattern. They are so dark that darkness does not allow to see anything, like ether, at all. It might be one of the worse theory ever announced. Science never proves the certainty of a hypothesis, but its falsity, and the "darkness" is as "false" and empty as the ether.

Conclusion

We don't need exotic and bizarre distributed matter to balance equations, no ghost forces, not even constants, not anthropocentric principles, beyond the fact that has been so many years without appearing in the detectors, even with restrictions of interacting only with weak WZ particles. Although the rotational speeds are apparently running away, Expansion will always be faster and galaxies will evolve while they get away one from another, decreasing its density and increasing its size until fuzzy star clouds, increasingly distances one from each other, switching off, leaving a tail of stars and dust in its track. Sad and dull destination or maybe there is another way.

References

1. Weinstein S (2008). Multiple Time Dimensions. Cornel University Library.
2. Canuto VM (1981) The Earth's radius and the G variation. Nature 290: 739-744.
3. Nodland B, Ralston JP (1997)Indication of Anisotropy in Electromagnetic Propagation over Cosmological Distances. Phys Rev Lett 78: 3043.
4. Longo MJ (2007) Does the Universe has a Handedness? Astrophysics and Astronomy
5. Kashlinsky A, Atrio-Barandela F, Kocevski D, Ebeling H (2008) A measurement of large-scale peculiar velocities of clusters of galaxies: results and cosmological implications. Astrophys J 686: 49-52.
6. Albrecht A, Magueijo J (1999) A time varying speed of light as a solution to cosmological puzzles. Phys Rev D59: 043516.
7. Carroll SM, Field GB (1997) Is there evidence for cosmic anisotropy in the polarization of distant radio sources? Phys Rev Lett 79: 2394.
8. Eisenstein DJ, Bunn EF (1997) Comment on the Appropriate Null Hypothesis for Cosmological Birefringence. Phys Rev Lett Reply Phys Rev Lett 79:1958.
9. Masreliez CJ (2000) The Expanding Spacetime Theory. Nu Inc Corvallis.
10. Bergson H (1922) Duration and Simultaneity: Bergson and the Einsteinian Universe. Clinamen Press Ltd.
11. Lanczos C (1924) Zeitschrift für Physik A Hadrons and Nuclei. Springer.
12. Gamov G (1946) Rotating Universe? Nature 158: 549.
13. Gödel K (1949) An example of a new type of cosmological solution of Einstein's field equations of gravitation. Rev Mod Phys 21: 447-450.
14. Ozsváth I, Schücking E (1962) Finite Rotating Universe. Nature 193: 1168.
15. Hawking S (1969) On the rotation of the universe. Mon Not R Astron Soc 142:129-141.
16. Valdes F, Tyson JA, Jarvis JF (1983). Alignment of faint galaxy images: Cosmological distortion and rotation. Ap J 271: 431.
17. Birch P (1982) Is the universe rotating? Nature 298: 451
18. Phinney ES, Webster RL (1983) Is there evidence for universal rotation? Nature 301: 735
19. Barrow JD, Juskiewicz R, Sonoda DH (1985) Universal rotation: how large can it be Mon Not R Astron Soc 213: 917-943.
20. Obukhov YN, Ivanenko DD, Korotky VA (1986) On the Rotation of the Universe. Astron Circ Acad Sci USSR 1458: 1-10.
21. Kühne RW (1997) On the Cosmic Rotation Axis. Mod Phys Lett A.
22. Penrose R (1967) Twistor Algebra. Journal of Mathematical Physics 8: 345-366.
23. Hamilton WR (1844) On quaternions, or on a new system of imaginaries in algebra. Philosophical Magazine 25: 489-495.
24. Teller E (1948) On the change of physical constants. Phys Rev 73: 801
25. Sato K (1990) The quest for the fundamental constants in cosmology: Proceedings of the XXIVth meeting Rencontre de Moriond, Les Arcs, Savoie, France.
26. Kochanek CS (1996) Is There a Cosmological Constant? Nasa Technical Reports.
27. Belinchón JA (2014) cosmological models anisótropos con Variable constants: an autosimilar approach, Complutense University of Madrid, Spain.
28. Guth AH (1981) Inflationary universe: A possible solution to the horizon and flatness problems. Physical Review D 23: 347-356.
29. Navarro JF, Frenk CS, White SM (1996) The Structure of Cold Dark Matter Halos. The Astrophysical Journal 462: 563-575.

30. Bidin CM, Smith R, Carraro G, Mendez RA, Moyano M (2014) On local dark matter density. Journal of Physics G: Nuclear and Particle Physics 41:0-6.
31. Clowe D (2006) A Direct Empirical Proof of the Existence of Dark Matter. The Astrophysical Journal Letters 648: L109-L113.
32. Brada M, Allen SW, Ebeling H, Massey R, Morris RG, et al. (2008) Revealing the Properties of Dark Matter in the Merging Cluster MACS J0025.4-1222. The Astrophysical Journal 687
33. Eckert D, Jauzac M, Shan HY, Kneib JP, Erben T, et al (2015). Warm-hot baryons comprise 5–10 per cent of filaments in the cosmic web. Cornell University Library.
34. Simon JD (2011) A Complete Spectroscopic Survey of the Milky Way Satellite Segue 1: The Darkest Galaxy. The Astrophysical Journal.
35. Azcorra iñigo (2014) Does inflation explain the deficit of galaxy matter? Ministry of paranormal affairs.
36. Dirac PAM (1937) The Cosmological Constants. Nature: 139: 323.
37. Zwicky F (1939) On the Theory and Observation of Highly Collapsed Stars. Phys Rev 55: 726.
38. Chandrasekhar S (1937) The Cosmological Constants. Nature 139: 757-758.
39. Sciama DW (1953) On the origin of inertia. Mon Not Royal Astronomical Society 113: 34-42.
40. Brans C, Dicke R (1961) Mach's Principle and a Relativistic Theory of Gravitation. Physical Review 124: 925-935.
41. Ferguson HC (1998) The Hubble Deep Field. Cambridge University Press.
42. Martinez-Manso J (2011) Velocity dispersions and stellar populations of the most compact and massive early type galaxies at redshift ~1. The Astrophysical Journal Letters 738: L22.
43. Genzel R, Förster Schreiber NM, Wilman D, Momcheva I, Tacchella S, et al. (2017) Strongly baryon-dominated disk galaxies at the peak of galaxy formation ten billion years ago. Nature 543: 397-401.
44. Jordan P (1971) The Expanding Earth: some consequences of Dirac's Gravitation Hypothesis. Pregamon Press Oxford.
45. Newman J (2005) Testing for Evolution in the Fine Structure Constant with DEEP2. American Physical Society
46. Shlyakhter AI (1982). Direct Test of the Time-Independence of Fundamental Nuclear Constants Using the Oklo Natural Reactor. Cornel University Library
47. Lamoreaux SK (2004) Neutron Moderation in the Oklo Natural Reactor and the Time Variation of alpha. Cornel University Library.
48. Webb JK, King JA, Murphy MT, Flambaum VV, Carswell RF, et al. (2011) Indications of a Spatial Variation of the Fine Structure Constant. Physical Review Letters 107 :191101.
49. Milgrom M (2007) The MOND Paradigm. Cornel University Library.
50. Moffat JW (2006) Scalar tensor vector gravity theory. Journal of Cosmology and Astroparticle Physics 3: 4-6.

Ergodic Theory and the Structure of Non-commutative Space-Time

Wang CHT*, James Moffat J and Oniga O

Department of Physics, University of Aberdeen, King's College, UK

Abstract

We develop further our fibre bundle construct of non-commutative space-time on a Minkowski base space. We assume space-time is non-commutative due to the existence of additional non-commutative algebraic structure at each point **x** of space-time, forming a quantum operator 'fibre algebra' A(x). This structure then corresponds to the single fibre of a fibre bundle. A gauge group acts on each fibre algebra locally, while a 'section' through this bundle is then a quantum field of the form $\{A(x); x \in M\}$ with *M* the underlying space-time manifold. In addition, we assume a local algebra *O(D)* corresponding to the algebra of sections of such a principal fibre bundle with base space a finite and bounded subset of space-time, *D*. The algebraic operations of addition and multiplication are assumed defined fibrewise for this algebra of sections.

Keywords: Quantum; Algebra; Entropy; Space

Introduction

We characterise 'ergodic' extremal quantum states of the fibre algebra invariant under the subgroup *T* of local translations of space-time of the Poincare group *P* in terms of a non-commutative extension of entropy applied to the subgroup *T*. We also characterise the existence of *T*-invariant states by generalizing to the non-commutative case Kakutani's work on wandering projections. This leads on to a classification of the structure of the local algebra *O(D)* by using a '*T*-Twisted' equivalence relation, including a full analysis of the *T*-type III case. In particular we show that *O(D)* is *T*-type III if and only if the crossed product algebra *O(D)* × *T* is type III in the sense of Murray-von Neumann.

Ergodic Theory in the Classical (Commutative) Case

In the commutative case, a general von Neumann algebra *R* is isomorphic to the set $L^\infty(Z,v)$ of essentially bounded, measurable, complex-valued functions on the locally compact set *Z* with *v* a positive regular borel measure. If *G* is a group of automorphism of *R* then *G* is by definition isomorphic to a group of automorphisms of $L^\infty(Z,v)$ which we also denote as *G*. Projections *P* in the algebra *R* become, under the isomorphism, characteristic functions of borel subsets of *Z*. If $g:P\rightarrow Q$ and *P* is isomorphic to χ_E for the borel set *E* then *Q* is also a projection thus is isomorphic to χ_F for some borel subset *F*. Thus the group *G* induces a group of transformations of the σ-ring B of borel subsets of *Z*.

Let *T* be such a transformation which is measure preserving, then for any borel set *E* in B, $v(T^{-1}(E)) = v(E)$ i.e., the measure v is *T*-invariant. *T* is defined to be ergodic if it mixes the space. i.e., $T^{-1}(E)$= *E* modulo a null set implies either *v(E)*=0 or *v(Z/E)*=0. If *T* is ergodic in this sense and the measure *v* is *T*-invariant, then *v* is defined to be an ergodic measure [1]. It then follows that if *Z* is compact, v is a probability measure on *Z* which is ergodic, and *T* and its inverse are continuous mappings, then this is equivalent to v being an extreme point of the invariant measures on Z. For if Ɛ denotes the set of such invariant probability measures, and v is an extreme point of Ɛ, then any measurable set *E* with $0 < \lambda = v(E) < 1$ allows the construction of measures $\mu_1(*) = \frac{1}{\lambda} v(* \cap E)$ and $\mu_2(*) = \frac{1}{1-\lambda} v(* \cap (Z \setminus E))$ such that $v = \lambda\mu_1 + (1-\lambda)\mu_2$; a contradiction. Conversely if v is a probability measure on Z which is ergodic, and 0<*μ*<*v*, then we have, for any Borel set *A*, $\mu(A) = \int f(x)dv(x)$ for some $f \in L^1(Z,v)$ and *f* is a *T*- invariant, positive function by the properties of probability measures. If *f* is not constant we can define Borel sets $S_1 = \{x \in Z; f(x) < t\}$ and $S_2 = \{x \in Z; f(x) > t\}$ for some positive real *t*. Both sets are invariant and non-trivial, thus they must both have measure 1; a contradiction. Hence *f* is constant and the measure is an extreme point since for any convex combination of measures, v dominates these measures and this leads to the tautology $v = \lambda v + (1-\lambda)v$ for some λ: $0 < \lambda < 1$.

Given a partition *p* of *Z* into measurable subsets $\{A_j, j=1,2,3\ldots\}$, If Z is compact and has total measure equal to 1 then we can interpret the value $\mu(A_j)$ as the probability of the set A_j. The expression $-log\mu(A_j)$ is then a measure of the description length or Kolmogorov Complexity of the partition subset A_j. A measure preserving transformation such as *T* transforms the partition *p* into the partition *Tp* which is $\{TA_j; j=1,2,3\}$. The entropy or, equivalently, the expected value of the Kolmogorov Complexity of the partition *p* is defined as $-\sum_j \mu(A_j)\log\mu(A_j)$. Since *T* is measure preserving, the partitions *p* and *Tp* have the same entropy.

Hopf [2] was interested in the question of when a measure representable as a Lebesgue integral is invariant under a measurable transformation *T*. Hopf considers a partition of the measurable set into subsets and the effect of multiples T^k of *T* acting on those subsets. In operator theoretic language we can express this as follows Stormer [3]. We replace measurable sets by projections and the group $\{T^k; k\in Z\}$ by a general discrete group *G* of automorphisms of a *commutative* von Neumann algebra *R* acting on a Hilbert space *H* which is implemented by the unitary representation $g \rightarrow U_g$ from *G* to the set of unitaries acting on *H*. Two projections *H* and *K* in *R* are *Hopf equivalent* if there is an orthogonal family of projections E_j in *R* and group elements $g_j \in$ G with $H = \sum_j E_j$ and $K = \sum_j U_{g_j}^* E_j U_{g_j}$. This equivalence leads to a partition based criterion, '*H*-finiteness', for the existence of a finite invariant measure. For this kind of orthogonal partition we can define the entropy as being derived from the partition weightings (all equal in this case).

***Corresponding author:** Wang CHT, Professor, Department of Physics, University of Aberdeen, King's College, UK, E-mail: c.wang@abdn.ac.uk

With this definition it is then clear that if the two projections H and K in R are Hopf equivalent then they have the same entropy relative to an invariant measure. This idea is easily extended to the noncommutative case, as we will see later.

Noncommutative (Quantum) ergodic theory

The low energy regime beyond the standard model can be represented, we postulate, by linearised gravity with matter on a flat space-time manifold M. For this regime, invariance of quantum states to the Poincare group is a key symmetry and new results were presented [4]. Within the resulting fibre bundle construct defined there, our focus is on the local fibre algebra A(x) and the subgroup T of the Poincare group consisting of translations of space-time as a group of automorphisms of A(x).

If A(x) is a fibre algebra and the group T of translations of space-time a subgroup of the Poincare group, we define $\alpha: g \rightarrow \alpha_g$ to be a representation of T as automorphisms of A(x). Let f be a faithful normal state of **A(x)** and define the 'induced' transformation.Since the $v_g(f)=f \circ \alpha_g$ this means that $V_g(f)$ is also a normal state and $V_g(f)(A)=f(\alpha_g(A))$ subgroup T is abelian, and the mapping $g \rightarrow v(g)$ is a group homomorphism, the set $\{v(g); g \in T\}$ is a continuous group of commuting transformations of the dual space A(x)*. If f is a state of the algebra then define E to be the weak* closed convex hull of the set $\{v(g)f; g \in T\}$. Then E is a weak* compact convex set and each $v(g)$: $\mathcal{E} \rightarrow \mathcal{E}$. By the Markov-Kakutani fixed point theorem [5] it follows that E has an invariant element. In other words, the group T has the fixed point property and thus is amenable. In summary, because T is an abelian group and locally compact it is an amenable group, since the closed convex hull of any quantum state of the system contains a T-invariant state [6]. This leads us to define the following;

Definition: Let A(x) be a fibre algebra and the group T of translations of space-time a subgroup of the Poincare group. Let $\alpha: g \rightarrow \alpha(g)$ be a representation of T as automorphisms of A(x). The group representation α acts ergodically on A(x) if given a projection E in A(x), $\alpha_g(E)= E\ \forall_g \in T$ implies that $E=0$ or $E=I$.

This definition is a direct generalisation of the commutative case where **A(x)** is the set of essentially bounded measurable functions on a locally compact space with a regular borel measure, discussed above. We can also have the following non- commutative generalisation of an ergodic probability measure as an extreme point of the set of invariant measures; as first pointed out by Segal [7].

Definition: Let A(x) be a fibre algebra and $\alpha: g \rightarrow \alpha(g)$ a representation of the translation subgroup T of the Poincare group P as automorphisms of A(x). A quantum state f of A(x) is α-invariant if $f(\alpha_g(A))= f(A))\ \forall\ A$ in A(x). If f is a normal (i.e., density matrix) state and an extreme point of the set of invariant states, then f is defined to be a T-ergodic state.

Note that the set of invariant states is a compact convex subset of the quantum state space of A(x) and is thus generated by its extreme points [8].There is a non-trivial invariant state for the amenable group T, as discussed earlier, thus there is an extremal T- invariant state of A(x). Since, by definition, the Hilbert space representation on which A(x) acts is separable, the algebra contains a faithful normal state and hence a T-invariant normal state. The norm limit of a set of normal states is again normal, and thus for a separable Hilbert space, the fibre algebra **A(x)** with the assumptions above always contains a T-ergodic state [9].

Definition: The fibre algebra **A(x)** is a quantum operator algebra and thus always contains an identity operator I. If f is a quantum state of **A(x)** then by definition, $f(I)=1$. The support of f is the unique smallest projection E in **A(x)** such that $f(E)=1$ and is denoted E_f.

We also require the following definition;

Definition: E is an α-invariant projection in **A(x)** if $\alpha_g(E)=E\ \forall_g \in T$

The following result is well known in the finite dimensional (matrix algebra) case. For completeness we give a general proof.

Theorem 2: Let **A(x)** be a fibre algebra and $: g \rightarrow \alpha(g)$ a representation of the translation subgroup T of the Poincare group P as (gauge) automorphisms of **A(x)**. Assume there exists at least one state f of **A(x)** which is α-invariant. Then the support of f, E_f, is an invariant projection and f is a normal and ergodic state if and only if the representation α acts ergodically on the cut down algebra $E_f A(x) E_f$

Proof: We start with the observation that assuming f is α-invariant implies that $f(\alpha_g(E_f)) = f(E_f) = 1\quad \forall g \in T$; we call such states 'symmetric'. By uniqueness of the support of f it follows that E_f is an α-invariant projection. Let π be the Gelfand-Naimark-Segal (GNS) representation of **A(x)** induced by the state f on the Hilbert space $H(f)$. We make the simplifying assumption for now that f is a faithful state; i.e., $E_f=I$, and revisit this assumption later. In this case the von Neumann algebra π(**A(x)**) has a separating-generating vector ξ and the representation π is a *-isomorphism. Define the unitary group $U_g \pi(A)\xi = \pi\left(\alpha_g(A)\right)\xi$ on a dense subset of $H(f)$, then U_g extends to a unitary on $H(f) = \{\pi(B)\xi; B\}^-$ in the fibre algebra where $\{\}^-$ denotes closure of the set in the norm topology. The mapping U:$g \rightarrow g$ is then a unitary representation of the translation group T and for B a quantum observable in the fibre algebra **A(x)** we have $U_g \pi(B) U_g^* = \pi(\alpha_g(B))\quad \forall g \in T$ i.e. the unitary representation U implements the automorphic representation α: $g \rightarrow \alpha(g)$.

Consider now the involution mapping on π(**A(x)**) defined as $A \rightarrow A$. * This induces an anti-linear mapping on a dense subset of the Hilbert space $H(f)$; S: $A\xi \rightarrow A^*\xi$. Moreover, this extends to a mapping with closed graph which we also denote by S. By the theorem of Tomita-Takesaki S has a polar decomposition $S = J\Delta^{\frac{1}{2}}$ such that $J(\mathbf{A(x)})J=\pi(\mathbf{A(x)})'$; the commutant of the fibre algebra $\pi(A(x))$ [9]. If $x=B\xi$ is in the domain of S, then it follows that $U_g x$ also lies in the domain of S, and we have the relationship;

$$U_g SB\xi = U_g B^*\xi = \alpha_g(B^*)\xi = \alpha_g(B)^*\xi = S\alpha_g(B)\xi = SU_g B\xi$$

This leads to the conclusion that, on the domain of S, we have $U_g S=SU_g$

Then we have;

$$S = U_g S U_g^* = U_g J\Delta^{\frac{1}{2}} U_g^* = U_g J U_g^* U_g \Delta^{\frac{1}{2}} U_g^*$$

By uniqueness of the polar decomposition, $J = U_g J U_g^*\ ; J$ from this we deduce J and U_g *commute* for all $g \in T$

$B \in \pi(\mathbf{A(x)}) \cap \{U_g; g \in T\}'$ implies that $JBJU_g=U_gJBJ$ for all $g \in T$

Thus $JBJ \in \{U_g; g \in T\}' \cap \pi(\mathbf{A(x)})'$

Conversely if $C \in \{U_g; g \in T\}' \cap \pi(\mathbf{A(x)})'$ then $C=JBJ$ *for some* $B \in \pi$ (A(x)) and

$BJU_g=U_gJBJ$ *implies* $JBU_gJ=JU_gBJ$ and thus $BU_g=U_gB$ so that $B \in \{U_g; g \in T\}'$

that; We conclude that $J\left\{\pi(A(x))\cap\{U_g; g\in T\}'\right\}J = \pi(A(x))'\cap\{U_g; g\in T\}'$

The automorphic representation α: g α_g of T acts ergodically if and only if $\pi(A(x))\cap\{U_g; g\in T\}'$ is trivial, containing only the projections 0 and I and thus consisting of the set of complex multiples of *I*. From the reasoning above it follows that the representation $\alpha: g\to\alpha_g$ of T acts ergodically if and only if $\pi(A(x))'\cap\{U_g; g\in T\}'$ invariant is also trivial.

If *E* is a projection in the set $\pi(A(x))'\cap\{U_g; g\in T\}'$ we can define a state $f_E(A)=\frac{\langle\xi, E\pi(A)\xi\rangle}{\langle\xi, E\xi\rangle}$ on the fibre algebra **A(x)**. Then $f_E=\omega_{E\xi}\circ\pi$ is a state dominated by $f=\omega_\xi\circ\pi$ and we have;

$$f(A)=\frac{\langle\xi,\pi(A)\xi\rangle}{\langle\xi,\xi\rangle}=\lambda\frac{\langle\xi,E\pi(A)\xi\rangle}{\langle\xi,E\xi\rangle}+(1-\lambda)\frac{\langle\xi,(I-E)\pi(A)\xi\rangle}{\langle\xi,(I-E)\xi\rangle}\quad\text{for } A\in\mathbf{A(x)},$$

$$\text{where } \lambda=\frac{\langle\xi,E\xi\rangle}{\langle\xi,\xi\rangle}=\frac{\|E\xi\|^2}{\|\xi\|^2}\text{ and } 1-\lambda=\frac{\|\xi\|^2-\|E\xi\|^2}{\|\xi\|^2}=\frac{\langle\xi,(I-E)\xi\rangle}{\|\xi\|^2}$$

Thus *f* is an extremal invariant state if and only if the projection *E=0 or I*. The result follows for the support of *f* equal to 1. Finally, we need to extend the result to a general invariant state *f* with support E_f, $0<E_f<I$. This follows from what we have already proved, since the restriction of *f* to E_fA(x)E_f is a faithful state, and a state extremal among the invariant states of the cut down algebra E_fA(x)E_f is also extremal among the invariant states of the full fibre algebra **A(x)**. This follows from the fact that if *f* is a convex combination of states from the full fibre algebra, then each of them has a support less than or equal to E_f. In the next section we develop and prove a noncommutative version of a well-known result in classical ergodic theory and use it to characterize the existence of such symmetric states.

Wandering Projections and Invariant Symmetry States

(Hajian and Kakutani, 1964) defined a wandering set as follows; [10]

Definition: Let (*X*, *B*, *μ*) be a measure space with finite measure; $\mu(X)<\infty$ and where *B* is the set of all measurable subsets of *X*. Let *T* be a bijective transformation of *X* such that both *T* and its inverse are measurable mappings. A wandering set for *T* is a measureable subset *S* of *X* such that the sets $\{T^{nk}(S)\}$ are disjoint, for some infinite sequence of integers *nk*.

Definition: Two measures *v* and *μ* on the measure space X are said to be equivalent if they share the same null sets. A measure *v* is *T*-invariant if *v(T(E))*=v(*E*) for all measurable subsets *E* of *X*.

With these definitions, showed that there is a finite *T*-invariant measure *v* on *X*, equivalent to *μ*, if and only if there are no wandering subsets of *X* [10].

If now we consider an abelian von Neumann algebra *R*, then *R* is isomorphic to *C (X)* with *X* a compact stonean space of finite measure, and the positive, normal, regular borel measures on *X* correspond to the normal states of *R*. By Dixmier we can characterise these normal measures as being equivalent to measures which annihilate each nowhere dense subset of *X* [11]. It follows that if measures *v* and *μ* on the measure space *X* are equivalent and measure *v* is normal, then measure *μ* is also normal.

If *θ* is a continuous automorphism of the abelian algebra *R*, isomorphic to *C (X)*, then we can define the homeomorphism *T* of *X* by $\theta f(x)=f(Tx)$ for $f\in C(X)$. By the result quoted above, if *μ* is a normal measure on *X* with support equal to *X*, there is a measure equivalent to *μ* which is *T*-invariant if and only if there are no wandering measurable subsets *E* of *X*.

If such a set *E* did exist, such that the sets $\{T^{nk}(E)\}$ are disjoint, for some infinite sequence of integers *nk*, then by regularity of *μ* we can assume that *E* is closed. Since *X* is a stonean space, *E* is both open and closed. Thus the characteristic function χ_E corresponds to a projection in the algebra *R* and the set of projections $\theta^{nk}(\chi_E)$ is an orthogonal set. From the algebraic perspective then we can say the following. Given an abelian von Neumann algebra *R*, an automorphism *θ* of *R* and a faithful normal state acting on *R*. Then there is a faithful normal *θ*-invariant state acting on *R* if and only if there are no non-trivial projections *E* in *R* such that for some infinite sequence of integers *nk*, the projections $\theta^{nk}(E)$ are mutually orthogonal. It can be easily shown that for a commutative algebra this condition on the set of projections is equivalent to the requirement that that there are no nonzero projections *E* with $\theta^{nk}(E)$ 0 in the ultraweak toplogy $nk\to\infty$ for some infinite sequence *nk* of integers. This new formulation now generalises easily to the non-commutative (quantum) case as follows.

Definition: Let *R* be a von Neumann algebra, *G* a group of auto morphisms of *R*. Then a nontrivial projection *E* in *R* is wandering if *E* is such that; $g_{nk}(E)\to 0$ for some infinite sequence g_{nk} in *G*. Convergence is defined in the weak operator topology.

If **A(x)** is a fibre algebra then it is a von Neumann algebra with trivial centre and is countably decomposable. Let $\alpha: g\to\alpha_g$ be a group representation of the translation subgroup *T* of the Poincare group which is ultraweakly continuous.

Theorem 3: There is a faithful normal translation invariant quantum state on the fibre algebra **A(x)** if and only if there are no wandering projections in **A(x)**.

Clearly if *E* is a projection in **A(x)** such that $g_{nk}(E)\to 0$ for some infinite sequence g_{nk} in *G* and *f* is a faithful, normal *α*-invariant state, then *f(E)*=0, thus *E*=0.

The proof of the converse is based on work by Takesaki on singular states [12]. We assume that there are no wandering projections in **A(x)**. The fibre algebra **A(x)** has a faithful normal state *f*. By the fixed point property, applied to the set; *E*=weak* closed convex hull of {*v*(*g*) *f*; *g*∈*T*}, **A(x)** has an invariant state which we denote as *h*. We need to show that *h* is both normal and faithful. By Takesaki *h* has a unique decomposition $h=h_n+h_s$ with h_n a normal positive linear functional and h_s a singular positive linear functional [13]. *By u*niqueness of the decomposition, both of these linear functionals are also *α*-invariant. Let *S* be the support of h_n so that $0\le S\le I$. If $S\neq I$ we can choose a projection *F* with $0<F<I-E$ $h_s(F)=0$ and $h_s(F)=0$ [12]

Let $\lambda=\inf_g\in_T(f\circ\alpha_g(F))$. Since $h=h_n+h_s$, we have $h(F)=0$. Therefore $\lambda=0$. Thus there is a sequence g_{nk} with. $f\circ\alpha_{gnk}(F)\to 0$ Since *f* is faithful and normal this implies that $\alpha_{gnk}(F)\to 0$ in the weak operator topology; i.e. *F* is a wandering projection. This contradiction shows that the support of h_n equals I and h_n is the required normal, faithful invariant state

The Structure of the Local Algebra *O(D)*

For each event point **x** in Minkowski space-time, we have a fibre algebra **A(x)** defined as a von Neumann algebra with trivial centre and a faithful representation as an algebra of operators acting on a separable Hilbert space. Thus *O(D)* is an associative principal fibre subbundle; *associative* in the sense that a Lie group (the translation subgroup of the Poincare group) acts on each fibre; a *subbundle* in the sense that only that subset of $\{\mathbf{A(x)};\mathbf{x}\in M\}$ with x∈*D* is of physical interest.

The local von Neumann algebra *O(D)* does not necessarily have a trivial centre; its structure is more complex in some ways. We assume that the quantum system it represents has an energy operator with discrete countable eigenstates and we thus assume also that *O(D)* is separable. We propose to use the ideas of noncommutative ergodic theory to gain insight into the structure of *O(D)*, as we now describe.

Von Neumann introduced the idea of equivalence of measurement 'projection' operators as a way of gaining traction on the structure of a general von Neumann algebra [9]. Much of this analysis centres around the question of whether or not the algebra possesses a finite trace, extending the idea of the trace of a finite matrix operator as the sum of its observable eigenvalues. This analysis was enhanced to take account of groups of (unitarily implemented) automorphisms of the algebra by Stormer [3]. This allows him to define a 'G-equivalence' of projections which generalises to the non-commutative quantum case the definition used in standard commutative ergodic theory [2]. One of us, extended this work to a characterisation of the tensor product of 'G-type III' algebras. As a result of this previous work we can now develop a classification of the structure of our local algebra *O(D)* [14]. We do this by applying these earlier results where the group concerned is now the subgroup *T* of local translations of space-time of the Poincare group *P*.

Definition: Let α: $g \to \alpha_g$ be a group representation of the translation subgroup *T* of the Poincare group as a discrete group acting on the von Neumann algebra *O(D)*. A representation π of *O(D)* acting on a Hilbert space *H* is covariant if there is a homomorphism $g \to U_g$ from *G* to the group of unitary operators on *H* with $\pi(\alpha_g(A)) = U_g \pi(A) U_g^* \quad \forall A \in O(D)$.

If φ is a normal state of *O(D)* then $\phi \circ \alpha_g$ is also a normal state since each automorphism preserves the algebraic structure and hence preserves complete additivity. If *S* denotes the set of all normal states of *O(D)* then the direct sum $\pi = \oplus\{\pi_\varphi; \varphi \in S\}$ of their Gelfand-Naimark-Segal (GNS) representations is a faithful representation of *O(D)* as a von Neumann algebra acting on a Hilbert space *H* which is the direct sum of the GNS Hilbert spaces. If we define

$U_g\left(\oplus_{\varphi\in S}\pi_\varphi(A_\varphi)x_\varphi\right) = \oplus_{\varphi\in S}\pi_\varphi(\alpha_g(A_{\varphi\circ\alpha_g}))x_\varphi$ as a mapping on each of the pre-Hilbert spaces for the GNS constructions, then U_g extends to a unitary operator on *H* and the representation $\{A_g; g \in T, A_g \in O(D)\}$ is a faithful normal representation of *O(D)*. We can therefore assume that *T* acting on *O(D)* as a discrete group of automorphisms is unitarily implemented, if necessary.

Definition: Let α: $g \to \alpha_g$ be a group representation of the translation subgroup *T* of the Poincare group as a discrete group acting on the von Neumann algebra *O(D)*. If *E* and *F* are projections in *O(D)* we say that *E* and *F* are *T*-equivalent if there is a set of operators $\{A_g; g \in T, A_g \in O(D)\}$ with $E = \sum_g A_g^* A_g$ and $F = \sum_g \alpha_g(A_g A_g^*)$.

Definition: We write this *T*-equivalence as $E \approx F$ and call it a *T-twisted equivalence*. In the special case that each A_g is a projection, this definition is a direct non-commutative generalisation of Hopf equivalence.

Definition: A projection *F* is defined to be *T-finite* if *F* contains no proper sub-projections which are *T*-equivalent to *F*. The algebra *O(D)* is defined to be *T-finite, or T-Type II(1),* if the identity of *O(D)* is a *T*-finite projection. *O(D)* is *T-semifinite, or T-Type II(∞)* if every projection in O(D) dominates a *T*-finite projection. *O(D)* is *T-purely infinite, or T-Type III,* if *O(D)* does not contain any *T*-finite projections.

The *T*-type III case is the most difficult to analyse. In the *T*-type III case there is not even the 'shadow' of a trace. A *T*-invariant trace is a bounded faithful normal linear mapping τ: *O(D)*→*C* with;

$$\tau(AB) = \tau(\alpha_g(AB)) = \tau(BA) \quad \forall g \in T; A, B \in O(D).$$

If τ is a trace, then by the earlier remarks we can assume that the group representation of *T*, as a discrete group, is unitarily implemented by the unitary representation U:$g \to U_g$ so that τ is automatically *T*-invariant. Further, if *F* is a *T*-finite projection and *E*~*F* in the sense of Murray and von Neumann then $E \leq F$ and *E*~*F* imply that $E \lesssim F$ and $E \approx F$ (using only the identity of the group). Thus *E*=*F* and *F* being *T*-finite implies *F* is finite.

Stormer established that *O(D)* is *T*-semifinite if and only if there is a faithful normal semifinite *T*-invariant trace on *O(D)*.

The Crossed Product Algebra of O(D)

Assume (by taking a faithful representation if necessary) that *O(D)* acts on a Hilbert space *H*. Define the Dirac function ε_g to take the value 1 at g and zero elsewhere on *T*. Then $\{\varepsilon_g; g \in T\}$ is an orthonormal basis for the Hilbert space $l^2(T)$ Given $l^2(T)$ and H we can form the tensor product Hilbert space $H \otimes l^2(T)$. Define;

$$U_h(x \otimes \varepsilon_g) = x \otimes \varepsilon_{gh^{-1}} \text{ for } xH\ g,h \in T$$

$$A)(x \otimes \varepsilon_g) = \alpha_g(A)x \otimes \varepsilon_g \text{ for } A \in O(D), g \in T$$

Then U_h extends to a unitary operator on $H \otimes l^2(T)$and the mapping $h \to U_h$ is a group homomorphism from the translation group T into the group of unitaries acting on $H \otimes l^2(T)$.

Similarly $\Phi(A)$ extends to a bounded linear operator on $H \otimes l^2(T)$ for all *A* in *O(D)* and the mapping $h \to U_h$ implements the automorphic representation $h \to \alpha_h$. *O(D1)*⊗ *O(D2)*

The transformation Φ is an ultraweakly continuous *isomorphism of *O(D)* and it follows that *Φ(O(D))* is a von Neumann algebra. Finite sums $\sum_j U_{g_j}\Phi(A_j)$ form a *algebra denoted $(O(D)\times T)_0$ which contains *Φ(O(D))*. The cross product algebra *O(D)*×T is defined as the closure of the *algebra $(O(D)\times T)_0$for the ultraweak operator topology. The crossed product algebra *O(D)*×T can be used to prove the following structural result.

Assume *O(D1)* and *O(D2)* are local von Neumann algebras in space-time regions *D1* and *D2* which are not space-like separated. Let *G* and *H* be discrete representations of the translation subgroup of the Poincare group as automorphisms of *O(D1)* and *O(D2)* respectively. Then if either *O(D1)* or *O(D2)* is *G/H*-purely infinite (*G/H*-Type III), the joint algebra *O(D1)*⊗ *O(D2)* is *G*×*H*-purely infinite (equivalently *G*×*H*-type III) under the action of the joint representation G×H of the translation group. If both O(D1) and O(D2) are G/H finite or G/H semifinite, then the same applies to the joint algebra *O(D1)* ⊗ *O(D2)*. These results follow from the fact that; *O(D1)* ×*G)* ⊗ *O(D2)* × *H)* is spatially *isomorphic to (OD1 ⊗ OD2) × (G × H)[14].

A Symmetry of Types

In this part of our analysis of the structure of *O(D)* we find a pleasing symmetry for purely infinite type III algebras between the *T*-type of *O(D)* and the corresponding Murray-von Neumann type of its cross product algebra *O(D)*×*T*. First we have to prove the following key result. Recall that, by construction, the crossed product von Neumann algebra O(D) × T contains the embedded closed sub-algebra *Φ(O(D))*, isomorphic to *O(D)*.

Theorem 4: There is an ultraweakly continuous mapping, denoted Γ, from O(D) × T to O(D) such that the restriction $\Gamma|_{\Phi(O(D))} = \Phi^{-1}$, the inverse of the embedding of the algebra O(D); and the composite map $\Gamma \circ \Phi : O(D) \times T \to \Phi(O(D))$ is a continuous projection of norm one.

Proof: Continuing with the notation introduced earlier; the map $x \to x \otimes \varepsilon_g : H \to H_g$, where H_g is a closed linear subspace of $H \otimes l^2(T)$, is both isometric and linear. Since the set $\{\varepsilon_g; g \in T\}$ is orthonormal, the Hilbert space $K=H \otimes l^2(T)$ is the direct sum of the H_g's and every element x of K can be represented as $x = \sum_{g \in T} x_g \otimes \varepsilon_g$ with $\|x\|^2 = \sum_{g \in T} \|x_g\|^2 < \infty$.

If E_g is the projection from K onto H_g, and $B = \sum_g U_g \Phi(A_g)$ is an element of $(O(D) \times T)_0$ then straightforward arguments show that $E_s B E_t = E_s U_{s^{-1}t} \Phi(A_{s^{-1}t}) E_t$ Taking the weak closure, we have $B \in O(D) \times T$ with $B = \lim \{B^\alpha; E_s B^\alpha E_t = E_s U_{s^{-1}t} \Phi(D^\alpha_{s^{-1}t}) E_t\}$. From the Kaplansky density theorem we can choose B^α with $\|B^\alpha\| \le \|B\|$ and the net $D^\alpha_{s^{-1}t}$ is then a bounded net in the ultraweakly compact ball of radius $\|B\|$ [9]. It thus has a subnet converging to an element D_{s-1t} of *O(D)*. From this we have the following expression;

$$E_s B E_t = E_s U_{s^{-1}t} \Phi(D_{s^{-1}t}) E_t \quad (1)$$

In particular we have $E_e B E_e = E_e \Phi(D_e) E_e$. If we define *Γ(B)=De* then clearly the mapping Γ is linear, and $\Gamma|_{\Phi(O(D))} = \Phi^{-1}$. Finally if $B^\alpha \to B$ ultraweakly then $E_e B^\alpha E_e \to E_e B E_e$ and the mapping Γ is ultraweakly continuous.

If *B* is in the kernel of Γ then $D_e=0$ from equation (1) above this implies that $E_s B E_s=0$ ∀*s* and thus *B*=0; the kernel of Γ is {0} and Γ is a faithful mapping. This shows that Γ has the required properties, and completes the proof.

This allows us to now prove the following key structural result.

Theorem 5: *O(D)* is *T*-type III if and only if the crossed product algebra *O(D)* × *T* is type III in the sense of Murray-von Neumann.

Proof: If *O(D)* is not *T*-type III then it contains a non-trivial *T*-finite projection E. Then it follows that if Φ is the identification of *O(D)* within the crossed product algebra *O(D)* × *T* then Φ(E) is finite in the sense of Murray-von Neumann. Thus *O(D)*×*T* is not type III. Conversely assume the crossed product algebra *O(D)* × *T* is not type III. From Theorem 4 we know that there is a faithful normal projection Γ of norm one from *O(D)* × *T* onto *O(D)*. From Sakai [15] it follows that *O(D)* cannot be type III. Thus *O(D)Z* is semifinite for some projection Z in the centre of *O(D)*. From Stormer [16,17] *O(D)Z* is *T*-semifinite thus *O(D)* cannot be *T*-type III. This completes the proof.

Conclusion

Through the experiment we explained fibre bundle construct of non-commutative space-time on a Minkowski base space.

References

1. Richard HP (1956) Lectures on ergodic theory. American Mathematical Soc.
2. Hopf E (1932) Theory of measure and invariant integrals. Transactions of the American Mathematical Society 34: 373-393.
3. Stormer E (1973) Automorphisms and equivalence in von Neumann algebras. Pacific Journal of Mathematics 44: 371-383.
4. Moffat J, Oniga J, Charles HT, Wang (2016) Group cohomology of the Poincare group and invariant quantum states. Cornell Universty library.
5. Reed M, Barry S (1980) Methods of modern mathematical physics.
6. Jean-Paul P (1984) Amenable locally compact groups. Wiley-Interscience.
7. Segal IE (1951) A class of operator algebras which are determined by groups. Duke Mathematical Journal 18: 221-265.
8. Krein M, David M (1940) On extreme points of regular convex sets. Studia Mathematica 9: 133-138.
9. Kadison RV, John RR (1983) Fundamentals of the theory of operator algebras. Springer.
10. Hajian AB, Kakutani S (1964) Weakly wandering sets and invariant measures. Transactions of the American Mathematical Society 110: 136-151.
11. Dixmier, Jacques D (1951) Sur certains espaces considérés par MH Stone. Instituto Brasileiro de Educação.
12. Takesaki M (1959) On the singularity of a positive linear functional on operator algebra. Proceedings of the Japan Academy 35: 365-366.
13. Takesaki M (1958) On the conjugate space of operator algebra. Tohoku Mathematical Journal 10: 194-203.
14. Moffat J (1974) On groups of automorphisms of the tensor product of von Neumann algebras. Mathematica Scandinavica 34: 226-230.
15. Sakai S (1998) C*-algebras and W*-algebras. Springer Science.
16. Størmer E (1970) States and invariant maps of operator algebras. Journal of Functional Analysis 5: 44-65.
17. Moffat J (1977) On groups of automorphisms of operator algebras. Mathematical Proceedings of the Cambridge Philosophical Society. Cambridge University Press.

Scalar Field Theory for Mass Determination

Panicaud B*

Department of Physics, University of Technology of Troyes, France

Abstract

Matter is actually under numerous investigations because of our misunderstanding on some observed phenomena, especially at the astronomic and cosmological scales. In the present article, a scalar field theory is investigated to explain the mass of particles from a global point of view. An universal mechanism is developed and a general relation is eventually proposed that enables to make some accurate predictions for the mass of composite particles. Numerical values are provided including predictions for existing particles, with discrete energy spectrum in relation to large-scale phenomena.

Keywords: Mass determination; Mass prediction; Dark energy; Quantum oscillator; Baryons masses

Introduction

Nowadays, the mass concept seems to be well understood and is mainly associated to particles of matter. However, several problems still remain unexplained. First, astrophysical observations [1,2] lead to the conclusion that some dark matter could exist to explain, for example, the rotation curve of galaxies as pro- posed on galaxy clusters by Zwicky [3] and proved on several observed galaxies by Rubin [4]. This dark matter has several explanations either baryonic or non-baryonic, but none is definitively accepted, meaning that matter is eventually not well known. In the same way, the concept of dark energy has no more convincing explanation. Second, mass can be explained by different theories that will be reminded in the present article. However, few of them lead to predictions that can be experimentally tested. Some mechanisms are nowadays identified and experimentally verified, such as the mass acquisition of massive bosons Z and W through the Higgs process [5,6]. Last but not least, mass is still the last physical unit based on a physical standard bulk material subjected to inherent difficulties [7,8]. We can definitively wonder what mass means. In the present article, the possible origins of the mass of physical systems are reminded with different explanations according to the different physical theories. Too exotic theories are not considered (such as negative mass). Therefore, the proposed list is not exhaustive but shows the main explanations, through some examples. Leading to a no-way road, we suggest and build another theory. This paper is thus an attempt to calculate the masses of both fundamental and composite particles from a master formula containing several parameters. The developments are based on a Lagrangian approach with a variational principle. It is consistent with the developments in physics over the last 50 years. Indeed, field theory is the based of actual physics, for which the present theory is derived using the classical tools of differential geometry. This theory is based on arguments already considered but not in the same way as presently. We base our theory on few simple assumptions. That theory for mass calculation is then described and more specifically the assumed interaction of matter with a real scalar field. The parameters of this model are then investigated. The use of that model on fundamental and composite particles enables to quantify explicitly these parameters. Relations between those parameters lead to predictions of particles masses. The investigation of one of those parameters according to the energy level of different families of particles leads to some observable energy density, which is eventually interpreted.

Review of Possible Mass Explanations

Mass is mainly associated to matter. Viewing at the classification of particles, we can see that fermions and bosons may have a mass. It is worth noting that all fermions have a mass, including neutrinos, whereas only some bosons have. For physical body, mass is usually separated in active gravitational mass, passive gravitational mass and inertial mass [9]. Reciprocity of the gravitational action leads to the equality between the active and passive masses [9]. The strong equivalence principle leads to the equality between the gravitational and inertial masses according to the Einstein's theory of General Relativity explained [10,11]. In addition, Einstein proposes the equivalence between the total mass and total energy in the framework of Special Relativity [12]. This last equivalence will be systematically used further. Besides, mass definition is clear for closed systems; whereas for open systems where energetic interactions take place with the surrounding environment through boundaries, it is always more difficult to define it clearly.

Global energy balance

Mass may be defined through the balance of total energy of a physical system (according to the equivalence between the total mass and total energy). It means a global definition of the corresponding mass. There are several examples. First, let us consider the relativistic scattering of particles. When energy thresholds are reached, particles may be created with specific masses. These thresholds and masses are defined according to the global balance of energy [11]. This "mechanism" applies during collisions in labs, stars or during the Big Bang. The last one may be considered as the fabric of particles defining their masses once forever in specific conditions that are no more accessible. In such examples, each system is constituted of several particles without interaction (or weakly interacting) with the outside of the system. Second, during nuclear or chemical reactions, the separated and thus non-interacting constituents have a different energy to an interacting system when constituents are closer. Global balance of energy between both configurations (interacting and non-interacting) enables to calculate the bounding

***Corresponding author:** Panicaud B, Professor, Department of Physics, University of Technology of Troyes, France
E-mail: benoit.panicaud@utt.fr

energy [13]. This "mechanism" is much more important (106 times) for nuclear interactions than for chemical interactions, because of the amplitude and characteristic length of nuclear interactions compared to the electromagnetic ones. Third, Komar made an attempt to define mass in the context of General Relativity, especially when exists a gravity field [14]. Other mass definitions exist such as Bondi or ADM masses [15,16]. This approach can be interpreted as a generalization of the equivalence between total mass and total energy in the context of General Relativity, by using the global energy balance with the momentum-energy tensor of contravariant component $T_{\mu\nu}$ and the metric tensor of contravariant component $g_{\mu\nu}$. The Komar mass MK is a definition of the total mass of the system based on its total energetic content, according to the relation [14,10]

$$M_K = \frac{1}{c^4} \iiint_V \sqrt{g44}\left(2T^{\mu\nu} - g^{\mu\nu}\left(T^{\alpha\beta} g\alpha\beta\right)\right) u_\mu u_\nu dV \quad (1)$$

where c is the speed of light, V is the space volume, uµ is the covariant component of the quadrivector speed of particles such that *uµuµ=c2*, and *g44* is the covariant "time" component of the metric tensor. Greek subscript or superscript runs from 1 to 4 in the present article.

Total energy decompositions

There is another way to define the mass of physical systems: by decomposing their different parts due to different physical effects. In such a decomposition, it is assumed that the different parts are independent and can be associated additively. There are several examples. In Special Relativity, the link with the total mass is given by the equivalence $E=mc^2$, where E is the total energy and m is the total mass.

First, at non-relativistic limit, the total energy of a single and isolated particle without internal degrees of freedom is the sum of its energy at rest m_0c^2 and its kinetic energy. The total mass of this particle for an observer is thus the sum of its mass at rest m_0 plus a mass depending on the speed of the particle to the observer (supposed to be inertial/Galilean) (see eqn. 3). This decomposition is the first-order expansion of the general equivalence between mass and energy. It can be also interpreted as a consequence of the fourth-component (on time direction) of the relation between the quadrivector momentum-energy of con- travariant component p^μ and the quadrivector speed of contravariant component.

u^μ :

$$E/c = p^4 = m_0 u^4 = m_0 \gamma c = mc \quad (2)$$

$$\Rightarrow m = m_0\gamma = \frac{m_0}{\sqrt{1 - v^2/c^2}}$$

$$\approx m_0 + \frac{1}{2} m_0 v^2 / c^2 \quad (3)$$

where γ is the Lorentz factor [10] depending on the spatial norm of the speed v2. Second, when taking into account for internal degrees of freedom of a single and isolated particle, it is necessary to add an internal energy, for which a statistic assumption is often performed. We consider that internal thermalization is reached. Each internal degrees of freedom nd has thus the same equi probable energy [16]. It enables to decompose the total mass as the sum of a constant part m1 (different throughout this article), plus a contribution depending on temperature T:

$$m \approx m_0 = m_1 + \frac{n_d k_B T}{2c^2} \quad (4)$$

where k_B is the Boltzmann constant. A similar relation was proposed by De Broglie in 1955 to explain the mass from a domain at temperature T. It is quite easy to extend eqn. 4 for Special Relativity by using the Lorentz factor γ, provided that temperature is correctly coupled to γ [17]. It is more difficult to extend it for General Relativity. For example, a heated and massive gas leads to an increase of its mass. However, its numerical value depends on the chosen equation for mass definition (Komar or others). Moreover, when the system is dynamic (gas exploding...), these mass definitions are quite useless because the calculations are easy only for stationary metric tensor.

Third, when extending to open systems with external weak interactions, any physical effect can contribute to the total mass using a similar way. For example, if we consider the same single particle as previously with an internal structure and specific magnetic properties, then interaction with an external magnetic vector field of component B^k_{ext} can contribute to the total mass, such as:

$$m \approx m_0 = m_1 + \frac{n_d k_B T}{2c^2} + \frac{B^2_{ext} V}{2\mu c^2} \quad (5)$$

where V is the volume of the particle and µ is its magnetic permeability. Any other energy may contribute to the total mass. This last example ought to be placed in section 2.4 as a case of mass due to an interaction mechanism. We see that the contribution depends on the square of the inverse of the speed of light. Consequently, the numerical contributions of those effects to the total mass are often very small.

Last, another decomposition is possible when considering a composite system, based on the additivity of energy when independent parts of this system are assumed. Indeed, let us consider a complex system composed of n_p parts.

If the different parts do not interact, or weakly such as it can be neglected, then the total mass can me written as:

$$m = \sum_{i=1}^{np} m_i \approx \sum_{i=1}^{np} m_{0i} \quad (6)$$

Last approximation corresponds to the non-relativistic limit. For continuous media, the sums in eqn. 6 are replaced by integrals.

Quantum effects and interpretation

Quantum mechanics leads to some specific interpretation of the mass due to the duality between particles and waves. Indeed, the energy of a particle is related to the angular frequency of its wave; whereas the group speed vector of contravariant component is related to the variation of this energy E to the wavenumber vector of covariant component k_i. It leads to a quantum explanation of the rest mass of the wave group, according to (at non-relativistic limit) [18]:

$$m_0 = \frac{p^i}{v^i_g} = \frac{\hbar^2 k^i}{\frac{\partial E(k_l)}{\partial k_i}} \quad (7)$$

where $\hbar$ is the reduced Planck constant. This "mechanism" can be used, for example, to calculate the mass contribution in a lattice with electrical charges moving within. Indeed, we can calculate in a more general way the effective second-order mass tensor of particles useful in solid-state physics [19,20]:

$$m^{ij}_0 = \hbar^2 \left(\frac{\partial^2 E(k_l)}{\partial k_i \partial k_j}\right)^{-1} \quad (8)$$

Let us consider the case of a cubic lattice of lattice spacing a with an energy $E(k_l) = E_0 + \sum_{l=1}^{3} (Aa^2 (k_l)^2)$, where A is the amplitude of the energy of particles moving on this lattice. With equations 7 or 8, we obtain the same result for the isotropic effective mass of the particles:

these Relations can also be applied to the quantum tunnelling or to the cyclotron motion. It is also possible to have quantum contribution due to degenerate states. For example, we can mention the case of degenerate fermions gas and degenerate bosons gas for which quantum effects lead to a specific relation for the effective mass of those "almost ideal" gases [21]. For example, the degenerate Fermi gas with repulsive interaction of amplitude U_a and interaction characteristic Length

$$L_a = \frac{m_1 U_a}{4\pi\hbar^2}$$

$$m_0 = m_1 + 0.59 m_1 L_a^2 \left(\frac{3\zeta}{\pi}\right)^{2/3} \tag{9}$$

where ζ is the volumic concentration of particles. Quantum effects are required to explain the masses of subatomic particles and more precisely at a scale below the Compton wavelength. Some attempts to explain the light hadrons masses can be found [22]. The quantum theory offers an interesting framework for quantification of masses. Formulas based on the resonance theory of elementary particles as de Broglie waves have been proposed, such as for example m0,n ≈ 137 n me where n is the mode and me is the electron mass [23]. At the opposite, the masses of leptons (electron, muon and tau) may be described with an empirical relation, known as Koide formula [24]:

$$m_e + m_\mu + m_\tau = \frac{2}{3}\left(\sqrt{m_e} + \sqrt{m_\mu} + \sqrt{m_\tau}\right)^2 \tag{10}$$

Some authors have tried to applied it to the prediction of the neutrinos masses or extended it for other particles [25-27]. Eventually, we have also to focus on the seesaw mechanism as a possible explanation for masses of neutrinos [28]. Indeed, because of the quantum probabilities, each neutrino can transform into each other. It is known as neutrinos oscillation [29,30]. Their mass could be then explained thanks to a super-partner particle for which the mass is big enough to correlate with the small observed masses of neutrinos [28,30,31]. More explanations and details for the neutrino masses can be found [32].

Mechanisms of interaction

From the previous explanations, some cases correspond to local effect due to interactions with the different fields. There are several examples for such open systems. First, such an effect is not limited to microscopic scale and can be observed at macroscopic scale. We can mention the hydrodynamic interaction of a sphere of finite volume V accelerating in a fluid of mass density ρ [33]. The local interaction of such a finite solid with the environing fluid leads to a contribution of the mass as:

$$m \approx m_0 = m_1 + \frac{\rho V}{2} \tag{11}$$

Second, at microscopic field, similar mechanisms may be considered. For example, the mass of massive bosons Z and W can be explained through the local interaction with a complex scalar field [5,6]. This mechanism leads to the acquisition of particle mass thanks to the Higgs boson, through the broken symmetry in the scalar field [5,6]. Such a mechanism may be generalized to any particles (Goldstone model [34]). However it presents different limitations, such as the number of necessary coupling constants as numerous as the existing particles.

Third, there exists different possible couplings with scalar field. The present article focuses especially with a specific coupling between matter and a real scalar field (see in section 3).

Last, we can discuss the case of interacting systems through gravitation. Boratav and Kerner [35] have presented an interpretation of mass as the interaction with the far distant universe. This mechanism is consistent with the Mach principle that has oriented Einstein for the construction of the General Relativity. Different assumptions are assumed: finite radius of the universe; constant mass density of the universe ρU at the observed scale; only gravitational interaction (without magneto-gravitational effect, but with radiative gravitational terms); expansion of the universe according to the Hubble law. It can be then proved that mass emerges from the radiative gravitational force of the far distant universe on a test particle. The inertial mass and the gravitational mass are equal provided that $\rho U\, GH^{-2}=1$, where G is the Newtonian gravitational constant. Actually, the experimental evaluations of the Hubble constant $H \approx 73 km\cdot s^{-1} \cdot M\, pc^{-1}$ and of the mass density $\rho U \approx 10^{-26}$to $10^{-25}\, kg\, m^{-3}$ suggest that this relation could be numerically verified.

Mechanisms of self-interaction

Self-interaction is also a kind of interaction that could also explain the mass for open or closed systems. The most famous development has been made for electrodynamics. Several relations can be proposed based on different approximations: with or without relativistic assumption, with or without quantum assumption [36-39]. For some of those relations, the mass depends on the inverse of the length scale r:

$$m \approx m_0 = m_1 + Cste' \lim_{r \to 0}\left(\frac{m_0 L_T}{r}\right) \tag{12}$$

where L_T is the Thomson scattering length and $Cste'$ s a numerical parameter depending on the considered geometry. A strong divergence problem occurs for non-quantum approach (relativistic or not) [36,37] or for quantum and non- relativistic approach [38]. It is only with a simultaneous relativistic and quantum theory that the mass of particles can be renormalized with a logarithmic dependence on the length scale [39]:

$$m = \frac{m_1}{1 - \frac{3\alpha e}{2\pi} \lim_{\Lambda r \to \infty} \log\left(\frac{\Lambda_r}{m_1}\right)} \tag{13}$$

This process of renormalisation is done in quantum electrodynamics and can be performed for all order of the expansion of interaction terms with Feynman diagrams [40]. This is possible because: the theory depends on only one parameter Λr related to the length scale; the electromagnetic energy is local; the coupling constant αe ≈ 0.0073 is a small parameter to unity. Such a renormalisation explains mass as a result of the self-interaction, but does not enable to calculate a priori the observable mass.

Observable mass is the total mass. In the most general case, it corresponds to the Komar mass (eqn. 1), whose definition can also be applied to the calculation of the electromagnetic charge when taking into account of its local gravitational field; it leads at first order to the same relation as eqn. 12. We can wonder if such an explanation for mass by considering self-interaction could be experimentally checked. If we consider a macroscopic ball of electrical charge Q and finite radius r, the total mass is:

$$m \approx m_0 = m_1 + Cste'' \frac{Q^2 \mu_0}{4\pi r} \tag{14}$$

where $\mu_0 = \left(\epsilon_0\, c^2\right)^{-1} = 4\pi 10^7 USI$ is the vacuum permeability and $Cste''$ is a numerical parameter depending on the considered geometry ∈ [1/2; 3/5]. For an electrical charge Q of some Coulomb, a radius r of some mm and for an uncharged mass $m1$ around some mg, it should be possible to measure an effective increase of the charged mass $m0$.

However, because of the additional electrostatic interaction of the charge with the surrounding environment (long-distance and strong interaction), it is more difficult to define clearly the mass of this open system and what is really weighted with a weighing machine. In other words, it is not sure that eqn. 14 is still valid. Accurate experimental tests would be required to separate the different effects. Eventually, the same developments can be proposed for gravitation and quantum chromodynamics. However, it has been demonstrated that gravitational interaction cannot be renormalized [10, 41].

Thermodynamics mechanism

Thermodynamics arguments may also be used to define mass. There are several examples. First, let us start with the proposition of Verlinde based on entropic considerations [42]. He has proposed that gravitation is a consequence of the holographic principle due to Bekenstein and would be an entropic force [43]. The gravitational mass "emerges in the part of space surrounded by a screen where the energy is evenly distributed over the occupied bits of information" [42]. He proves then that the inertial mass may be related to the entropy S of this information as:

$$m \approx m_0 = n^i(\nabla_i S)\frac{\hbar}{2\pi k_B c} \tag{15}$$

Where n^i is the contravariant component of a unitary space vector and ∇i is the covariant component of the space differential operator. Second, we can also directly applied thermodynamics principles [16]. Let us propose a very simple but innovative example at microscopic scale for an atomic nucleus without detailing the inside nuclear interactions, using non-quantum and non-relativistic arguments. We suppose that a thermodynamic dissipation occurs at local scale because of the different particles inside the nucleus of global temperature T, where diffusion may occur. Indeed, thermal equilibrium is assumed, whereas "chemical" equilibrium is not within this nucleus. The created volumic heat power rv is dissipated for example through thermal radiation $\tilde{Q}_R$ at the surface. This volumic heat is related to the flux of particles, through a linear electrochemical-like coupling that is built to respect the Clausius Duhem inequality [16]. It is then supposed that diffusion of particles in the nucleus follows the Fick's first law. The Fick diffusion coefficient can be $D_F = \frac{k_B T}{6\pi\eta L}$. Electrical charges could be also taken into account through a factor (z+1) in the diffusion coefficient. Using eqn. 4 with m1=0, it leads then to the mass of the nucleus of characteristic length L.

$$\begin{aligned} & -j^i(\nabla_i\beta)L^3 = r_v L^3 = -\widetilde{Q_R} \\ & \Leftrightarrow \frac{k_B T D_F}{\zeta}(\nabla_i\zeta)^2 L^3 = \sigma_{SB} T^4 L^2 \\ & \Rightarrow m_0^2 = \frac{2.5 n_N^3 \hbar^3}{\pi^3 c^2 L^5 \eta} \\ & \qquad \Leftrightarrow m_0 \approx \frac{1.7 \cdot 10^{-22}}{\sqrt{\eta}}(kg) \end{aligned} \tag{16}$$

where β=β0 + kB T ln ζ is the "chemical" potential of the nuclear fluid ji and is the contravariant component of flux of particles relatively to the barycenter of the system. ζ is the volumic concentration of particles ≈ nN /L3 in this nuclear fluid. σSB is the Stefan-Boltzmann constant. First equality can also be derived from general Ohm's relation and Nernst-Einstein relation [16]. Advanced developments on viscosity of nuclear fluid can be found [44,45]. From bibliography, we can use either the viscosity in common nuclei around 2 10-8kg/(m · s) or in quark-gluon plasma around 5· 10-11 kg/(ms). Numerical calculations have been done for nN=3 and L=1 Fermi. It leads respectively to the masses of a nucleon 1.3 · 10-18 kg (in common nuclei) and 2.4 · 10-28 kg (in quark-gluonplasma). According to the considered assumptions, alculation is more valid for a gas than for a liquid. It seems thus logical to obtain a correct magnitude for the mass of a nucleon with the viscosity of quark-gluon plasma. For a viscosity around 10^{10} kg/(ms), we obtain a more accurate value for the mass. Mass and viscosity appear in general simultaneously with fundamental interactions [15], and seems here to be directly related.

From this short review, it has been shown that mass depends on kinematics quantity (through the speed), on thermodynamics quantities (with temperature or entropy), on scale transition (according to eqn. 6), on interactions or self- interactions. The mass is eventually difficult to understand because it involves modern physical theories that are difficult to merge. Moreover, the use of such an ultimate theory has interest only at the Planck scale. Without matter, the only reference mass is the Planck mass ≈ 2.176 · 10^{-8}*kg*=1.221 · 10^{19} *GeV /c2*, built with the system of units (*c, G,* ﬁ). However, the corresponding energy seems to have no practical interest for determining the mass of most of the "common" observed particles.

Nevertheless, one conclusion can be drawn: when the system is closed, the mass is intrinsic; when the system is open, part of the whole mass may be linked to interactions with the surrounding medium. Because none part of the universe is strictly isolated, then it could be extrapolated that mass of those parts are not intrinsic properties for any considered local matter. Only a global point of view should be thus considered to define clearly mass. Eventually, except relations between physical variables (*v, T, S*...) or general definitions (Komar, Verlinde), there are few mechanisms to explain mass: Higgs process for bosons *Z* and *W*, hypothetic seesaw process for neutrinos, hypothetic far distant gravitational interaction for all particles... This list is far to be exhaustive. Some other theories for mass generation can be found in literature, which involve gravity or not. None of them are really satisfying for all particles. Only one has been verified experimentally with huge difficulties (Higgs process). Consequently, it is appropriated to look for a more general mechanism.

Theory for Mass Calculation

In the present article, it is proposed to develop a general theory enabling to calculate the mass of all particles. As previously said, different contributions may be considered for this mass calculation. Only three contributions will be further considered based on some general assumptions.

Assumptions

The proposed model aims to predict and calculate the rest mass of physical systems m0 (corresponding to the total mass in a rest frame). In the following development, this mass is simply noted m. We assume that mass can be additively decomposed. This assumption is directly linked to an additive de-composition of the Lagrangean functional that is supposed to exist. In this decomposition, we consider three terms (not necessarily independent):

$$m = m_{ref} + m_Q + m_\Phi \tag{17}$$

For non-relativistic approximation, this theory is based on the following total Lagrangian functional:

$$L = L_{field} + L_{matter} = \frac{1}{2}\dot{\Phi}^2 + \frac{1}{2}m\upsilon^2$$

$$= \frac{1}{2}\dot{\Phi}^2 \frac{1}{2}\left(m_{ref} + m_Q + C_{ste}\dot{\Phi}\right)\upsilon^2$$

We suppose that *mref* + *mQ* do not depend explicitly on Φ, scalar field that will be detailed in section 3.2, or Φ˙, its mass derivative, and thus correspond to constant terms to these fields. Moreover, we assume that the parameter *Cste* does not depend on the total mass *m*.

First, we define a mass level denoted as reference mass *mref*, meaning that particles are associated to this energy level. Its value may a priori depend on the considered family of particles. The concept of family of particles will be detailed at the end of this section. Second, from a macroscopic point of view, particles seem to access all the positions of space. However, accessing small positions requires high energies ac- cording to quantum effects. Moreover, a threshold is presently supposed around the Planck limit. Particles cannot then access smaller distance than the Planck length. It means that space is no more continuous and can be regarded as a lattice with a step related to the Planck length L_p. Consequently, the quantum mass contribution m_Q can be calculated according to eqns. 7 and 8. For a cubic lattice of lattice spacing L_p , we have:

$$m_Q = C_Q \frac{\hbar^2}{L_P^2 mc^2} = C_Q \frac{\hbar c}{Gm}$$

where m is the mass, G is the Newtonian gravitational constant and $\tilde{\tilde{C}}_Q$ is a dimensionless parameter linked to the real length to the ideal Planck lattice. This parameter can be expressed as a function of other dimensionless parameters such as parameters of coupling. Indeed, at the Planck length, one expects that gravitational interaction occurs with a coupling constant associated to the reference mass $\alpha_G = \frac{Gm_{ref}^{m^2}}{\hbar c}$ that leads to the quantum mass contribution:

$$m_Q = \tilde{C}_Q \frac{Gm_{ref}^2}{\hbar c}\frac{\hbar c}{Gm} = \tilde{C}_Q \frac{m_{ref}^2}{m} = -\frac{1}{4\tilde{\tilde{C}}_Q^2}\frac{m_{ref}^2}{m} \tag{20}$$

The last equality is written to simplify the interpretation of the dimensionless parameter $\tilde{\tilde{C}}_Q$ as a ratio of a characteristic length to twice the Planck length. This parameter is further denoted as Planck lattice spacing. The ideal Planck lattice spacing is 0.5, meaning that the particle can access distance equal to the Planck length. This mass contribution is also supposed to be negative.

Third, we consider the existence of a real scalar field Φ with a contribution mass *m*Φ. Generally, physical fields are assumed to depend on space x^j and time *t*=*x*4/*c*. We suppose here that this field is uniform and stationary, i.e. its values are assumed to be constant for all time and space. For a given family of *N* particles, masses are distinct and can be related to a mass vector m^k associated to the ground state at rest of these *N* particles, where *k* runs from 1 to *N*. As an example to illustrate a family of particles, we have just to consider the electron, muon and tau particles. This scalar field is supposed to depend only on a mass vector variable, such that $\Phi(m^k)$.

Scalar field equation

The field equation can be obtained from a variation principle, especially the Lagrangean approach. Let us consider the simple case for a single particle of mass m, we have a Lagrangean functional

L_{field} (Φ, Φ˙, *m*) where $\dot{\Phi} = \frac{d\Phi}{dm}$. The scalar field corresponds to the functional that minimizes the action, with adapted boundary conditions, such that:

$$\delta\left(\int L_{field}\left(\Phi,\dot{\Phi},m\right)dm\right) = 0 \tag{21}$$

It leads to the Euler-Lagrange equation:

$$\frac{\partial L_{field}}{\partial \Phi} - \frac{d}{dm}\left(\frac{\partial L_{field}}{\partial \dot{\Phi}}\right) = 0 \tag{22}$$

Considering the chosen Lagrangean of Eq. 18, we assume the simplest form for the Lagrangean of the scalar field:

$$L_{\text{field}} = \frac{1}{2}\dot{\Phi}^2 + CsteL_1\dot{\Phi} + CsteL_2 \tag{23}$$

As previously said, we assume that the parameter *CsteL*1=0.5*Cstev*2 does not depend on the total mass *m*, leading to the equation:

$$\frac{d}{dm} = \left(\dot{\Phi}(m)\right) = \ddot{\Phi}(m) = 0 \tag{24}$$

For non-single particles, the scalar mass is replaced by a mass vector of contravariant components m_k. Therefore, the second-order differential equation is replaced by considering a covariant derivative in the mass space instead of a simple derivative:

$$\frac{D}{Dm_j}\left(\frac{D}{Dm^j}\Phi\left(m^k\right)\right) = 0 \tag{25}$$

The expression of the covariant derivative is interesting for some simple cases:

$\frac{D}{Dm^j} = \frac{1}{\sqrt{|g|}}\frac{\partial}{\partial m_j}$ and $\frac{D}{Dm^j} = \sqrt{|g|}\frac{\partial}{\partial m_j}$ and $|g|$ is the determinant of the metric tensor, when orthogonal coordinate systems in the mass space are considered as further.

This Laplacian equation in the mass space has to be solved to obtain the field dependence $\Phi(m^k)$. Assuming isotropy of this function to its argument $|m|$ and $g|m||m|=1$, we can replace equation 25 by the simplified one:

$$\frac{d^2\Phi(|m|)}{d|m|^2} + \frac{d\Phi(|m|)}{d|m|}\Gamma^k_{|m|k} \tag{26}$$

where $\Gamma^k_{|m|k}$ is the Christoffel symbol. For dimension *N*=2 ("polar" coordinates), its value is $\frac{2}{|m|}$ For dimension *N*=3 ("spherical" coordinates), its value is $\frac{2}{|m|}$. For higher dimension *N* ("hyper-spherical" coordinates), its value is in general $\frac{N-1}{|m|}$. It leads to the equivalent equations:

$$\frac{1}{|m|^{(N-1)}}\frac{d}{d|m|}\left(|m|^{(N-1)}\frac{d}{d|m|}\Phi\left(|m|\right)\right) = 0 \tag{27}$$

$$\Leftrightarrow \frac{N-1}{|m|}\frac{d}{d|m|}\Phi\left(|m|\right) = \frac{d^2}{d|m|^2}\Phi\left(|m|\right) = 0 \tag{28}$$

$$\Leftrightarrow \frac{N-1}{|m|}\dot{\Phi}\left(|m|\right) + \frac{d}{d|m|}\dot{\Phi}\left(|m|\right) = 0 \tag{29}$$

where $|m|$ is the Euclidean norm of m^k. The vector dimension is *N*. Because of this assumption of isotropy throughout the article, the norm will be simply written by using *m* instead of $|m|$. The solution of eqn. 29 is then:

$$\dot{\Phi}(m) = \frac{C_\Phi}{m^{(N-1)}} \tag{30}$$

where C_Φ is a parameter. However, we can wonder whether this parameter can depend on other parameters or is really a "universal" physical constant (for a given choice of the units system). For example, that parameter could depend on the configuration of the system, i.e., its scale, geometry or number of particles.

Matter coupling

It is now necessary to consider the coupling of Φ and/or Φ˙ with matter. This coupling can be performed through the definition of an adapted Lagrangean functional. The latter has not to depend on an undetermined level of field, but should only depend on its derivative. Thus we assume a coupling with Φ. This is the same argument as using electrical potential gradient instead of the electrical potential for the electrical force construction. For example, in a four-dimensional formalism, we have eventually to couple Φ˙ and the 4-speed uμ of a material point to obtain a scalar functional. The simplest coupling is performed by multiplying Φ˙ with Consequently, a contribution to the total mass is simply, according to eqn. 18

$$m_\Phi = Cste\ \dot{\Phi} \tag{31}$$

At this step of the derivation, a choice on the unit of Φ is possible. Consequently, the *Cste* parameter in eqn. 31 is chosen equal to $2C_\Phi\hbar^2$. For the specific solution of eqn. 30, the mass contribution is then:

$$m_\Phi = \frac{2C_\Phi^2\hbar^2}{m^{(N-1)}} \tag{32}$$

In the international units system, for all *N*, the scalar field ˙ is homogeneous to $kg^{N/2-1}$=2 m^{-2} .s^1 and the parameter $C_\Phi$is homogeneous $kg^{N/2-1}$=2 m^{-2}. It is possible to define a quantity homogeneous to a mass with the parameter C_Φ, and thus independent on *N*, as:

$$\mu_\Phi = (2C_\Phi^2\hbar^2)^{1/N} \tag{33}$$

$$\Rightarrow m_\Phi = \frac{\mu_\Phi^N}{m^{(N-1)}} \tag{34}$$

As said before, C_Φ could also depend on specific parameters of the system. For instance, we can assume that C_Φ may be related to *mref*. For a M-power dependence of this scalar charge, the scalar field parameter may be then expressed as:

$$C_\Phi = c_1 m_{ref}^M \tag{35}$$

$$\Leftrightarrow \mu_\Phi = c_2 m_{ref}^{(2M/N)} \tag{36}$$

In the present theory, because the field Φ does not depend on the space and time coordinated, the mass of particles can be explained everywhere and every when by the same mechanism with the same parameters leading to the same mass. From a space time point of view, the present theory is non-local. The proposed mechanism is global.

Solutions for identical particles

Identical particles mean here a priori for a family of particles with the same quantum numbers (total angular momentum, spin, magnetic momentum, electrical charge, leptonic number, baryonic number...), but with a different mass. It is supposed that these particles interact isotropically with the scalar field. By taking eqns. 32 and 20 in eqn. 17, we obtain the following relation:

$$m = m_{ref} - \frac{1}{4\tilde{C}_Q^2}\frac{m_{ref}^2}{m} + \frac{2C_\Phi^2\hbar^2}{m^{(N-1)}} \tag{37}$$

$$\Leftrightarrow m^N - mref^{m^{(N-1)}} + \frac{m_{ref}^2}{4\tilde{C}_Q^2} m^{(N-2)} - 2C_\Phi^2\hbar^2 = 0 \tag{38}$$

This last equation depends on 4 parameters N, $\tilde{C}_Q$, C_Φ the number of particles in a family, the Planck lattice spacing, the parameter of scalar interaction and the reference mass. The number of particles can be chosen a priori for a given family. For example, let us consider the case for 1, 2 or 3 particles. For the first case *N*=1, we have:

$$m - m_{ref} + \frac{m_{ref}^2}{4\tilde{C}_Q^2} m^{-1} - 2C_\Phi^2\hbar^2 = 0 \tag{39}$$

$$\Leftrightarrow m^2 - (m_{ref} + 2C_\Phi^2\hbar^2)m + \frac{m_{ref}^2}{4\tilde{C}_Q^2} = 0 \tag{40}$$

If $2C_\Phi^2\hbar^2$, one particle can exist if and only $\tilde{C}_Q$>1 Else, it is not possible to have strictly one single massive particle with the present theory. There exists another possibility if *mref*=0, for which the two solutions of Eq. 40 are *m* =0 and *m* =*μ*Φ(*N* =1)= $2C_\Phi^2\hbar^2$. One single massive particle can exist associated to a second one as a non-massive particle. For *N*=2 and *N*=3, the equations are respectively:

$$\Leftrightarrow m^2 - m_{ref} m \left(\frac{m_{ref}^2}{4\tilde{C}_Q^2} - 2C_\Phi^2\hbar^2 \right) = 0 \tag{41}$$

$$m^3 - m_{ref} m + \frac{m_{ref}^2}{4\tilde{C}_Q^2} m - 2C_\Phi^2\hbar^2 = 0 \tag{42}$$

For specific conditions on the different parameters, 2 or 3 massive particles are expected, corresponding respectively to the solutions of the second-order algebraic equation (eqn. 41) or the third-order algebraic equation (eqn. 42)

Simplified influence of the reference mass *mref*

Except N, three other parameters have to be considered that are supposed to be independent. The Planck lattice spacing $\tilde{C}_Q$ is supposed to have values more or less around unity. However, each family of particles will strictly have a different value of this parameter because of their nature. With the present theory, it is not a priori possible to explain and calculate accurately these values. The parameter of scalar interaction C_Φ may have a value for the considered particles depending a priori on the number of particles N and/or other parameters. Here for simplicity and illustration, C_Φ is supposed to be independent of *mref*. For the present calculation of this subsection, a constant value of $10^{-9}kg^{-N/2} \cdot m^{-2} \cdot s^1$ is thus chosen. The only parameter that remains to be investigated is the reference mass m_{ref}. It depends on the nature of particles and on the number of particles *N* in a family. The influence of the reference mass is illustrated respectively for *N*=2, *N*=3, *N*=4 and *N*=5 (Figures 1-4).

It is worth noting that these diagrams present different behaviours. For *N*=2 (Figure 1), for low reference mass (below ≈ 2 · 10^{-15}*mP r*) the number of particles is asymptotically 1, whereas for high reference mass (above ≈ 2 · 10^{-15}*mP r*) the number of particles is 2. For *N*=3 in Figure 2, for low reference mass (below ≈ 0.5*mP r*) there are strictly 2 particles, where as for high reference mass (above ≈ 0.5*mP r*) the number of particles is 3. For *N*=4 in Figure 3, for low reference mass (below ≈ 7 · 106*mP r*) there are strictly 3 particles, whereas for intermediary reference mass (between ≈ 7 · 10^6*mP r* and ≈ 10^7*mP r*) the number of particles is 4, and for high reference mass (above

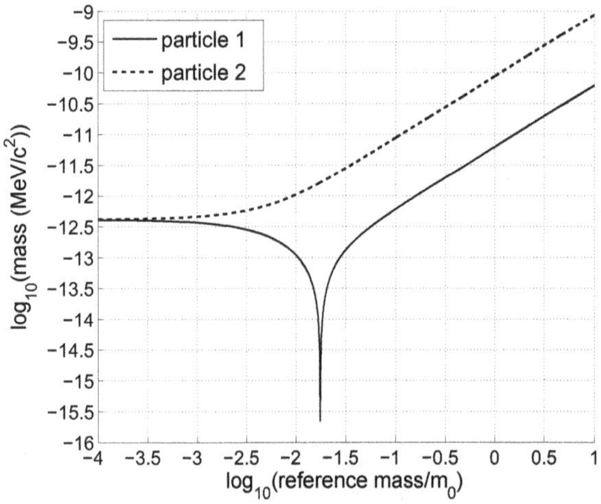

Figure 1: Influence of the reference mass for the case *N*=2 corresponding to the solutions of Eq.41; C_{Φ}=5· 10^{-9} $kg^{-1}\cdot m^{-2}\cdot s^{1}$and $\tilde{\tilde{C}}_{Q}$ =2; the normalization mass m_0 has been arbitrarily chosen to $10^{-13}mP\ r$.

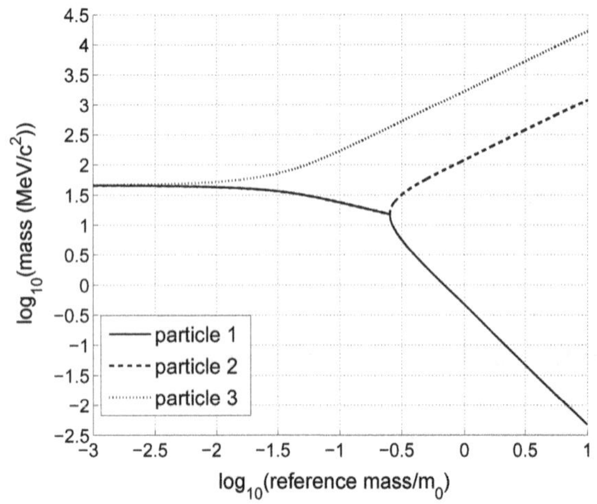

Figure 2: Influence of the reference mass for the case *N*=3 corresponding to the solutions of Eq.42;C_{Φ}=5 · $10^{-9}kg^{-3/2}\cdot m^{-2}\cdot s^{1}$ and $\tilde{\tilde{C}}_{Q}$ =2; the normalization mass m_0 has been arbitrarily chosen to $2mP\ r$.

≈ $107mP\ r$) the number of particles tends asymptotically to 3. For *N*=5, the behavior is similar to *N*=4 because two of the solutions are identical. Differences only occur on the limit of ranges of the reference mass. First we can conclude that a family of 5 particles is not possible for the considered values of the parameters of this theory. Second we can suggest that a family of 4 particles is possible to be observed only in a restrictive domain of reference mass corresponding to very high values to the proton mass $mP\ r$. Therefore, for particles of "common" reference mass (i.e., around the proton mass), only families of 2 or 3 massive particles should be theoretically observed. This is relevant with experimental observations. In the present theory, it is also worth noting that the mechanism to obtain mass for particles is a kind of seesaw mechanism (for *N*=1).

Application to Massive Constituents

Now the present theory is going to be applied to different kinds of particles either fundamentals (leptons, quarks, massive bosons) or composites (baryons, mesons).

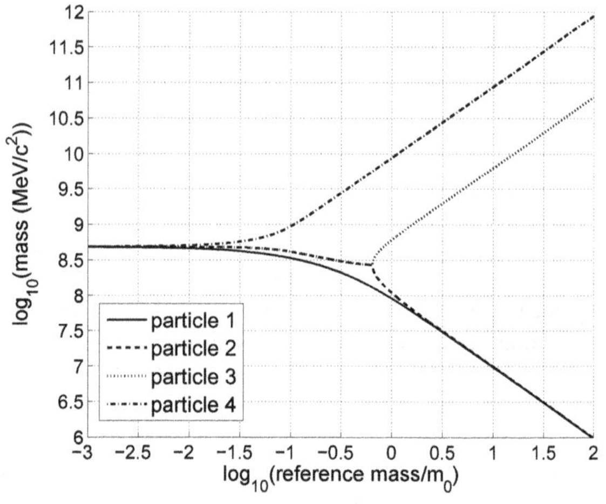

Figure 3: Influence of the reference mass for the case *N*=4 corresponding to the solutions of Eq.38 for *N*=4; C_{Φ}=5·$10^{-9}kg^{-2}\cdot m^{-2}\cdot s^{1}$ and $\tilde{\tilde{C}}_{Q}$ =2; the normalization mass m_0 has been arbitrarily chosen to $10^{7}mP\ r$.

Assumptions

We assume that for any kind of particles the theory of section 3 can be applied according to the assumptions (additivity of mass contributions, existence of a quantum mass contribution linked to a lattice at the Planck scale, existence of a scalar field contribution). We consider groups of N=2 or 3 particles, denoted as family. For the different kinds of particles (fundamentals or composites), there can exist differences such as the parameter CΦ, which may be related to the reference mass m_{ref}. This can be easily understood by doing analogy with electromagnetic field theories where interaction parameter depends on the number of electrical charge and their relative geometries. This dependence is supposed to be related to the number of particles N and to the scale of description (fundamental or composite particles), but not on the quantum numbers of the system (total angular momentum, electrical charge...). It is also important to emphasize that we expect to obtain a general theory. However, with triplet of particles, it is always possible to obtain a set of three parameters from a third-order algebraic equation. It is only if some links between those parameters and/or if predictions may be done that the present theory can be assessed to experimental observations.

Leptons masses

Two families of leptons should be described: electronic and neutrino ones. The triplets are easy to identify. First, we consider the family constituted of electron, muon and tau particles. In this triplet, all the particles have the same quan- tum numbers [46]. Moreover, the experimental masses are quite well-known. It is first interesting to question how such leptons are created. For example, for electrons, they can be created from neutrons by beta decay or from gamma radiation annihilation, for which nucleons are intermediaries [13]. Consequently, the reference mass for this kind of reference should be a multiple of the nucleon mass. Considering the order of magnitude of the tau, we expect to have the reference mass around 2 m Pr. Because of the slight difference between the neutron and proton masses, it is better to let vary this parameter. It is thus possible to identify the different parameters of eqn. 42, which are related in Table 1.

As expected, the ratio of reference mass to proton mass is very close to 2. This parameter ought to be considered as known. Moreover, the

Particles	Electron	Muon	Tau
Experimental mass [46] (MeV/c^2)	0.510998928 ± 0.000000011	106.6583715 ± 0.0000035	17776.82 ± 0.16
Calculation (MeV/c^2)	0.510998928	106.	1777.
Parameters	$\frac{m_{ref}}{m_{Pr}}$	$\approx C_Q$	C_Φ($kg^{1/2}. m^{-2}. s^1$)
Fitted values	2.00687 ± 0.00017	2.16738 ± 0.00009	4.943062. 10^{-9} ± 2.23. 10^{-13}

Table 1: Input and output data for electronic family.

Particles	Electron	Muon	Tau
Experimental mass [46] (MeV/c^2)	4.8 ± 0.8	95 ± 5	4180 ± 30
Calculation (MeV/c^2)	4.8	95	4180
Parameters	$\frac{m_{ref}}{m_{Pr}}$	$\approx C_Q$	C_Φ($kg^{1/2}. m^{-2}. s^1$)
Fitted values	4.561 ± 0.038	3.311 ± 0.0081	2.203.10^{-8} ± 2.50.10^{-9}

Table 2: Input and output data for the first quark family.

Particles	Up	Charm	Top
Experimental mass [47] (MeV/c^2)	2.3 ± 1.2	1275 ± 25	173210 ± 1220
Calculation (MeV/c^2)	2.3	1275	173210
Parameters	$\frac{m_{ref}}{m_{Pr}}$	$\approx C_Q$	C_Φ($kg^{1/2}. m^{-2}. s^1$)
Fitted values	185.967 ± 1.328	5.86 ± 0.039	3.516. 10^{-7} ± 1.021. 10^{-7}

Table 3: Input and output data for the second quark family.

Planck lattice spacing $\tilde{\tilde{C}}_Q$ very small, meaning that this family may access distance close to the Planck length (≈ 4.3LP). The last parameter C_Φ may depend on the particles family and especially on the mass of reference. This will be investigated further in section 4.5. Second, to complete it should be theoretically possible to express the masses of neutrinos with the present theory. One main problem is that some of the parameters change to the electronic family, such as the reference mass. Moreover, the experimental values of masses of those particles present huge uncertainties. Consequently, identification of the parameters that depend on the considered family would have a too weak accuracy. This is consistent with the explanations proposed in section 2. Indeed, neutrinos can be seen roughly as electrons without electrical charge; thus renormalisation may also be used to explain the link between the masses of those two families, according to the relation:

$$m_{e,\mu,T} - v = m_{e,\mu,T} - v + \frac{e^2}{L_{e,\mu,T} \, c^2} \tag{43}$$

where Le,μ,τ is the characteristic length of self-interaction (proportional to the length scale r used in Eq. 12) and $e2=Q2/(4\pi E0)$. We expect those lengths are strictly positive. However, for each particles of those families (electron and e- neutrino, muon and μ-neutrino, tau and τ-neutrino), there is a priori a different value for Le,μ,τ. In other words, there are always three unknown parameters Le, $L\mu$, $L\tau$ as in the proposed theory, which are difficult to identify because of the huge uncertainties of the experimental values for the neutrinos masses.

Quarks masses

Two families of quarks could be described. First, we consider the family con- stituted of down, strange and bottom particles. In this triplet, all the particles have not strictly the same quantum numbers: the isospin, strangeness and bottomness are not equal [47]. We suppose that these properties do not affect the mass of quarks at first order of approximation. Contrarily to leptons, it is harder to induce the reference mass for quarks because they do not exist freely [13]. It is nevertheless possible to identify the different parameters of Eq. 42, with a higher uncertainties, which are related in Table 2.

The ratio of reference mass to proton mass is around 4.56. Its interpretation is not trivial. Similarly to the electronic family, the parameter $\tilde{\tilde{C}}_Q$ is very small, meaning that this family may access distance close to the Planck length. The last parameter $C\Phi$ may depend on the particles family and especially on the mass of reference. This will be investigated further in section 4.5. Second, it is also possible to describe the other quark family constituted of up, charm and top quarks. In this triplet, all the particles have again not strictly the same quantum numbers: the isospin, charm and topness are not equal [47]. We suppose again that these properties do not affect the mass of quarks at first order of approximation. Similarly to the first quark family, it is harder to induce the reference mass for quarks because they do not exist freely [13]. It is nevertheless possible to identify the different parameters of eqn. 42, with a higher uncertainties, which are related in Table 3.

The ratio of reference mass to proton mass is around 186. Its interpretation is definitively not trivial. Similarly to the electronic family, the parameter $\tilde{\tilde{C}}_Q$ is very small, meaning that this family may access distance close to the Planck length. The higher the reference mass is, the higher this parameter is. For the second quark family, it is twice the value found for the first quark family. The last parameter $C\Phi$ may depend on the particles family and especially on the mass of reference. This will be investigated further in section 4.5. Its uncertainty is quite important, because of the bigger uncertainties for the masses of that family.

Massive bosons

Nowadays, it has been clearly evidence the masses of bosons W and Z of the electroweak theory, as discussed in introduction of this article. Let us imagine that the Higgs mechanism is not a significant part of the mass contribution of these bosons (meaning that the Higgs mass could not to be as important as expected). Let us also suppose that the difference of electrical charge effect between this two bosons is neglected, meaning that no renormalisation process contributes significantly to the mass of these particles. The latter is a strong assumption only to see if the present theory is able to describe consistently the mass of those two particles. However, because W + and W – are usually assumed to have the same mass, the latter depends only on even power of the electrical charge, so that it should be a second order effect to the total mass provided that this contribution is a small parameter. Two possibilities may be considered for this family, either N=3 or directly N=2. For the first case, the reference mass should be inferior to 0.5 mP r, but this is not possible because values for the boson masses are superior. Consequently, the solution has to be searched directly for the case N=2 supposing that W and Z form a doublet. It is possible to identify the different parameters of eqn. 41, which are related in Table 4.

Particles	Z°	W+ or W	
Experimental mass [48](GeV/c^2)	91.1876 ± 0.0021	80.385 ± 0.015	
Calculation (GeV/c^2)	91.1876	80.385	
Parameters	$\frac{m_{ref}}{m_{Pr}}$	$\tilde{\tilde{C}}_Q$	C_Φ(m^{-2}. s^1)
Fitted values	182.861 ± 0.018	1.00199 ± 0.00001	$5.3632.10^{-8}$ ± 3.15. 10^{-10}

Table 4: Input and output data for the massive bosons.

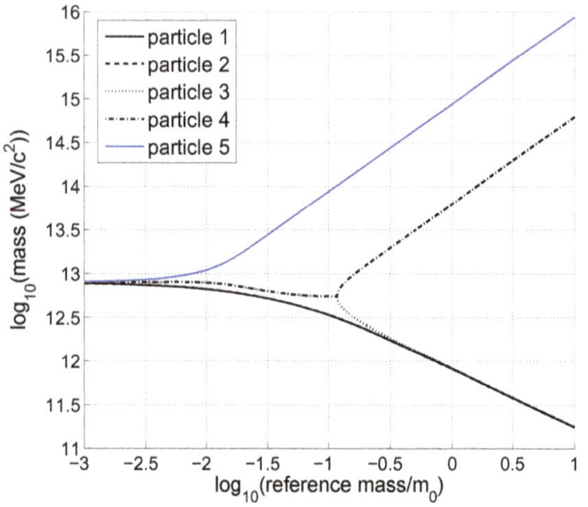

Figure 4: Influence of the reference mass for the case *N*=5 corresponding to the solutions of Eq.38 for *N*=5; C_Φ=5·$10^{-9}kg^{-5/2}\cdot m^{-2}\cdot s^1$ and $\tilde{\tilde{C}}_Q$=2; the normalization mass m_0 has been arbitrarily chosen to $10^{12}mPr$.

The ratio of reference mass to proton mass is around 182.9. Its interpretation is once more not trivial. This value is close to the one obtained for the second quark family. More than the electronic and quarks families, the parameter $\tilde{\tilde{C}}_Q$ is very close to unity (1.002), meaning that this family moves quasi strictly on a Planck lattice of twice the Planck length. It seems logical that bosons may access distance values smaller than fermions, because the exclusion principle does not limit the position of bosons but only the one of fermions. For the parameter of scalar field, the value is not the same as the ones found for the quark and electronic families. This last parameter depends definitively on the particles family and/or on the mass of reference.

Possible link between fundamental particles masses

First, we are trying to link the scalar field interaction to the reference mass. Indeed, the parameter CΦ does not appear as an universal constant. From Tables 1-3, we can plot μΦ versus the reference mass. This quantity is more interesting to plot, because its unit does not depend on N (eqn. 33). The massive bosons (N=2) have not been considered, mainly because it does not correspond to the same case as the others (N=3). The graph is presented in Figure 5.

The points are roughly aligned (in log scale) according to the relation:

$$\log 10\mu_\Phi = -30.40 + 0.71\log_{10} m_{ref} \tag{44}$$

$$\Leftrightarrow \mu_\Phi = 3.97.10^{-31}\, m_{ref}^{0.71} \tag{45}$$

$$\Leftrightarrow C_\Phi = 1.68.10^{-12}\, m_{ref}^{1.06} \tag{46}$$

When fitting the data without the error bars, the linear relation (in log scale) is slightly better and the fitted parameter are slightly different.

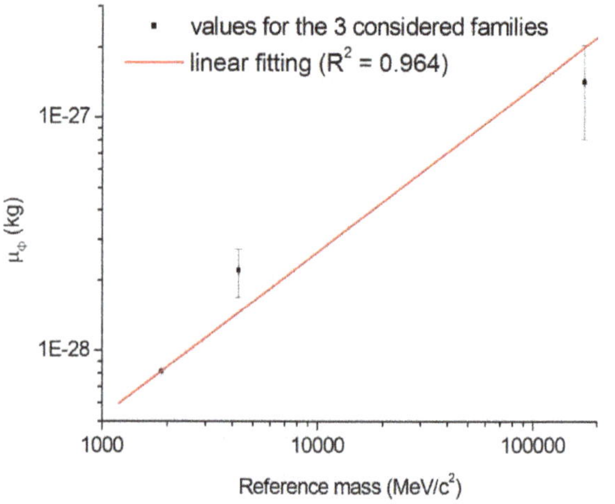

Figure 5: μ_Φ versus the reference mass m_{ref} for the three fundamental considered families (electrons, first quarks and second quarks families).

Errors are 0.63 and 0.19 respectively on the intercept and on the slope (in log scale). This non- linear relation links the reference mass with a coupling parameter according to Eq. 35, with M=1.06. C_Φ is thus quasi linear to m_{ref}. In this case, we may rewrite the equation for mass determination as:

$$m^3 - m_{ref}m + \frac{m_{ref}^2}{4\tilde{\tilde{C}}_Q^2}m - \mu_\Phi^3 = 0 \tag{47}$$

$$m^3 - m_{ref}m^2 + \frac{m_{ref}^2}{4\tilde{\tilde{C}}_Q^2}m - \mu_\Phi^3 = 0 \tag{48}$$

Second, we may think that it would be possible to use this relation to extrapolate the values for the neutrinos family. It reduces the number of unknown because of the dependence. Only two parameters are now required: the reference mass mref and the Planck lattice spacing $\tilde{\tilde{C}}_Q$ The latter could be numerically taken equal for both leptonic families (≈ 2.17). Only the reference mass remains then eventually unknown, but it is difficult to choose a value leading to results compatible with the experimental conditions imposed to those particles. This aspect remains to be investigated.

In conclusion of this section, we have obtained a theory that can describe the mass of fundamental particles. For the three considered families for N=3, there exists a priori 9 unknown parameters. It is possible to give a power law relation for the scalar mass of coupling as a function of the reference mass. Moreover, the reference mass of electronic family can be a priori given (=2 m Pr). Consequently, there exist only 7 unknown parameters to calculate the 9 masses of those particles with the considered universal mechanism.

Composite particles masses

It is now interesting to look for the masses of composite particles. Indeed, the present theory should and can be adapted to any particles: baryons, mesons. For baryons, it is quite easy to define the different families. A given family consists on a common root added to one of the fundamental quark family or the other. For mesons, it is more difficult to define a priori a family. We will look for the following considered families of N=3 particles, where J is the is the total angular momentum and Q is the electrical charge. For a given family (mesons or baryons), we assume to have only the same Q and J. This is consistent with previous assumptions considered for the quark families. We work with:

- Proton, Sigma, bottom Sigma, for J=1/2 and Q=1

In this triplet of baryons, all the particles have not strictly the same quantum numbers: the isospin, strangeness and bottomness are not equal. We suppose that these properties do not affect the mass of composites at first order of approximation. The results are summarized in Table 5.

- Delta, Sigma, bottom Sigma, for J=3/2 and Q=1

In this triplet of baryons, all the particles have not strictly the same quantum numbers: the isospin, strangeness and bottomness are not equal. We suppose that these properties do not affect the mass of composites at first order of approximation. The results are summarized in Table 6. In this triplet of baryons, all the particles have not strictly the same quantum numbers: the isospin, strangeness and bottomness are not equal. We suppose that these properties do not affect the mass of composites at first order of approximation. The results are summarized in Table 7.

neutron, Lambda, bottom Lambda, for J=1/2 and Q=0 In this triplet of baryons, all the particles have not strictly the same quantum numbers: the isospin, strangeness and bottomness are not equal. We suppose that these properties do not affect the mass of composites at first order of approximation. The results are summarized in Table 8.

- π0, η, ηl, for J=0 and Q=0

This regroupment has been considered according to their specific quantum numbers and because they belongs to the same octet. In this triplet of mesons, all the particles have not strictly the same quantum numbers: the isospin is not equal. We suppose that these properties do not affect the mass of composites at first order of approximation. The results are summarized in Table 9.

- ρ0, ω, φ, for J=1 and Q=0

This regroupment has been considered according to the same reason as previously. In this triplet of mesons, all the particles have not strictly the same quantum numbers: the isospin is not equal. We suppose that these properties do not affect the mass of composites at first order of approximation. The results are summarized in Table 10.

We have considered that any triplet can be evaluated by the present theory, providing the electrical charge Q and the total angular momentum J are similar per family. At first order of approximation, we consider that the flavor quantum numbers does not affect the mass value. Whatever the family, the present theory is able to reproduce the experimental results with consistent values of the three parameters.

Particles quarks content	**Proton p^+ *u- u- d***	**SigmaΣ^+ *u-u-s***	**Bottom Sigma Σ_b^+ *u-u-b***
Experimental mass [46] (*MeV/c²*)	938.272046 ± 0.000021	1189.37 ± 0.07	5811.3 ± 1.9
Calculation (*MeV/c²*)	938.	1189.	5811.
Parameters	$\frac{m_{ref}}{m_{Pr}}$	$\overset{\approx}{C}_Q$	$C_\Phi(kg^{1/2}.\ m^{-2}.\ s^1)$
Fitted values	8.4613 ± 0.0021	1.08114 ± 0.00009	$1.28520 - 10^{-6} \pm 2.5.\ 10^{-10}$

Table 5: Input and output data for the u-u-x family, for *J*=1/2 and Q=1.

Particles quarks content	**Delta Δ+*u - u - d***	**Sigma Σ^*+*u-u-s***	**Bottom Sigma Σ_b^+ *u - u - b***
Experimental mass [46] (*MeV/c²*)	1232 ± 2	1382.80 ± 0.35	5832.1 ± 1.9
Calculation (*MeV/c²*)	1232	1382.80	5832.
Parameters	$\frac{m_{ref}}{m_{Pr}}$	$\overset{\approx}{C}_Q$	$C_\Phi(kg^{1/2}.\ m^{-2}.\ s^1)$
Fitted values	9.0026 ± 0.0045	1.02574 ± 0.00015	$1.59078.\ 10^{-6} \pm 1.75.\ 10^{-9}$

Table 6: Input and output data for the u-u-x family, for *J*=3/2 and Q=1.

Particles quarks content	**Delta Δ^- *d-d-d***	**Sigma Σ^{*-} *d-d-s***	**Bottom Sigma *d - d - b***
Experimental mass [46] (*MeV/c²*)	1232 ± 2	1387.2 ± 0.5	5835.1 ± 1.9
Calculation (*MeV/c²*)	1232	1387.2	5835.1
Parameters	$\frac{m_{ref}}{m_{Pr}}$	$\overset{\approx}{C}_Q$	$C_\Phi(kg^{1/2}.\ m^{-2}.\ s^1)$
Fitted values	9. 0105 ± 0.0047	1.02547 ± 0.0002	$1.59372.\ 10^{-6} \pm 1.84.\ 10^{-9}$

Table 7: Input and output data for the d-d-x family, for *J*=3/2 and Q=¡1.

Particles quarks content	**neutron n° *u - d - d***	**Lambda Λ° *u-d-s***	**Bottom Lambda Λ_b^0 *u- d-b***
Experimental mass [46] (*MeV/c²*)	939.565379 ± 0.000021	1115.683 ± 0.006	5619.4 ± 0.6
Calculation (*MeV/c²*)	939.565379	1115.683	5619.4
Parameters	$\frac{m_{ref}}{m_{Pr}}$	$\overset{\approx}{C}_Q$	$C_\Phi(kg^{1/2}.\ m^{-2}.\ s^1)$
Fitted values	8.1796 ± 0.0006	1.08115 ± 0.00003	$1.22487.\ 10^{-6} \pm 6.87.\ 10^{-11}$

Table 8: Input and output data for the u-d-x family, for J=1/2 and Q=0.

Particles quarks content	π^o uū — dđ	η uū+ dđ - 2ss	η' uū + dđ + ss
Experimental mass [47] (MeV/c^2)	134.9766 ± 0.0006	547.862 ± 0.018	957.78 ± 0.06
Calculation (MeV/c^2)	134.9766	547.862	957.78
Parameters	$\frac{m_{ref}}{m_{Pr}}$	$\tilde{C}_Q$	$C_\Phi(kg^{1/2}.\ m^{-2}.\ s^1)$
Fitted values	1.74856 ± 0.00008	0.96145 ± 0.00001	1.343105. 10^{-7} ± 6.7. 10^{-12}

Table 9: Input and output data for a mesons family, for J=0 and Q=0

Particles quarks content	ρ° uū - dđ	ω uū+ dđ	Φ $s\bar{s}$
Experimental mass [47] (MeV/c^2)	775.26 ± 0.25	782.65 ± 0.12	1019.461 ± 0.019
Calculation (MeV/c^2)	775.26	782.65	1019.461
Parameters	$\frac{m_{ref}}{m_{Pr}}$	$\tilde{C}_Q$	$C_\Phi(kg^{1/2}.\ m^{-2}.\ s^1)$
Fitted values	2.74694 ± 0.00042	0.86982 ± 0.00001	3.96922. 10^{-7} ± 9.8. 10^{-11}

Table 10: Input and output data for a mesons family, for J=1 and Q=0.

The results for composite particles show that the reference masses ratios are between 8 and 9 for baryons, whereas for mesons smaller values are obtained between 1 and 3. The parameter C_Q is very close to unity (between 1.025 and 1.082) for baryons, meaning that these families may access distance close to the Planck length. It is interesting to note that mesons families have a value for this parameter slightly inferior to unity (0.961 and 0.870), and thus closer to Planck length than baryons. The last parameter CΦ may depend on the particles family and especially on the mass of reference. This will be investigated further in section 4.7

Possible link between fundamental particles masses

Because the experimental data are not necessarily available for all the triplets, it is impossible to perform identification for all the existing families of particles. But the identifications performed in 4.6 are sufficient to find a possible relation between CΦ and the reference mass mref. Indeed, the parameter CΦ does not appear as an universal constant. From Tables 5-9, we can plot μ_Φ versus the reference mass. Only the family of mesons with the less experimental uncertainties have been used for the further step. The quantity μ_Φ is more interesting to plot because its unit does not depend on N. The graph is presented in Figure 6.

The points are correctly aligned (in log scale) according to the relation:

$$\log 10\mu_\Phi = -30.30 + 0.98\log_{10} m_{ref} \quad (49)$$

$$\Leftrightarrow \mu_\Phi = 5.03.10^{-31}\, m_{ref}^{0.98} \quad (50)$$

$$\Leftrightarrow C_\Phi = 2.39.10^{-12}\, m_{ref}^{1.48} \quad (51)$$

When fitting the data without the error bars, the linear relation (in log scale) is similar and the fitted parameter are the same. Errors are 0.12 and 0.03 respectively on the intercept and on the slope (in log scale). This non-linear relation links the reference mass with a coupling parameter according to the equation 35, with M=1.48. In this case, we may rewrite the equation for mass determination as:

$$m^3 - m_{ref}m^2 + \frac{m_{ref}^2}{4\tilde{C}_Q^2}m - \mu_\Phi^3 = 0 \quad (52)$$

$$\Leftrightarrow m^3 - m_{ref}m^2 + \frac{m_{ref}^2}{4\tilde{C}_Q^2}m - 127.26.10^{-93}\, m_{ref}^{2.95} = 0 \quad (53)$$

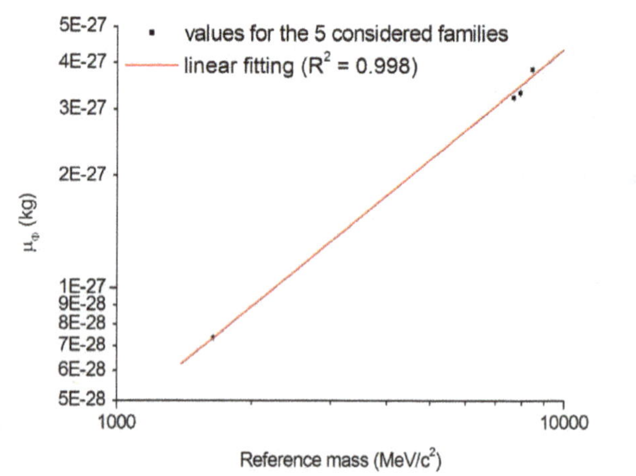

Figure 6: μ_Φ versus the reference mass m_{ref} for the five considered families (composite particles).

Predictions for composite particles masses

Even if it not possible to identify the parameters for all the particles, it is interesting to make some prediction based on triplets, for which only 2 masses are known. However, it can be done only if one parameter can be a priori calculated. This can be done by using eqn. 51 during the identification process. This has been performed for the following list of families:

Neutron, charmed Sigma, top Sigma, for J=1/2 and Q=0

Calculations for the uncertainties of C_Φ do not take into account for a possible uncertainty on the exponent of the power law. All other origins for the uncertainties have been transported. The results for those composite particles show that the references masses ratios are between 9 and 16. The Planck lattice spacing C-Q is very close to unity (between 1.017 and 1.058), meaning that this family may access distance close to twice the Planck length. Those results are consistent with the ones obtained in section 4.6. Consequently, we can have good confidence. As a consequence of this identification, Tables 11-17 proposes a predicted value for the third member of each family that remains now to be experimentally verified.

Masses at macroscopic scale

At macroscopic scale, the mass should be calculated according to

Particles quarks content	Neutron n^o $d—d—u$	Charmed sigma $\sum_c^0$ $d—d—c$	Top sigma $\sum_t^0 d—d—t$
Experimental mass [46] (MeV/c²)	939.565379 ± 0.000021	2453.74 ± 0.16	Unknown
Calculation (MeV/c^2)	939.565379	2453.74	
Prediction (MeV/c^2)			6971 ± 86
Parameters	$\frac{m_{ref}}{m_{Pr}}$	$\overset{\approx}{C}_Q$	$C_\Phi(kg^{1/2}.\ m^{-2}.\ s^1)$
Fitted values with C_Φ= f(m_{ref})	11.046 ± 0.092	1.0171 ± 0.0027	1.1301. 10^{-6} ± 2.06. 10^{-8}

Table 11: Input and output data for the d-d-x family, for J=1/2 and Q=0.

Particles quarks content	Xi Ξ^o s - s - u	Charmed omega Ω^o_c s - s - c	Top omega Ω^o_t s - s - t
Experimental mass [46] (MeV/c^2)	1314.86 ± 0.2	2695.2 ± 1.7	unknown
Calculation (MeV/c^2)	1314.86	2695.2	
Prediction (MeV/c^2)			9122 ± 107
Parameters	$\frac{m_{ref}}{m_{Pr}}$	$\overset{\approx}{C}_Q$	$C_\Phi(kg^{1/2}.\ m^{-2}.\ s^1)$
Fitted values with C_Φ=f(m_{ref})	14.00 ± 0.12	1.0366 ± 0.0029	1.6808. 10^{-6} ± 2.92. 10^{-8}

Table 12: Input and output data for the s-s-x family, for J=1/2 and Q=0.

Particles quarks content	Delta Δ^o d - d - u	Charmed Sigma $\sum^{*0}_c$ d - d - c	Top sigma $\sum^{*0}_t$ d - d - t
Experimental mass [46] (MeV/c²)	1232.0 ± 2	2518.8 ± 0.6	unknown
Calculation (MeV/c^2)	1232.0	2518.8	
Predictions (MeV/c^2)			8514 ± 104
Parameters	$\frac{m_{ref}}{m_{Pr}}$	$\overset{\approx}{C}_Q$	$C_\Phi(kg^{1/2}.\ m^{-2}.\ s^1)$
Fitted values with C_Φ=f(m_{ref})	13.07 ± 0.11	1.0360 ± 0.0029	1.5181. 10^{-6} ± 2.73. 10^{-8}

Table 13: Input and output data for the d-d-x family, for J=3/2 and Q=0.

Particles quarks content	Xi Ξ^{*0} s - s - u	charmed Omega Ω^{*0}_c s - s - c	top Omega Ω^{*0}_t s - s – t
Experimental mass [46] (MeV/c²)	1531.8 ± 0.32	2765.9 ± 2	Unknown
Calculation (MeV/c^2)	1531.8	2765.9	
Prediction (MeV/c^2)			10183 ± 117
Parameters	$\frac{m_{ref}}{m_{Pr}}$	$\overset{\approx}{C}_Q$	$C_\Phi(kg^{1/2}.\ m^{-2}.\ s^1)$
Fitted values with C_Φ=f(m_{ref})	15.43 ± 0.13	1.0451 ± 0.0029	1.9773. 10^{-6} ± 3.37. 10^{-8}

Table 14: Input and output data for the s-s-x family, for J=3/2 and Q=0.

Particles quarks content	Delta Δ^{++} $u — u — u$	charmed Sigma $\sum^*_c$++ $u — u — c$	top Sigma $\sum^*_t$++ $u — u — t$
Experimental mass [46] (MeV/c²)	1232 ± 2	2517.9 ± 0.6	unknown
Calculation (MeV/c^2)	1232	2517.9	
Prediction (MeV/c^2)			8513 ± 104
Parameters	$\frac{m_{ref}}{m_{Pr}}$	$\overset{\approx}{C}_Q$	$C_\Phi(kg^{1/2}.\ m^{-2}.\ s^1)$
Fitted values C_Φ=f(m_{ref})	13.07 ± 0.11	1.0360 ± 0.0029	1.5177. 10^{-6} ± 2.73. 10^{-8}

Table 15: Input and output data for the u-u-x family, for J=3/2 and Q=2.

Particles quarks content	Xi E*- $s—s—d$	Omega $s—s—s$	Bottom Omega $s — s — b$
Experimental mass [46] WWI c2)	1535.0 ± 0.6	1672.45 ± 0.29	unknown
Calculation (MeV/c^2)	1535.0	1672.45	
Prediction (MeV/c^2)			8104 ± 90
Parameters	$\frac{m_{ref}}{m_{Pr}}$	$\overset{\approx}{C}_Q$	$C_\Phi(kg^{1/2}.\ m^{-2}.\ s^1)$
Fitted values with az,=f(m_{ref})	12.06 ± 0.096	1.0583 ± 0.0030	1.4113. 10-6 ± 2.31. 10-8

Table 16: Input and output data for the u-d-x family, for J=3=2 and Q=0.

eqn. 37. However, the second and third terms of the right-member of this equation vanished when masses are big enough, which corresponds for example to the cases of macroscopic systems. Consequently, at those scales, the mass of physical systems is equal to the reference mass that corresponds to the mass obtained with a weighing machine. Another consequence is that additivity of masses of a complex and composed

Particles quarks content	Delta $\Delta\,°u—d—d$	Sigma $\Sigma^{*}°u—d—s$	Bottom Sigma $\sum_{b}^{*0}\ u—d—b$
Experimental mass (MeV' e2) [46]	1232 ± 2	1383.7 ± 1	unknown
Calculation (MeV/c^2)	1232	1383.7	
Prediction (MeV/c^2)			6542 ± 76
Parameters	$\frac{m_{ref}}{m_{Pr}}$	$\tilde{C}_Q$	C_Φ($kg^{1/2}.\ m^{-2}.\ s^{1}$)
Fitted values with C_Φ,=$f(m_{ref})$	9.761 ± 0.082	1.0556 ± 0.0030	$1.0290.\ 10^{-6} \pm 1.76.\ 10^{-8}$

Table 17: Input and output data for the u-d-x family, for J=3/2 and Q=0.

system is strictly correct only for macroscopic scales (eqn. 6). For microscopic systems, it is not necessarily true (with or without taking into account for interactions between the particles of systems).

Discussion on the scalar field

In section 3.3, we have introduced different masses linked to the scalar field. According to eqn. 17, mΦ is the contribution part of the particle mass corresponding to its interaction with the scalar field. μ_Φ is also related to the scalar field. It is denoted as the scalar mass of coupling/interaction. Equations 44 and 49 show that this mass may be directly related to the reference mass m_{ref}, which would play the role of a source for the scalar field. It is interesting to investigate the different μ_Φ, calculated from the different identified C_Φ (eqn. 33).

As an energy, we can wonder how is distributed the spectrum of this quantity? At first approximation, we can try to compare it to a harmonic quantum oscillator because the values are discrete. One-dimensional, two-dimensional and three-dimensional harmonic oscillators have been tested. Only the one- dimensional harmonic oscillator leads to consistent results. We have calculated the different ratios μ_Φ/mP r for all the families considered in sections 4.2, 4.3, 4.6 and 4.8. The massive bosons (N=2) have not been considered once again, mainly because it does not correspond to the same case as the others (N=3). For the different families, we have also calculated the quantity:

$$n = \left(\frac{\mu\Phi, n}{m_{Pr}} - \frac{1}{2}\frac{E0}{m_{Pr}C^2}\right) / \frac{E_0}{m_{Pr}C^2} \tag{54}$$

where the zero-level energy E_0 is identified for the electronic family, for which μ_Φ is the smallest. The value of E_0 is thus 2μelec=91.556M eV/c2. This value enables to plot the energy for the different n calculated with eqn. 54, according to:

$$\frac{E_n}{m_{Pr}C^2} = \frac{E0}{m_{Pr}C^2}\left(\frac{1}{2} + n\right) \tag{55}$$

It can be also compared to the theoretical values obtained with Eq. 55 for the same *E*0 with integer values for *n* running from 0 to 27. We have plotted both results in Figure 7. When comparing the two sets of values, we obtained a good correlation. Indeed, for *n*=0 the relative error is null. For *n*=1, the relative error is maximum: 14% for *n* and 9% for *En*. For other *n*, the relative error is less than 1.3% for *n* and *En*. With good confidence, we can conclude to the existence of the mass quantification according to the one-dimensional harmonic quantum oscillator, whose interaction/coupling with scalar field leads to a part of the mass of matter. This conclusion is independent of the influence of the reference mass, because of two reasons. First the influence of *mref* on *Cφ* is not the same for all kinds of families (cf. eqns. 46 and 51). Second, the quantum oscillator has been identified using all the families (for *N*=3). This oscillator leads then to the quantification of the different scalar masses of interaction *μ*Φ,*n* that leads to the quantification of the particles masses *m*, thanks to the dependence to the reference mass of the scalar field interaction (with specific coupling "constant" for each family of particles).

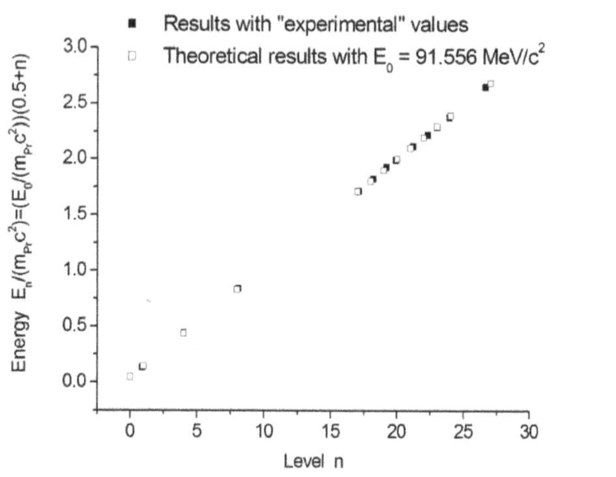

Figure 7: Normalized energy $E_n/(mP_rc^2)$ versus level n for the different considered families (fundamental and composite particles) for N=3.

The zero-level energy of this oscillator may be a good candidate to solve cosmological problems such as dark energy. Its value corresponds to a scalar mass of coupling *μ*Φ,0=45.778*M eV*/*c*2 and is equivalent to a fundamental angular frequency *ω*0=*E*0/fi=1.391 · 10²³*rad*/*s*. Indeed, it is possible to calculate the density of this energy in the universe. Considering a number of baryons of *nB*,*U*=10⁸¹ with the age of the universe of *tU*=13.819 · 10⁹ years, we can calculate the density as follows [49]:

$$\begin{aligned} \rho\Phi &= 3n_{B,\,U}E_0 / (4\pi R_U^3) \\ &= 3n_{B},_{U}E_0 / (4\pi(ct_U)^3 \\ &\approx 1.74.10^{-26}\,kg / m^3 \end{aligned} \tag{56}$$

where R_U is the radius of the universe considered as a sphere and c the speed of light. The calculus has to be compared with the value given in the specialized literature around 10^{-26}kg/m³ [49]. It is obvious that the previous calculation suffers from different approximations:

- The universe is not necessarily and simply a sphere with isotropic expansion;
- The number of baryons is uncertain; moreover, the leptons and mesons have not been taken into account in the calculation; however the present calculation corresponds to an admissible maximum value for *nB*,*U*;

From the present theory the baryons are not in the minimum of energy, but correspond to modes with

n /= 0, so that it cannot be the observed baryons that eventually participate to this density; it should be the same quantity as the observed one: we suggest thus that it could be anti-baryons, for which actual observations are lacking.

However, taking care of these effects will not drastically change the numerical result. The dark energy could be thus interpreted as anti-particles at the zero- level energy and not converted in direct observable matter. This proposition does not explain this asymmetry between particles and anti-particles, but suggests that the ones have been converted in matter through the scalar coupling, while the others have not and would thus constitute the "observed" dark energy

Conclusion

The present paper is an attempt to calculate the masses of both fundamental and composite particles from a master formula containing several parameters. The developments are based on a Lagrangian approach with a variational principle. It is consistent with the developments in physics over the last 50 years. Indeed, we propose a theory explaining mass of the different particles as an interaction/ coupling with a real scalar field based on physical arguments. Other arguments such as symmetries remain still to be investigated. This theory is based on quantum arguments, at first order of approximation, with non-relativistic assumption, in a global approach, assuming that mass can be additively separated. Mass may be thus composed of three parts depending differently on a reference mass: the reference mass itself;

a quantum correction linked to the Planck scale; an interaction with a real scalar field. No self-interactions are eventually taken into account. The interaction with the scalar field may be compared to the Higgs mechanism. Indeed, each family of particles interacts differently with this scalar field and corresponds to a different coupling constant. These "constants" of coupling are less numerous in the present approach. The different parameters of the corresponding equation (mainly used as a third-order algebraic equation) have been identified for different particles: fundamental or composite ones. For fundamental ones, only neutrinos have not been used because of the experimental uncertainties of their masses. The latter may be equivalently interpreted through the renormalisation mechanism. For composite ones, several baryons families have been used. It should be extended to more families of mesons. Moreover, the present theory is able to make predictions for the mass of unknown particles belonging to families, for which at least two masses are known. Indeed, it is possible to predict the interaction parameters, because C_Φ (or μ_Φ) is related to the reference mass m_{ref} for a given scale and a given number N of particles per family, independently of other quantum numbers (Q, J ...). Moreover, the interaction with a scalar field seems to follow a specific scheme related to a quantum harmonic oscillator of zero-level energy/mass 91.556 MeV/c^2.

Its energy/mass density is evaluated around 2 · 10-26 kg/m^3. It may be interpreted as the dark energy linked to the whole content of anti-particles (antibaryons...) of the universe that would remain as fundamental oscillations in the mass space at the zero-level energy and have not been converted into matter through this scalar interaction. This asymmetry between particles and anti- particles in the mass space remains yet to be investigated and will be done in a future work. The quantum and non-relativistic 1D harmonic oscillator predicts the scalar mass of interaction μΦ for each family of particles (baryons, mesons...) whatever their quantum numbers, scale and could be also independent of N. This last point requires also more investigations from doublets of particles.

The present theory proposes thus a general description of family of N particles from one to any expected integer value. However, only the cases N=2 and N=3 seem to correspond to the common energy around the proton mass mP r. The reference mass could be predicted according to the relation with the scalar mass of interaction μΦ that seems to be related for a given family to a mode of oscillation of a quantum harmonic oscillator. The latter can be seen as a source for the scalar field, which enables the oscillations ("germs" of matter) in the mass space to acquire mass through the coupling with the scalar field. It would lead also to the possibility to predict masses for family for which only one mass is known. Moreover, the present theory could be also applied for anisotropic state of the scalar field.

Besides, we propose a quantized lattice structure for space at the Planck scale, for which consequences of the present theory are experimentally checkable, with some predictions. This proposed lattice structure is based on quantum principles and physical considerations at Planck energy. Eventually, the theory enables to predict a minimum distance related to the Planck length that each family can access. This simple theory is thus powerful to explain and calculate the mass of physical systems, especially at microscopic scale, for which some contributions have influence on large large-scale phenomena (dark energy). The predicted masses require now to be experimentally tested.

References

1. Rubin VC, Ford Jr WK, Thonnard N (1980) Rotational properties of 21 SC galaxies with a large range of luminosities and radii. Astrophysical Journal 238: 471-487.
2. Gaitskell RJ (2004) Direct Detection of Dark Matter. Annual Review of Nuclear and Particle Science 54: 315-359.
3. Zwicky F (1937) On the Masses of Nebulae and of Clusters of Nebulae. Astrophysical Journal 86: 217.
4. Rubin V, Ford Jr WK (1970) Rotation of the Andromeda Nebula from a Spectroscopic Survey of Emission Regions. Astrophysical Journal 159: 379.
5. Higgs PW (1964) Broken Symmetries and the Masses of Gauge Bosons. Physical Review Letters 13/16: 508-509.
6. Brout R, Englert F (1964) Broken Symmetry and the Mass of Gauge Vector Mesons. Physical Review Letters 13/9: 321-323.
7. Bava E, Kuhne M, Rossi AM (2013) Metrology and Physical Constants. IOS Press.
8. Davis RS (2015) What Is a Kilogram in the Revised International System of Units (SI)? J Chem Educ 92/10: 16041609.
9. Jammer M (2009) Concepts of Mass in Contemporary Physics and Philosophy. Princeton University Press.
10. Weinberg S (1972) Gravitation and cosmology: principles and applications of the general theory of relativity. Wiley.
11. Landau L, Lifchitz E (1980) The classical theory of fields, 4th Edition. LevButterworth-Heinemann.
12. Einstein A (1905) Ist die Tr̈agheit eines K¨orpers von seinem Energieinhalt abh¨angig? in Annalen der Physik 1: 639.
13. Stacey WM (2007) Nuclear Reactor Physics, 2nd Edition. Wiley VCH.
14. Komar A (1963) Positive-Definite Energy Density and Global Consequences for General Relativity. Physical Review 129/4: 18731876.
15. Blanchet L, Spallicci A, Whiting B (2011) Mass and Motion General Relativity. Springer.
16. Balian R (2006) From Microphysics to Macrophysics: Methods and Applications of Statistical Physics. Springer.
17. Ott H (1963) Zeitschrift fur̈ Physik 175: 70.
18. Landau L, Lifchitz E (1965) Quantum mechanics: non-relativistic theory, 2th edition. Pergamon Press.
19. Kittel C (2004) Introduction to Solid State Physics, 8th Edition. John Wiley and Sons.
20. Ashcroft NW, Mermin N (1976) Solid State Physics. Brooks Cole.
21. Landau L, Lifchitz E (1969) Statistical Physics, 2th edition. Pergamon Press.

22. Du̇rr S, Fodor Z, Frison J, Hoelbling C, Hoffmann R, et al. (2008) Ab Initio Determination of Light Hadron Masses. Science 322/5905: 1224-1227.

23. Nambu Y (1952) An empirical mass spectrum of elementary particles. Prog. Theor. Phys. 7: 595-596.

24. Sumimo Y (2009) Family gauge symmetry and Koides mass formula. Physics Letters B 671: 477480.

25. Li N, Ma BQ (2005) Estimate of neutrino masses from Koides relation. Physics Letters B 609: 309316.

26. Xing, Zhang H (2006) On the Koide-like relations for the running masses of charged leptons, neutrinos and quarks. Physics Letters B 635: 107111.

27. Ma E (2007) Lepton family symmetry and possible application to the Koide mass formula. Physics Letters B 649: 287291.

28. King SF (2016) Neutrino mass and mixing in the seesaw playground. Nuclear Physics B 908: 456-466.

29. Goswani S (2016) Neutrino Phenomenology: Highlights of Oscillation Results and Future Prospects. Nuclear and Particle Physics Proceedings 273-275: 100-109.

30. Miranda OG, Valle JWF (2016) Neutrino oscillations and the seesaw origin of neutrino mass. Nuclear Physics B 908: 436-455.

31. Cai Y, Chao W (2015) The Higgs seesaw induced neutrino masses and dark matter. Physics Letters B 749: 458463.

32. King SF (2003) Neutrino Mass Models. Corenll University Library.

33. Guyon E, Hulin JP, Petit L (1996) Hydrodynamique physique. EDP Sciences.

34. Goldstone J, Salam A, Weinberg S (1962) Broken Symmetries. Physical Review 127: 965970.

35. Boratav M, Kerner R (1998) Relativity. Ellipses

36. Feynman RP, Leighton RB, Sands M (2011) The Feynman Lectures on Physics. Basic Books.

37. Babin A (2015) Explicit mass renormalization and consistent derivation of radiative response of classical electron. Annals of Physics 361: 190214

38. Cohen-Tannoudji C (1977) Quantum Mechanics. Wiley VCH.

39. Heitler W (1984) The Quantum Theory of Radiation. Dover Publications Inc, USA.

40. Mattuck RD (1992) A Guide to Feynman Diagrams in the Many-Body Problem. 2th Edition, Dover Publications Inc, USA.

41. Feynman RP (1999) Feynman Lectures on Gravitation. Penguin Books Ltd

42. Verlinde E (1992) On the origin of gravity and the laws of Newton. Journal of High Energy Physics.

43. Bekenstein JD (1973) Black holes and entropy. Phys Rev D 7: 2333.

44. Parihar V, Widom A, Drosdo D, Srivastava YN (2008) Viscosity of high energy nuclear filuids. Journal of Physics G: Nuclear and Particle Physics.

45. Auerbach N, Shlomo S (2009) η/s Ratio in Finite Nuclei. Phys Rev Letters 103: 172501.

46. Beringer J (2012) Particle Data Group. PR D86, 010001.

47. Olive KA (2014) Particle Data Group. Chin Phys.

48. Yao WM (2006) Particle Data Group. J Phys G 33: 1-6.

49. Spergel DN (2007) Wilkinson Microwave Anisotropy Probe (WMAP) Three Year Results: Implications for Cosmology. The Astrophysical Journal Supplement Series 170: 377-408.

Permissions

All chapters in this book were first published in JPM, by OMICS International; hereby published with permission under the Creative Commons Attribution License or equivalent. Every chapter published in this book has been scrutinized by our experts. Their significance has been extensively debated. The topics covered herein carry significant findings which will fuel the growth of the discipline. They may even be implemented as practical applications or may be referred to as a beginning point for another development.

The contributors of this book come from diverse backgrounds, making this book a truly international effort. This book will bring forth new frontiers with its revolutionizing research information and detailed analysis of the nascent developments around the world.

We would like to thank all the contributing authors for lending their expertise to make the book truly unique. They have played a crucial role in the development of this book. Without their invaluable contributions this book wouldn't have been possible. They have made vital efforts to compile up to date information on the varied aspects of this subject to make this book a valuable addition to the collection of many professionals and students.

This book was conceptualized with the vision of imparting up-to-date information and advanced data in this field. To ensure the same, a matchless editorial board was set up. Every individual on the board went through rigorous rounds of assessment to prove their worth. After which they invested a large part of their time researching and compiling the most relevant data for our readers.

The editorial board has been involved in producing this book since its inception. They have spent rigorous hours researching and exploring the diverse topics which have resulted in the successful publishing of this book. They have passed on their knowledge of decades through this book. To expedite this challenging task, the publisher supported the team at every step. A small team of assistant editors was also appointed to further simplify the editing procedure and attain best results for the readers.

Apart from the editorial board, the designing team has also invested a significant amount of their time in understanding the subject and creating the most relevant covers. They scrutinized every image to scout for the most suitable representation of the subject and create an appropriate cover for the book.

The publishing team has been an ardent support to the editorial, designing and production team. Their endless efforts to recruit the best for this project, has resulted in the accomplishment of this book. They are a veteran in the field of academics and their pool of knowledge is as vast as their experience in printing. Their expertise and guidance has proved useful at every step. Their uncompromising quality standards have made this book an exceptional effort. Their encouragement from time to time has been an inspiration for everyone.

The publisher and the editorial board hope that this book will prove to be a valuable piece of knowledge for researchers, students, practitioners and scholars across the globe.

List of Contributors

Michael Lawrence
Maldwyn Centre for Theoretical Physics, Cranfield Park, Burstall, Suffolk, UK

Michaud A
SRP Inc Educational Research Service Quebec, Canada

Tim Tarver
Department of Mathematics, Bethune-Cookman University, USA

YinYue Sha
Dongling Engineering Center, Ningbo Institute of Technology, Zhejiang University, China

Cusack P
Brealey Drive, Peterborough, 77432, Infrobright, Canada

André Michaud
Senior Researcher, Canada

Guerra V and Abreu R
Department of Physics, Instituto Superior Técnico, University of Lisbon, Portugal

Dr. Andreas Bacher
Faculty of Mathematics and Natural Sciences, University of Cologne, Albertus-Magnus-Platz, Germany

Masahiko Makanae
Representative Free Web College, Nishikasai, Edogawa-ku, Tokyo 134-0088, Japan

Stelian Liviu B
Independent Researcher, Israel

Kalinowski MW
Bioinformatics Laboratory, Medical Research Centre, Polish Academy of Sciences, Poland

Abdelrahman MAE and Moaaz O
Department of Mathematics, Faculty of Science, Mansoura University, Egypt

Alexandris NG
Independent Researcher, Greece

Ghanam R
Department of Liberal Arts and Sciences, Virginia Commonwealth University in Qatar, Qatar

Basim Mustafa B
Department of Mathematics, An-Najah National University, Palestine

Mustafa MT
Department of Mathematics, Statistics and Physics, Qatar University, Qatar

Thompson G
Department of Mathematics, University of Toledo, USA

Makanae M
Independent Researcher, Representative Free Web College, Nishikasai, Edogawa-ku, Tokyo 134 0088, Japan

Kohut P
Researcher, Czech republic

Murad PA
Morningstar Applied Physics, LLC Vienna, VA 22182, USA

Moffat J, Oniga T and Wang CHT
Department of Physics, University of Aberdeen, King's College, Aberdeen AB24 3UE, UK

Murad PA
Morningstar Applied Physics, LLC Vienna, VA 22182, USA

Matoog RT
Department of Mathematics, Faculty Applied Sciences, Umm Al-Qura University, Kingdom of Saudi

Moffat J, Oniga T and Wang CHT
Department of Physics, University of Aberdeen, UK

Asokan S.P.
Independent Researcher, New No 9, Old No 2, T.S.V Koil Street, Mylapore, Chennai, Tamil Nadu, India

Pons-Rullán B
The Stenbock House, Rahukohtu 3,15161 Tallinn, Estonia

Wang CHT, James Moffat J and Oniga O
Department of Physics, University of Aberdeen, King's College, UK

Panicaud B
Department of Physics, University of Technology of Troyes, France

Index